Soil Processes and Quality Analysis

Soil Processes and Quality Analysis

Edited by **Henry Wang**

Callisto Reference
New York

Published by Callisto Reference,
106 Park Avenue, Suite 200,
New York, NY 10016, USA
www.callistoreference.com

Soil Processes and Quality Analysis
Edited by Henry Wang

International Standard Book Number: 978-1-63239-565-8 (Hardback)

Printed in China.

Contents

Preface

Every book is initially just a concept; it takes months of research and hard work to give it the final shape in which the readers receive it. In its early stages, this book also went through rigorous reviewing. The notable contributions made by experts from across the globe were first molded into patterned chapters and then arranged in a sensibly sequential manner to bring out the best results.

This book discusses the various soil processes and also emphasizes on the topic of quality analysis. Soil quality and properties are constantly affected by human activities and natural processes. A number of researches are being conducted to assess soil quality in various environmental scenarios because of the growing need for soil protection. The estimation of soil quality determines the capability of soil to carry out its functions in current scenarios and also how it can be preserved for future land use. There are certain indicators currently being employed to estimate soil quality such as biological, chemical and physical properties along with various soil processes. Current trends and soil processes employed in assessment of soil quality provide wide depiction of current researches in soil quality assessment comprising of general studies on soil processes, evaluation of significant indicators of soil quality such as soil organic matter dynamic and soil-plant interaction along with elaboration on various methodologies for maintaining soil health.

It has been my immense pleasure to be a part of this project and to contribute my years of learning in such a meaningful form. I would like to take this opportunity to thank all the people who have been associated with the completion of this book at any step.

Editor

Section 1

Soil Quality Assessment

Chapter 1

Solute Transport in Soil

S.E.A.T.M. van der Zee and A. Leijnse

Additional information is available at the end of the chapter

1. Introduction

1.1. Classification of solutes, pollutants and subsurface pollution

Solute transport is of importance in view of the movement of nutrient elements, e.g. towards the plant root system, and because of a broad range of pollutants. Pollution of the subsurface is often considered to be either point source pollution or diffuse source pollution. Point source pollution covers a limited area, and is often caused by accidental (or illegal) spills (e.g. leaking pipes, tanks, mine tailings, etc.). Diffuse source pollution covers a large area and is in general caused by large-scale application of both beneficial and hazardous compounds at the soil surface (manure and fertilizer, pesticides, atmospheric deposition of acids and radio nuclides, etc.). Pollution is not necessarily man induced, but may be due to geological or geohydrological causes, e.g. in the cases of pollution with arsenic, and salt.

For the polluting species, a distinction can be made between dissolved and immiscible, and between conservative and reactive. Dissolved pollutants (aqueous phase pollutants) will spread with the groundwater due to groundwater flow, diffusion and dispersion. Immiscible pollutants will spread as a separate phase (non-aqueous phase liquids, NAPL). They will contain components with very low solubility in the water phase. They constitute a long-term source for pollution.

Conservative pollutants are those that do not react with the solid soil material, do not react with other pollutants and will not be degraded by biological activity. Reactive solutes may enter or leave the water phase through adsorption/desorption, chemical reactions, dissolution/ precipitation and/or biodegradation.

1.2. Some basic definitions

Advection : the spreading of a pollutant by groundwater flow.

Diffusion : the spreading of a species dissolved in the water phase by the Brownian motion of the ions (molecules).

Dispersion : the spreading of a species dissolved in the water phase by local variations in the water velocity.

Adsorption/desorption : interaction of species dissolved in the water phase with the solid matrix. This process can be physically based or chemically based, reversible or irreversible.

Chemical reactions : reactions of species dissolved in the water phase with other species, resulting in the occurrence of different species altogether.

Biodegradation : the degradation of species dissolved in the water phase by bacteria.

Radioactive decay : the degradation of species by radioactivity.

Concentrations of species in the water phase C_i (including pure water itself) are defined as the mass of the species per unit volume: *kg/m*^{3}, *g/l, mg/l*, etc.

The density of a multi-component fluid, consisting of N components, is then given as:

$$\rho = \sum_{i=1}^{N} C_i \tag{1}$$

Mass fractions ω of the components (mass per unit of mass: *kg/kg, g/g, etc.)* are defined as:

$$\omega_i = \frac{C_i}{\rho} \text{ such that } \sum_{i=1}^{N} \omega_i = 1 \tag{2}$$

For dilute solutions (tracer concentrations) all mass fractions $\omega_i \ll 1$, except for the pure water. This means that the density of the fluid is close to the density of pure water, and can be assumed to be constant.

Water density is a function of pressure, temperature and composition. This last dependence is only important at high concentrations. E.g. in case of seawater intrusion, or in deep saline aquifers which are sometimes used to store waste or to produce energy. In these deep aquifers salt concentrations can be as high as 300 *g/l*, resulting in a water density of 1200 *g/l* (giving a salt mass fraction of 0.25). Water density fluctuations will also play a role in the subsurface storage of heat.

Water viscosity is a function of pressure, temperature and composition. This influences the hydraulic conductivity (see next section). The dependence on the temperature is by far the most important. Hence, this dependence must be taken into account in the analysis of subsurface storage of heat.

1.3. Groundwater flow

Groundwater flow is described by Darcy's law. Darcy's law is in principle the form of the momentum balance (Navier-Stokes equation), averaged over a large number of pores. It also follows from a balance of forces on water flowing through a porous medium.

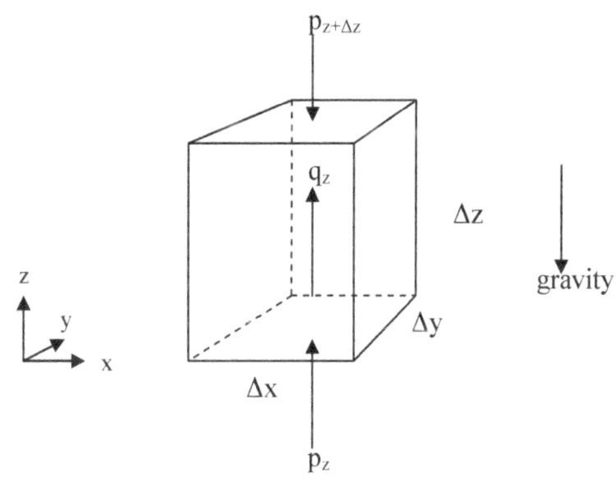

Figure 1. Balance of forces on water in a porous medium

Consider flow in the z-direction (see Figure 1). The net forces (positive upward) working on a body of water with dimensions Δx, Δy and Δz are:

Pressure forces: $(p(z)-p(z+\Delta z))\Delta x\Delta y$

Gravity forces: $-\rho g\Delta x\Delta y\Delta z$

Friction forces: $-R'q_z\Delta x\Delta y\Delta z$

where R' is the resistance factor and q_z is the specific discharge (Darcy velocity) in the z-direction.

For the pressure forces, we can make the following approximation by using a first order Taylor series expansion:

$$p_z - p_{z+\Delta z} = p_z - \left(p_z + \frac{\partial p}{\partial z}\Delta z\right) = -\frac{\partial p}{\partial z}\Delta z \tag{3}$$

Then setting up the force balance, we find:

$$q_z = -\frac{1}{R'}\left(\frac{\partial p}{\partial z} + \rho g\right) = -\frac{\kappa_z}{\mu}\left(\frac{\partial p}{\partial z} + \rho g\right) \tag{4}$$

where we have assumed that the resistance factor R' is proportional to the liquid viscosity μ. κ_z is the intrinsic permeability in the z-direction (L^2), which is assumed to be a property of the porous medium. The intrinsic permeability of a porous medium is largely determined by the pore sizes and shapes. A strong correlation between permeability and porosity exists. Similar expressions can be obtained for the flow in x and y direction:

$$q_x = -\frac{\kappa_x}{\mu}\frac{\partial p}{\partial x} \qquad q_y = -\frac{\kappa_y}{\mu}\frac{\partial p}{\partial y} \tag{5}$$

Basic assumptions in this derivation are that the acceleration of the water can be neglected, and that the friction forces are linear dependent on the velocity.

The latter is not always true (especially at high water velocities, e.g. close to an abstraction or infiltration well), in which case Darcy's law is not valid, but should be replaced by Forchheimer's equation:

$$q_x + \beta q_x^2 = -\frac{\kappa_x}{\mu}\frac{\partial p}{\partial x} \qquad q_y + \beta q_y^2 = -\frac{\kappa_y}{\mu}\frac{\partial p}{\partial y} \qquad q_z + \beta q_z^2 = -\frac{\kappa_z}{\mu}\left(\frac{\partial p}{\partial z} + \rho g\right) \tag{6}$$

where β is again a property of the porous medium.

Define a piezometric head h as:

$$h = \frac{p}{\rho g} + z \quad or \quad p = \rho g\left(h - z\right) \tag{7}$$

Basically, the piezometric head consists of a pressure head $p/\rho g$ and the vertical position z with respect to the reference level. It is the position of the top of the water column in an observation well with respect to the reference level (usually mean sea level). This is different from unsaturated flow, which is formulated in terms of the pressure head.

Substitution of equation (7) in equation (4), assuming that the density ρ is constant then gives Darcy's law in terms of the groundwater head h:

$$q_z = -\frac{\kappa_z}{\mu}\left(\frac{\partial p}{\partial z} + \rho g\right) = -\frac{\kappa_z}{\mu}\left(\rho g\left(\frac{\partial h}{\partial z} - 1\right) + \rho g\right) = -\frac{\kappa_z \rho g}{\mu}\frac{\partial h}{\partial z} \tag{8}$$

and similar expressions can be obtained for q_x and q_y.

Consequently, if the density ρ and the viscosity μ are constant, we can define hydraulic conductivities as:

$$k_x = \frac{\kappa_x \rho g}{\mu} \qquad q_x = -k_x \frac{\partial h}{\partial x} \tag{9}$$

and the same for the y and z direction. This shows that the hydraulic conductivity k (L/T) is dependent on the fluid properties.

The groundwater flow equation follows from a mass balance for the complete water phase (including all dissolved species). Consider the element as depicted in Figure 2 with dimensions $\Delta x, \Delta y$ and Δz. The net mass influx over a period Δt in the x-direction is given by:

$$\begin{aligned}&\left(\rho q_x(x) - \rho q_x(x+\Delta x)\right)\Delta y \Delta z \Delta t \approx \left(\rho q_x(x) - \left(\rho q_x(x) + \frac{\partial(\rho q_x)}{\partial x}\Delta x\right)\right)\Delta y \Delta z \Delta t \\ &= -\frac{\partial(\rho q_x)}{\partial x}\Delta x \Delta y \Delta z \Delta t\end{aligned} \tag{10}$$

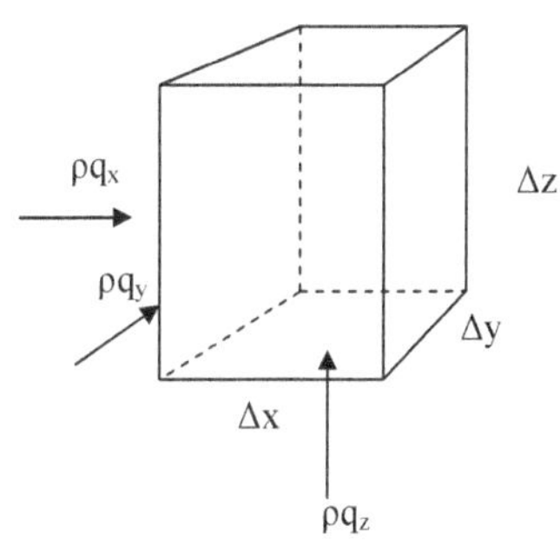

Figure 2. Water balance in a porous medium

A similar expression can be obtained for the net mass influx in the y and z directions. The change in the total mass in the element is given by:

$$\left(n\rho(t+\Delta t) - n\rho(t)\right)\Delta x \Delta y \Delta z \approx \frac{\partial}{\partial t}(n\rho)\Delta x \Delta y \Delta z \Delta t \tag{11}$$

where n is the porosity.

Equating the net mass influx and the change in mass gives the mass balance equation for the liquid phase:

$$\frac{\partial}{\partial t}(n\rho) + \frac{\partial}{\partial x}(\rho q_x) + \frac{\partial}{\partial y}(\rho q_y) + \frac{\partial}{\partial z}(\rho q_z) = 0 \tag{12}$$

For situations with varying fluid properties (salt water intrusion, storage of heat, etc.) this equation together with the pressure formulation of Darcy's law (equations (4) and (5)) should be used. Note, that the flow equation in that case is non-linear. Note also, that in these cases, even though a piezometric head can be defined, it will not be the driving force for groundwater flow.

If the density of the liquid ρ and the porosity n is assumed to be dependent on the pressure p only, the time derivative in the mass balance equation can be written as:

$$\frac{\partial}{\partial t}(n\rho) = \left[\rho\frac{\partial n}{\partial p} + n\frac{\partial \rho}{\partial p}\right]\frac{\partial p}{\partial t} = \rho g\left[\rho\frac{\partial n}{\partial p} + n\frac{\partial \rho}{\partial p}\right]\frac{\partial h}{\partial t} = \rho S_s\frac{\partial h}{\partial t} \qquad (13)$$

where S_s is the specific storage. Combining this equation with the piezometric head formulation of Darcy's law (equations (8) and (9)) and division by the (constant) density gives the well-known groundwater flow equation:

$$S_s\frac{\partial h}{\partial t} - \frac{\partial}{\partial x}\left(k_x\frac{\partial h}{\partial x}\right) - \frac{\partial}{\partial y}\left(k_y\frac{\partial h}{\partial y}\right) - \frac{\partial}{\partial z}\left(k_z\frac{\partial h}{\partial z}\right) = 0 \qquad (14)$$

Note, that the average pore water velocity v is different from the specific discharge q :

$v = q/n$.

2. Simplified description of processes in reactive transport

2.1. General

Similar to the water balance, we can derive a general form for the mass balance of a dissolved component in groundwater. Assume that the mass fluxes in x, y and z-directions are given by F_x, F_y and F_z (M/L^{2}T) respectively (see Figure 3).

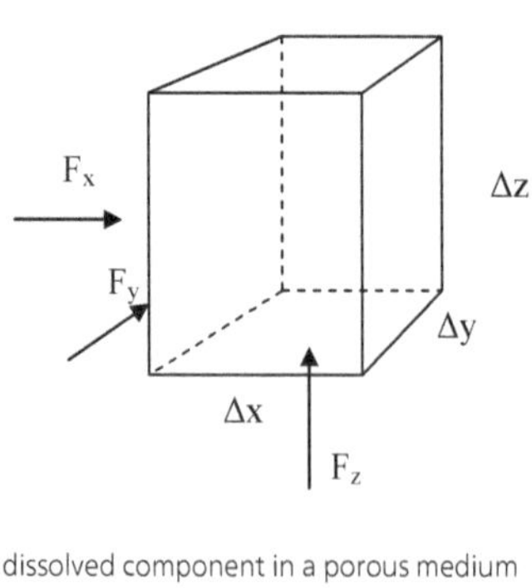

Figure 3. General mass balance for a dissolved component in a porous medium

The net mass influx in the x-direction over a period Δt is then given by:

$$\begin{aligned}&\left(F_x(x)-F_x(x+\Delta x)\right)\Delta y\Delta z\Delta t\approx\left(F_x(x)-\left(F_x(x)+\frac{\partial F_x}{\partial x}\Delta x\right)\right)\Delta y\Delta z\Delta t\\&=-\frac{\partial F_x}{\partial x}\Delta x\Delta y\Delta z\Delta t\end{aligned}\tag{15}$$

and similar expressions can be obtained for the net mass influx in the y and z-directions.

The change in mass of the component in the element over a period Δt is given by:

$$\left(nC\left(t+\Delta t\right)-nC\left(t\right)\right)\Delta x\Delta y\Delta z\approx\frac{\partial}{\partial t}\left(nC\right)\Delta x\Delta y\Delta z\Delta t\tag{16}$$

Due to the different processes occurring, mass of a component can be produced or lost in a period, e.g. because of adsorption/desorption, chemical reactions, decay, etc. The loss of mass due to these processes per unit volume and unit time will be indicated by I (M/L^3T). Note, that I can be either positive (loss of mass) or negative (gain of mass). Combining the different terms then gives the following general mass balance equation:

$$\frac{\partial}{\partial t}\left(nC\right)+\frac{\partial}{\partial x}\left(F_x\right)+\frac{\partial}{\partial y}\left(F_y\right)+\frac{\partial}{\partial z}\left(F_z\right)+I=0\tag{17}$$

In the following, the mass fluxes and/or the mass production associated with the different processes playing a role will be given. For the time being, simplified (linear) expressions will be given, which will result in a mass balance equation in the form of the classical Advection-Dispersion (or Convection Dispersion) equation, CDE. Later, more complicated expressions will be covered.

2.2. Advection

Advection (or convection) is the transport of dissolved components by flowing groundwater. The mass transport per unit area of porous medium of a dissolved component by flowing groundwater is given by:

$$F_x=q_xC\tag{18}$$

where F_x is the mass flux of the component in the x-direction (M/L^2T), q_x is the specific discharge of water (Darcy velocity) in the x-direction (L^3/L^2T) and C is the concentration of the component in the water phase (M/L^3). No mass is produced or lost, hence, $I=0$.

The underlying assumption is that the average velocity of the ions or molecules of the dissolved substance is the same as the average water velocity: if we move one liter of water over a certain distance, also all chemicals in that liter will have moved that distance. In most cases, this will be true, but there are exceptions. These exceptions occur e.g. when the molecules of the dissolved substance are very large (colloids, virus). If we consider the flow of water in a capillary, the water velocity v at a distance r from the centre is given by:

$$v = 2v_{avg}\left(1 - \frac{r^2}{r_0^2}\right) \tag{19}$$

where v_{avg} is the average water velocity and r_0 is the radius of the capillary. For large molecules, only part of the capillary is available for transport. That can be caused by either the size of the molecules or by the electrical charges on the surface. As a result, the average velocity of such particles in a capillary will exceed the average velocity of the water itself. If the radius of such particles is given by r_c, it can easily be inferred that the average velocity of the particles compared to the average water velocity of the water is given by:

$$\frac{v_c}{v_{avg}} = 1 + 2\alpha + \alpha^2 \qquad \alpha = \frac{r_c}{r_0} \tag{20}$$

(find the average water velocity by integrating the water velocity from 0 to r_0, and find the average particle velocity by integrating the water velocity from 0 to r_0-r_c). For instance, for particles with a size of 20% of the capillary diameter, the average velocity is some 40% larger than the water velocity. These effects have been observed in virus and colloid transport.

If only advective transport takes place, the mass balance for a component follows from (17) and (18):

$$\frac{\partial}{\partial t}(nC) + \frac{\partial}{\partial x}(q_x C) + \frac{\partial}{\partial y}(q_y C) + \frac{\partial}{\partial z}(q_z C) = 0 \tag{21}$$

Now consider the 1-dimensional mass balance equation with constant porosity n and constant specific discharge q. Division by n then gives the following equation:

$$\frac{\partial}{\partial t}(C) + \frac{q}{n}\frac{\partial}{\partial x}(C) = \frac{\partial}{\partial t}(C) + v\frac{\partial}{\partial x}(C) = \frac{DC}{Dt} = 0 \tag{22}$$

where DC/Dt is the material derivative, i.e. the change in concentration when moving along with a water particle. Since this derivative is zero, advective transport only results in a displacement of the initial concentration distribution by vt, where t is the elapsed time. This is also true in 3 dimensions.

Equation (22) can be written in dimensionless form by defining the following dimensionless variables:

$$C_d = \frac{C}{C_r} \quad t_d = \frac{t}{t_r} \quad x_d = \frac{x}{L_r} \tag{23}$$

where C_r, t_r and L_r are reference or characteristic values for the system considered. Substitution of these dimensionless variables in the mass balance equation (22) gives:

$$\frac{\partial C_d}{\partial t_d} + \frac{vt_r}{L_r}\frac{\partial C_d}{\partial x_d} = 0 \tag{24}$$

We can now choose any one of the characteristic values t_r or L_r such that the coefficient in front of the spatial derivative in (24) is 1. This means that for a given characteristic time t_r, the characteristic length is given by vt_r, while for a given characteristic length L_r, the characteristic time is given by L_r/v. These characteristic values obviously are related to respectively the travel distance and the travel time of water particles.

The specific discharge that is required to quantify the advective fluxes follows from the mass balance equation for the water phase in combination with Darcy's law. Basically, this means that (local) information about the value of the permeability (or hydraulic conductivity) is required.

2.3. Diffusion

Diffusion is the spreading of a component dissolved in the water phase by the Brownian motion of the molecules/ions. In open water, the mass flux due to diffusion is given by Fick's first law:

$$F_x = -D_m \frac{\partial C}{\partial x} \tag{25}$$

where D_m is the molecular diffusion coefficient (L^2/T), which is typical for the component considered. In a porous medium, the mass flux due to diffusion is given by a similar expression:

$$F_x = -n\frac{D_m}{\tau}\frac{\partial C}{\partial x} = -nD_{eff}\frac{\partial C}{\partial x} \tag{26}$$

The porosity n enters the equation to account for the area that is effectively available for mass transport. τ is the tortuosity of the porous medium (-), which accounts for the fact that the length of the path molecules or ions have to take in a porous medium to travel from one position to another is larger than the distance between these positions. For normal porous media, τ has a value in the order of 1.6 to 1.7. No mass is produced or lost due to diffusion, hence $I=0$.

In case only diffusion occurs, the mass balance equation reads:

$$\frac{\partial}{\partial t}(nC)-\frac{\partial}{\partial x}\left(nD_{eff}\frac{\partial C}{\partial x}\right)-\frac{\partial}{\partial y}\left(nD_{eff}\frac{\partial C}{\partial y}\right)-\frac{\partial}{\partial z}\left(nD_{eff}\frac{\partial C}{\partial z}\right)=0 \tag{27}$$

Now consider a 1-dimensional mass balance equation with constant porosity n and diffusion coefficient D_{eff}:

$$\frac{\partial C}{\partial t}-D_{eff}\frac{\partial^2 C}{\partial x^2}=0 \tag{28}$$

For this equation, numerous analytical solutions dependent on the boundary conditions are known, both in Cartesian and radial coordinate systems. The steady state solution in a Cartesian coordinate system is:

$$C = Ax + B \tag{29}$$

where A and B are determined by the boundary conditions. Note, that this solution is not dependent on the effective diffusion coefficient.

Equation (28) can be written in dimensionless for by defining appropriate characteristic values for the concentration and the time and length scales:

$$\frac{\partial C_d}{\partial t_d}-\frac{D_{eff}t_r}{L_r^2}\frac{\partial^2 C_d}{\partial x_d^2}=0 \tag{30}$$

Setting the coefficient in front of the spatial derivative to 1, characteristic values for the time and the length are found. For a given characteristic time t_r, the characteristic length is given by $L_r=\sqrt{(D_{eff}t_r)}$, and for a given characteristic length L_r, the characteristic time is given by $t_r=L_r^2/D_{eff}$.

In general, molecular diffusion will not play an important role in porous media transport, unless the groundwater velocities are very small (which is e.g. the case for transport through very low permeable clays).

Measurement of effective diffusion coefficients in a porous medium is usually done in the laboratory by performing time dependent experiments

2.4. Dispersion

Dispersion is the spreading of a dissolved component due to local variations in the ground-water velocity. In general, we distinguish mechanical and hydrodynamic dispersion.

Mechanical dispersion takes place on the pore scale, and is caused by velocity variations across the cross section of the capillaries (or pores). Usually the groundwater velocities are so small, as are the pore diameters, that molecular diffusion is fast enough to balance concentration differences in the direction perpendicular to the flow (i.e. across the pores).

Hydrodynamic dispersion is the sum of molecular diffusion and mechanical dispersion. It usually occurs on a larger scale than a single pore, and is caused by all variations in the average groundwater velocity (i.e., averaged over a large number of pores) that we did not account for explicitly, including diffusion (Figure 4). Thus, if we consider layers with different values of the hydraulic conductivity (or permeability), this variation does not necessarily give rise to hydrodynamic dispersion. However, it is clear that such variation certainly may lead to variation of the rate of displacement of chemicals and to true mixing, if it combines with diffusion. This is commonly called macro or mega dispersion (see e.g. Dagan, 1987), and considered later in this chapter (section 7).

For both mechanic and hydrodynamic dispersion, the mass fluxes are assumed to be given by the following form of Fick's first law:

$$F_x = -D_{xx}\frac{\partial C}{\partial x} - D_{xy}\frac{\partial C}{\partial y} - D_{xz}\frac{\partial C}{\partial z} \qquad (31)$$

where D_{xx}, D_{xy} and D_{xz} are elements of the dispersion tensor (L^2/T). Similar expressions are valid for the mass fluxes in y and z-direction. The dispersion tensor is symmetric, and consists of 6 different numbers, D_{xx}, D_{xy}=D_{yx}, D_{xz}=D_{zx}, D_{yy}, D_{yz}=D_{zy} and D_{zz}. The elements of the dispersion tensor are dependent on the groundwater velocity v, such that the dispersion coefficients in the direction of the flow and perpendicular to the flow are given by:

$$D(//v) = \alpha_l |v| \qquad D(\perp v) = \alpha_t |v| \qquad (32)$$

where α_l and α_t are the longitudinal and transversal dispersivities (L) respectively. These are assumed to be properties of the porous medium, and indicate the size of heterogeneities in the system that is not accounted for by variations in the (average) groundwater velocity. Because mass transport by hydrodynamic dispersion and by molecular diffusion is described by the same law, they are usually combined.

In a fully 3-dimensional system, with velocity components v_x, v_y and v_z respectively, the elements of the hydrodynamic dispersion (including molecular diffusion) are given by:

$$
\begin{aligned}
D_{xx} &= D_{eff} + \alpha_t |v| + (\alpha_l - \alpha_t)\frac{v_x^2}{|v|} \\
D_{yy} &= D_{eff} + \alpha_t |v| + (\alpha_l - \alpha_t)\frac{v_y^2}{|v|} \\
D_{zz} &= D_{eff} + \alpha_t |v| + (\alpha_l - \alpha_t)\frac{v_z^2}{|v|} \\
D_{xz} &= D_{zx} = (\alpha_l - \alpha_t)\frac{v_x v_z}{|v|} \\
D_{yz} &= D_{zy} = (\alpha_l - \alpha_t)\frac{v_y v_z}{|v|} \\
D_{xy} &= D_{yx} = (\alpha_l - \alpha_t)\frac{v_x v_y}{|v|}
\end{aligned}
\tag{33}
$$

Because hydrodynamic dispersion occurs only in combination with groundwater flow, a mass balance for a component follows from the combination of mass fluxes as defined by equations (18) and (31), with the elements of the dispersion tensor given by equation (33). Using a short hand notation, this mass balance is given by:

$$\frac{\partial}{\partial t}(nC) + \nabla \cdot (qC) - \nabla \cdot (nD \cdot \nabla C) = 0 \tag{34}$$

Now, consider a one-dimensional system with constant porosity, velocity and dispersion coefficient:

$$\frac{\partial C}{\partial t} + v\frac{\partial C}{\partial x} - D\frac{\partial^2 C}{\partial x^2} = 0 \tag{35}$$

Making this equation dimensionless by choosing appropriate characteristic values for the concentration, time and length gives:

$$\frac{\partial C_d}{\partial t_d} + \frac{v t_r}{L_r}\frac{\partial C_d}{\partial x_d} - \frac{D t_r}{L_r^2}\frac{\partial^2 C_d}{\partial x_d^2} = 0 \tag{36}$$

Choosing the characteristic time $t_r = L_r / v$ then gives the following dimensionless mass balance equation:

$$\frac{\partial C_d}{\partial t_d} + \frac{\partial C_d}{\partial x_d} - \frac{1}{Pe}\frac{\partial^2 C_d}{\partial x_d^2} = 0 \qquad Pe = \frac{vL_r}{D} \tag{37}$$

where *Pe* is the Peclet number. This number is characteristic for the ratio of advective transport and dispersive transport. The solution of (57) depends only on *Pe*, and large Peclet numbers indicate that advection dominates; small Peclet numbers indicate that dispersion dominates.

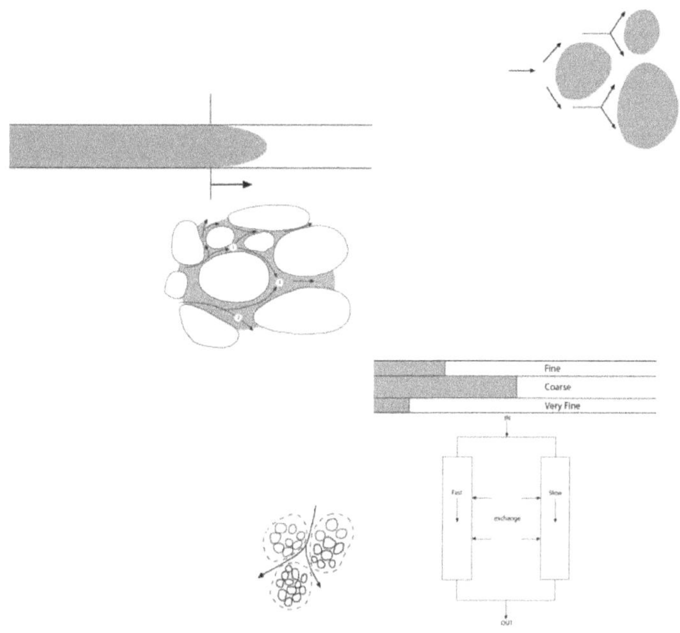

Figure 4. Illustration of different mechanisms of dispersion. Mixing occurs due to velocity variations within pores, and between pores, in combination with diffusional mixing at locations where water flows with different concentrations meet. In addition, flow velocities may be aligned in the mean flow direction, but it can also have components that are at an angle with this direction. Also larger scale variations in flow velocity, due to aggregation and layering, may lead to enhanced mixing.

For mechanical dispersion, the longitudinal dispersivity α_l is in the order of the pore sizes (mm scale). For hydrodynamic dispersion, the longitudinal dispersivity is dependent on the scale of the problem. For laboratory experiments in columns, values of less than 1 mm to values larger than 1 cm have been reported. For field scale experiments values larger than 10 m have been reported, dependent on the size of the experiment and the heterogeneity of the aquifer in which the experiment was performed.

A cautioning remark is needed with regard to such large dispersivities, as such large values cannot possibly be due to complete true mixing of water of different composition in porous media. Such large values are commonly obtained based on several methodological complications: (i) the equipment used to measure 'local' concentrations (e.g. observation wells, geophysical methods) are themselves responsible for mixing, and (ii) modelling with large spatiotemporal discretization in view of computational efficiency may lead to numerical mixing, and (iii) dispersivities may be 'fitted' using a relatively simple transport equation (e.g. a one dimensional version of the transport equation), which leads to artifacts.

Based on field experiments, an empirical relation for the longitudinal dispersivity has proposed as:

$$\alpha_l \approx 0.0175L^{1.46} \tag{38}$$

where L is a characteristic length for the domain of interest. For large scales (large values of L), an upper bound for α_l is reached. Note, that relation (38) is based on the evaluation of a number of field experiments, and gives an estimate of the dispersivities only. Also note, that these estimates are based on the assumption that the aquifer permeability is homogeneous over the domain of interest.

As a guideline, the transversal dispersivity is commonly assumed to be 5 to 10% of the longitudinal dispersivity.

2.5. Adsorption/desorption

Adsorption/desorption creates a sink/source term for a component in the water phase. Two processes take place at the same time: molecules/ions will attach to the solid material, and attached molecules/ions can be released from the solid to the water phase. For the time being, we will adopt a linear description of the process, corresponding with case A of Figure 5:

$$I = k_aC - k_dC_s = k_d\left(\frac{k_a}{k_d}C - C_s\right) = k_d\left(K_dC - C_s\right) \tag{39}$$

where k_a and k_d are the attachment (1/T) and detachment (M/L^3T) coefficients respectively, and C_s is the concentration of the component adsorbed (M/M). k_a and k_d have different units because of the different units for C and C_s respectively. In this formulation, no equilibrium has been assumed. K_d is the distribution coefficient. In case of equilibrium, the expression in (39) between brackets is 0, hence:

$$C_s = K_dC \tag{40}$$

which defines a linear adsorption isotherm. In principle, such a relation is only valid if the concentrations are very low, and if equilibrium between the water phase and the solid material exists.

If advection, dispersion and linear adsorption/desorption occur, the mass balance equation for a component in the water phase can be given as:

$$\frac{\partial}{\partial t}(nC)+\nabla\cdot(qC)-\nabla\cdot(nD\cdot\nabla C)+k_aC-k_dC_s=0 \quad (41)$$

Note, that this equation is not based on the assumption of local equilibrium. Also, two unknowns are present in this equation: C and C_s. The other equation required to solve for the concentrations is given by a mass balance equation for the adsorbed component:

$$\frac{\partial}{\partial t}\left((1-n)\rho_sC_s\right)-k_aC+k_dC_s=0 \quad (42)$$

where ρ_s is the density of the solid material. Basically, this mass balance equation is comparable to equation (41), because the adsorbed component is not transported (no advection and dispersion), and that the source term due to adsorption/desorption has the opposite sign.

Adding equations (41) and (42) gives the total mass balance:

$$\frac{\partial}{\partial t}(nC)+\frac{\partial}{\partial t}\left((1-n)\rho_sC_s\right)+\nabla\cdot(qC)-\nabla\cdot(nD\cdot\nabla C)=0 \quad (43)$$

If we now assume equilibrium, equation (40) can be used to eliminate C_s from equation (43):

$$\frac{\partial}{\partial t}(nRC)+\nabla\cdot(qC)-\nabla\cdot(nD\cdot\nabla C)=0 \quad R=1+\frac{1-n}{n}\rho_sK_d \quad (44)$$

where R is the retardation factor. Note, that for non-reactive solutes (no adsorption/desorption) R=1. It is clear from this equation, that the retardation factor is only found in the time-derivative term. For this reason, it got its name, as this factor R implies that both ad/convection and dispersion are R times slower: they are retarded by a factor R.

Now consider a one-dimensional form of this mass balance, with constant porosity, velocity, dispersion coefficient and retardation factor:

$$\frac{\partial C}{\partial t}+\frac{v}{R}\frac{\partial C}{\partial x}-\frac{D}{R}\frac{\partial^2 C}{\partial x^2}=0 \quad (45)$$

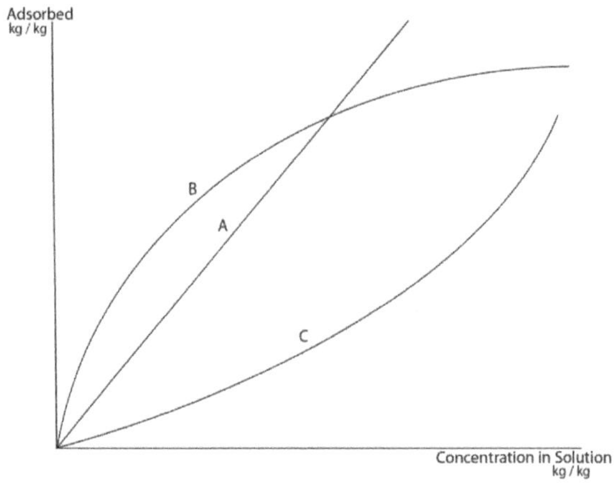

Figure 5. linear (A) and nonlinear adsorption, with B similar to Freundlich and Langmuir type equations and C resembling precipitation controlled reactions, if the nonlinearity is much more distinct than in this figure.

Note that this equation is identical to equation (35), the mass balance equation for a non-reactive component, with both the velocity v and the dispersion D scaled by a factor R. If the molecular diffusion can be neglected, the dispersion coefficient is proportional to the velocity v, and equation (45) will give the same results as the mass balance equation for a non-reactive component with a velocity that is decreased by a factor R.

Now consider the one-dimensional form of equation (41) (non-equilibrium) with constant porosity, velocity, dispersion coefficient and attachment and detachment constants. This equation can be made dimensionless by choosing appropriate values for the characteristic concentration, time and length. If (as done before) we define the characteristic time $t_r=L_r/v$, the non-dimensional equation is given by:

$$\frac{\partial C_d}{\partial t_d}+\frac{\partial C_d}{\partial x_d}-\frac{1}{Pe}\frac{\partial^2 C}{\partial x^2}+\frac{k_a L_r}{nv}C_d-\frac{k_d L_r}{nv}C_{sd}=0 \tag{46}$$

The last two coefficients in this equation are two forms of the dimensionless Damkohler number. This number gives the ratio of the groundwater travel time and the time required to reach equilibrium. Large Damkohler numbers indicate that the assumption of local equilibrium is appropriate, while small Damkohler numbers indicate that adsorption/desorption should be described as a non-equilibrium process.

Measurement of the adsorption distribution coefficient K_d is commonly done in a laboratory batch experiment. A soil sample is mixed with water that contains a dissolved component at a certain concentration. This mixture is stirred gently for a long time in order to assure that equilibrium between the water phase and the soil is establihed. From a measurement of the resulting concentration in the water phase at equilibrium, the amount adsorbed can be determined, and the distribution coefficient calculated:

Assume we have a mass of soil M_s. We add a volume of water V which has dissolved in it a component at concentration C_i. At equilibrium, the concentration in the water phase is measured as C_{eq}. The amount of mass of the component added to the system is VC_i. At equilibrium, the total mass of the component in the water phase is VC_{eq}. Consequently, the total mass adsorbed is $V(C_i\text{-}C_{eq})$, and the concentration of adsorbed component is $C_s=V(C_i\text{-}C_{eq})/M_s$ while the concentration in the water phase is C_{eq}. The value of the distribution coefficient follows directly from equation (40).

Measurement of the attachment and detachment coefficients can be done in batch experiments by measuring the concentration in the water phase as a function of time.

Interaction of solutes do not only take place with the solid material, but can also exist with colloïdal particles (e.g. natural organic material), which in itself are mobile. Consequently, a competition between adsorption on the solid matrix and adsorption on colloïds may occur. This may lead to an enhanced transport of species (e.g. heavy metals) that may otherwise be considered to be highly retarded.

2.6. Decay

For the simplified description given in this chapter, we will assume that the decay due to chemical reactions, biological activity and/or radioactivity is given by a first order expression:

$$I = n\lambda C \tag{47}$$

where λ is the decay/degradation constant. λ is related to the half life $t_{1/2}$ of the component by:

$$\lambda = \frac{\ln(2)}{t_{1/2}} \tag{48}$$

The halflife $t_{1/2}$ is commonly measured in batch experiments by mixing a sample of soil material with water which has the component dissolved in it. Measuring the concentration in the water phase as a function of time will give an estimate of the decay. In these experiments, adsorption/desorption should also be taken into account. The process of first order decay/degradation is of great importance for much of the transport theory. As may be already apparent, radionuclides decay proportional to the total decaying mass present. For instance the recent tsunami accident with the Fukushima nuclear plant in Japan may have resulted in soil contamination with radionuclides, where the decay rate determines the period for which radiation problems may be acute. Likewise, the Chernobyl melt down resulted in continental scale contamination with radionuclides by different elements, that move towards groundwater with different rates, and different degradation rates. To appreciate the hazard for life, the rate of downward movement of chemicals in relation with the decay rate of hazardous radiation is a typical transport problem.

The first order degradation rate law is also the most commonly used rate law for describing the degradation of contaminants such as pesticides, nutrient chemicals such as nitrate, contaminants such as PAHs, BTEX, chlorinated hydrocarbons (the last under anaerobic conditions), and other contaminants (Keijzer et al., 1999, Jaesche et al., 2006, French et al., 2009), despite that it ignores that transformation products may be hazardous too.

The importance of degradation can be appreciated from an example of groundwater (rather than soil) contamination. About, say, a decade ago, the concept of natural attenuation has been developed. This concept proposes that the subsoil environment is able to cause natural degradation of contaminants, e.g. due to the intrinsic activity of microbial populations. Although dispersional mixing and dilution, as well as volatilization of chemicals may contribute to natural attenuation, degradation is a major process in this concept. The concept as such is important as it diminishes the environmental hazards of soil contamination, and therefore has become a major issue in soil and groundwater contamination strategies, management, and decision making.

2.7. Full simplified mass balance equation

A full mass balance equation assuming linear equilibrium adsorption and first order decay can now be written as:

$$\frac{\partial}{\partial t}(nRC)+\nabla\cdot(qC)-\nabla\cdot(nD\cdot\nabla C)+n\lambda RC=0 \tag{49}$$

where it has also been assumed that the component dissolved in the water phase as well as the component adsorbed onto the solid phase can both decay with the same decay constant. In reality, this is not necessarily true (Beltman et al., 2008).

This equation is the foundation of most software aimed at modelling soil and groundwater contamination, such as MODFLOW/MT3D and related models. As such, this equation is the core of much scientific as well as management supporting investigations done at international, national and local levels. Few, if any, predictions and prognoses are made on the fate of contaminants, that are not based on equation (49).

For some one-dimensional and simple two- or een three-dimensional problems, analytical solutions exist for equation (49). A number of these solutions have been programmed and are available at the web site:

www.cee.uiuc.edu/transport

With those solutions, it is easy to obtain an impression of the effects of the different processes and their parameters on the transport behaviour of dissolved chemicals.

3. Some effects in the numerical solution of the transport equation

3.1. General

In many cases, analytical solutions of the simplified transport equation are not available. That is e.g. the case for heterogeneous systems or complicated boundary conditions. In those cases one has to resort to the numerical solution of the groundwater flow equation and the solute transport equation.

In order to analyse the behaviour of the numerical solution of the partial differential equation describing the transport of a solute, we consider a simplified 1-dimensional system with constant porosity n, constant specific discharge q and constant dispersion coefficient D:

$$\frac{\partial C}{\partial t}+v\frac{\partial C}{\partial x}-D\frac{\partial^2 C}{\partial x^2}=0 \tag{50}$$

For the evaluation of the spatial derivatives in a finite difference approach, we use the following Taylor series expansion:

$$\begin{aligned}C(x+\Delta x)&=C(x)+\Delta x\frac{\partial C}{\partial x}+\frac{(\Delta x)^2}{2}\frac{\partial^2 C}{\partial x^2}+\frac{(\Delta x)^3}{6}\frac{\partial^3 C}{\partial x^3}+....\\ C(x-\Delta x)&=C(x)-\Delta x\frac{\partial C}{\partial x}+\frac{(\Delta x)^2}{2}\frac{\partial^2 C}{\partial x^2}-\frac{(\Delta x)^3}{6}\frac{\partial^3 C}{\partial x^3}+....\end{aligned} \tag{51}$$

where Δx is the blocksize in the x-direction.

The first order derivative of C with x can now be approximated in two ways. A backward finite difference approximation follows from the second equation given by (51):

$$\frac{\partial C}{\partial x}\approx\frac{C(x)-C(x-\Delta x)}{\Delta x}+\frac{\Delta x}{2}\frac{\partial^2 C}{\partial x^2}-...... \tag{52}$$

Another approximation for the first order derivative can be obtained by taking the difference of the two equations given in (51):

$$\frac{\partial C}{\partial x}\approx\frac{C(x+\Delta x)-C(x-\Delta x)}{2\Delta x}+\frac{(\Delta x)^2}{6}\frac{\partial^3 C}{\partial x^3}+...... \tag{53}$$

An approximation for the second order spatial derivative is obtained by adding the two equations (51). After some manipulation, the following approximation is obtained:

$$\frac{\partial^2 C}{\partial x^2} \approx \frac{C(x+\Delta x)-2C(x)+C(x-\Delta x)}{(\Delta x)^2} + \frac{(\Delta x)^2}{24}\frac{\partial^4 C}{\partial x^4} + ... \quad (54)$$

where a higher order term not given in equation (51) has been taken into account. The truncation error for the approximation of the first order derivative is for the backward difference (52) of the order Δx, and for the central difference of the order $(\Delta x)^2$. In other words, the central difference approximation is more accurate. The truncation error for the approximation of the second order derivative is of the order $(\Delta x)^2$.

For the time derivative the following approximation can be obtained:

$$\frac{\partial C}{\partial t} \approx \frac{C(t)-C(t-\Delta t)}{\Delta t} + \frac{\Delta t}{2}\frac{\partial^2 C}{\partial t^2} - ...... \quad (55)$$

where Δt is the time step and the derivative in the higher order term is evaluated at time $t+\Delta t$ (i.e. at the end of the time step). Note, that the truncation errors in the approximation for the spatial and temporal derivatives indicate, that small grid blocks (small Δx) should be used where the second order derivative of the concentration with respect to x is large, and small time steps should be used when the second order derivative with respect to t is large.

In the following, we will adopt the notation: $C_i=C(x)$, $C_{i-1}=C(x-\Delta x)$, $C_{i+1}=C(x+\Delta x)$, while values evaluated at the beginning of a time step will have a superscript o, and values evaluated at the end of a time step will have a superscript n.

If we evaluate the spatial derivatives at a time level between the beginning of the time step and the end of the time step, the discretised mass balance equation can then be written as:

$$\frac{C_i^n - C_i^o}{\Delta t} + v\left[\theta\frac{C_i^n - C_{i-1}^n}{\Delta x} + (1-\theta)\frac{C_i^o - C_{i-1}^o}{\Delta x}\right] - D\left[\theta\frac{C_{i+1}^n - 2C_i^n + C_{i-1}^n}{(\Delta x)^2} + (1-\theta)\frac{C_{i+1}^o - 2C_i^o + C_{i-1}^o}{(\Delta x)^2}\right] = 0 \quad (56)$$

where θ is a factor between 0 and 1. For $\theta=1$, all spatial derivatives are evaluated at the new time level (the end of the time step). This is a fully implicit scheme. For $\theta=0$, all spatial derivatives are evaluated at the old time level (the beginning of a timestep). This is a fully explicit scheme. A mixed scheme (known as the Crank-Nicholsen scheme) is obtained by setting $\theta=0.5$.

Note that in equation (56) a backward difference for the advective term is used. A similar expression can be obtained for a central difference approximation for the advective term.

3.2. Numerical dispersion

Numerical dispersion is an extra dispersion in the numerical solution of the transport equation which is caused by the discretisation of the advective term.

Consider the discretised equation (56), and evaluate all spatial derivatives at the end of the time step (fully implicit, θ=1):

$$\frac{C_i^n - C_i^o}{\Delta t} + v\frac{C_i^n - C_{i-1}^n}{\Delta x} - D\frac{C_{i+1}^n - 2C_i^n + C_{i-1}^n}{(\Delta x)^2} = 0 \tag{57}$$

One could now ask the question which partial differential equation is approximately solved by these equations if higher order terms are taken into account. Using equations (52), (54) and (55), keeping all terms with derivatives of order 2 then gives the following partial differential equation:

$$\frac{\partial C}{\partial t} - \frac{\Delta t}{2}\frac{\partial^2 C}{\partial t^2} + v\frac{\partial C}{\partial x} - \frac{v\Delta x}{2}\frac{\partial^2 C}{\partial x^2} - D\frac{\partial^2 C}{\partial x^2} = 0 \tag{58}$$

In order to obtain an expression for the second order derivative with respect to time, we will differentiate equation (50) with respect to time:

$$\frac{\partial^2 C}{\partial t^2} = -v\frac{\partial^2 C}{\partial x \partial t} + D\frac{\partial^3 C}{\partial x^2 \partial t} \tag{59}$$

An expression for the cross derivative term is obtained by differentiating equation (50) with respect to x:

$$\frac{\partial^2 C}{\partial x \partial t} = -v\frac{\partial^2 C}{\partial x^2} + D\frac{\partial^3 C}{\partial x^3} \tag{60}$$

Substitution of equation (60) in (59) then gives:

$$\frac{\partial^2 C}{\partial t^2} = v^2\frac{\partial^2 C}{\partial x^2} - vD\frac{\partial^3 C}{\partial x^3} + D\frac{\partial^3 C}{\partial x^2 \partial t} \tag{61}$$

Substitution of (61) in (58), collecting the second order derivative terms and neglecting higher order terms then gives:

$$\frac{\partial C}{\partial t}+v\frac{\partial C}{\partial x}-\left[D+\frac{v\Delta x}{2}+\frac{v^2\Delta t}{2}\right]\frac{\partial^2 C}{\partial x^2}=0 \tag{62}$$

Equation (62) is up to second order terms identical to the discretised equation (57). In other words, the discretised equation is an approximation to a solute transport equation with enhanced dispersion (cf. the term between brackets). Note, that in the analysis higher order terms are neglected. As a consequence, the discretised equation (57) is not completely identical to equation (62).

The extra dispersion $v\Delta x/2+v^2\Delta t/2$ is called the numerical dispersion. If we assume that molecular diffusion can be neglected, the physical dipersion is given by $D=\alpha v$, where α is the dispersivity of the medium. The numerical dispersion can now be neglected if the block size Δx and the time step Δt are chosen such that:

$$\alpha >> \frac{\Delta x}{2}+\frac{v\Delta t}{2} \tag{63}$$

Carrying out the same analysis with the advective term in equation (57) approximated by a central difference, will show that the numerical dispersion in that case is $v^2\Delta t/2$, which is smaller than the one for the backward difference of the advective term. There can be, however, reasons for adopting the backward difference approximation (see next section).

There are a number of ways in which we can neutralize the effect of numerical dispersion. The most obvious way is to correct the dispersivity for the numerical dispersion. Suppose the physical dispersivity is given by α_f, while the dispersivity defined for the numerical calculations is given by α_m. For a finite difference approximation with a backward difference for the advective term, the following choice:

$$\alpha_m=\alpha_f-\frac{\Delta x}{2}-\frac{v\Delta t}{2} \tag{64}$$

will result in a total dispersion in the numerical calculations equal to the physical dispersion. This approach can, however, only be adopted if the model dispersivity α_m remains positive because negative values of the model dispersivity will generate instabilities in the numerical solution.

More complicated ways to minimise numerical dispersion in the numerical solution of the solute transport equation can be thought of. In all cases one should consider the fact that numerical dispersion is solely caused by the first order spatial derivative (the advective term).

One way of avoiding numerical dispersion is by what is called operator splitting. In that approach the change in concentration is split in two parts: one part due to advective transport, and one part due to dispersion:

$$\left(\frac{\partial C}{\partial t}\right)_a = -v\frac{\partial C}{\partial x}$$
$$\left(\frac{\partial C}{\partial t}\right)_d = D\frac{\partial^2 C}{\partial x^2} \tag{65}$$

and adding the two contribution gives the full transport. The first part in equation (65), the advective part is now solved by a characteristic method. In this method, water particles are followed as they are transported with velocity v. Each water particle represents a certain mass of solute, which in a time step Δt is transported over a distance $v\Delta t$. Once the advective transport has been solved, the dispersion has to be added. That again can be done in a number of ways, the most simple being a finite difference approximation. Another way is the random walk method, where the effect of dispersion is simulated by random displacements of the water particles around the mean displacement given by the velocity v. These random displacements are related to the dispersion coefficient.

Using a characteristic method to simulate the advective transport has the disadvantage that only a discrete number of particles can be followed, and that interpolation is required to transform mass per particle to concentration distribution. A smooth concentration distribution from the distribution of the particles can only be obtained if a very large number of particles is used. This is especially true for the relative low concentration contours.

A method that strongly resembles the characteristic method is the Eulerian-Lagrangian method. In this method, the time derivative is approximated by taking the difference in concentration not at the same place, but at different places. Using a Taylor series expansion, it can be shown that the expression:

$$\frac{C(x,t+\Delta t)-C(x-v\Delta t,t)}{\Delta t} \approx \frac{\partial C}{\partial t}+v\frac{\partial C}{\partial x} \tag{66}$$

is second order correct, i.e. neglected derivatives are of third order or higher. Consequently, no numerical dispersion is generated by this method. In a standard finite difference method the second term in the left side of (66) is evaluated by evaluating the concentration at position $x\text{-}v\Delta t$ at the beginning of a time step. That can usually be one by interpolation, although special precautions have to be taken close to boundaries.

Each of the methods mentioned here can easily be extended to two or three dimensions.

3.3. Oscillations in the solution

Some of the finite difference schemes can, under certain conditions, generate oscillations in the solution, i.e. negative concentrations or concentrations larger than the maximum value defined by the initial and boundary conditions may occur. It should be pointed out that these are not instabilities, where errors in the solution can grow unbounded.

Consider the discretised, implicit equation, where the advective term is approximated by a backward difference (equation (57)). With some manipulation, this equation can be written as:

$$C_i^n = \beta_i C_i^o + \beta_{i-1} C_{i-1}^n + \beta_{i+1} C_{i+1}^n$$
$$\beta_i = \frac{1}{1 + \frac{v\Delta t}{\Delta x} + \frac{2D\Delta t}{(\Delta x)^2}} \quad \beta_{i-1} = \frac{\frac{v\Delta t}{\Delta x} + \frac{D\Delta t}{(\Delta x)^2}}{1 + \frac{v\Delta t}{\Delta x} + \frac{2D\Delta t}{(\Delta x)^2}} \tag{67}$$
$$\beta_{i+1} = \frac{\frac{D\Delta t}{(\Delta x)^2}}{1 + \frac{v\Delta t}{\Delta x} + \frac{2D\Delta t}{(\Delta x)^2}}$$

Equation (67) shows that the concentration at the new time level is a weighted average of the concentration at the old time level and the concentrations in the adjacent grid blocks. Inspection of the weighing coefficients β shows that:

$$\beta_i + \beta_{i-1} + \beta_{i+1} = 1 \quad and \quad \beta_i, \beta_{i-1}, \beta_{i+1} > 0 \tag{68}$$

As a consequence C obeys a maximum principle, i.e. it can never become smaller that the smallest value given in the initial and boundary conditions, or larger than the largest value given in the initial and boundary conditions.

If we would have used a central difference for the advective term an expression similar to equation (67) can be written, however, with weighting coefficients:

$$\beta_i = \frac{1}{1 + \frac{2D\Delta t}{(\Delta x)^2}}$$
$$\beta_{i-1} = \frac{\frac{D\Delta t}{(\Delta x)^2} + \frac{v\Delta t}{2\Delta x}}{1 + \frac{2D\Delta t}{(\Delta x)^2}} \tag{69}$$
$$\beta_{i+1} = \frac{\frac{D\Delta t}{(\Delta x)^2} - \frac{v\Delta t}{2\Delta x}}{1 + \frac{2D\Delta t}{(\Delta x)^2}}$$

Inspection now reveals that the sum of the weighting factors β again equals 1, but that all weighting factors are positive only under the condition:

$$\frac{D\Delta t}{(\Delta x)^2} > \frac{v\Delta t}{2\Delta x} \quad or \quad D > \frac{v\Delta x}{2} \quad or \quad \alpha > \frac{\Delta x}{2} \tag{70}$$

Condition (70) is often given in a slightly different form:

$$Pe_{cell} = \frac{v\Delta x}{D} < 2 \tag{71}$$

where Pe_{cell} is called the cell Peclet number (cf. equation (37)).

Basically, a backward difference for the advective term generates more numerical dispersion than a central difference. However, a central difference approximation might result in oscillations in the solution, which will not occur for a backward difference.

It should be pointed out that the global mass balance will in all cases by perfect, irrespective of numerical dispersion or oscillations. This also indicates that damping oscillations in a solution by simply not allowing the concentration to become larger than a predefined value or smaller than another predefined value will ultimately result in mass balance errors.

3.4. Stability of the explicit solution

The explicit formulation of the discretised equations has the advantage that an explicit expression for the concentrations in each grid block is obtained, which therefore does not require matrix manipulation to obtain the solution. However, under certain conditions, such an explicit formulation may become unstable, i.e. small errors in the solution may grow uncontrolled and unbounded in time, resulting in very large positive and negative concentrations.

Appendix A gives the derivation of the criteria for the explicit formulation to be stable. In general, conditions (139) and (140) are given in a slightly different form. Condition (139) can be written as:

$$2D < \frac{(\Delta x)^2}{\Delta t} \tag{72}$$

Substitution of this relation in condition (140) then gives:

$$\Delta t < \frac{2D}{v^2} < \frac{(\Delta x)^2}{v^2 \Delta t} \quad or \quad \frac{v^2 (\Delta t)^2}{(\Delta x)^2} < 1 \tag{73}$$

which can then be given as the well known Courant condition:

$$\frac{v\Delta t}{\Delta x} < 1 \tag{74}$$

and the Neumann condition:

$$\frac{D\Delta t}{(\Delta x)^2} < 0.5 \tag{75}$$

Both conditions have a physical interpretation. The Courant condition states that in one time step, a water particle cannot travel further than the length of a grid block. The Neumann condition relates the characteristic length associated with the dispersion over a time step to the block size.

For a stable solution it is required that both conditions are satisfied, which means that the most restrictive condition determines the time step size that will still result in a stable solution.

Even an instable solution will give a perfect global mass balance (provided we are able to calculate the mass balance with enough significant digits). Consequently, a perfect mass balance (very small errors) is no guarantee for a good solution. However, large errors in the global mass balance is a guarantee for errors in the concentration distribution.

4. Initial and boundary conditions

In order to be able to model the transport of (reactive) solutes in groundwater it is necessary to define both the initial and boundary conditions. Initial conditions are (mathematically) only required for transient or time-dependent problems.

For local pollution problems, the initial conditions (a clean soil) can usually be given for the time before the pollution or spill occurred. For diffuse sources of pollution, the initial condition (present day situation in cases predictions have to be made) are derived by simulating long periods before the present day, using known (or estimated) mass inflow.

Boundary conditions for solute transport can be defined similar to the boundary conditions for groundwater flow. We distinguish three types of boundary conditions:

Dirichlet: the concentration on the boundary is fixed. Although mathematically this is a valid boundary condition, it is physically almost always impossible to create such a boundary condition. Nevertheless it is often applied at inflow boundaries, where the concentration of the solute in the water phase is known.

Neumann: the total mass flow of a solute across a boundary is defined. This type of boundary condition (also called mass loading) is often applied to inflow boundaries in e.g. experiments, but also in field situations dealing with point sources of pollution where the total mass of solute entering the groundwater is known (or can be estimated). For outflow boundaries, this type of boundary condition is physically not possible, because at outflow boundaries the total mass leaving the system is unknown (dependent on the concentration in the aquifer).

Cauchy: the mass flux of solute across the boundary is dependent on the concentration in the water phase itself. This type of boundary condition is often applied at outflow boundaries, defining the total mass flux as qC, where q is the water flux and C is the concentration of the solute in the water. Implicit in this definition of the boundary condition is the assumption that the dispersive flux across the boundary is zero.

For local point source pollution problems, we choose boundaries far enough away from the point source to make sure that the boundary conditions do not influence the concentration distribution.

5. Non-linear, non-equilibrium processes

5.1. General

Diffusion and radio-active decay have already been described in section 2, and do not need further elaboration.

5.2. Advection and dispersion

In heterogeneous systems, advection and dispersion are closely related. Dispersion is used to describe the transport of a contaminant due to variations in the groundwater flow which are not described by our "model" (where model can mean anything from complex numerical systems to the assumption of uniform flow). These variations are caused by local heterogeneities which have not been taken into account.

Basically, this means that the more information on convective or advective transport (in fact the variation in hydraulic conductivity) are taken into account, the smaller the dispersion will be.

From large numbers of field measurements, it is known that within one geological formation, the hydraulic conductivity may show a log-normal distribution with for instance an exponential covariance function defining the spatial correlation:

$$Cov(Y_1, Y_2) = \sigma_Y^2 e^{-\frac{r}{l_Y}} \tag{76}$$

where $Y=ln(k)$, σ_Y is the standard deviation in Y, r is the distance between the points where Y_1 and Y_2 were measured, and l_Y is the correlation length. The basic assumption is that the covariance is dependent on the distance between the measurement points only. This relation can be extended to account for direction dependence. For instance, the fact that sedimentation in geological formation have taken place in a certain direction will generate this direction dependence (correlation length in the z-direction will be smaller than the correlation length in the x and y-directions). Further, more practical aspects are discussed later in this chapter.

If we consider the formation to be homogeneous, all heterogeneities in the formation will have to be described by dispersion. In such a system, the dispersivity will be related to the correlation length (relation dependent on the flow pattern). That is, however, only true if the plume "has seen" all heterogeneities, i.e. if the size of the plume is (much) larger than the correlation length. In that case we will call the plume "ergodic". For small plumes (non-ergodic), that is not the case, and the behaviour of such a plume cannot be described on a large scale by global dispersion. For the behaviour of such plumes it is necessary to incorporate information on local heterogeneities.

Note, that hydrodynamic dispersion generates a spreading in the average values of the concentrations, where the averaging volume is determined by the scale on which we assume the system to be homogeneous. If that scale is large, we are in principle not allowed to make a comparison of calculated concentrations with local measurements. Another reason to use spatial moments.

5.3. Adsorption/desorption

Equilibrium adsorption/desorption, especially for larger concentrations, can often be described by either a Freundlich equation (Figure 5, cases B and C)

$$C_s = K(C)^p \tag{77}$$

where K is a constant, and $p<1$, or by a Langmuir isotherm:

$$C_s = \frac{C_{s\max} kC}{1+kC} \tag{78}$$

where C_{smax} the maximum amount is that can be adsorbed (occurs when $kC>>1$). For low concentrations (where $kC<<1$) relation (78) becomes linear. The Langmuir isotherm is typical for soils and solid surfaces that have a limited number of sites available for adsorption.

For the non-linear adsorption, we can still define a retardation factor:

$$R = 1 + \frac{1-n}{n}\rho_s \frac{dC_s}{dC} \tag{79}$$

which will, however, be dependent on the concentration C. The mass balance equation is then given by:

$$R\frac{\partial C}{\partial t}+\nabla\cdot F=0 \tag{80}$$

where F is the mass flux (by advection and dispersion).

Inspection of the Freundlich and Langmuir isotherms shows that the retardation factor becomes smaller for larger concentrations. For the displacement of a front of solute, this means that the higher concentrations are less retarded than the lower concentrations. In other words, the higher concentrations try to "overtake" the lower concentrations. This effect is counteracted by dispersion. After some time, equilibrium will occur between these competing processes and a travelling wave will develop (Bosma and Van der Zee, 1993). An example of such displacement is shown in Figure 6.

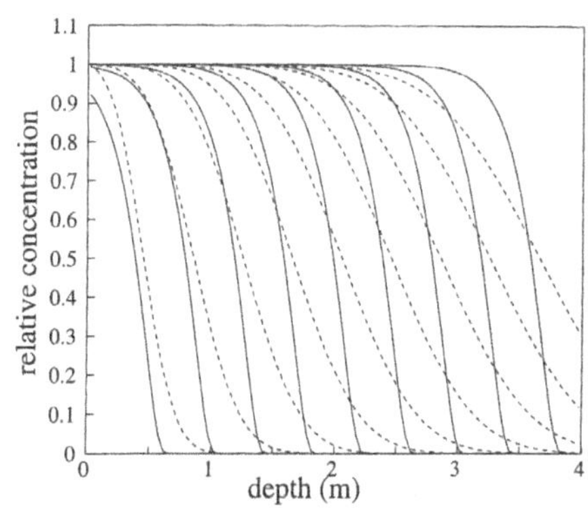

Figure 6. Traveling wave type of displacement, with concentration fronts given by solid lines, compared with linear convective-dispersive transport, with fronts given by dashed lines

In case of non-equilibrium, the adsorption/desorption may also be non-linear due to the limited number of sites available. For non-equilibrium adsorption/desorption we also need to solve for a mass balance of the adsorbed solute. For instance, a Langmuir type non-equilibrium interaction takes the form:

$$I=k_a\left(1-\theta\right)C-k_dC_s \tag{81}$$

where θ is the fraction of the adsorption sites that is occupied (Van der Zee et al., 1987). Non-equilibrium often results in tailing in breakthrough curves.

If more solutes are present, competition for the adsorption sites may occur. One of the effects that may occur, is, that some solute adsorbs fast, but will later be replaced by a solute that adsorbs slower, but has a higher affinity for the adsorption sites.

5.4. Chemical reactions

Chemical reactions are usually described by equilibrium reactions. These are in general highly non-linear, while a large number of species play a role. However, they are local in nature (there is no spatial partial derivative in these equations). Solving for chemical reactions in combination with transport can be done in two ways:

i. Operator splitting: solve for the transport of all species required. The result is a redistribution of the concentrations. Starting with these concentrations, calculate the new chemical equilibrium, etc. Advantage is that the number of equations is limited. Disadvantage is that the time step size is limited.

ii. Combine the equations for chemical equilibrium with the transport equations, and solve simultaneously. Advantage is that the time step size is less limited than in the operator splitting method. Disadvantage is the large system of non-linear equations that need to be solved.

Note, that negative concentrations (however small they are) cannot be allowed when dealing with chemical reactions.

Dissolution/precipitation reactions play a slightly different role. These reactions are dependent on threshold values. Consider a mineral the dissolves in the water phase. As long as the mineral is present, the concentration in the water phase is constant, and the mass transfer from the solid to the liquid phase is unknown. When the mineral is not present, it will act in the water phase as any other species. Until the concentration becomes large, and precipitation starts to occur.

One of the biggest problems in chemical reactions is to limit the number of species (and hence the number of mass balances that need to be considered) that have to be taken into account (Schroeder, 2005).

5.5. Biodegradation

Biodegradation needs at least, beside the present of bacteria, also the presence of a carbon source and the presence of an electron acceptor (if degradation is aerobic). The degradation can also be anaerobic (e.g. of inflammable NAPLs such as chlorinated hydrocarbons), in which a chemical is needed that supplies electrons for the degradation process. In general, aerobic degradation is faster than anaerobic degradation.

Biodegradation can be assumed to take place in the water phase only. Dependent on the number of bacteria present, we can distinguish between:

- single bacteria in the water phase; these can also be transported with the flowing groundwater;
- colonies on the solid matrix; these bacteria are not transported, but they still have access to all species in the water phase;

- biofilms on the solid matrix; for bacteria in biofilms, mass transport from the free water phase to the biofilm is usually diffusion controlled, and this (slow) mass transfer has to be taken into account when describing biodegradation.

If a carbon source is available, the biomass (number of bacteria) will grow, and in time we can have a transition from free bacteria to colonies to biofilm. Two other effects play a role in biodegradation: 1) bacteria die, which can usually be described by a first order decay, and 2) species may be present that inhibit the biodegradation.

If a first order description of biodegradation is not sufficient, a typical way to describe biodegradation then is e.g. by Monod type relations:

$$I = I_{\max}\left(\frac{C}{k_C + C}\right)\left(\frac{O}{k_O + O}\right)\left(1 - \frac{I}{k_I + I}\right) \tag{82}$$

where I_{max} is the maximum amount that can be degraded, C is the concentration of the carbon source, O the concentration of the electron acceptor, I the concentration of the inhibitor, and k are constants. Equation (82) can be extended to include more species that play a role in the biodegradation. Note, that for each of these species a mass balance needs to be solved. Equation (82) does not include the biomass, but that can be done in the same way, provided a mass balance for the biomass is solved as well.

If biodegradation takes place, the products of the degradation may in itself be degraded again by the same (type of) bacteria. In these cases it is required to analyse (or predict) the transport of multiple solutes, where the degradation of one solute results in a source term for the daughter product. These multiple solutes will in general have different adsorption/desorption characteristics and different decay characteristics.

5.6. Non Aqueous Phase Liquids (NAPL)

Non-aqueous phase liquids or oil does not mix with water. Infiltration of these liquids in the subsurface (e.g. through leaking tanks or pipes) will create multiphase systems.

If a spill occurs, oil will be transported through the unsaturated zone, leaving behind a residual oil saturation (around 20-30% of the pore volume), which is not mobile due to capillary forces.

When the oil reaches the water table the oil will float on the water table if it is lighter than water (LNAPL). Such floating lenses have been found on many places worldwide, often due to spills of gasoline at gasoline stations. The analysis of such lenses has been studied experimentally (Wipfler et al., 2004) but also numerically and analytically (Van Dijke and Van der Zee, 1998). If the NAPL is denser than water (DNAPL), it will move downwards through the groundwater until it reaches an impermeable layer. The transport of DNAPL through the saturated zone is controlled by instabilities (local heterogeneities), and it is therefore very difficult to predict where exactly the DNAPL will be present. However, residual oil will be present at those locations where the DNAPL has passed through.

Components of the oil will dissolve in the water phase, although generally in very small quantities. If the oil phase is in equilibrium with the water phase, the concentration of the oil components in the water phase can be described by Raoult's law:

$$C = C_{\max} X \tag{83}$$

where C_{max} the maximum concentration is if the water phase is in contact with the pure component, and X is the mol fraction of the component in the oil phase.

For a layer of the pure oil, there is in general no equilibrium between the water phase and the oil phase. In these cases, the mass transfer is diffusion controlled, and described by a first order relation:

$$I = k\left(C - C_{eq}\right) \tag{84}$$

where k is the mass transfer coefficient, and C_{eq} is the equilibrium concentration, which is given by equation (83). The mass transfer coefficient is dependent on a.o. the diffusion coefficient, the oil saturation, average pore diameter and the groundwater velocity. A number of empirical relations exist that define these relations.

If a free oil phase is present, the transport of oil components requires both the solution of the oil and water flow equations and the transport equation of each species in both phases. Note, that the oil and water flow are coupled due to the capillary forces, and the dependence of the water hydraulic conductivity on the oil saturation.

6. Density dependent flow and transport

6.1. General

In many cases, the properties of groundwater, and in particular the density, are influenced by the concentration of dissolved species. That is e.g. the case for sea water intrusion in coastal aquifers, or for the infiltration of leachate from a landfill into a fresh water aquifer. In many of these cases, we may assume that the density is the only property of the water phase that is influenced by the concentrations of the dissolved species. In other cases, we have to take into account the fact that the viscosity of the water phase is also influenced by the concentration of the dissolved species. That is in particular true for very high solute concentrations (or for large temperature differences). For instance, if we consider the disposal of waste in deep saline aquifers, the effect of a changing viscosity cannot be neglected. The viscosity of water varies by a factor of 2 from fresh water to concentrated brine. These situations differ from those of NAPLs, as the different types of groundwater are miscible, whereas NAPL and water are immiscible.

In case of changing fluid properties, the governing flow equation cannot be formulated in terms of groundwater potential but have to be given in terms of pressures and a gravity term. In the following, the governing equations for density dependent flow and transport will be given, followed by a simplified set of equations.

6.2. Basic equations

For density dependent flow, the general form of Darcy's law needs to be used:

$$q_x = -\frac{\kappa_x}{\mu}\frac{\partial p}{\partial x} \quad q_y = -\frac{\kappa_y}{\mu}\frac{\partial p}{\partial y} \quad q_z = -\frac{\kappa_z}{\mu}\left(\frac{\partial p}{\partial z} + \rho g\right) \tag{85}$$

The mass balance equation for the water phase is given by:

$$\frac{\partial}{\partial t}(n\rho) + \nabla \cdot (\rho q) = 0 \tag{86}$$

and the mass balance for the solute by:

$$\frac{\partial}{\partial t}(n\rho\omega) + \nabla \cdot (\rho\omega q) - \nabla \cdot (n\rho D \cdot \nabla \omega) = 0 \tag{87}$$

where it has been assumed that there are no external sources and sinks, and that the solute creating the density differences is a conservative one.

We can now define a "fresh water potential" as:

$$h_f = \frac{p}{\rho_f g} + z \tag{88}$$

where ρ_f is the density of fresh water (ω=0) and z is the vertical position with respect to a reference level. Substitution of (88) in (85) gives the following form of Darcy's law:

$$q_x = -\frac{\kappa_x \rho_f g}{\mu}\frac{\partial h_f}{\partial x} \quad q_y = -\frac{\kappa_y \rho_f g}{\mu}\frac{\partial h_f}{\partial y} \quad q_z = -\frac{\kappa_z \rho_f g}{\mu}\left(\frac{\partial h_f}{\partial z} + \frac{\rho - \rho_f}{\rho_f}\right) \tag{89}$$

which shows that the hydraulic conductivities are dependent on the fluid properties. It is also obvious that the fresh groundwater potential is not the only driving force for the groundwater flow. The system of equations (85) or (89) with (86) and (87) have to be supplemented by equations of state for both the density and the viscosity of the water.

The mass balance equations with Darcy's law form a set of coupled, non-linear equations. The coupling is a two-way coupling: the flow is dependent on the concentration distribution (through the dependence of the density and the viscosity on the concentration), while the concentration distribution is dependent on the flow through the advective term in equation (87).

6.3. Simplified equations

In many cases, the equations governing the density dependent flow and transport can be simplified by making a number of assumptions. First of all, for relatively low concentrations (e.g. the salt concentration in seawater) we may assume that the changes in the viscosity with the concentration are negligible. Furthermore, the density can be assumed to be linear dependent on the salt mass fraction. That is certainly the case for salt water:

$$\rho = \rho_f\left(1+\gamma\omega\right) \tag{90}$$

where for salt, γ has a value of 0.7.

Secondly, we will assume the (temporal) changes in the porosity can be neglected.

Finally, we can adopt Boussinesq's approximation, which states that the variations in the liquid density can be neglected everywhere, with the exception in the gravity term of Darcy's law. With all these assumptions, the governing equations can be written as:

Mass balance of the water phase:

$$\nabla \cdot q = 0 \tag{91}$$

Mass balance of the salt:

$$\frac{\partial\left(n\omega\right)}{\partial t} + \nabla\cdot\left(\omega q\right) - \nabla\cdot\left(nD\cdot\nabla\omega\right) = 0 \tag{92}$$

Darcy's law:

$$q = -\frac{\kappa}{\mu}\left(\nabla p + \rho g e_z\right) \tag{93}$$

Equation of state:

$$\rho = \rho_f\left(1+\gamma\omega\right) \tag{94}$$

where e_z is a unit vector in the z-direction (positive upward).

Substitution of the equation of state (94) in Darcy's law (93), and assuming that the porosity and the dispersion coefficients are constants, even further simplifies the set of equations:

$$\nabla \cdot q = 0 \tag{95}$$

$$n\frac{\partial\left(\omega\right)}{\partial t} + \nabla \cdot \left(\omega q\right) - nD\nabla^2\omega = 0 \tag{96}$$

$$q = -\frac{\kappa}{\mu}\left(\nabla\left(p+\rho_f gz\right) + \rho_f \gamma\omega g e_z\right) \tag{97}$$

These equations can be made dimensionless by choosing appropriate reference values for the different variables in the equations:

$$q_d = \frac{q}{q_r} \quad x_d = \frac{x}{L} \quad t_d = \frac{t}{t_r} \quad \omega_d = \frac{\omega}{\omega_r} \quad p_d = \frac{p+\rho_f gz}{p_r} \tag{98}$$

where L is a characteristic dimension of the system considered.

Now, if we choose the following reference values:

$$q_r = \frac{nD}{L} \quad t_r = \frac{L^2}{D} \quad \omega_r = \omega_{\max} \quad p_r = \frac{\mu L^2}{\kappa n D} \tag{99}$$

the system of equations reduces to:

$$\nabla \cdot q = 0 \tag{100}$$

$$\frac{\partial \omega}{\partial t} + \nabla \cdot \left(\omega q\right) - \nabla^2\omega = 0 \tag{101}$$

$$q = -\nabla p - A\omega e_z \tag{102}$$

where it is understood that all variables are dimensionless (subscript *d* has been omitted), and:

$$A = \frac{n\kappa D\Delta\rho_{\max}}{\mu L} \tag{103}$$

is the Rayleigh number. This number defines the ratio of gravity and dispersive forces. Together with the initial and boundary conditions it fully controls the solution of equations (100) through (102).

For the stable situation, where fresh water (derived from rainfall) is situated above saline groundwater, Eeman et al. (2011, 2012) provided analyses for the behaviour of such rainfall lenses in coastal delta areas. Experimental evidence of such lenses was given by Lebbe et al. (2008), Vandenbohede et al. (2008) and Eeman et al. (2011, 2012). Their work was preceded by over a century of investigations of the dynamics of fresh water pockets in coastal dunes (Herzberg, 1901, Van der Veer, 1977, Maas, 2007), barrier dunes, and atols (see Eeman et al. 2012).

For systems that are possibly instable, i.e. systems where fresh water is overlain by water with a higher density, the Rayleigh number controls whether such instabilities will occur. These situations arise e.g. under landfills, where the (heavier) leachate infiltrates in a fresh water aquifer, or at transgression of the coast line by rise of the seawater level.

For reasonable simple geometries, a perturbation analysis can show us when instabilities may occur. For the Rayleigh-Bernard problem, i.e. a rectangular vertical slab, with no-flow boundary conditions at the sides, and ω=1 at the top boundary and ω=0 at the bottom boundary, it can be shown that instabilities will occur for $A>4\pi^2$.

7. Transport in heterogeneous media

7.1. General considerations

If we wish to convey our understanding of transport of solutes to real soils, we have to account for spatiotemporal variability. After all, it is well known that soils vary spatially (layers, horizons, pores size distribution), and as a function of time, e.g. due to time varying weather, groundwater flow and many other causes. Such variability leads to rather complex behaviour in space and time. Instead of the rather simple, smooth concentration profiles and break-through curves, that we obtain for homogeneous soil and simple initial and boundary conditions, quite involved transport trajectories and concentration distributions result.

These are difficult to convey to the stakeholders, that have to base decisions on such results. For such reasons, spatial and temporal moment theory can be used to capture the transport phenomena in relatively robust terms. The moment theory is explained and for the case of spatial moments, it is presented mathematically. This is not done for temporal moments, because the mathematical details are completely in analogy to the spatial moments, and therefore obsolete here. Before going into details, first a qualitative impression is given of the moments.

The first spatial moment is related to the solute velocity, or, in case of a conservative solute, the groundwater velocity. For an instantaneous release of a non-reactive tracer in a steady state groundwater flow field, the first spatial moments as a function of time tell us exactly where the tracer is located and therefore also what the groundwater velocity is. Comparison of the first spatial moment of a reactive solute with the first spatial moment of a non-reactive tracer gives an estimate of the retardation factor. For a solute plume, these first moments characterize therefore the velocity of the entire plume, not of individual solute particles.

Following the zero'th spatial moment of a solute in time gives information about the degradation of the reactive solute. It is even better to compare these moments with the zero'th moment of a non-reactive (inert) tracer. If we indicate the properties of the reactive tracer with superscript *r* and the properties of the non-reactive tracer with *n*, we can define:

$$F^r = \frac{M_0^r(t)}{M_0^r(0)} \qquad F^n = \frac{M_0^n(t)}{M_0^n(0)} \tag{104}$$

The decay constant λ of the reactive solute is then given by:

$$\frac{F^r}{F^n} = e^{-\lambda t} \tag{105}$$

We could have used the zero'th spatial moment of the reactive solute only (as a function of time). However, comparing it with the zero'th spatial moment of a non-reactive solute may give a better answer because possible errors due to limited available information may cancel if the concentrations required for the determination of the zero'th spatial moment are measured at the same locations.

The second central spatial moment is related to the dispersion on the scale of the plume. For an instantaneous release of a solute in a steady state groundwater flow field, one of the elements of the dispersion coefficient is given by:

$$D_{xx} = \frac{1}{2}\frac{d}{dt}\left(\sigma_{xx}^2\right) \tag{106}$$

and similar expressions can be given for the other 5 elements of the dispersion tensor (cf equation (33)).

The accuracy with which the spatial moments can be determined is of course dependent on the amount of data available. Higher order spatial moments are often less accurate than the lower order spatial moments.

Characterizing the behaviour of a plume of contaminant on the basis of spatial moments is usually more robust than trying to characterize it by individual measurements of concentrations. The latter are very dependent on local heterogeneities. Apart from that, decision makers are usually interested in global measures like the transport of the plume (first spatial moment) and the spread around the mean travel distance (second central spatial moment) rather than in local concentrations.

7.2. Moment theory

As flow and transport conditions are spatiotemporally variable, complex patterns can develop, as illustrated in Figure 7 for the distribution of a chemical that moves through soil. This pattern is difficult to describe to someone who cannot see it. However, often that is not really important, because we are only interested in simpler information. Simpler information, that is also statistically robust, are the so-called moments. Moments can be described for any property that is distributed (in space, in time). For instance, if we consider the concentration of a solute as in Figure 7, that is distributed spatially, we can regard this distribution as a probability density function pdf (or: frequency function), in particular: a pdf of travelled distances, in this case. Its discrete analogue is a histogram, provided the lengths of all bars are normalized in such a way that their sum is equal to 1. A pdf f is related with a cumulative distribution (P) according to:

$$P = \int_{-\infty}^{x} f(\xi)d\xi \tag{107}$$

and for integration to $x=+\infty$, P becomes one: it is the probability that the value of x lies between the lower and the upper boundary, and this probability is one for this $x=+\infty$ upper boundary. Therefore, if we deal with a concentration distribution, we first have to make a pdf out of it. We do so with the zero'th moment.

Instead of doing so, we will illustrate moments on mass instead of concentration. A concentration is a mass (of solute) divided by a mass or volume of water. You can of course calculate what happens if you add one glass of water with a particular concentration to another glass with another concentration. However, physically it is much better to go to more basic properties: how much mass of solute is in each glass, how much water, and what is the final mass of solute and of water if both are combined? Hence: never calculate moments of concentrations, or water fractions. Instead, calculate moments of quantities (kg, volume,...). For the present purpose, we calculate moments of solute mass in solution ($\theta(x, y, z, t).C(x, y, z, t)$) instead of concentration.

The zero'th moment of the solute mass distribution in solution is given by:

$$M_0(t) = \int_{-\infty}^{\infty} \theta(x, t)C(x, t)dx \tag{108}$$

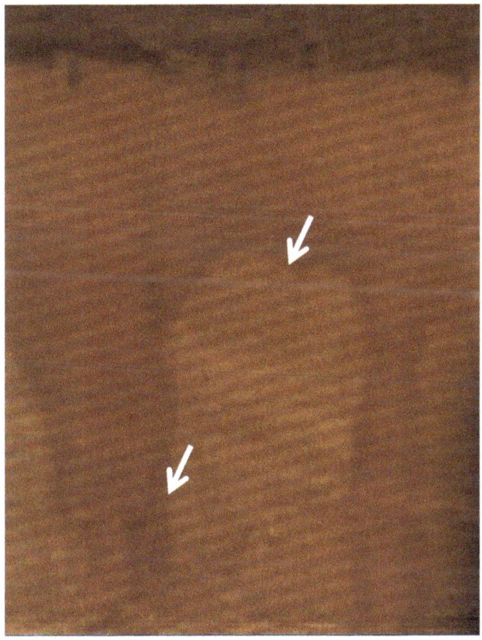

Figure 7. Evidence of spatial differences in the rate of propagation of a solute front. Part of a photograph kindly provided by R.G. (Gary) Kachanoski (Canada) of a loamy soil profile in Ontario, Canada. Front depth (white arrows) is visible because the lime rich subsoil is lighter coloured than the topsoil, where lime has dissolved and leached due to natural and anthropogenic acid inputs during the Holocene. Topsoil pH about 6.5, subsoil pH about 8.2.

if water fraction and concentration are only distributed in the x-direction (otherwise, we have to integrate for x, y, and z). The zero'th moment is also known as the mass of the distribution, and is a 'normalizing' property: division of higher order moments by the mass, renders the distribution of water fraction times concentration a pdf!

The first moment is given by:

$$M_1(t)=\int_{-\infty}^{\infty} x\theta(x,t)C(x,t)dx=\int_{-\infty}^{\infty} x f_{\theta C}(x,t)dx \tag{109}$$

and written this way, illustrates that division by the 0th moment is needed to obtain a pdf. The first spatial moment divided by the 0th spatial moment is known as the average position of the plume of solute (x_{av}), whereas the first temporal moment divided by the 0th temporal moment is the mean breakthrough time. The second moment is obtained by multiplying with x^2 instead of x. More common to use is the second central moment, given by:

$$M_2^c(t)=\int_{-\infty}^{\infty}(x-x_{av})^2\theta(x,t)C(x,t)dx=\int_{-\infty}^{\infty}(x-x_{av})^2 f_{\theta C}(x,t)dx \tag{110}$$

from which the variance can be obtained as:

$$\sigma_{xx}^2(t)=\frac{M_2^c(t)}{M_0(t)} \tag{111}$$

This variance is a good measure for the spreading in space (here the x-direction), and is related with the concept of dispersion, as mentioned before. To describe the transport behaviour of solutes, the moments up to the variance are not always sufficient. The main reason is that the transport velocities, i.e., including retardation effects, of parcels of solute may not be symmetrically (e.g. normally) distributed. In that case, higher order moments may be needed, or the assumption that the pdf has another than normal shape is required. An important alternative to the normal (symmetric) pdf is the lognormal pdf.

7.3. The lognormal pdf

The normal pdf of X is given by:

$$f(X)=\frac{1}{s_X\sqrt{2\pi}}e^{\left(\frac{X-m_X}{s_X\sqrt{2}}\right)^2} \tag{112}$$

where s_X is the standard deviation and m_X is the mean. The lognormal pdf of Y is very similar, namely:

$$f(Y)=\frac{1}{Y\,s_Y\sqrt{2\pi}}e^{\left(\frac{Y-m_Y}{\sigma\sqrt{2}}\right)^2} \tag{113}$$

where the mean m_Y and standard deviation s_Y are those of the lognormally transformed parameter X=ln(Y). That means that if Y is lognormally distributed, then X is normally distributed, and the pdf of lognormally distributed Y is characterized with the statistics of X. The following equation shows that this is correct:

$$f_Y dY=\frac{1}{Y}f_{lnY}dY=f_{lnY}d(lnY)=f_X dX \tag{114}$$

So, if we wish to characterize the lognormal pdf, we first have to log-transform, to calculate the statistics, and then we know the pdf. There is another way, as the statistics of both pfd's are related. The following relationships can be derived if Y is lognormally distributed and if X=ln(Y):

$$m_X=\ln(m_Y)-\frac{1}{2}s_X^2$$

$$s_X^2=\ln\left(1+\left(\frac{s_Y^2}{m_Y^2}\right)\right) \tag{115}$$

So, to calculate the mean, you first have to calculate the variance (of X), from the statistics (mean and variance) of Y.

Median and mode for the normally distributed X are equal to the mean.

If you know the statistics of X (for instance log(hydraulic conductivity)), but not those of hydraulic conductivity itself (Y), then you can calculate those easily by inverting the above equations. The result is given by:

$$m_Y = e^{\left(m_X + \frac{1}{2} s_X^2\right)}$$

$$s_Y^2 = e^{\left(2m_X + s_X^2\right)}\left(e^{s_X^2} - 1\right) \qquad (116)$$

and for completeness, also the median and modus:

$$\begin{aligned} median_Y &= e^{m_X} \\ mode_Y &= e^{\left(m_X - s_X^2\right)} \end{aligned} \qquad (117)$$

The lognormal pdf is very attractive for several reasons. An important reason is that many properties appear to be well lognormally distributed, whereas they are not normally distributed. For instance, the saturated hydraulic conductivity K_s has been found experimentally to be often lognormally distributed, and Miller and Miller (1955) also give a theoretical basis with the similar media or similitude theory.

7.4. Approaches to model transport in heterogeneous media

Modeling of heterogeneous media can be done in several ways, that vary in complexity. In all cases, an impression of the heterogeneity is gained experimentally. This gives us information regarding the statistics of the important soil properties, such as hydraulic conductivity. This information can be directly used as input for any model. However, it is also possible to generate a field of hydraulic conductivities and use that for input. This is attractive, because often we have data only for a very limited number of positions, leaving large volumes for which we do not have data and for which we have to find a good strategy to parameterize.

If we generate random fields, we can do a calculation for each generated field. If we do so repeatedly, for a designated set of statistics, then the calculations will be similar but different: just as when you take a picture of a meadow every five minutes: the meadow will be the same, but each picture is different. Repeated calculations for the same statistics is what we call Monte Carlo simulation. With this technique, we can determine the uncertainty that is intrinsic to our model results.

Monte Carlo simulation is a numerical approach, with which we can check whether analytical solutions are sound. But Monte Carlo simulation may also be done using analytical solutions, if they are available. Besides Monte Carlo, a common way to deal with heterogeneity is in the context of a GIS system, where each cell or pixel gets its own value, depending on overlay maps and transfer functions. Then, for each cell or pixel the same set of calculations is done, only with different parameters. An example of the output obtained this way is shown in Figure 8 for pesticide leaching. In these calculations, geospatial data of soil type, weather conditions, geohydrology, land use (which pesticides, which crops, which growing season and pesticide application date), and other information can be linked. Clearly, then it is possible to determine which regions have a major hazard of leaching a particular pesticide (or more generally, a contaminant of interest).

This type of modelling is very important for risk analysis and screening of pesticides in the admission of these chemicals. Whether or not a pesticide is admitted to the market, or should be taken out of the market (as happened about 1990 with quite a number of pesticides in the EU), may depend on the outcome of fate calculations with models as described here. For instance, the pesticide screening in EU uses the so-called FOCUS protocols, as mentioned by Beltman et al. (2008).

In a relatively simple approach, the spatial configuration is neglected. This is suitable, if the interest is not where something happens but what its effect is on the entire system. An example is leaching of solute from a field: instead of wishing to know where the leaching occurs, it may be sufficient to know what the average leaching is. This is the essence of the parallel stream tube PST model.

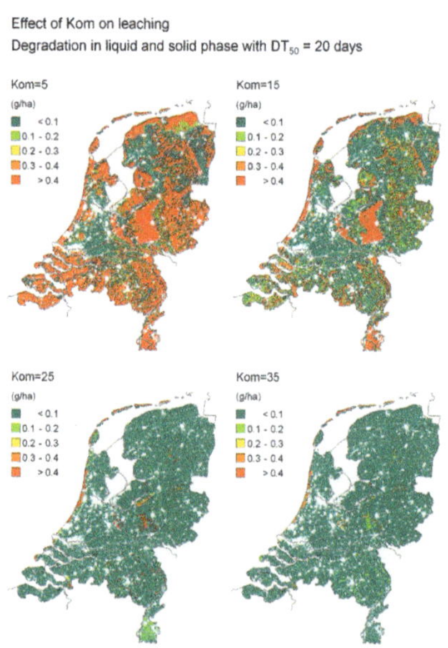

Figure 8. Leached pesticide into groundwater, as calculated with the EU pesticide screening model GEO-PEARL, for four K_{om}-values and a half life of 20 days. Calculations and figure kindly provided by A. Tiktak (PBL, Bilthoven, Netherlands). The assumption has been made that only dissolved pesticide may degrade (see Beltman et al., 2008)

7.5. Parallel stream tube model

This approach has been developed by Bresler and Dagan in two highly innovative papers in 1979, and their analyses for a conservative tracer was later extended to reacting chemicals (Van der Zee and Van Riemsdijk, 1987, Destouni and Cvetkovic, 1992). It is very effective to capture the effect that fronts do not move with the same velocity for all places. It is used more for leaching in unsaturated soil, than for groundwater transport, because stream tubes are more unidirectional in soil (vertical) than in groundwater.

For the vertical transport of solute, the simplest model is the purely convective model. For the case of linear sorption, it is given by:

$$z_c = \frac{vt}{R} \tag{118}$$

for each concentration that at time 0 was located at depth (z) equal to zero. For gravity leaching, the velocity is determined by the unit gradient and the hydraulic conductivity, where the latter has been demonstrated to be strongly spatially variable. The retardation factor is likewise known to vary spatially. This implies, that even if all concentrations 'start' at time 0 at the same position (say z=0), then after a certain time, they will have moved more in some stream tubes than in others. This is illustrated in Figure 9. Due to spatial variability of flow velocity and retardation factor, also the front moves with a distributed velocity!

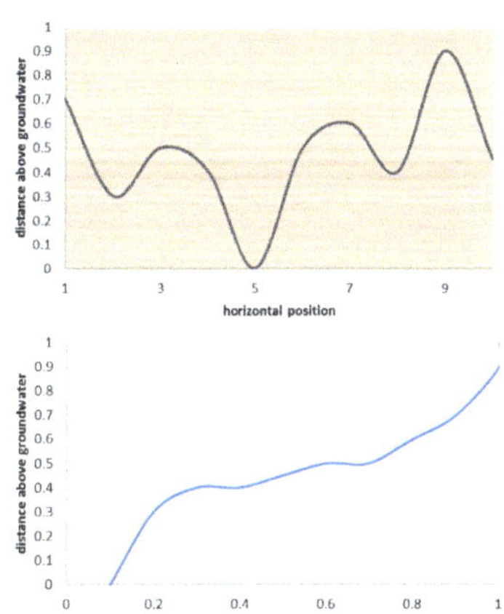

Figure 9. Illustration of the PST model for unsaturated soil, and Heaviside input of solute. The top panel was obtained by calculating the front position for many horizontal positions, giving a front position with above a relative concentration of 1 and below of 0. The bottom panel was obtained by calculating the fraction of the horizontal plane, for which the relative concentration is 1. This fraction is then the horizontally averaged (relative) concentration.

Now, it is important to recognize with which complexity we wish to describe such spatiotemporal variable transport. In case of the PST model, we are satisfied if we know how the transport is for the entire area (for instance field) on average, and we are less interested in which part of the field the transport is fast or slow. In that case, the configuration of the solute front in x and y is of less importance, and we consider the field average front. For that front, we can be interested in the mean front position, its distribution, and its variance. In this chapter, we focus on the mean position and the distribution around this mean.

The field average front can be easily determined using Monte Carlo simulation, where random numbers are drawn for the distributed parameters, combined in the convective model, to determine the distributed front position at designated time (Van der Zee and Van Riemsdijk, 1987). However, it is also possible to determine the field average front analytically, in case that the parameters v and R are lognormally distributed. This analytical procedure is as follows, for the simple case of a Heaviside solute input at the soil surface.

First, we experimentally or theoretically determine the mean (m) and standard deviation (s) of the lognormally transformed v and R, i.e. of ln(v) and ln(R). The so-called reproductive properties of the lognormal distribution are for the purely convective transport model of use to determine the statistics of front depth: if $z=vt/R$, then:

$$m_{lnZ} = m_{lnv} + ln(t) - m_{lnR}$$

$$s^2_{lnZ} = s^2_{lnv} + s^2_{lnR} \qquad (119)$$

As for the Heaviside input, the concentration at the soil surface changes abruptly from one value to another, we obtain different results depending on the initial and boundary conditions. As a first step, we determine, for a distributed ln(z) (for front depth), what is the probability that the front has not passed a certain depth. This probability is given by:

$$Pr\{z<z^*\} = \int_{-\infty}^{z^*} f_z dz \qquad (120)$$

with the pdf given in section 8.3. Inserting the lognormal pdf, with statistics as just determined, we obtain:

$$Pr\{z<z^*\} = \frac{1}{2} erf\left\{\frac{ln(z^*) - m_{lnz}}{s_{lnz}\sqrt{2}}\right\} \qquad (121)$$

This probability is the probability that the front has not passed the depth z^*. To interpret that as a concentration, we have to reason what that means. If the front has not passed a certain depth, at that depth we still find the original (initial) concentration. Hence, Pr represents the fraction of the area, where we still find the initial concentration at depth $z=z^*$. We now consider two simplified systems. In the first case, the initial concentration is 1 and the incoming concentration is 0. In that case, the areally averaged concentration at depth z^* is:

$$\langle C(z^*)\rangle = Pr\{z<z^*\} \qquad (122)$$

If, on the other hand, the initial concentration is equal to 0 and the incoming concentration is 1, then:

$$\langle C(z^*)\rangle = 1 - Pr\{z < z^*\} = \frac{1}{2} erfc\left\{\frac{In(z^*) - m_{lnz}}{s_{lnz}\sqrt{2}}\right\} \tag{123}$$

More generally, if the initial and final (or input) concentrations are not equal to 0 or 1, the mean concentration can be transformed in the real value, using transformations such as:

$$\langle C\rangle = \frac{C(z^*) - C_{initial}}{C_{final} - C_{initial}} \tag{124}$$

Due to the logarithmic transforms, the mean concentration profile commonly looks as in Figure 10.

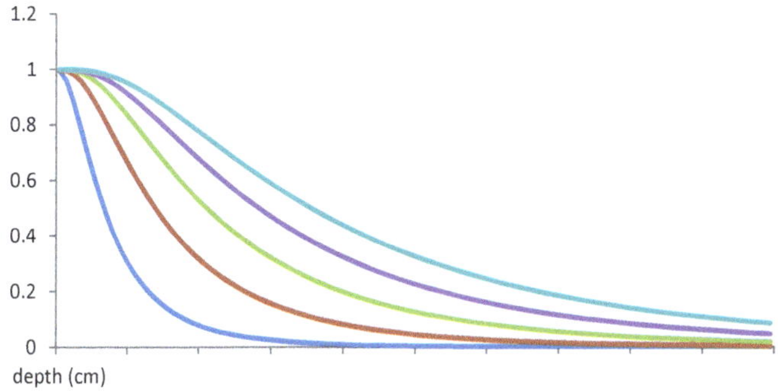

Figure 10. Areally averaged concentration (vertical axis) as a function of depth (horizontal), for different times: initial concentration 0 and final concentration 1.

In the presentation so far, we assumed that the important properties are lognormally distributed, i.e., taking the logarithmic transform, we obtain numbers that are normally distributed. Often, this assumption is appropriate (Van der Zee et al., 1988), but also convenient, if the convective PST model is adopted: the reproductive properties hold for multiplication and division in case of a lognormal pdf, whereas they are additive/subtractive for the normal pdf: see Eq. (119). Also, lognormally distributed parameters cannot be negative, whereas normally distributed parameters can. This is important, because for instance the hydraulic conductivity and the retardation factor cannot be negative.

Although transport in the water unsaturated soil is predominantly vertically downward or upward, some horizontal displacement always happens. This transversal transport can also be important, for instance, if a slug of contaminant moves downwards, but dispersional mixing and movement in the transversal directions occurs for a contamination event of limited areal extent (in the horizontal x and y directions). In that case, the explained theory is valid but spatial moments have to be calculated in the other directions than only depths, as in the previous sections.

7.6. Multidimensional stochastic modelling

In the period 1980-1995, stochastic groundwater hydrology made major advances. The basis of those advances was that the saturated hydraulic conductivity was a lognormally distributed random space function RSF, that could be characterized with e.g. an exponential autocorrelation function. For this assumption, analytical expressions were derived for the hydraulic head and the specific discharge RSF. Water flow understanding was compiled in the milestone book by Gedeon Dagan (1987), and an insightful paper on water flow in unsaturated soil was written by Kurt Roth (1995), with attractive illustrations of water channeling due to hydraulic conductivity variability, under steady state flow.

Whereas numerical and analytical work has been done with regard to solute transport in 2D heterogeneous conductivity fields, only in 1993 two papers came out that addressed transport of solutes that adsorb linearly, hence with RSF for the hydraulic conductivity as well as for the (linear) adsorption coefficient, and retardation factor (Bellin et al., 1993, Bosma et al., 1993). Results were produced in terms of second central moments of the displacement distance.

At the moment, the techniques to generate RSF for hydraulic parameters has advanced enormously compared to the early days of stochastic groundwater hydrology and contaminant hydrology. Therefore, it has become possible to look at real cases of soil and groundwater contamination, taking random variability of properties into account. This versatility is made a bit more modest, as it remains a critical, often subjective and much debated issue of which values to give to core properties, such as the dispersivities.

8. Conclusion

Solute transport has originally been considered for the local scale, where small scale heterogeneity resulted in a diffusion-dispersion type of mixing that at the scale of interest (often columns of porous media, soil between soil surface and groundwater level, or aquifers) resulted in a normally distributed concentration in space and time. Because of this normal distribution, this mixing could be described by a normally (or Gaussian, Fickian) distributed process, which is in full agreement with a linear parabolic partial differential equation.

In the past few decades, particularly between 1979, with the pioneering work of Bresler and Dagan (1979), and today, it has become apparent, that also at larger scales than the sand grain, heterogeneity leads to variability in flow and transport. This variability has been approached predominantly from a stochastic point of view, by considering it as a spatially distributed stochastic process. This approach appeared to be a rewarding one to address both real mixing and uncertainty at scales of interest: the decision and management scale. At the same time, it is still debatable which values of the dispersivities are appropriate, and in view of the theory provided in this chapter, section 3, this also holds for the discretization used in numerical modelling. For instance, the stochastic theory for water flow and solute transport resulted in equations for the macro dispersivities. However, these dispersivities do not necessarily represent real mixing. Briefly, this issue is discussed by both Janssen et al. (2006) and Eeman et al. (2012).

Whereas conventional solute transport often considered only one scale of heterogeneity (the grain or soil sample scale), the stochastic approach addressed so far mainly two scales, the microscopic and one larger scale, the last characterized by statistics of macroscopic properties such as the hydraulic conductivity or the retardation factor. In reality, we have to deal with a whole hierarchy of scales, from grain, to sample, horizon/layer, geological strata, to watershed. Comprehensive theory cannot address all these scales in a simple theory, so it considers those that are deemed most important.

Scientifically, further advances are being made. Whereas the mentioned work in section 7 particularly addresses spatial variability, in the research on transport of tracers and reactive solutes in soil and groundwater, also the variability in time is getting more interest. A prominent example in this respect has been that since 2000, eco-hydrological theory has been added to the spectrum, that emphasizes variability as a function of time, particularly through atmospheric forcing: rainfall varies erratically as a function of time (Rodriguez-Iturbe and Porporato, 2004, Vervoort and Van der Zee, 2008). In near future, these developments of, on one side, spatial, and on the other side temporal variability, will be no doubt combined.

In soil and groundwater management, modelling as discussed in this chapter is essential to prioritize decision making. Recently, a paper considered policy making in the Netherlands (Witte et al., 2012), and was quite influential nationally and internationally. This paper emphasized local variations of several conditions that affect the fate of factors that influence ecological impacts. It is a clear example of the potential and limitations of transport modelling.

Meanwhile, throughout the world, this type of modelling is central in risk assessments of soil and groundwater contamination, in dimensioning soil and groundwater remediation, and in evaluations of how policy decisions work out in practice for the cases of changed and unchanged policies. For these reasons, awareness and in some cases experience, with models as described in this chapter, is important.

Appendix A: Stability analysis of explicit finite difference transport equation

Consider the explicit finite difference approximation of the 1-dimensional solute transport equation, where the advective term has been approximated by a central difference:

$$\frac{C_i^n - C_i^o}{\Delta t} + v\frac{C_{i+1}^o - C_{i-1}^o}{2\Delta x} - D\frac{C_{i+1}^o - 2C_i^o + C_{i-1}^o}{(\Delta x)^2} = 0 \qquad (125)$$

Collecting terms, this can be rewritten as:

$$C_i^n = \left(1 - \frac{2D\Delta t}{(\Delta x)^2}\right) C_i^o + \left(\frac{v\Delta t}{2\Delta x} + \frac{D\Delta t}{(\Delta x)^2}\right) C_{i-1}^o + \left(-\frac{v\Delta t}{2\Delta x} + \frac{D\Delta t}{(\Delta x)^2}\right) C_{i+1}^o = \\ (1-\beta_2) C_i^o + \left(\beta_1 + \frac{\beta_2}{2}\right) C_{i-1}^o + \left(-\beta_1 + \frac{\beta_2}{2}\right) C_{i+1}^o \tag{126}$$

where:

$$\beta_1 = \frac{v\Delta t}{2\Delta x} \qquad \beta_2 = \frac{2D\Delta t}{(\Delta x)^2} \tag{127}$$

Now assume that at a certain time a small error ε (perturbation) is introduced in the solution to the equations. That can e.g. be caused by roundoff errors in a computer calculation. If we indicate the "correct" solution with C, subtitution of the perturbed solution $\underline{C}$ in the finite differencd equation (126) leads to:

$$\underline{C}_i^n + \varepsilon_i^n = (1-\beta_2)\underline{C}_i^o + \left(\beta_1 + \frac{\beta_2}{2}\right)\underline{C}_{i-1}^o + \left(-\beta_1 + \frac{\beta_2}{2}\right)\underline{C}_{i+1}^o + \\ (1-\beta_2)\varepsilon_i^o + \left(\beta_1 + \frac{\beta_2}{2}\right)\varepsilon_{i-1}^o + \left(-\beta_1 + \frac{\beta_2}{2}\right)\varepsilon_{i+1}^o \tag{128}$$

Since the correct solution $\underline{C}$ obeys equation (126), we obtain the following equation for the perturbation ε:

$$\varepsilon_i^n = (1-\beta_2)\varepsilon_i^o + \left(\beta_1 + \frac{\beta_2}{2}\right)\varepsilon_{i-1}^o + \left(-\beta_1 + \frac{\beta_2}{2}\right)\varepsilon_{i+1}^o \tag{129}$$

The fact that the perturbation ε is given by the same equation as the correct value of C is caused by the fact that the equation in C is a linear one.

Now consider one Fourrier component of the perturbation given by:

$$\varepsilon_i = \lambda e^{i\omega x} \tag{130}$$

where ω is the wavenumber, and λ the time dependent amplification factor. In order to have a stable solution, we will require that any perturbation ε in the solution will decrease with time for any wave number. Basically this means that λ will decrease with time.

Substitution of (130) in (129) gives:

$$\lambda^n e^{i\omega x} = (1-\beta_2)\lambda^o e^{i\omega x} + \left(\beta_1 + \frac{\beta_2}{2}\right)\lambda^o e^{i\omega(x-\Delta x)} + \left(-\beta_1 + \frac{\beta_2}{2}\right)\lambda^o e^{i\omega(x+\Delta x)} \tag{131}$$

Division by $\lambda^n e^{i\omega x}$ then gives:

$$\begin{aligned} &\frac{\lambda^n}{\lambda^o} = (1-\beta_2) + \left(\beta_1 + \frac{\beta_2}{2}\right)e^{-i\omega\Delta x} + \left(-\beta_1 + \frac{\beta_2}{2}\right)\lambda e^{i\omega\Delta x} = \\ &1-\beta_2-\beta_1\left(e^{i\omega\Delta x} - e^{-i\omega\Delta x}\right) + \frac{\beta_2}{2}\left(e^{i\omega\Delta x} + e^{-i\omega\Delta x}\right) \end{aligned} \tag{132}$$

Now we will use the following relations:

$$\begin{aligned} e^{i\omega\Delta x} - e^{-i\omega\Delta x} &= 2i\sin(\omega\Delta x) \\ e^{i\omega\Delta x} + e^{-i\omega\Delta x} &= 2\cos(\omega\Delta x) \end{aligned} \tag{133}$$

Substitution in equation (132) then leads to:

$$\begin{aligned} &\frac{\lambda^n}{\lambda^o} = 1-\beta_2-2\beta_1 i\sin(\omega\Delta x) + \beta_2\cos(\omega\Delta x) = \\ &1-\beta_2\left(1-\cos(\omega\Delta x)\right) - 2\beta_1 i\sin(\omega\Delta x) = \\ &1-2\beta_2\sin^2\left(\frac{\omega\Delta x}{2}\right) - 2\beta_1 i\sin(\omega\Delta x) \end{aligned} \tag{134}$$

If we now take the abolute value, and substitute for shorthand $\theta=\omega\Delta x$, the following relation is obtained:

$$\left|\frac{\lambda^n}{\lambda^o}\right| = \left(1-2\beta_2\sin^2\frac{\theta}{2}\right)^2 + 4\beta_1^2\sin^2\theta \tag{135}$$

From well known goniometric relations:

$$\sin^2\theta = 1 - \cos^2\theta$$
$$\cos\theta = 1 - 2\sin^2\frac{\theta}{2} \tag{136}$$
$$\sin^2\theta = 4\left(\sin^2\frac{\theta}{2} - \sin^4\frac{\theta}{2}\right)$$

Substitution in equation (135) gives, after rearranging terms:

$$\left|\frac{\lambda^n}{\lambda^o}\right| = 1 - \left(4\beta_2 - 16\beta_1^2\right)\sin^2\frac{\theta}{2} + \left(4\beta_2^2 - 16\beta_1^2\right)\sin^4\frac{\theta}{2} \tag{137}$$

In order to have a stable solution, the absolute value of the ratio of λ^n/λ^o should be <1. Bearing in mind that both $\sin^2$ and $\sin^4$ are always positive, it is sufficient to require that:

$$4\beta_2 - 16\beta_1^2 > 0 \qquad 4\beta_2 - 16\beta_1^2 > 4\beta_2^2 - 16\beta_1^2 \tag{138}$$

The second relation in (138) can be written as:

$$\beta_2 < 1 \quad or \quad \Delta t < \frac{(\Delta x)^2}{2D} \tag{139}$$

and the first relation as:

$$\beta_2 > 4\beta_1^2 \quad or \quad \Delta t < \frac{2D}{v^2} \tag{140}$$

Note, that in order to obtain a stable solution, both requirements (A-15) and (A-16) need to be fulfilled.

Acknowledgements

We are grateful for funding by the Dutch research program Knowledge for Climate (themes 2 and 3), Stichting Retourschip in Wassenaar (Netherlands) and the EU project SoilCAM (Topic ENV.2007.3.1.2.2, grant nr. 212663).

Author details

S.E.A.T.M. van der Zee and A. Leijnse

Wageningen University, Environmental Sciences, Wageningen, Netherlands

References

[1] Appels, W. M, Bogaart, P. W, & Van Der Zee, S. E. A. T. M. (2011). Influence of spatial variations of microtopography and infiltration on surface runoff and field scale hydrological connectivity, Advances in *Water Resources*, doi:10.1016/j.advwatres.

[2] Bear, J. Hydraulics of Groundwater, New York, Elsevier, (1982).

[3] Bear, J. Dynamics of fluids in porous media. New York, Elsevier, (1972).

[4] Bellin, A, Rinaldo, A, Bosma, W. J. P, Van Der Zee, S. E. A. T. M, & Rubin, Y. (1993). Linear equilibrium adsorbing solute transport in physically and chemically heterogeneous porous formations: 1. Analytical solutions, Water Resour. Res., , 29, 4019-4030.

[5] Beltman, W. H. J, Boesten, J. J. T. I, & Van Der Zee, S. E. A. T. M. Spatial moment analysis of transport of nonlinearly adsorbing pesticides using analytical approximations, Water Resour. Res., W05417, doi:10.1029/2007WR006436,(2008). , 44

[6] Boekhold, A. E, & Van Der Zee, S. E. A. T. M. (1991). Spatial patterns of cadmium contents related to soil heterogeneity, *Water, Air, Soil Pollution*, 57/58, 479-488, 1991

[7] Bolt, G. H. Soil Chemistry B, Physico-Chemical Models, Elsevier, (1982).

[8] Bosma, W. J. B, & Van Der Zee, S. E. A. T. M. Water Resources Research, (1993).

[9] Bresler, E, & Dagan, G. Solute dispersion in unsaturated heterogeneous soil at field scale, Soil Science Society America Journal, (1979).

[10] Cirpka, O. A, & Kitanidis, P. K. (2000). An advective-dispersive stream tube approach for the transfer of conservative-tracer data to reactive transport, Water Resour. Res., , 36, 1209-1220.

[11] Dagan, G, Bresler, E, Dagan, G, & Bresler, E. (1979). Solute dispersion in unsaturated heterogeneous soil at field scale: Theory. *Soil Sci. Soc. Am. J.*, , 43, 461-467.

[12] Dagan, G, & Neuman, S. P. (1997). Subsurface Flow and Transport: A Stochastic Approach. 1st edn. Cambridge Univ. Press, Cambridge, 241 pp.

[13] Dagan, G. Flow and Transport in Porous Formations, Springer Verlag, Berlin, (1987).

[14] De Louw PGBOude Essink GHP, Stuyfzand PJ, van der Zee SEATM. Upward groundwater flow in boils as the dominant mechanism of salinization in deep polders. The Netherlands J Hydrol (2010). , 394, 494-506.

[15] Eeman, S, Van Der Zee, S. E. A. T. M, Leijnse, A, De Louw, P. G. B, & Maas, C. (2012). Response to recharge variation of thin lenses and their mixing zone with underlying saline groundwater Hydrol. Earth Syst. Sci. Discuss., 16, p. 3535 – 3549, doi:10.5194/hess-16-3535-2012 , 9, 1435-1480.

[16] Eeman, S, Leijnse, A, Raats, P. A. C, & Van Der Zee, S. E. A. T. M. Analysis of the thickness of a fresh water lens and of the transition zone between this lens and upwelling saline water, Water Resour. Res., Advances in Water Resources doi:10.1016/j.advwatres.2010.12.001,(2011). , 34, 291-302.

[17] Fiori, A, Berglund, S, Cvetkovic, V. D, & Dagan, G. analysis of solute flux statistics in aquifers: the combined effect of pore-scale dispersion, sampling, and linear sorption statistics, *Water Resour. Res.*, 38, doi:10.1029/2001WR000678

[18] FOCUS ((2000). FOCUS groundwater scenarios in the EU review of active substances, Report of the FOCUS groundwater scenarios workgroup, EC document Sanco/321/2000 rev. 2. Available at http://viso.ei.jrc.it/focus/gw/docs/FOCUS_GW_Report_Main.pdf

[19] French, H. K, Swensen, B, Englund, J, Meyer, O, & Van Der Zee, K. -F. S.E.A.T.M. ((1994). A lysimeter trench for reactive pollutant transport studies. In: Soveri, J. and Suokko, T. (eds.) Future Groundwater Resources at Risk. International Association of Hydrogeological Science (IAHS), June, Helsinki, Finland IAHS Publication (1994) 222: 131-138., 13-16.

[20] French, H. K. and Van der Zee ((1999). Field scale observations of small scale spatial variability of snowmelt drainage and infiltration, *Nordic Hydrology*, , 30, 166-176.

[21] French, H. K, Van Der Zee, S. E. A. T. M, & Leijnse, A. (1999). Differences in gravity dominated unsaturated flow during autumn rains and snowmelt, *Hydrological processes* 13, 17: 2783-2800

[22] French, H. K, Van Der Zee, S. E. A. T. M, & Leijnse, A. (2001). Transport and degradation of propyleneglycol and potassium acetate in the unsaturated zone. *Journal of Contaminant Hydrology*, , 49, 23-48.

[23] French, H. K, Van Der Zee, S. E. A. T. M, & Meju, M. (2009). SoilCAM: Soil contamination: advanced integrated characterisation and time-lapse monitoring, *Rev. Environ. Sci. Biotechnol.*, , 8, 125-130.

[24] Gao, X, Schroder, T. J, Hoffland, E, Zou, C, Zang, F, & Van Der Zee, S. E. A. T. M. Geochemical Modeling of Zinc Bioavailability for Rice, Soil Sci. Soc. Am. J. doi: 10.2136/sssaj2008.0060,(2009). , 74, 301-309.

[25] Herzberg, A. Die Wasserversorgung einiger Nordseebäder. J Gasbeleucht Wasserversorgung (1901). in German]., 44, 815-9.

[26] Jaesche, P. J, Totsche, K. U, & Kögel-knabner, I. (2006). Transport and anaerobic biodegradation of propylene glycol in gravel-rich soil materials, *Journal of Contaminant Hydrology*, , 85, 271-286.

[27] Janssen, G. M. C. M, Cirpka, O. A, & Van Der Zee, S. E. A. T. M. (2006). Stochastic analysis of nonlinear biodegradation in regimes controlled by both chromatographic and dispersive mixing, *Water Resour. Res.*, 42, doiWR004042

[28] Janssen, G. M. C. M, Valstar, J. R, & Van Der Zee, S. E. A. T. M. Measurement network design including travel time determinations to minimize model prediction uncertainty, Water Resour. Res., 44, W02405, doi:WR005462, (2008).

[29] Keijzer, H, Van Der Zee, S. E. A. T. M, & Leijnse, A. (1998). Characteristic regimes for in-situ bioremediation of aquifers by injecting water containing an electron acceptor, *Comput. Geosci.*, , 2, 1-22.

[30] Keijzer, H, Schotting, R. J, & Van Der Zee, S. E. A. T. M. (1999). Semi-analytical traveling wave solution of one-dimensional aquifer bioremediations, *Communications on Applied Nonlinear Analysis*, , 7, 1-20.

[31] Lebbe, L, Van Meir, N, & Viaene, P. Potential implications of sea-level rise for Belgium. J Coast Res (2008). , 24(2), 358-66.

[32] Leistra, M, Van Der Linden, A. M. A, Boesten, J. J. T. I, & Tiktak, A. van den Berg, F. ((2000). PEARL model for pesticide behaviour and emissions in soil-plant systems. Description of processes. Alterra report 13, RIVM report 711401009, Alterra, Wageningen, 107 pp.

[33] Maas, K. Influence of climate change and sea level rise on a Ghijben Herzberg lens. J Hydrol (2007). , 347, 223-8.

[34] Miller and MillerMiller, E. E. and Miller, R. D. ((1956). Physical theory for capillary flow phenomena. *J.Appl. Phys.*, , 27, 324-332.

[35] Roth, K. (1995). Steady state flow in an unsaturated, two-dimensional macroscopically homogeneous miller-similar medium. *Water Resources Research*, , 31, 2121-2140.

[36] Suweis, S, Rinaldo, A, Van Der Zee, S. E. A. T. M, Daly, E, Maritan, A, & Porporato, A. Stochastic modeling of soil salinity, Geophysical Research Letters, L07404, doi: 10.1029/2010GL042495,(2010). , 37

[37] Van Der Velde, Y, Torfs, P. J. J. F, Van Der Zee, S. E. A. T. M, & Uijlenhoet, R. (2012). Quantifying catchment-scale mixing and its effect on time-varying travel time distributions, Water Resources Research, Vol. 48, W06536, doi:10.1029/2011WR011310

[38] Rode, M, Arhonditsis, G, Balin, D, Kebede, T, Krysanova, V, Van Griensven, A, & Van Der Zee, A. , a. n. d S. E. A. T. M. (2010). New challenges in integrated water quality modelling, *Hydrological processes*, DOI:10.1002/hyp.7766,

[39] Vandenbohede, A, Luyten, K, & Lebbe, L. Effects of global change on heterogeneous coastal aquifers: a case study in Belgium. J Coast Res (2008). , 24, 160-70.

[40] Van Der Zee, S. E. A. T. M, & Boesten, J. J. T. I. (1991). Effects of soil heterogeneity on pesticide leaching to groundwater, Water Resour. Res., 27(12), 3051- 3063

[41] Van Der Veer, P. Analytical solution for steady interface flow in a coastal aquifer involving a phreatic surface with precipitation. J Hydrol (1977). , 34, 1-11.

[42] Van Der Zee, S. E. A. T. M, & Van Riemsdijk, W. H. (1987). Transport of reactive solute in spatially variable soil systems, Water Resour. Res., , 23, 2059-2069.

[43] Van Der Zee, S. E. A. T. M, Fokkink, L. G. J, & Van Riemsdijk, W. H. (1987). A new techinique for assessment of reversibly adsorbed phosphate, Soil Sci. Soc. Am. J. , 51, 599-604.

[44] Van Dijke, M. I. J, & Van Der Zee, S. E. A. T. M. (1998). Analysis of oil lens removal by extraction through a seepage face, Comput. Geosci., , 2, 47-72.

[45] Vervoort, R. W, & Van Der Zee, S. E. A. T. M. Simulating the effect of capillary flux on the soil water balance in a stochastic ecohydrological framework, Water Resour. Res., W08425, doi:10.1029/2008WR006889,(2008). , 44

[46] Van Der Zee, S. E. A. T. M, Van Uffelen, C. J, Shah, H. S, & Raats, P. A. C. and N. dal Ferro, Soil sodicity as a result of periodical drought, Agricultural Water Management doi:10.1016/j.agwat.2009.08.009,(2010). , 97, 41-49.

[47] Van Der Zee, S. E. A. T. M. (1990). Analytical traveling wave solutions for transport of solutes with nonlinear and nonequilibrium adsorption, Water Resour. Res., 26(10), 2563-2578

[48] Vervoort, R. W, & Van Der Zee, S. E. A. T. M. Stochastic soil water dynamics of phreatophyte vegetation with dimorphic root systems, Water Resources Research, W10439, doi:10.1029/2008WR007245,(2009). , 45

[49] Witte, J. P. M, Runhaar, J, Van Ek, R, Van Der Hoek, D. C. J, Bartholomeus, R. P, Batelaan, O, Van Bodegom, P. M, Wassen, M. J, & Van Der Zee, S. E. A. T. M. An ecohydrological sketch of climate change impacts on water and natural ecosystems for The Netherlands: bridging the gap between science and society, Hydrol. Earth Syst. Sci. Discuss., 9, 6311-6344, 2012, doi:10.5194/hessd-9-6311-2012,(2012).

Chapter 2

Physical Indicators of Soil Quality in Oxisols Under Brazilian Cerrado

Diego Antonio França de Freitas,
Marx Leandro Naves Silva, Nilton Curi,
Mayesse Aparecida da Silva,
Anna Hoffmann Oliveira and
Sérgio Henrique Godinho Silva

Additional information is available at the end of the chapter

1. Introduction

The Brazilian Cerrado makes up one of the most biodiverse savannas in the world and it harbors a mosaic of plant physiognomies that include from open forms (grasslands) to forest (dense woodlands), possessing high structural, functional and life forms diversity. Little valued traditionally, the Cerrado has been neglected in most of the conservationist initiatives because its vegetation is considered sparse and of low value.

The Brazilian Cerrado comprises an area of 2,036,448 km^2 [1]. Its largest part is within the Aw type of Köpen climatic classification (tropical seasonal savanna), with a rainy period, from October to March, followed by a dry period, from April to September. In this environment, the irregular distribution of the rain and the existence of short droughts constitute serious limitation for farming in the absence of irrigation.

The main soils of the Cerrado area are Latosols (Oxisols) that correspond to 46%, followed by Neosols (Entisols) with 16% and Argisols (Ultisols) with 15%. Latosols occupy a flat to gentle rolling topography in the landscape, which facilitates the mechanized management, those soils being of high potential for the production of annual and perennial crops and also pasture.

In recent decades the Cerrado has undergone various transformations as to its land use, mainly due to the high investments in soil correctives, fertilizers and various crop varieties adapted to this biome. This generated a disordered occupation of the land, with a rampant increase of deforestation that contributed to the loss of species diversity and, concomitantly, some inadequate soils management techniques propitiated the fast degradation of that resource [2], erosion, aquifer pollution, ecosystem degradation, alteration of the soil physical, chemical and biological attributes and consequent reduction of the soil quality.

In that context, according to [3], the soil quality is expressed when the soil works within the limits of a natural ecosystem, so as to sustain biological production, promote animal and plant health, and to maintain the quality of the environment. It is usually determined by a group of physical, chemical and biological attributes, that represents the different soil characteristics and that influences its various functions. Each one of these edaphic attributes, in turn, may or may not perform well, which will influence agricultural production in a significant way.

The direct evaluation of the soil properties seems to be the most appropriate way to measure or to monitor its conservation or any degradation process underway [4]. Thus, the evaluation of the soil quality has been increasingly proposed as an integrated indicator of the environmental quality and sustainability of agricultural systems. The quantification of soil attribute alterations, due to the intensification of production systems or natural systems exploration, besides being useful in the evaluation of anthropic interference in the environment because it considers the relationship between the soil and the other aspects of the ecosystem, supplies important subsidies to the definition of rational management systems, contributing to making the soil less susceptible to the loss of productive capacity and, fundamentally, to the environmental conservation [5].

The soil quality indicator attributes are defined as measurable properties that have influence on the capacity of the soil to produce crops or on the performance of environmental functions [6]. For those attributes to be capable of indicating soil quality alterations, they should be well correlated with processes within the ecosystem; be applied in a relatively easy manner under field conditions and be appraised not only by specialists but also by producers; they should be sensitive to variations in management and climate, reflecting soil quality changes, without being influenced by accidental alterations and be components of previously existent databases [6]. For reference [7] no individual indicator is able to describe and quantify all of the soil quality aspects and not even an isolated soil function is enough, since it should have a relationship among all its attributes.

Among the various soil attributes responsible for its quality, the physical attributes stand out, the soil structure being of one of the most important indicators for plant growth, since it has a direct influence on the densification, compaction, crusting, water infiltration and soil susceptibility to erosion conditions [8]. The structure can be evaluated through the soil density, macro and microporosity, aggregate stability, and resistance to water penetration and infiltration in the soil. These indicators show the effect of the soil management, being easy to measure, with fast and reasonably precise responses [3].

Knowing the soil density is an important indicator of the soil management conditions and its value will reflect in the characteristics of the soil pore system, so as to hinder the water and oxygen supply, limiting the plant development and organisms activity [9], thus influencing various fundamental processes that the soil exercises in its function [10]. The porosity is the volumetric fraction of the soil occupied with air and/or water and it is empirically divided in macroporosity (pores diameter > 0.05 mm) and microporosity (pores diameter < 0.05 mm). The phenomena of water infiltration in the soil (descending flow) occur mainly via the macropores, while the storage (retention) of water occurs in the micropores. The soil compaction tends to mainly reduce the macroporosity values, the reason why there is water infiltration reduction and, consequently, an increased erosion risk.

The aggregate stability varies with the inherent soil characteristics and with the management systems. The intense soil tillage provokes the aggregate breakup, being able to drastically reduce its stability. With the breakage of the aggregates, the organic matter which was in its interior becomes unprotected, accelerating its decomposition process, reducing both resistance of these aggregates and the soil quality [10].

As such, due to the importance of the physical attributes for the soil quality maintenance, the objective of the present study was to analyze and characterize the physical attributes indicative of the quality of the Latosols in native, agricultural, pasture and planted forest environments, in the Cerrado biome.

2. Location, climate, soil and management systems

In order to conduct this study, it was constructed a physical attribute database of the Latosols located in areas under Cerrado. The database in question was prepared through the selection of information and data contained in the files of the Soil and Water Conservation Sector, Soil Science Department, Federal University of Lavras, gathering information from [11- 16]. The studied physical attributes indicators of soil quality were the soil density, total pore volume, macroporosity, microporosity, average geometric diameter, hydraulic conductivity of the saturated soil and soil resistance to penetration.

3. Study areas

The study was conducted in several areas of the Brazilian Cerrado, with samplings in the State of Minas Gerais (Campos das Vertentes, Vale do Rio Doce - Guanhães, Noroeste, Vale do Rio Doce - Belo Oriente, Central) and the State of Goiás (South). Figure 1 presents the location of the environments under study.

3.1. Campos das Vertentes, MG

Campos das Vertentes is located in the Alto Rio Grande basin (20° 21'-21° 42' S; 43° 16'-44° 42' WGr), in the South-Central of Minas Gerais State. According to the Köppen climatic classification, the predominant climate in the area is the Cwa type (mild-temperate mesothermal), that is characterized by having an average temperature of the coldest month under 18°C and the average of the hottest month over 22°C, with rainy summers and dry winters. The average annual precipitation is 1435 mm, concentrated between December and April.

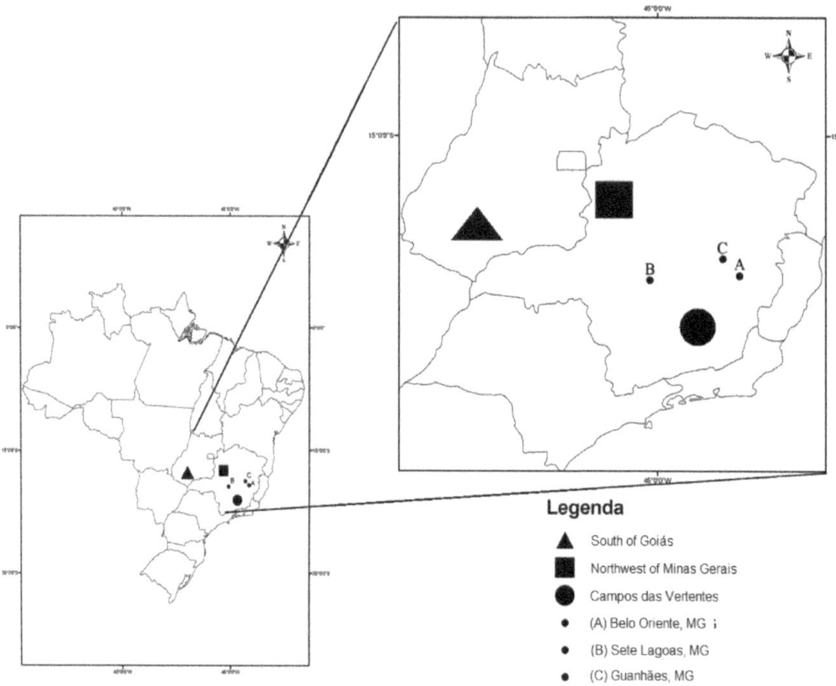

Figure 1. Location of the studied areas.

The soil of the mentioned area is classified as typical acric Red-Yellow Latosol (LVA_1) of clayey texture (EMBRAPA, 2006), developed over a geological substratum corresponding to poor metapelitic rocks of the São João del Rei (phyllite) and Andrelândia (micaschist) groups. The selection of the studied management systems was conducted was conducted attempting to reach a better representativeness of the area the area over that soil class, as shown in Table 1.

3.2. Vale do Rio Doce, Guanhães, MG

The Vale do Rio Doce region, municipal district of Guanhães (18° 46' S; 42° 55' WGr), State of Minas Gerais, presents the Cwa climate (dry winter and rainy summer) with temperature of the coldest month under 18°C and the one of the hottest month passing 22°C, according to the Köppen climatic classification. The dry season occurs between the months of April and September, the average annual precipitation of the area being 1,180.8mm and an average altitude of 850m.

Systems	Symbol	Characteristic of the systems
Native Cerrado	NC1	Primary vegetation represented by the sub deciduous tropical Cerrado and Savanna woodland.
Conventional cultivation with potato	CCP	Conventional cultivation with potato, sampled postharvest.
Conventional cultivation with potato and oat	CCPO	Conventional cultivation with potato, followed by oat, after a postharvest subsoiling of potato and rotated with corn, sampled 15 days after the corn planting.
Conventional cultivation with corn	CCC	Conventional cultivation with corn
No-till with corn	DPC	No-till with corn after the conventional cultivation with potato in the harvest, rice and conventional corn, sampled 52 days after planting.
Conventional eucalyptus	CE1	Conventional eucalyptus cultivation, without application of subsequent management practices.

Source: [16] modified.

Table 1. Characterization of the management systems in typic acric Red-Yellow Latosol in the Campos das Vertentes - MG region.

The soil of the area under studies is classified as typical dystrophic Red Latosol (LV_1), very clayey texture, prominent A horizon, alic, kaolinitic-oxidic, mesoferric, wavy relief, gneiss and granite-gneiss substrate. The selected study areas are presented in Table 2.

Systems	Symbol	Characteristic of the systems
Native savanna	NC2	Original condition
Native pasture	NP	Degraded native pasture of long duration.
Uncovered soil	UCS1	System where the soil does not possess plant covering.
Eucalyptus planted following the slope, with burning of crop debris	ESq	Eucalyptus planted in line accompanying the slope of the land, burning of crop debris being carried out.
Eucalyptus planted following the slope, without burning of crop debris	ES	Eucalyptus planted in line accompanying the slope of the land, the crop debris being left on the surface.
Eucalyptus planted cross slope	ECS1	Eucalyptus planting line is perpendicular to the slope direction.

Source: [12] modified.

Table 2. Characterization of the management systems in typical dystrophic Red Latosol (LV_1) Vale do Rio Doce region, town of Guanhães – MG.

3.3. Northwest Minas Gerais

The Northwest region of the State of Minas Gerais is located between latitudes 16° 10' and 18° 42' S, and longitudes 44° 24' and 47° 44' WGr, the climate of the area being the Cwa type, characterized by the temperature of the coldest month under 18° C, and the precipitation of the driest month less than 60 mm with annual averages varying from 1,300 to 1,400 mm. The soil is classified as typical dystrophic Red Latosol (LV_2) [17]. Seven production systems were studied in this area, according to Table 3.

Systems	Symbol	Characteristic of the systems
Native Cerrado	NC3	Typical Cerrado vegetation, without reports of human interference and agricultural use.
Eucalyptus + rice	ER	Eucalyptus intercropped with rice, eucalyptus being 4 months of age.
Eucalyptus + soybeans	ESy	Eucalyptus intercropped with soybeans. On the date of the sampling, the eucalyptus was 1 year and 4 months old.
Eucalyptus + pasture	EP	Eucalyptus intercropped with planted pasture. On the date of the sampling, the eucalyptus was 3 years and 4 months old.
Eucalyptus + pasture + cattle	EPC	Eucalyptus intercropped with planted pasture. On the date of the sampling, the eucalyptus was 7 years and 4 months old.
Conventional pasture	CP	Conventional pasture
Conventional eucalyptus	CE2	Conventional eucalyptus (3x2 spacing)

Source: [14] modified.

Table 3. Characterization of the management systems in a typical dystrophic Red Latosol (LV_2), Northwestern Minas Gerais region.

3.4. Vale do Rio Doce, Belo Oriente, MG

In the municipal district of Belo Oriente (19° 17′ S; 42° 23′ WGr), in the Rio Doce region, State of Minas Gerais, the predominant climate is the Aw type, in other words, tropical, with dry winters and rainy season in the summer, according to the Köppen classification, presenting an average annual temperature varying between 22° and 27°C, the maximum temperature being 32° C and the lowest, 18° C with an average annual precipitation varying from 701 to 1500 mm at an altitude of 233m. The dry season occurs between the months of May to September.

The soil of this area was classified as typical dystrophic Red-Yellow Latosol (LVA_2), very clayey texture. The geological formation is granite-gneisse bedrock of the Pre-Cambrian period, and the material of origin are gnaisse alterations. The study areas were made up of six soil use systems, as presented in Table 4.

Systems	Symbol	Characteristic of the systems
Native Cerrado	NC4	Reference system in equilibrium
Planted pasture	PP1	Pasture of *Brachiaria sp.* Soil use reference in the study area.
Eucalyptus planted cross slope	ECS2	Planting line of the eucalyptus is perpendicular to the slope direction.
Conventional eucalyptus with burning of crop debris	ECq	Eucalyptus planted in line accompanying the slope of the land, burning of crop debris being carried out.
Conventional eucalyptus	CE3	Eucalyptus planted in line accompanying the slope of the land, the crop debris being left on the surface.
Uncovered soil	UCS2	System Referencial in degradation process.

Source: [15] modified.

Table 4. Characterization of the management systems in a dystrophic Red-Yellow Latosol (LVA_2) in the Vale do Rio Doce region, town of Belo Oriente – MG.

3.5. Central region of Minas Gerais

The samples were collected in the city of Sete Lagoas, MG, located at 19° 25′south and 44°15′ west at an altitude of 732m. The average annual temperature in the area is 22.1°C and the average annual precipitation is 1340 mm. According to the Köppen climatic classification, the climate is Aw (seasonal tropical savanna).

The soil is an alic Red Latosol (LV_3), with moderate A horizon, very clayey texture, Cerrado tropical semideciduous phase and gentle undulated relief, derived from pelitic rocks of the Late Proterozoic Bambuí group. For the Sete Lagoas area, five production systems and a reference environment were studied, according to Table 5.

Systems	Symbol	Characteristic of the systems
Native Cerrado	NC5	Environment without anthropic interference
Disk harrow and corn cultivation	DHCC	Conventional preparation with disk harrow and continuous cultivation with corn
Disk plow and cultivation with corn	DPICC	Conventional preparation with disk plow and continuous cultivation with corn
Disk plow and cultivation in rotation with corn and beans	DPIRCB	Conventional preparation with disk plow and cultivation in rotation with corn and beans.
No-till and corn	DPCntC	No-till and continuous cultivation with corn
No-till and cultivation with rotation with corn and beans	DPRCB	No-till and cultivation with rotation with corn and beans

Source: [11] modified.

Table 5. Characterization of the management systems in an alic Red Latosol (LV_3) in the Sete Lagoas region.

3.6. Southern region of Goiás

The work was developed in agricultural properties in the municipal districts of Morrinhos and Caldas Novas in the Southern area of the State of Goiás, located in the Central Goiano Plateau geomorphological unit, Lowered Goiânia Plateau sub-unit.

The soil was classified as typical dystrophic Red Latosol (LV_4), according to [17]. Being this soil developed from Pleistocenic lateritic-residues covering on micaxists of the group Araxá of Proterozóico Inferior. The studied areas were constituted of eight systems of use of the soil as presented in Table 6.

System	Symbol	Characteristic of the systems
Native Cerrado	NC6	Environment without anthropic interference
Pasture	PP2	Planted pasture of long use, without fertility management in the last 10 years and under continuous cattle grazing.
Irrigated No-till	IDP 1	System under central pivot in the last 5 years, with corn cultivation in rotation with beans, with subsoiling to 15 cm of depth 2 years before the harvest.
Irrigated No-till	IDP 2	No-till under central pivot in the previous 5 years, with corn, beans and rice cultivation and a harvest of industrial tomato with surface harrowing from 0 to 1 cm depth, 2 years before.
Dryland No-till	DDP	No-till planting with soybean cultivation over millet straw in the previous 7 years, after conventional system (soybeans).
Irrigated conventional planting	ICP 1	Conventional system, with use of heavy harrowing, irrigated under central pivot in the previous 2 years, with corn after more than 15 years of dryland soybeans-corn succession.
Irrigated conventional planting	ICP 2	Recently irrigated conventional system, with use of heavy harrowing, under central pivot in the previous 2 years, with squash/beans/sweet corn rotation, after more than 10 years as pasture.
Dryland conventional planting	DCP	Conventional dryland system, with use of heavy harrowing for the soil preparation and soybeans-corn succession for more than 15 years.

Source: [13] modified.

Table 6. Characterization of the management systems in a typical dystrophic Red Latosol (LV_4) in the South of State of Goiás.

4. Determination in the field

Samples with undisturbed structures were collected with the use of the Uhland sampler, in cylinders with average dimensions of 8.25 cm of height by 6.90 cm of internal diameter. The samples with deformed structure were collected in the depth of 0-20 cm, being air-dried and sieved in a 2 mm mesh sieve (fine soil) for analyses.

The soil resistance to penetration was determined in the field using an impact penetrometer (IAA/PLANALSUCAR STOLF model), according to the methodology of [18]. The values obtained in Kgf cm^{-2} were multiplied by a factor of 0.098 to be expressed in MPa.

5. Laboratory determinations

The textural analysis was conducted by the pipette method [19], using NaOH 1 mol L^{-1} as chemical dispersant and fast agitation (12,000 rpm), for 10 minutes. The soil density was determined according to [20]. The total pore volume was determined according to the expression recommended by [21]. The pore size distribution was determined using a porous filter plate funnel in the suction unit with 60 cm of water column height for macro and microporosity separation in samples previously saturated for 48 hours. In this situation, the water volume retained in the samples after the equilibrium corresponds to the microporosity, the macroporosity being obtained by difference between the total pore volume and the microporosity [22].

The aggregates with diameter from 4.76 to 7.93 mm were obtained by soil sieving, and the aggregate stability determined through sieving in water after slow pre-wetting of the aggregates by capillarity for 24 hours [23-24]. Sieves with 2.00; 1.00; 0.50; 0.25 and 0.105 mm meshes were used for separation of the aggregate size classes [25]. Concerning the aggregate stability and size distribution, the procedure involves a known weight of soil mass which is submitted to slow wetting and sieving, so the average geometric diameter could be used as an index of the aggregate size distribution [25].

The soil water permeability was evaluated in laboratory, from samples previously saturated by capillarity and using a constant-head permeameter, adapted for elimination of the percolated water close to the cylinder walls, following methodology described by [26].

6. Granulometric characterization of soil

The granulometric characteristics and the particle density (PD) of the studied Latosols are presented in Table 7.

In relation to the relative particle size proportion, we verified high clay fraction and low of silt content and a silt/clay ratio lower than 0.7. The particle density values of presented variations from 2.29 to 2.65 kg dm^{-3}, and the particle density is not influenced by mechanical alterations, but by the organic matter content in the soils.

Local	Soil	Silt	Clay	Sand	Silt / Clay	PD
		--------- g kg^{-1} -------				kg dm^{-3}
Campos das Vertentes – MG	LVA1	155	627	218	0.25	2.43
Guanhães – MG	LV1	71	598	331	0.12	2.56
Northwest – MG	LV2	141	681	178	0.21	2.29
Belo Oriente – MG	LVA2	109	425	466	0.26	2.50
Central – MG	LV3	234	582	184	0.40	2.65
South – Goiás	LV4	198	335	467	0.59	2.52

Table 7. Silt, clay, sand content and particle density for the Latosols under Cerrado [27].

7. Mineralogical properties

The tropical soils present high degree of weathering, with the clay fraction mineralogy dominated by silicate minerals of the 1:1 type and iron and aluminum oxides [28]. Latosols are what best represents the pedogenetic tendencies of the tropical soils, being defined as those that present a latosolic mineral subsurface B horizon, that evidences an advanced of weathering stage, as shown by the complete or almost total alteration and decomposition of easily weathered minerals, the high depth and by the low cationic exchange capacity [29].

Latosols present clay fraction mineralogy basically dominated by kaolinite, gibbsite, goethite and hematite, besides poorly crystallized iron and aluminum oxides. Although the predominant mineralogical composition in the tropical soils can be considered simple, variations that can occur within and among the mineral groups as to particle size and specific surface, exposed faces, degrees of isomorphic substitution and crystallinity can provide high soil behavior variability within a same class [30].

The mineralogical properties of the studied Latosols are presented in Table 8

Local	Symbol	Gb	Kt	Gb/ Gb+Kt	SiO_2	Al_2O_3	Fe_2O_3	TiO_2	ki	kr
		mg kg^{-1}			--------g kg^{-1}-------					
Campos das Vertentes - MG	LVA_1	290	350	0.45	161	260	145	10.8	1.05	0.78
Guanhães - MG	LV_1	162	364	0.31	177	235	69	13.6	1.28	1.08
Northwest - MG	LV_2	480	260	0.65	175	252	72	6.3	1.18	1.00
Belo Oriente - MG	LVA_2	117	380	0.24	116	173	77	16.6	1.14	0.89
Central - MG	LV_3	160	310	0.34	234	319	120	5	1.24	1.08
South - Goiás	LV_4	335	188	0.64	123	174	140	1.96	1.2	0.8

Gb = gibbsite; Kt = kaolinite; Ki = SiO_2/Al_2O_3 molecular ratio; Kr = $SiO_2/(Al_2O_3 + Fe_2O_3)$ molecular ratio.

Table 8. Mineralogical characteristics of Latosols under Cerrado [27].

8. Soil density

The soil density (SD) increased in the management systems that underwent anthropic interference, three lowest values being the found in natural environments, represented by the native Cerrado (NC5) of the Central Minas Gerais region and native Cerrado (NC2 and NC4), both from Vale do Rio Doce, MG, with values of 0.83, 0.87 and 0.93 kg dm^{-3}, respectively (Table 9).

The highest soil density values were observed for the use systems installed in the South of Goiás, the three highest values being for the irrigated conventional planting, conventional dryland and direct dryland planting (ICP2, DCP and DDP), with values of 1.36, 1.35 and 1.31 kg dm^{-3}, respectively. The maintenance of the soil in the uncovered forms (UCS1 and UCS2) caused increases of 41 and 39% in the soil density in relation to the native environment of the same area, because in these areas the direct impact of the rain drops occurs, which provides elevation of the soils density. The density results found for Latosols under study were below the value of 1.40 kg $dm\text{-}^{3}$, a value that, according to [31], is restrictive to the plants root growth in clayey soils.

Alterations in the cultivated soil density values in relation to the natural condition have been reported by several references [5, 32]. The soil density in non-cultivated environments is a physical property that depends on pedogenetic factors and processes. The lowest soil density value for native areas and that did not undergo anthropic interference results from a higher accumulation of plant residues incorporated into the soil, associated to the non-disturbance of the structure by the machines and agricultural implements traffic, animal trampling, intensive cultivation and inadequate management systems [33].

According to [34], it becomes difficult to affirm under which use systems the soil density increase would tend to be harmful to other functions and the soil quality in reason of the differences in the granulometric composition, the chemical and mineralogical nature of the soil, the time of use of the management systems and the resilience and resistance inherent to each soil class, among other factors.

Land use	Ds	Pt	Macro	Micro	Ks	AGD
	Kg dm^{-3}	----------------	m^3m^{-3}	------------	mm h^{-1}	mm
Campos das Vertentes – MG - LVA_1						
CCP	1.17	0.52	0.09	0.43	42.6	4.67
CCPO	1.22	0.51	0.10	0.41	29.8	4.85
CCC	1.05	0.56	0.10	0.46	49.4	4.68
DPC	1.18	0.52	0.05	0.47	13.8	4.64
CE1	1.15	0.53	0.07	0.46	41.3	4.77
NC1	1.11	0.55	0.18	0.37	38.0	4.87

Land use	Ds	Pt	Macro	Micro	Ks	AGD
	Kg dm^{-3}		---------------- m^{3}m^{-3} ------------		mm h^{-1}	mm
Vale do Rio Doce – Guanhães – MG - LV$_1$						
NC2	0.87	0.65	0.33	0.32	230.0	4.90
NP	1.08	0.56	0.19	0.38	90.0	4.71
UCS1	1.23	0.49	0.10	0.40	183.0	4.79
ESq	1.14	0.54	0.17	0.37	174.0	4.92
ES	1.13	0.55	0.18	0.37	160.0	4.92
ECS1	1.18	0.54	0.17	0.37	75.0	4.89
Northwest – MG – LV$_2$						
NC3	1.07	0.54	0.24	0.30	733.7	4.39
ER	1.06	0.54	0.23	0.31	136.4	4.05
ESy	1.01	0.55	0.19	0.36	348.6	4.17
EP	1.15	0.54	0.23	0.31	74.95	3.92
EPC	1.13	0.56	0.27	0.29	128.6	4.00
CP	0.99	0.60	0.21	0.39	240.9	4.03
CE2	1.14	0.52	0.28	0.24	214.5	4.33
Vale do Rio Doce – Belo Oriente – MG – LVA$_2$						
NC4	0.93	0.60	0.30	0.30	191.0	4.55
PP1	1.20	0.50	0.15	0.35	81.8	4.36
ECS2	1.13	0.55	0.26	0.29	180.0	4.57
ECq	1.21	0.51	0.19	0.32	161.0	4.46
CE3	1.19	0.52	0.21	0.3	152.0	4.35
UCS2	1.29	0.47	0.11	0.36	70.0	1.95
Central – MG – LV$_3$						
DHCC	1.11	0.58	0.16	0.42	6.1	2.44
DPICC	1.10	0.59	0.16	0.43	13.5	1.93
DPIRCB	0.98	0.62	0.21	0.41	14.2	3.87
DPCntC	1.11	0.58	0.17	0.41	6.7	3.87
DPRCB	0.97	0.63	0.21	0.42	6.7	2.71
NC5	0.83	0.68	0.29	0.39	95.0	4.42
South – Goiás – LV$_4$						
NC6	1.27	0.52	0.19	0.35	174.5	4.62
PP2	1.14	0.56	0.21	0.35	340.8	4.42

Land use	Ds	Pt	Macro	Micro	Ks	AGD
	Kg dm^{-3}	----------------	m^{3}m^{-3}	-------------	mm h^{-1}	mm
IDP1	1.21	0.51	0.17	0.34	65.1	4.38
IDP2	1.18	0.56	0.21	0.35	145.1	4.38
DDP	1.31	0.49	0.17	0.32	76.3	3.67
ICP1	1.19	0.56	0.23	0.33	253.6	3.15
ICP2	1.36	0.50	0.13	0.37	114.8	4.56
DCP	1.35	0.50	0.23	0.27	159.9	2.63
			Descriptive statistics			
Minimum	0.83	0.47	0.05	0.24	6.1	1.93
Lower quartile	1.08	0.52	0.16	0.32	46.0	3.96
Average	1.13	0.56	0.21	0.38	136.2	4.07
Median	1.14	0.54	0.19	0.36	114.8	4.39
Upper quartile	1.20	0.56	0.23	0.40	177.2	4.68
Maximum	1.36	0.68	0.33	0.47	733.7	4.92

Table 9. Soil density (Ds), total porosity (Pt), macroporosity, microporosity, hydraulic conductivity (Ks) and average geometric diameter (AGD) for land use systems in the Cerrado [27].

8.1. Total porosity and pore distribution per size

Considering all of the management systems, the values of total porosity varied between 0.47 and 0.68 m^{3}m^{-3} (Table 9). Among the soil use systems without anthropic interference, two systems presented the largest total porosity values, the native Cerrado of the Central area (NC5) being the system that presented the highest value, followed by the native Cerrado (NC2) of Guanhães, respectively 0.68 and 0.65 m^{3}m^{-3}, whereas No-till plus crop rotation with corn and beans (DPRCB) and planting with conventional preparation with disk plow plus crop rotation with corn and beans (DPIRCB), both in the Vale do Rio Doce of Minas Gerais, presented total porosity values of 0.63 and 0.62 m^{3}m^{-3}, respectively.

The reduction in the total porosity in areas under agricultural management in relation to the native areas is in agreement with the observations of [9, 35] and the latter verified a reduction of up to 24% in the total porosity, when compared with areas that did not undergo anthropic action. The trampling by animals, agricultural machines and inadequate management lead to interferences in the soil structure, promoting reduction in the total porosity. According to [36], crop rotation systems can increase the total porosity of the soil when implanted in agricultural areas, an effect being confirmed in this study.

The lowest total porosity values were found in the uncovered soils (UCS2 and UCS1), followed by the dryland No-till (DDP), irrigated conventional planting (ICP2) and dryland conventional planting (DCP), presenting values of 0.47, 0.49, 0.49, 0.50 and 0.50 m^{3}m^{-3}, respectively. The fact that the area with uncovered soil presented the lowest total porosity values can be related to the absence of the crop root systems, because after the decomposition of the roots, a soil pore increase occurs, and in these areas old pores can be obstructed due to reorganization of the surface after the removal of the plant covering.

The highest macroporosity values were found for areas without anthropic interference (Table 9), and the native Cerrado (NC2) in Guanhães, MG, the native Cerrado (NC4) in Belo Oriente and the native Cerrado (NC5) in the Central area of MG, presented macroporosity values equal to 0.32; 0.30 and 0.29 $m^3 m^{-3}$, respectively. These data show that a soil macropore reduction tendency exists when native areas are transformed into agricultural or forest areas.

The lowest macroporosity values were found in the systems installed in Campos das Vertentes (MG). According to [37], the low macropore presence tends to occur in the same area, because this attribute is related to the soil texture. In the Campos das Vertentes area the macropores presented low values, on the order of 0.05; 0.07; 0.09; 0.10 and 0.10 m^3m^{-3} for the DPC; CE1; CCP; CCPO and CCC systems, respectively. For this area, the native Cerrado (NC1) presented a macroporosity value above that of the agricultural systems, 0.18 m^3m^{-3}, which demonstrates the sensitivity of this attribute in the detection of the alterations imposed by the different management systems under natural conditions.

The highest microporosity values were found in the area of Campos das Vertentes, in Minas Gerais (Table 9). This occurs by the same explanation given to the macroporosity in the area, because, for the same total porosity, an increase in the macroporosity causes the reduction of the microporosity.

The Northwest area of MG uses an agrosilvopastoral system and it presented the lowest microporosity values, among them the conventional eucalyptus system (CE2) and eucalyptus + pasture + cattle (EPC) stand out with values of 0.24 and 0.29 m^3m^{-3}, respectively.

9. Permeability of the soil to water

The permeability of the soil to water, appraised through the hydraulic conductivity of the saturated soil, presented accentuated difference among the management systems used in Latosols (Table 9). The lowest soil permeability values were found in the Central area of Minas Gerais, and the systems of agricultural management underwent reductions between 85 and 93.3% of its permeability when compared to the native Cerrado of the same area. In the other environments, the system that uses No-till with corn (DPC) in the Campo das Vertentes, MG, presented the lowest permeability value, 13.8 mm h^{-1}. This value is justified since this system presented the lowest macroporosity value found in the studied soils, according to Table 9.

The highest permeability values were found in the native Cerrado (NC3) and eucalyptus + soybeans (ESy) systems of the Northwest area, MG, followed by the pasture (PP2) and irrigated conventional planting (ICP 1) of the South of Goiás and pasture of the Northwest area, MG, that presented values of 733.77; 348.6; 340.8; 253.6 and 240.92 mm h^{-1}.

According to the permeability classes adapted from the [38] and presented in Table 10, 71.8% of the soils were classified with the permeability varying between the moderate and fast classes, being this high permeability one of the characteristics of Latosols.

Class	Permeability (mm h^{-1})
Fast	> 254.00
Moderate to fast	254.00 – 127.00
Moderate	127.00 – 63.50
Slow to moderate	63.50 – 20.00
Slow	20.00 – 5.00
Very slow	< 5.00

Table 10. Classes of soil permeability to water [38].

9.1. Aggregate stability

The Average Geometric Diameter (AGD) represents an estimate of the most frequent aggregate size and demonstrates the stability of the structure facing the disaggregation action of the water, and may indicate the susceptibility degree of soil to hydric erosion [39].

As such, the native systems inside each area and the systems that possess eucalyptus in the area of Guanhães and Belo Oriente, MG, by not presenting constant tillage of the soil, low machine traffic and animal trampling, are the systems that possess the largest size of aggregates (Table 9).

Among the five lowest AGD values, three were found in soil use systems installed in the Central region of MG, they being conventional soil preparation with disk plow for corn planting (DPlCC), conventional preparation with disk harrow for corn planting (DHCC), and no-till with corn and beans rotation with (DPRCB), that present AGD values of 1.93, 2.44 and 2.71 mm, respectively. The other two systems are the uncovered soil (UCS2) of Belo Oriente, MG, and the dryland conventional planting (DCP) in the South area of the State of Goiás. A characteristic of these systems, except for DPRCB, was the high soil tillage operations for the soil preparation that fractioned the larger aggregates into smaller ones. For DPRCB, a possible explanation for the low AGD indexes presented would be the short transformation time from the conventional planting to no-till, thus the system still maintains AGD indexes from when the soil was tilled.

In general, a very high aggregate stability was verified for Latosols of the Cerrado area, since most of the aggregates diameter was superior to 4 mm, as suggested in Table 11. This high aggregate stability is a characteristic of Latosols, which enables the installation of intensive farming without even greater damage to the environment [40].

Class	Average geometric diameter - mm
Very high	> 4
High	4 – 3
Moderate	3 – 2
Low	2 – 1
Very low	< 1

Source: Summary of bibliographic research – DCS/UFLA

Table 11. Aggregate stability classes.

10. Soil resistance to penetration

The penetration resistance until the depth of 60 cm for the management systems studied in Cerrado Latosols are presented in FIGURES 2 to 7.

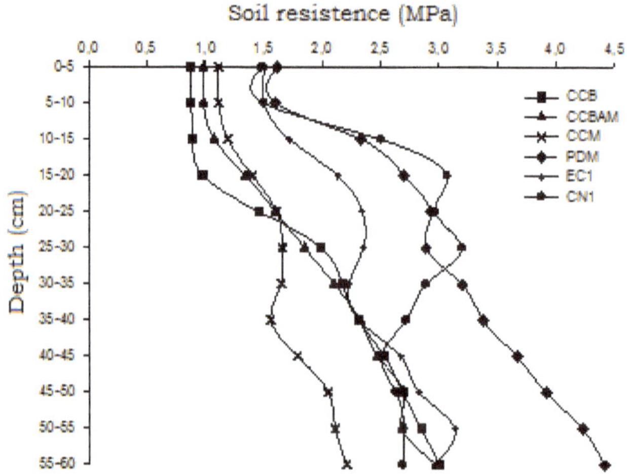

Figure 2. Penetration resistance for LVA_1, located in Campos das Vertentes, MG [27].

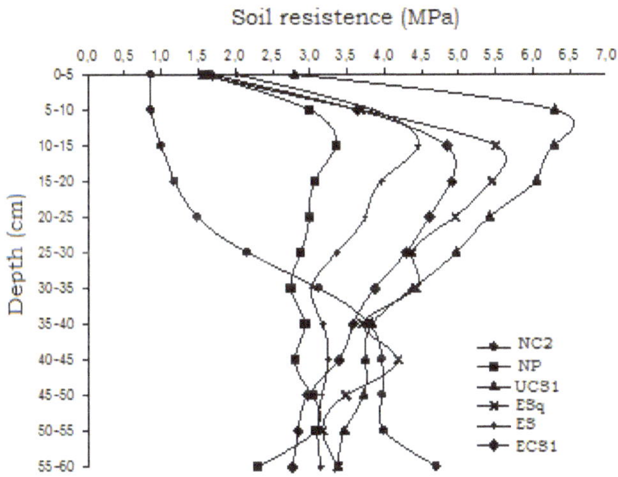

Figure 3. Soil penetration resistance for LV_1, located in Vale do Rio Doce, Guanhães, MG [27].

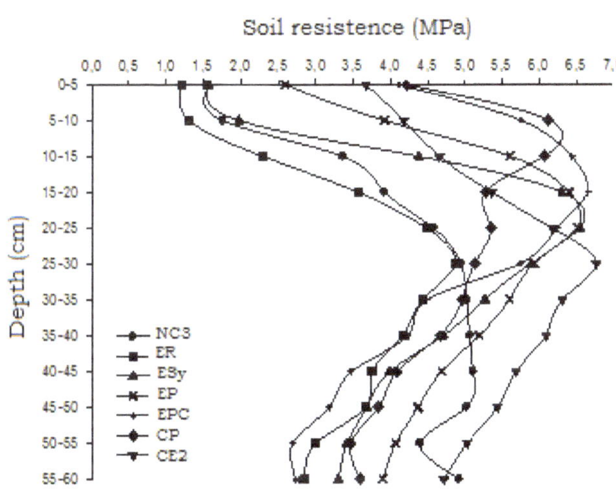

Figure 4. Penetration resistance of LV_2, Northwest region, MG [27].

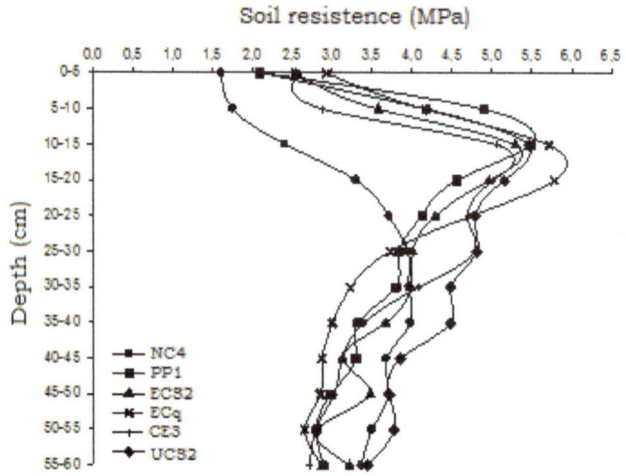

Figure 5. Soil penetration resistance for LVA_2, Vale do Rio Doce, Belo Oriente, MG [27].

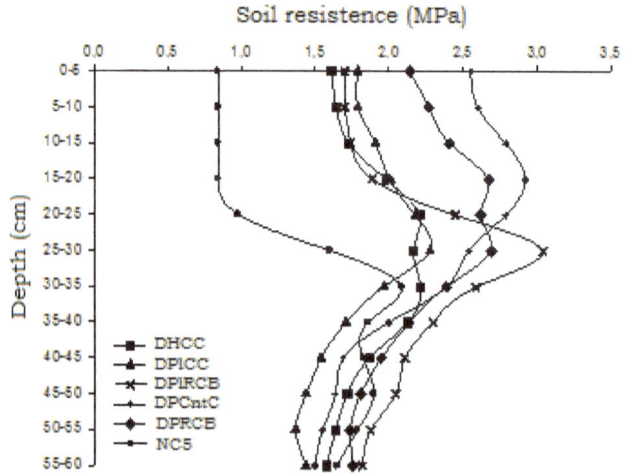

Figure 6. Soil penetration resistance for LV_3, Central MG region [27].

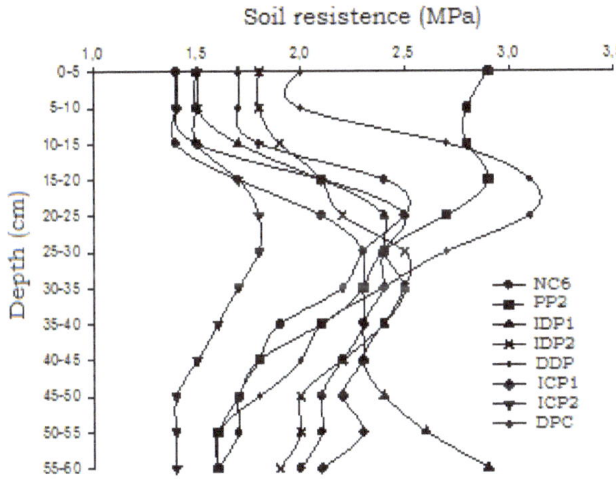

Figure 7. Soil penetration resistance for LV_4, South Goiás region [27].

The penetration resistance values varied from 0.84 MPa to 6.77 MPa (Table 12). In the native systems, except for the native Cerrado (NC2) of Guanhães, the average soil penetration resistance increases considerably in the sublayers, reaching values of 5.11 MPa in the native Cerrado of the Northwest area of Minas Gerais in the 40-45 cm depth. As such, a natural densification tendency is verified in Latosols located under the Cerrado biome, that can reach the average to high penetration resistance classes according to the classification contained in Table 13.

Depth	Soil penetration resistance, MPa					
	Lowest	Lower Quartile	**Average**	**Median**	Upper Quartile	**Highest**
0-10	0.84	1.50	2.24	1.75	2.80	6.30
10-20	0.84	1.82	3.27	2.79	5.00	6.66
20-30	0.98	2.31	3.41	2.93	4.68	6.77
30-40	1.56	2.30	3.15	2.89	3.98	6.32
40-50	1.40	2.01	2.88	2.83	3.68	5.69
50-60	1.37	1.89	2.73	2.79	3.35	5.03
0-60	0.84	1.90	2.96	2.7	3.83	6.77

Table 12. Descriptive statistics for the soil penetration resistance in the Cerrado.

Class	Penetration resistance (MPa)	Root growth limitation
Very low	< 1.1	Without limitations
Low	1.1 – 2.5	Few limitations
Moderate	2.6 – 5.0	Some limitations
High	5.1 – 10.0	Serious limitations
Very high	10.1 – 15.0	Roots practically do not grow
Extremely high	"/> 15.0	Roots do not grow

Source: Camargo & Alleoni (1997).

Table 13. Classes of soil mechanical penetration resistance and degree of root growth limitation [41].

The 0-10 cm depth was the one which presented the lowest average soil penetration resistance, and the lowest values were found in environments without anthropic interference, as in the native Cerrado of the Central area of Minas (NC5) and native Cerrado of Guanhães (NC2), that presented values of 0.84 and 0.86 MPa, respectively. In this depth the uncovered soil of Guanhães (UCS1) and of Belo Oriente (UCS2) presented high penetration resistance, with values of 6.30 and 4.18 MPa, respectively, a characteristic that can hinder the plant growth due to these values being classified in the average to high penetration resistance classes (Table 13). Even at this depth, the conventional pasture (CP) and eucalyptus + pasture + cattle (EPC) systems, both from the Northwest area of Minas Gerais, and planted pasture of Belo Oriente (PP1) presented values of 6.13; 5.75 and 4.9 MPa, that can be due to the animal trampling that can cause compaction, mainly in the first centimeters of the soil, proven for higher increases in the 0-10 cm depth [42].

In the 10-20 cm depth the tendency is continued for the layer above this, and the native Cerrado of the Central of Minas Gerais region (NC5) and the native Cerrado of Guanhães (NC2) systems present the lowest penetration resistance values. In this depth the systems that use potato planting (CCP and CCPO) in the Campo das Vertentes, MG, present low penetration resistance, due to the potatoes harvest process that provokes a high tillage of the soil.

The depths over 20 cm presented high soil penetration resistance values, which cannot only be demonstrated by the animal trampling in the pasture systems, because this effect is only limited to the surface layer of the soil. According to [43], the compaction and higher soil penetration resistance can also be the result of particle settling, a consequence of the pore blockage by the finer particles, as well as the wetting and drying cycles of the soil.

For [31] the root growth of annual cultures undergoes restriction in penetration resistance values over 2.0 MPa, and according to Table 12, above 2.6 MPa there are some restrictions to the root growth, and with this, for the depths over 20 cm, the development of roots in the studied areas can be compromised, because most of the soils presented penetration resistance over 2.5 MPa.

11. General overview on the variations among the physical indicators

Analyzing the soil physical attributes as indicators of soil quality in native and antrophic systems, it is verified that the land use for agricultural purposes provokes soil physical atttributes quality reduction. This degradation varies according to the geographical region, soil types and soil use and management practices. However, the use of conservationist systems coupled with little soil revolving tends to improve the soil physical attributes.

The systems that provoke soil revolving are more likely to compact the soil and increase water erosion, being this the main cause for the Brazilian soils degradation. Systems with reduced revolving or even the ones with no revolving (no-till) maintain the soil structure. Thus, those systems allow rapid water infiltration in the soil and, hence, reduce the water erosion.

12. Conclusions

The physical attributes analyzed in Latosols of Cerrado were sensitive to the reduction of the soil quality due to the substitution of native areas by agricultural areas, mainly in the planting systems that highly upturn the soil.

In most cases, the exploitation of the native soil caused an increase of the soil density and soil penetration resistance and it reduced the total porosity, macroporosity, hydraulic conductivity of the saturated soil and the average geometric diameter of the aggregates.

Systems that use direct planting and eucalyptus reforestation without burning present higher organic matter content, total organic carbon and carbon storage than agricultural systems.

The impact of the findings presented from this research is high considering the geographical extension of the Brazilian Cerrado region (one of the latest world agricultural frontiers) and their potential for developing land use and management policies is highly significant.

Author details

Diego Antonio França de Freitas, Marx Leandro Naves Silva, Nilton Curi,
Mayesse Aparecida da Silva, Anna Hoffmann Oliveira and Sérgio Henrique Godinho Silva

Federal University of Lavras, Brazil

References

[1] Instituto Brasileiro de Geografia e Estatística. Biomas e vegetações do Brasil: Rio de Janeiro. http://www.ibge.gov.br/ (accessed 23 jan 2010).

[2] Resck, D. V. S.; Ferreira, E. A.; Figueiredo, C. C.; Zinn, Y. L. Dinâmica da matéria orgânica no Cerrado. In: SANTOS, G. de A.; SILVA, L. S. da; CANELLAS, L. P.; CAMARGO, F. A. O. (ed.). Fundamentos da matéria orgânica do solo: ecossistemas tropicais & subtropicais. Porto Alegre: Metrópole; 2008. p359-417.

[3] Doran, J. W.; Parkin, T. B. Defining and assessing soil quality. In: Doran, J. W.; Coleman, D. C.; Bezdicek, D. F.; Stewart, B. A. (ed.) Defining soil quality for sustainable environment. Madison: Soil Science Society of America Proceedings; 1994. p3-21.

[4] Burger, J. A. Limitations of bioassays for monitoring forest soil productivity: rationale and exaple. Soil Science Society of American Journal 1996; 60(6) 1674-1678.

[5] Cardoso, E. L. Qualidade do solo em sistemas de pastagens cultivada e nativa na subregião da Nhecolândia, Pantanal Sul Mato-Grossense. PhD thesis. Universidade Federal de Lavras; 2008.

[6] Doran, J. W.; Parkin, T. B. Quantitative indications of soil quality: a minimum data set. In: Doran, J. W.; Jones, A. J. (ed.) Methods for assessing soil quality. Madison: Soil Science Society of America; 1996. p25-37.

[7] Stenberg, B. Monitoring soil quality of arable land: microbiological indicators. Soil and Plant Science 1999; 49(1) 1-24.

[8] Campos, B. C.; Reinert, D. J.; Nicolodi, R.; Ruedell, J.; Petrere, C. Estabilidade estrutural de um Latossolo Vermelho-Escuro distrófico após sete anos de rotação de culturas e sistemas de manejo de solo. Revista Brasileira de Ciência do Solo 1995;19(1) 121-126.

[9] Tormena, C. A.; Silva, A. P.; Libarde, P. L. Caracterização do intervalo hídrico ótimo de um Latossolo Roxo sob plantio direto. Revista Brasileira de Ciência do Solo 1998;22(4) 573-581.

[10] Aguiar, M. I. Qualidade física do solo em sistemas agroflorestais. Dissertation. Universidade Federal de Viçosa; 2008.

[11] Beutler, A. N. Produtividade de culturas e atributos físicos de Latossolo Vermelho-escuro fase Cerrado sob diferentes sistemas de manejo. Dissertation. Universidade Federal de Lavras; 1999.

[12] Brito, L. F. Erosão hídrica de Latossolo Vermelho distrófico típico em áreas de pós-plantio de eucalipto na região de Guanhães (MG). Dissertation. Universidade Federal de Lavras; 2004.

[13] D'Andrea, A. F. Atributos indicadores da qualidade do solo em sistemas de manejo no sul de Goiás. Dissertation. Universidade Federal de Lavras; 2001.

[14] Neves, C. M. N. Atributos indicadores da qualidade do solo em sistema agrossilvopastoril, no noroeste do estado de Minas Gerais. Dissertation. Universidade Federal de Lavras; 2001.

[15] Pires, L. S. Sistemas de manejo de eucalipto e erosão hídrica em Latossolo Vermelho-Amarelo muito argiloso na região de Belo Oriente (MG). Dissertation. Universidade Federal de Lavras; 2004.

[16] Silva, R. R. Qualidade do solo em função dos diferentes sistemas de manejo na região Campo das Vertentes, Bacia Alto Rio Grande-MG. Dissertation. Universidade Federal de Lavras; 2001.

[17] Empresa Brasileira de Pesquisa Agrop0ecuária. Sistema brasileiro de classificação de solos. 2.ed. Rio de Janeiro: Embrapa; 2006

[18] Stolf, R.; Fernandes, J.; Furlani Neto, V. L. Recomendação para uso do penetrômetro de impacto, modelo IAA/Planalsucar - Stolf. São Paulo: Planalsucar; 1983.

[19] Day, P. R. Particle fractionation and particle-size analysis. In: Black, C. A. (ed) Methods of soil analysis. Madison: American Society of Agronomy; 1965. p.545-566.

[20] Blake, G. R.; Hartge, K. H. Bulk density. In: Klute, A. (ed.) Methods of soil analysis. Madison: American Society of Agronomy; 1986. p.363-375.

[21] Danielson, R. E.; Sutherland, P. L. Porosity. In: Klute, A. (ed.) Methods of soil analysis. Madison: American Society of Agronomy; 1986. p443-461.

[22] Grohmann, F. Distribuição do tamanho de poros em três tipos de solo do Estado de São Paulo. Bragantia 1960;19(21) 319-328.

[23] Oliveira, M.; Curi, N.; Freire, J. C. Influência do cultivo na agregação de um Podzólico Vermelho-Amarelo textura média/argilosa da região de Lavras (MG). Revista Brasileira de Ciência do Solo 1983;7(3) 317-322.

[24] Kemper, W. D.; Rosenau, R. C. Aggregate stability and size distribution. In: Klute, A. (ed.). Methods of soil analysis: physical and mineralogical methods. Madison: American Society of Agronomy; 1986. p425-442.

[25] Empresa Brasileira de Pesquisa Agropecuária. Manual de análises de solo. 2. ed. Rio de Janeiro: Embrapa; 1997

[26] Lima, J. M.; Curi, N.; Resende, M.; Santana, D. P. Dispersão do material de solo em água para avaliação indireta da erodibilidade de latossolos. Revista Brasileira de Ciência do Solo 1990;14(1) 85-90.

[27] Freitas, D. A. F. Qualidade do solo em sistemas de manejo em Latossolos sob Cerrado. Dissertation. Universidade Federal de Lavras; 2010.

[28] Resende, M.; Curi, N.; Rezende, S. B.; Corrêa, G. F. Pedologia: base para distinção de ambientes. 5. ed. Lavras: UFLA; 2007

[29] Camargo, O. A. de; Jacomine, P. K. T.; Carvalho, A. P.; Olmos, I. L. The Brazilian classification of latosols. In: International soil classification workshop (ed.). Rio de Janeiro: EMBRAPA; 1988. p.190-199.

[30] Netto, A. R. Influência da mineralogia da fração argila sobre propriedades físicoquímicas de solos brasileiros. Dissertation. Universidade Federal de Viçosa; 1996.

[31] Arshad, M. A.; Lowery, B.; Grossman, R. Physical test for monitoring soil quality. In: Doran, J. W.; Jones, A. J. (ed.). Methods for assessing soil quality. Madison: Soil Science Society of America 1996. p.123-141.

[32] Costa, F. S.; Albuquerque, J. A.; Bayer, C.; Fontoura, S. M. V.; Wobeto, C. Propriedades físicas de um Latossolo Bruno afetadas pelos sistemas de plantio direto e convencional. Revista Brasileira de Ciência do Solo 2003;27(3) 527-535.

[33] Hamza, M. A.; Anderson, W. K. Soil compaction in cropping systems: A review of the nature, causes and possible solutions. Soil and Tillage Research 2005;82(2) 121-145.

[34] Araujo, E. A. Qualidade do solo em ecossistemas de mata nativa e pastagens na região leste do Acre, Amazônia Ocidental, PhD thesis. Universidade Federal de Viçosa; 2008

[35] Carneiro, M. A. C.; Souza, E. D.; Reis, E. F.; Pereira, H. S.; Azevedo, W. R. Atributos físicos, químicos e biológicos de solo de cerrado sob diferentes sistemas de uso e manejo. Revista Brasileira de Ciência do Solo 2009;33(1) 147-157.

[36] Stone, L. F.; Silveira, P. M. Efeitos do sistema de preparo e da rotação de culturas na porosidade e densidade do solo. Revista Brasileira de Ciência do Solo 2001;25(2) 395-401.

[37] Thurler, A. M. Estimativa da macro e microporosidade através da granulometria e densidade de partículas e do solo. Dissertation. Escola Superior de Agricultura Luiz de Queiroz; 1989.

[38] Soil survey staff. Soil survey manual. Washington: United States Government; 1993

[39] Bertol, I.; Albuquerque, J. A.; Leite, D.; Amaral, A. J.; Zoldan Junior, W. A. Propriedades físicas do solo sob preparo convencional e semeadura direta em rotação e sucessão de culturas comparadas às de campo nativo. Revista Brasileira de Ciência do Solo 2004;28(1) 155-163.

[40] Costa, L. M.; Nacif, P. G. S.; Costa, O. V.; Olszevski, N. Manejo dos solos da região dos Cerrados. In: Araújo, Q. R. (ed.). 500 anos de uso do solo no Brasil. Ilhéus: Editus; 2002. p.201-218.

[41] Camargo, O. A. de; Alleoni, L. R. F. Compactação do solo e o desenvolvimento das plantas. Piracicaba: Esalq; 1997

[42] Kondo, M. K.; Dias Junior, M. S. Compressibilidade de três Latossolos em função da umidade e uso. Revista Brasileira de Ciência do Solo 1999;23(2) 211-218.

[43] Araujo, E. A.; Lani, J. L.; Amaral, E. F.; Guerra, A. Uso da terra e propriedades físicas e químicas de Argissolo Amarelo distrófico na Amazônia Ocidental. Revista Brasileira de Ciência do Solo 2004;28(2) 307-315.

Chapter 3

The Role of Soil Mineralogy, Geochemistry and Grain Size in the Development of Mediterranean Badlands: A Review

Vito Summa and Maria Luigia Giannossi

Additional information is available at the end of the chapter

1. Introduction

It has long been recognized that the local Mediterranean climate, tectonics and human impact interact to determine the gross morphology and surface conditions of this landscape. However, attention has recently been given to the explanatory role of lithology, in particular sediment size and clay mineralogy, in explaining the badland formation [1-9].

For instance, on *biancane* sites, Battaglia et al. [10] found clay fractions to be significantly high. These sites have been reported to possess also high percentage of clay minerals in particular in the smectitic group.

Additionally, for these clay minerals, high exchangeable sodium on the exchange complex promotes dispersion (deflocculation) of the clays. The exchangeable sodium percentage (ESP), sodium adsorption ratio (SAR), sodium percentage (PS) and total dissolved salts (TDS) are commonly used to measure the dispersive state.

This chapter aims to contribute to the international framework of research on water erosion processes, and to identify critical emerging erosional risk factors. It focuses particularly on experimental research on material properties that could be the promoter of soil erosion processes.

Results show that many components of soil erosional response, such as soil dispersivity, badlands development or surface and subsurface processes like crusting or pipes, are strongly affected by spatially variable and temporally dynamic soil properties.

2. Soil degradation

When land is degraded, its productivity is reduced and many other eco system services are deleteriously affected. Land degradation may be primarily caused by natural processes, related to the characteristics of the given land resources and ecosystems. However, human activities often accelerate these degradation processes, leading to a rapid decline in the quality and quantity of the land resources and the ecosystem services flowing from these. Drylands are fragile and particularly susceptible to land degradation.

The United Nations Convention to Combat Desertification (UNCCD) defines land degradation in the context of drylands as: "a reduction or loss, in arid and semi-arid and dry subhumid areas, of the biological or economic productivity and complexity of rainfed cropland, irrigated cropland, or range, pasture, forest and woodlands resulting from land uses or from a process or combination of processes, including processes arising from human activities and habitation patterns" [11].

Land degradation is caused by a variety of complex interrelated degradation processes. These can be grouped into three major land degradation types, each of which can be subdivided according to a specific sub-set of degradation processes, namely:

1. Soil degradation;
2. Vegetation degradation;
3. Water resources degradation.

Vegetation degradation involves a combination of processes that may be natural, notably climate change which may lead to a loss of certain species and habitats, reduced biomass due to reduced moisture availability, or encroachment by invasive species. However, vegetation degradation is generally induced by human activity, through the over use or mis-management of forests, grazing and croplands, uncontrolled burning or introduction of pests and diseases.

Degradation of water resources in terms of quantity, quality and flow regime will lead to reduced productivity of the aquatic system in terms of fish and other useful aquatic species and products. It also affects the availability of clean drinking water for consumption by humans, livestock and wildlife.

Soil degradation is defined as the decline in soil quality caused through its misuse by human activity [12]. Degradation or decline of soil quality may occur due to physical or chemical processes triggered off by natural phenomena, or induced by humans through misuse of land resources. Processes such as soil erosion, nutrient run-off, water logging, desertification or compaction, may give examples of physical degradation processes, while acidification, organic matter loss, salinization, nutrient depletion by leaching, or toxicants accumulation, are all processes that can be classified as being agents and indicators of chemical degradation of soil.

2.1. Soil degradation types and processes

Soil degradation occurs when there is a decline in the productive capacity of the soil as a result of adverse changes in its biological, chemical, physical and hydrological properties and/or attributed to the removal of soil through erosion by water or wind or by mass movement. Sheet, rill and gully erosion by water, also the scouring and re-deposition of soil by wind and landslides are some of the most visible symptoms of soil degradation, but other less visible forms of degradation of soil properties are even more widespread and sometimes more serious, notably depletion of nutrients and soil organic matter decline.

The key processes that are responsible for soil degradation are listed in Table 1 [1, 11-12].

Degradation of soil biological properties	Degradation of soil chemical properties	Degradation of soil physical properties	Degradation of soil hydrological properties	Soil erosion	Soil pollution
• Increase in numbers and activity of harmful soil organisms • Reduction in numbers and activity of beneficial soil organisms	• Decline in number and availability of soil nutrients • Chemical imbalances and toxicities • Changes in soil pH (acidification or alkalinisation) • Salinization and sodicity • Chemical pollution	• Surface crusting and compaction and burning • Sub-soil compaction • Reduced soil rooting depth (erosion) • Loss of topsoil structure • Loss of soil fines (erosion of silts and clay) leaving sandier and stonier soils	• Waterlogging • Aridification • Reduced plant water uptake due to soil salinization	• Soil erosion by water (splash, sheet, rill and gully erosion) • Soil erosion by wind (removal and re-deposition of soil particles, abrasion by transported materials and formation of mobile sand dunes) • Gravitational erosion (mass movement through landslides, slumps, earth flows and debris avalanches) • Freeze/thaw erosion	• Soil chemical imbalances and nutrient toxicities • Build up of inorganic pollutants in the soil • Accumulation of pollutants/ toxicities of organic origin following the planting of certain crops • Emissions of toxic chemicals

Table 1. Soil degradation types

3. Soil erosion

Soil erosion is a major form of land degradation. It comprises various processes that are described separately below. However, any one of these processes may occur in the same locality, either in combination or at different times of year.

Soil erosion by water is often quite widespread and can occur in all parts of drylands where rainfall is sufficiently intense for surface runoff to occur. This category includes processes such as splash, sheet, rill and gully erosion. Splash erosion is commonly the first stage of water erosion and occurs when rain drops fall onto the bare soil surface. Their impact can break up surface soil aggregates and splash particles into the air. As water runs over the soil surface it has the power to pick up particles released by splash erosion and the capacity to detach particles from the soil surface. This may result in sheet erosion, where soil particles are removed from the whole soil surface on a fairly uniform basis.

Where runoff becomes concentrated into channels, rill and gully erosion may result. Rills are small rivulets of such a size that they can be ploughed over with farm machinery. Gullies are much deeper (often being several metres deep and wide) and form a physical impediment to the movement across the slope of farm machinery, even people and livestock [13].

Soil erosion by wind is also widespread throughout drylands that are exposed to strong winds. It includes both the removal and re-deposition of soil particles by wind action and the abrasive effects of moving particles as they are transported. In areas with extensive loose, sandy material, wind erosion can lead to the formation of mobile sand dunes that cause considerable economic losses through engulfing adjacent farm land, pastures, settlements, roads and other infrastructure [14].

Gravitational erosion tends to be more localised in regions with steep, rocky slopes and in mountain ranges. On sloping land when soil is saturated, its weight increases and the downward forces of gravity will induce a relatively large down-slope movement of soil and / or rocks (e.g. landslides, slumps, earth flows and debris avalanches). This mass movement of material may be very rapid and involve large volumes of soil, but is usually limited to isolated and localised events. Landslides may be natural events, however, their frequency and severity is likely to greatly increase following deterioration or loss of the natural vegetative cover by logging, overgrazing and / or clearing for cultivation [5, 15-17].

Freeze/thaw erosion is restricted to high altitude areas and areas with cold climates. It occurs when water in the topsoil initially freezes and expands, then melts, damaging topsoil structure and enabling loosened surface soil particles to be carried away in melt water runoff. It is primarily a natural process rather than one which is accelerated by particular human activities [15, 17-18].

This chapter covers only the assessment of soil erosion by water.

3.1. Soil erosion landscape: Badlands

The term badlands is currently used for areas of unconsolidated sediments or poorly consolidated bedrock, with little or no vegetation. They are useless for agriculture because of their intensely dissected landscape.

They appear to offer in a miniature spatial scale and a shortened temporal scale many of the processes and landforms exhibited by more normal fluvial landscapes, including a variety of slope forms, bedrock or alluvium-floored rills and washes, and flat alluvial expanses similar to large-scale pediments.

Although badlands evoke an arid image, they can develop in nearly any climate in soft sediments where vegetation is absent or disturbed. General reviews of badlands and badland processes are provided by Campbell [19] and Bryan and Yair [20], including discussions of the climatic, geologic, and geographic setting of badlands, sediment yields, host rock and regolith variations among badlands and field measurements of processes.

Badland landscapes are typically asymmetrical (Figure 1). The sunny aspects show impoverished or null vegetation cover because of the strong control on water availability effected by radiation, whereas the shady aspects may bear a vegetation cover close to 100% [21-23]. Steep slopes and gullies do not allow the formation of a developed soil because erosion processes are either frequent and/or intense.

Figure 1. Typical badland form at Pisticci (Basilicata Southern Italy) with southern eroded side and northern vegetated side

Calanchi and *biancane* are considered peculiar forms of badlands evolution [10].

The term *calanchi* describes the dendritic network of slope forms created on a single hillslope scarp (Figure 2). An individual *calanco* is defined by knife-edged ridges, separating small hydrographic drainage networks with horseshoe-shaped headwalls [24-25]. *Biancane* are small, conical or dome-shaped forms up to 20 m high (Figure 2), which may occur singly or in groups [9, 26-28].

Calanchi is the result of a rill erosion. Rill erosion is the removal of soil by water from very small but well defined, visible channels or streamlets where there is a concentration of overland flow [29]. In general, rill erosion is more serious than sheet erosion, and it is most accentuated when intense storms occur in watersheds or sites with high runoff-producing characteristics, loose, and grading operations.

Rill erosion is often described as the intermediate stage between sheet and gully erosion, and occurs by a concentration of runoff or low points through the soil.

Gully erosion could be considered as an advanced stage of rill erosion, where surface channel gullies (intermittent stream channels larger than rills) have been eroded to the point where they cannot be smoothed over by normal tillage operations.

Underground (groundwater) erosion is the removal of soil caused by groundwater seepage or movement towards a free face. It is also known as piping and occurs as a result of bank drainage or, in general, when seepage forces exceed intergranular stresses or cohesive forces [29]. Pipes can form in the downstream side of earth dams, gully heads, streambanks, and slopes where water exits from the ground. Once a cavity (pipe) forms, it is able to enlarge quickly since the flow follows the path of low flow resistance.

Figure 2. Typical morphological features of the landscape forms in Mediterranean area (Aliano, Basilicata – Southern Italy)

3.2. Compositional controls of badlands occurrence

The main factor controlling badland formation is the particular character of the rocks or other materials which form the base for the interaction of weathering and erosion processes [19]. However, the existence of other risk factors such as climatic condition, human activities, geomorphological exposition, structural features, encourages the intensification of erosion and development of morphological features of the landscape forms [27, 30-36].

Soils in badlands deserve special attention, because soils are the inter-phase between the lithosphere and the atmosphere, and so constitute one of the key elements either favoring or restricting the initiation of badland formation.

When soils are resilient against erosion processes, gullies do not form; however, when soils, either because of their particular ground cover, i.e. sparse vegetation, and/or intrinsic properties, cannot withstand erosive forces, the topsoil is eroded and deep gullies develop, which may give rise to badlands if the underlying material is also erosion-sensitive.

Consequently, the characteristics of the materials underlying soils are crucial for the development of true badlands.

Lithology is a major factor for badland production, and is probably of greater importance than tectonics, climate, topography or land use [4, 6, 19, 37-38].

The general characteristics of a soil, regolith or geological formation that favours badland relief are the unconsolidated or very poorly cemented material of clay and silt, sometimes with soluble minerals such as gypsum or halite [39]. Specific characteristics, like structure, mineralogy, physical and chemical properties, may play either a primary or secondary role in material disintegration and badland development.

In the Mediterranean area most parent materials are essentially silt-dominant, with clay as the second particle size, while sand is generally very poorly represented.

Texture depends on four factors: particle-size distribution, grain shape, degree of crystallinity and relationship among grains [40]. Of these, particle-size distribution plays the key role in susceptibility for material disintegration and erosion: the larger the range of particle sizes, the higher the degree of packing, and hence the greater resistance to breakdown processes.

Conversely, the narrower the particle-size distribution, the higher the susceptibility for material disintegration, piping and, consequently, for badland development [40-41].

Besides textural properties, porosity is the second most important physical property [42-43]. If suitable macropores are available within the material for enlargement, dispersion can encourage the rapid enlargement of subsurface pipes [44-45], a process sometimes referred to as piping or tunneling.

Infiltration is defined as the process by which water enters the soil. Its rate depends on soil type, soil structure and soil water content [46]. Infiltration is important for reducing run-off and consequent erosion. Increased soil compaction and loss of surface structure (reduced aggregation) are the main factors in reducing water infiltration rates in soils. Such rates are normally dependent upon the occurrence of large pores occupying the upper surfaces of the soil; therefore they depend on soil texture in the first place [47].

Soil bulk density is defined as the mass of soil per unit volume in its natural field state, including air space and mineral matter, plus organic substance. High values of bulk density may restrict the movement of surface waters through the soil, leading to a loss of nutrients by leaching. It may also increase erosion rates. Bulk density measurements are very important for assessing soil quality, since root growth and penetration of soil, together with the ease of soil aeration, are largely controlled by this factor [48].

Geotechnical properties provide another important control for erosion: Atterberg limits (for consistency), swelling, and slaking behavior are considered in many badland studies [10, 49-50].

Certain minerals play an essential role in the breakdown of some rocks at near surface conditions.

Some minerals are important because they may become soluble, like all soluble salts (halite), but also moderately soluble like sulphates (gypsum) or carbonates (calcite and dolomite), especially when they can be dissolved because of the small size of their constitutive particles and/or solvent characteristics [51].

Some other minerals; like clay minerals (smectite in particular), can absorb water in amounts several times their dry weight, with consequent volume increases. Wetting-drying alternations in materials with expandable clays cause the formation of nets of deep cracks and may also lead to the formation of a shallow layer of loose expanded regolith fragments, usually called popcorn [5].

It was found that the percentage presence of the swelling clays in the overall material mass is very important. Where clay percentages are high, the material mass is rendered impermeable on swelling, encouraging surface wash erosion and reducing infiltration. Where clay percentages are low, the deflocculation of the clay fraction merely destructures a material already lacking in other sources of cohesion, encouraging subsurface erosion. Given the presence of a suitable hydraulic gradient through a site, this distinction will separate materials that are dispersive but which do not develop large pipes from those that do.

Clay dispersion is a physico-chemical process relevant to erosion processes, particularly to the development of pipes. Materials (soils, regoliths or rocks) with a potential to disperse are those which contain a high exchangeable sodium percentage (ESP), saturating part of the exchangeable cations of their clays. This percentage is considered to be critical when higher than 13.

To predict the tendency of materials to pipe, Faulkner et al. [52-55] explored the effectiveness of the relationship between electrical conductivity (EC) and SAR (sodium adsorption ratio), originally used by Rengasamy et al. [54]. Whilst this improves diagnosis over the use of ESP or SAR values alone, it seems that this analysis is also insufficient in itself to distinguish between badland surfaces in terms of their morphology.

The relationship between pH and SAR can be used to indicate the extent of material buffering as dispersivity changes.

3.3. Erosion prediction

In the analysis of the processes connected to the land degradation, many models were developed in the past with the objective to give a qualitative and quantitative solution to the problem of the soil loss and erosion. This estimation of those phenomena is particularly difficult due to the number of variables to consider and their typologies including both natural, i.e. soil nature, vegetation and rainfall, and anthropic ones, as the many options for management practices and land use. Models about this evaluation are often based on empirical or process-based analysis and the synthetic equations used to describe the phenomenon are necessarily complex because they have to include the interactions of all the parameters. Besides, the complexity of the erosion processes, and the need for huge data banks to compile the many algorithms which are included in the models, are also technical problems to consider in the analysis plan. Anyway many erosion prediction models are available: event-based or long-term models, empirical or physically based models, on a basin or plot scale, which have been improved in the last few decades. One of the best examples in the estimation of long-term average annual soil loss from arable lands, is the Universal Soil Loss Equation (USLE) model.

The Universal Soil Loss Equation (USLE), developed by Wischmeier, Smith, and others in the 1960, predicts the long term average annual rate of erosion on a field slope based on rainfall pattern, soil type, topography, crop system and management practices.

In the applications related to the analysis of erosion processes, the USLE equation was replaced by the Revised Universal Soil Loss Equation (RUSLE), which has a similar structure (that is a black-box factor empirical model), but with more sophisticated inputs and it is designed for operation on personal computers

Because there is a wide discrepancy between predicted and observed erosion rates, models are better as research tools than as public policy and regulatory instruments or for prescriptive design measures for constructed landforms. But some models may provide useful guidance for the design engineer if adequately calibrated and verified for local conditions and if the design accounts for the uncertainty.

4. Influence of soil features on developing *calanchi* erosional landforms in Southern Italy (Basilicata)

The soil erosion risk is widespread in the Mediterranean. Some areas of Italy are an excellent example of soil erosion risk. Soil erosion vulnerability in the Basilicata region (Southern Apennines - Italy) is mostly represented by water erosion forms. *Calanchi* and *biancane* are two typical erosion landforms present in the Basilicata region and this area is a key reference for the international studies of water erosion processes [3, 8, 30, 56-58].

4.1. Geological and climatic settings

The studied sites are part of a well-known area of present desertification [1, 7, 20, 27, 30, 56-65], and is located in the far south of the Apennines (Figure 3).

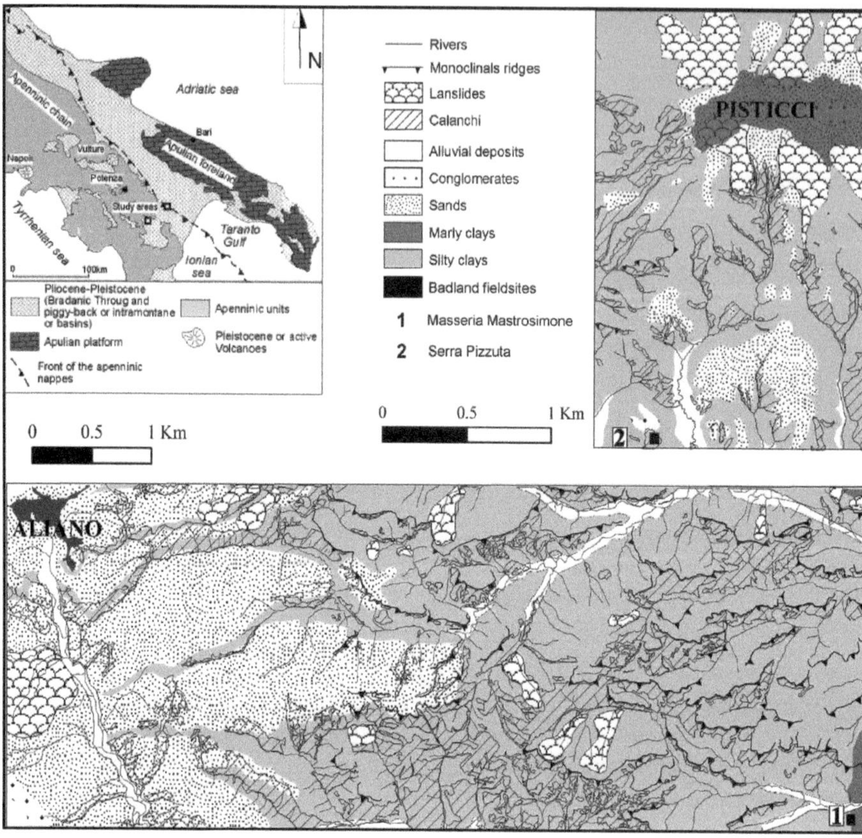

Figure 3. Geologic map showing the location of the study areas [adapted from Piccarreta et al., 2006]

The Pisticci site is a hilly area between the river Basento and the seasonal stream Salandrella, both deeply carved in the late Pliocene-Calabrian Sub-Apennine clays of the Bradanic Fore-deep [61].

The Sub-Apennine clays are clays and silty clays, sometimes sandy clays, typically blue in colour. The lithological uniformity of the pelitic facies is attenuated by thin sand-silt and tuff layers (centimetric scale).

The Aliano site is situated in the heart of the Basilicata Appenines and is part of the north-eastern area of the Plio-Pleistocene Sant' Arcangelo basin.

The formations were affected by the uplift of the eastern margin of the Apennine chain during the upper Pliocene and Post-Calabrian ages [66]. The whole area shows tectonic aspects due to movements which have prevailing vertical components. This is clearly caused by the features of the sedimentary layers, from sub-horizontal to gently dipping. Generally speaking the stratification is still not well-distinguished. The stratigraphic sequence and the basin structure have been studied in depth [67-74].

In the study areas the Plio-Pleistocene clays are the most diffused lithology to be found. These are 500 to 900 m thick [27, 75] and consist of marly and silty clays with a middle–high plasticity [27, 71].

Pieri et al. [76] redefined the Plio-Pleistocene successions by subdividing them into four depositional sequences, such as the Late Pliocene – Middle Pleistocene on the basis of the stratigraphic, sedimentologic and structural features of the deposits outcropping in the northern basin. Each cycle, several hundred metres thick, represents one or more depositional systems (alluvial, marine-deltaic and lacustrine).

Patacca and Scandone [73] suggest a new structural architecture for the Southern Appenines, especially for the Plio-Pleistocene Foredeep/thrust-sheet-top deposits. According to these authors, the Plio-Pleistocene thrust-sheet-top and Foredeep deposits were subdivided into two depositional sequences both of which are governed by tectonic processes active in the mountain chain (P1-2 - lower-upper Pliocene - and Q1-2 – lower-middle Pleistocene - thrust-related depositional sequences).

Details about the thrust-related depositional sequence in the Southern Appenines and the relative systems tracts, including the characteristic stratigraphic signatures, are supplied in Patacca and Scandone [72].

The area studied is characterized by extremely widespread erosion mainly affected by the lithological features of their soils. The overall geomorphological development of this area resulted from periodic intensive erosion, which began in the Late Pleistocene and continued during the Holocene due to tectonic movements, climatic changes and related sea-level fluctuations [77].

These two sites have been chosen because they are represent by two different badlands areas.

In the Pisticci area there are "typical" badlands, a representation of the usual morphology of the semi-arid Mediterranean area – characterized both by an eroded slope facing south and a non-eroded (covered) slope facing north [56, 58, 61, 65]. Erosion was studied on sediments of Sub-Apennine clays, unvegetated slopes (SE-facing scarp slopes up to 35°-40°) with high rates of erosion, labelled "Pisticci, eroded", and opposite vegetated slopes (NW-facing scarp slopes up to 20°) labelled "Pisticci, non-eroded" (Figure 4).

Also in the Aliano area there are slopes with features of erosion common to clayey-silty rocks exposed to the south-east as well as adjacent slopes having the same exposure (south-east), but showing a different erosional action. A partially-vegetated covering can be found (Figure 5).

The annual average rainfall for the Pisticci area 1923–2000, is about 645 mm [78-79]. The most abundant precipitation is in Autumn and Winter; Summer is the driest season [80].

Figure 4. Photos from Pisticci studied area: "typical" badlands on sediments of Sub-Apennine clays, characterized both by an eroded slope facing south (scarp slopes up to 35°-40°) and a non-eroded (covered) slope facing north (scarp slopes up to 20°).

Figure 5. Photos from Aliano studied area: slopes with clayey-silty rocks exposed to the south-east as well as adjacent slopes having the same exposure (south-east), but showing a different erosional action.

In the Aliano area, from 1955 to 2000, the annual mean precipitation was 738 mm (st.dev. 174mm), mainly concentrated from the months of October to January. In the same area the minimum and maximum values recorded over a period of 46 years are, respectively, 367mm and 1090 mm [34].

In both areas, the climate is typically Mediterranean, characterized by warm and dry Summers with temperatures averaging 26-27°C with a maximum as high as 39°C, and cold and rainy Winters with temperatures averaging 8-10°C in January.

4.2. Materials and methods

For each slope, eroded and non eroded (in all case studies), samples were collected in order to represent the several litho-pedological levels. Since vegetated soils resist breakdown and crusting [58], within the eroded slope, the crust was only differentiated with respect to the substrate and was defined as existing at 0-2 cm depth. Below the crust, samples were labelled "substrate". For each eroded and non-eroded slope, three different profiles were sampled: top, middle, and bottom (Figure 6).

Detailed grain size analyses were carried out by laser diffraction, a Malvern MasterSizerE laser particle-sizer with a 100-mm lens, which identifies grain-size intervals from 0.5 to 100 μm. For mineralogical analysis the clay fraction (<2 μm) was separated by means of fractioned sedimentation in accordance with Stokes' law.

Mineralogical analyses were carried out by X-ray diffraction (XRD) on a Rigaku D/Max-2200/ Pc powder diffractometer (theta-theta configuration, Cu Kα radiation). Quantitative mineralogical data were obtained according to Barahona [81], and the results were checked by means of a comparison with chemical data.

Chemical bulk-rock elements were measured by X-ray fluorescence (XRF) on a Philips PW 1480/10 spectrometer with Cr radiation. Recommendations made by Franzini et al. [82] were applied in order to correct matrix effects by using international geological standards.

pH measurements were made according to the procedure indicated in Italian law no.79 [83]. Dried samples were mixed with distilled water (ratio 1:2.5) and the mixture was then stirred and pH measured.

Figure 6. Sampling profile

Selected soluble salt concentrations (Na^+, K^+, Mg^{2+}, and Ca^{2+}) were measured by ion chromatography [52], and sodium adsorption ratio (SAR) and Exchangeable Sodium Percentage (ESP) calculated according to the formula:

$$SAR = \frac{Na}{\sqrt{\frac{Mg + Ca}{2}}}$$

$$ESP = \frac{Na}{CEC} 100$$

Pearson's correlation coefficients and Student's T-test were calculate to quantify the relationship between variables.

4.3. Results and discussion

The mineralogical, geochemical and grain-size composition features of these slopes has been determined to find common risk factors for the different areas.

Only a few grain-size parameters, mineralogical and geochemical features discriminate the eroded and non eroded substrates [8]. The water erosion phenomena is present where the fine fraction is abundant (more evident in Aliano than in Pisticci). This can be explained by a reduction of permeability in eroded soils while the non eroded ones are more stable with respect to the weathering phenomena, as they are more permeable.

Crusts represent the more weathered and modified part of eroded sides, but their grain size and chemical features resemble non eroded materials better than their own substrate. Such a similarity can be depicted as an auto-stabilization process of superficial portion of eroded slopes [e.g. 53, 84]. Chemical data enable discrimination between eroded and non-eroded slopes in all case studies.

pH, SAR (sodium adsorption ratio), TDS (total dissolved salts) and PS (percentage of sodium) are distinctive parameters for both eroded and non-eroded slopes. On average, eroded substrates are higher in pH, SAR and PS than non-eroded ones. The ESP (exchangeable sodium percentage) of the eroded slope has a higher value than the non-eroded one [8].

The results of this study show that, even if geological and geomorphological differences exist between the two areas, common erosion risk factors can be characterized.

4.3.1. Geomorphological and structural observations

In both study areas, the topography has a gentle dip and morphology is expressed as a typical monoclinal landscape. However, the causes for the monoclinal topography differ in the two regions (Figure 7).

The Aliano site has been interpreted as a simple monoclinal system, whereas at the Pisticci site, landslides are particularly widespread on the South-East facing hillslopes [33-34, 85].

Although the monoclinal morphology has differing origins in the two areas, in both settings, the existing primary and secondary network of fractures and joints appears to influence the genesis and development of surface drainage [33-34, 61].

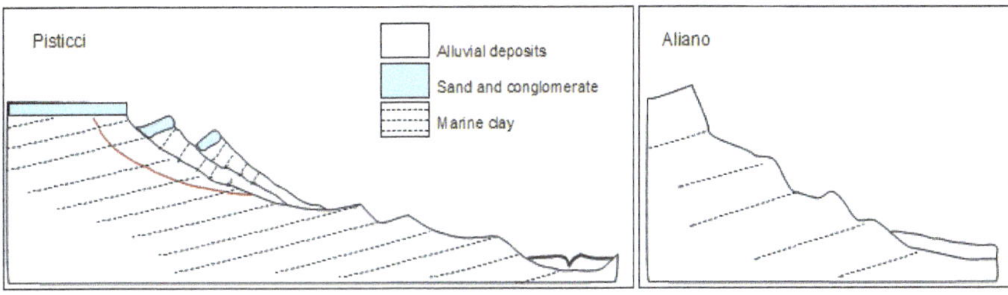

Figure 7. Schematic diagrams of the studied sites [33, modified]

In Aliano, down-valley structural features have been focussed around the weaker parts of the structural sequence of marine clays, although in the case of Pisticci, since these are failure planes not lithological features, these lineaments are more discontinuous in their down-valley pattern as might have been imagined.

In both settings, the rapidity of the geomorphic processes on the relatively steeper scarp slopes generally prohibits vegetation from securing a stable function.

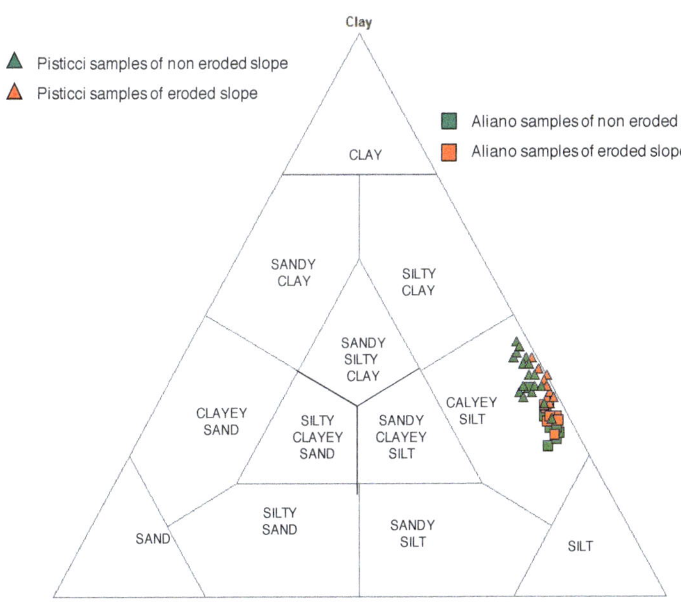

Figure 8. Granulometrical classification diagram

4.3.2. Granulometrical and mineralogical properties

After grain size analyses, all the substrate samples studied were found to be of the clayey-silt type, which is a typical substrate facilitating the formation of *calanchi*, as suggested by Battaglia et al. [10]. The grain size diagram [86] does not distinguish the eroded from the non-eroded slopes (Figure 8) neither does the soil erodibility nor the soil-quality diagram, from the CORINE Land cover [87]. However, more detailed grain-size distribution is shown in Figures 4 and 5, and gives further information, as it deals with the lower coarse fraction (>63 μm) of eroded substrates in all profiles ($r=0.783$, $p<0.000$). Comparing the profiles of the two slopes, further grain size discrimination is achieved due to the fact that the non-eroded profile of Aliano has larger course fraction (>16 μm, Figure 9) instead the non eroded profile of Pisticci are enriched in 4–63 μm fraction (Figure 10).

In both cases, the granulometric characteristics of the crust of the eroded slope are comparable with those of the substrate not eroded, as demonstrated by a linear correlation coefficient R close to 1 ($p<0.000$). This means that after erosion the most delicate part of the slope (the crust) becomes less dispersive as a sort of auto-stabilization process.

The micromorphological information on some samples of the eroded side show three distinct domains (Figure 11).

Below 20 cm the fine-grained dense sub-zone displays a massive structure and is relatively impermeable. Immediately above this dense sub-zone, a zone of isorientate structure with low porosity is present (2–20 cm).

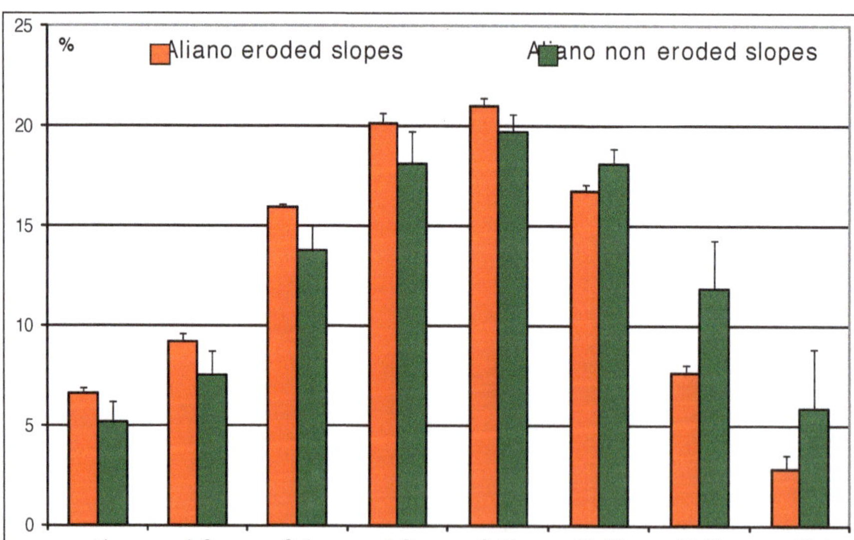

Figure 9. Average granulometrical composition of eroded and non eroded samples of Aliano

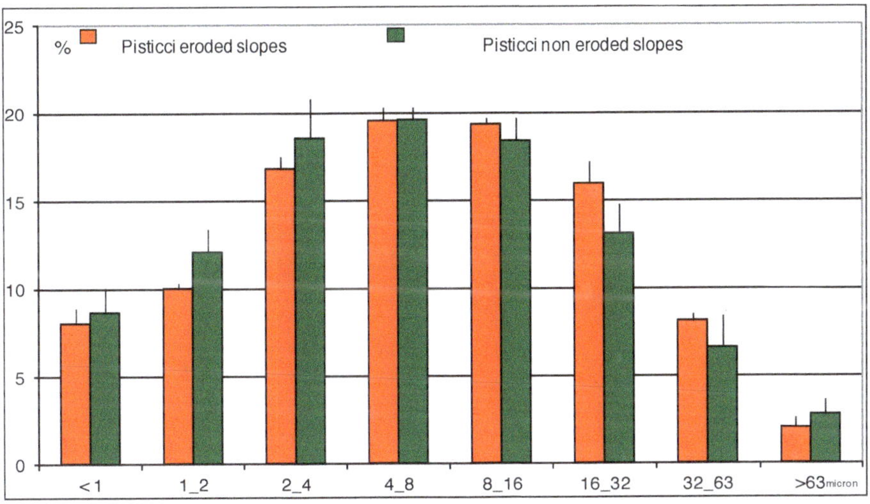

Figure 10. Average granulometrical composition of eroded and non eroded samples of Pisticci

The top 2 cm of the profile appears dispersive with high porosity. Infiltration process dominates, due to high permeability. The high porosity of the top 2 cm of the soil suggests that the most important hydraulic activity is restricted under this depth, at the intermediate level.

Both the eroded and non-eroded slopes in Aliano and Pisticci show a fairly comparable mineralogical composition of the samples on account of their mineralogical phases and quantity. So, it was not possible to define a systematic trend distinguishing bulk rock mineralogy with erosive features.

The mineralogical assemblage of the samples consists of quartz, calcite and feldspars (Table 2), with some difference in quantity for the two sites. Dolomite is always present but in lower concentrations; traces of gypsum and hematite occasionally occur at low levels. Among the clay minerals, illite is the most abundant (on average 50% of the clay fraction), while chlorite, kaolinite and mixed-layer illite–smectite generally having lower concentrations (Tab. 2). The amount of kaolinite is higher in the eroded slopes ($r=0.829$) than in the non-eroded ones ($r=-0.703$). The quantity of illite in the eroded slope is lower than that in the non-eroded one.

4.3.3. Chemical properties

Some parameters are found to be higher in the eroded substrates than in the non-eroded ones (Figure 12) ($p<0.000$). Also the crusts of eroded slopes differ from the substrates and the values of these three parameters increase with depth.

The total dissolved salts (TDS) can turn out to be quite distinctive in a comparison between the eroded and non-eroded sides. The sodium adsorption ratio (SAR) is a slightly different

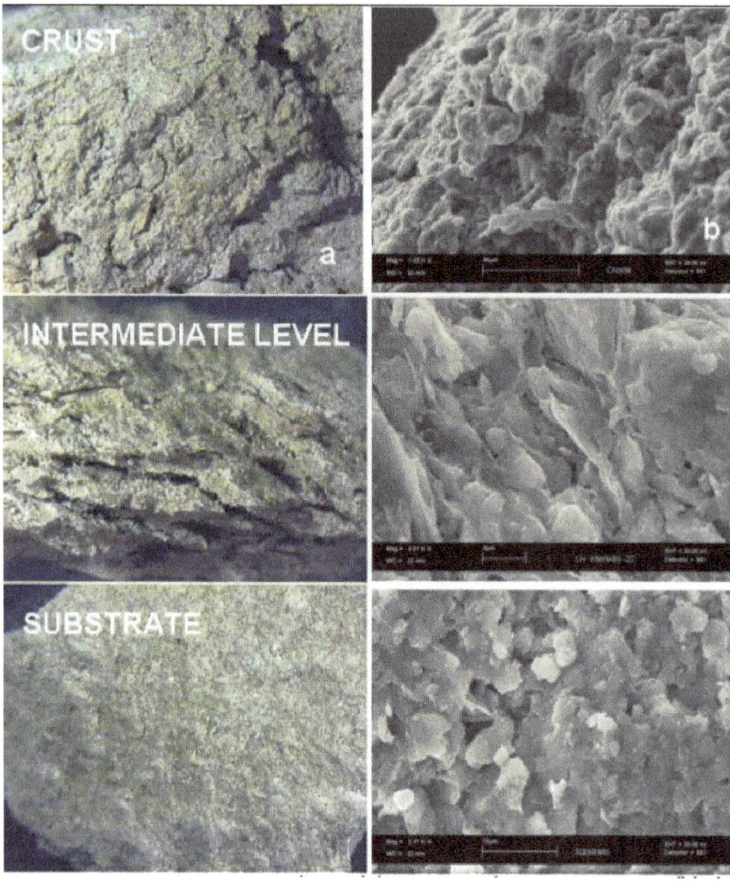

Figure 11. Sample of the eroded side at optical (a) and scanning electron microscopy (b

expression of the importance of Na in the soluble cation composition with respect to PS. pH values are alkaline (from 8.0 to 9.1).

The sodium effect of clay stability is often expressed as ESP (exchangeable sodium percentage, [3, 8, 55]), and is closely related to the Na available for cation-exchange from the clays. ESP–depth diagrams between the eroded and non-eroded slopes turned out to be extremely differentiating (Figure 13).

The various relations between these chemical parameters, which define soil susceptibility to dispersion, can also predict the performance of the surface layers and subsoil [10, 53-55, 88-89]. According to this approach the substrate of the eroded and non eroded slopes can be

Locality	Aliano				Pisticci			
Type	non eroded		eroded		non eroded		eroded	
	Average	std. dev.	average	std. dev.	average	std. dev.	average	std. dev.
Mineralogical composition of the bulk rock								
Sheet silicates %	36,0	4,7	32,6	4,3	43,4	3,5	39,5	7,5
Quartz	28,4	2,7	27,8	2,6	21,8	2,1	21,7	2,1
Calcite	17,1	1,8	20,9	2,3	23,2	2,9	21,6	1,9
Dolomite	3,4	0,8	3,7	1,1	2,3	0,5	3,8	1,2
Feldspars	14,8	3,0	14,8	3,1	9,2	1,2	12,7	3,1
Gypsum	0,1	0,3	0,0	0,0	0,0	0,0	0,0	0,0
Hematite	0,0	0,0	0,0	0,0	0,0	0,0	0,2	0,4
Mineralogical composition of the clay fraction (< 2 μm)								
Kaolinite %	12,0	2,6	15,4	3,3	22,7	3,9	26,8	1,9
Chlorite	20,1	2,6	20,7	5,1	17,3	3,1	22,7	2,1
Illite-smectite	9,8	4,9	6,6	4,0	17,4	14,5	18,0	7,9
Illite	58,0	6,6	57,4	9,7	42,7	10,0	32,5	8,0

Table 2. Mineralogical composition of sample of Aliano and Pisticci area

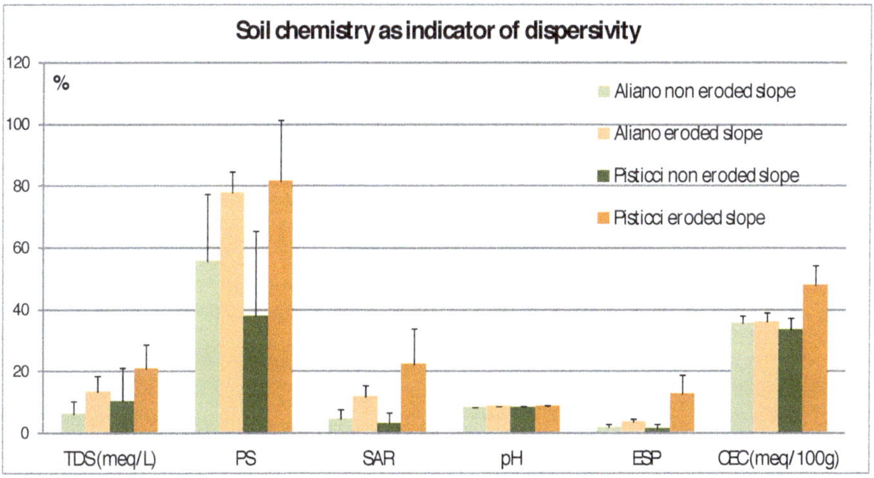

Figure 12. Comparison among some indicators of dispersivity (TDS= Total dissolved salts; PS= Sodium percentage; SAR= Sodium Adsorption ratio; ESP= Exchangeable sodium percentage; CEC= Cation exchange capability)

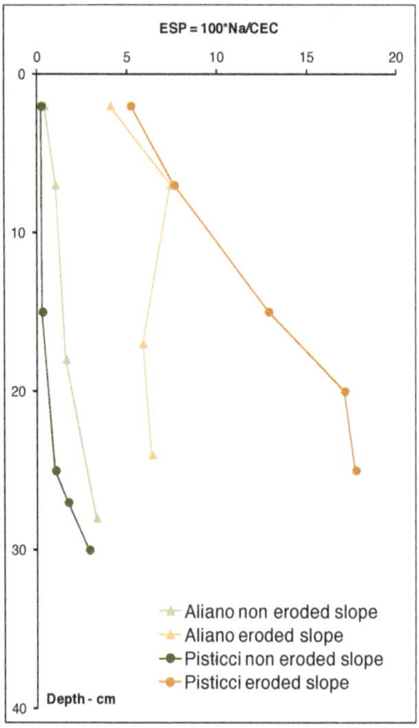

Figure 13. Exchangeable sodium percentage (ESP) versus soil depth

tentatively distinguished by using the general subdivision reported in the PS–TDS–SAR diagram of Figure 14 [8, 10, 89-90].

As shown in the diagram (Figure 14), all the samples of the non-eroded slopes have a high degree of variability as they are to be found in all the three classification zones.

The eroded slopes are mainly included in a dispersive zone, except for some samples included in the overlap zone. The dispersive properties of the latter samples are not so clear. In some cases, this shows a tendency of some portions of eroded slopes, which generally correspond to the topmost part of the slope, towards geochemical stabilization.

Other diagrams such as ESP-pH (Figure 15) can also be effective for distinguishing the eroded from non-eroded slopes [3, 8, 90]. The reason why ESP is a better discriminator may be due to the fact that the composition of the exchange complex is an intrinsic soil property.

An interesting feature that arises from many reported diagrams (Figures 14–15) is the anomalous plotting of crust samples in the eroded slope compared with other eroded profiles (white symbol in the non dispersive zone). Comparing the crusts with eroded substrates, crusts are clearly characterized by lower dispersivity parameters. The uppermost substrate samples of the non-eroded slope also follow the same trend as shown by crusts, suggesting

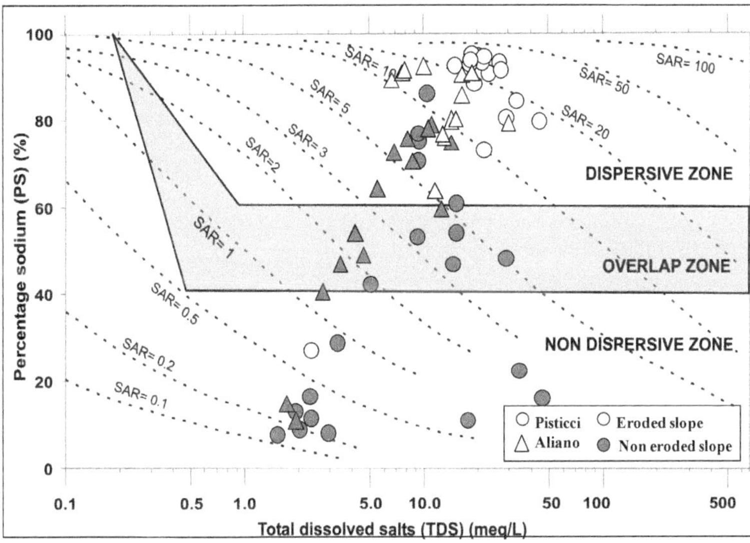

Figure 14. Relationships between sediment dispersivity and pore water composition (expressed through the PS, TDS and SAR parameters), as established by Sherard et al.[87]

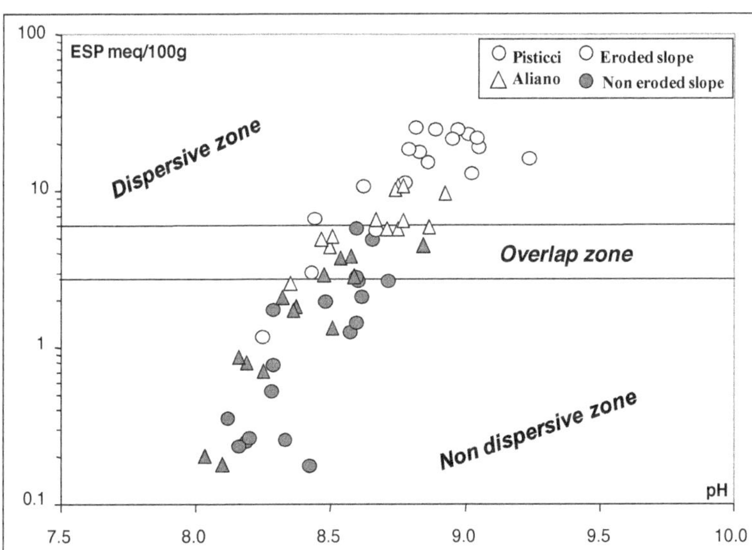

Figure 15. Correlation between ESP and pH for the prediction of dispersive behaviour of soil

that severely weathered portions of slope tend to reach a stable condition due to strong decreases in SAR, PS and ESP.

This effect, as mentioned by Faulkner et al. [53], is severe for the upper profile of the eroded slope, so that crust samples often plot in the non-dispersive field. This sort of 'auto-stabilization' process has been mentioned by several authors [e.g. 53, 84, 91] dealing with Na leaching.

As can be observed the eroded slope of Pisticci and that of Aliano have similar compositional characters with chemical characteristics associated with highly dispersive soils, regardless of exposure, or other geomorphological or climate factors.

4.4. Conclusion

This research has demonstrated that certain physico-chemical properties of the local sodic Plio-Pleistocene clays influence the different erosional processes in the two study slopes in Basilicata in fundamental ways.

Clay materials in the middle and base of the slopes retain a dispersive character. Only a few grain-size parameters discriminate eroded from non-eroded substrates. The water erosion phenomena is present where the fine fraction is abundant. This can be explained with a reduction of permeability in eroded soils while the non-eroded ones are more stable.

Erosional risk factors can be found in granulometrical features, in particular an higher fine fraction (< 16 μm for Aliano slopes and more abundant 8-16 μm for Pisticci slopes) promote erosional phenomena.

From a chemical perspective, a higher value of pH, SAR, PS parameters and above all ESP are an indication of intensive erosional processes.

The substrate of eroded and non eroded slopes can be discriminated by classification diagrams using some chemical parameters (SAR, PS and TDS) as dispersivity descriptors.

A better separation of substrates is obtained using other diagrams such as the ESP-pH.

If for the Pisticci site the exposure has always been considered one of the main erosion factors, by comparison with the slopes of Aliano we can understand how the intrinsic characteristics of the soil are crucial for the development of the erosion process.

The granulometric, mineralogical and chemical characters of the non-eroded slopes of Aliano (facing South) are comparable with those of the slopes of Pisticci (facing North).

The two study sites also have another common feature that it is possible to extend to all badlands domains: the auto-stabilization process.

This process has been identified thanks to the physico-chemical properties of the two monitored badland sites and the pH / SAR relationship that shows the tendency of the crust to auto-stabilise, confirming they are really effective signature sites.

Also the granulometrical similar composition of crusts and non-eroded substrates can be interpreted as an auto-stabilization process of superficial portion of eroded slopes.

5. Concluding remarks

Badlands are a typical landform of greatly dissected fine-grained materials. *Calanchi* and *biancane* are considered peculiar forms of badlands evolution. Their formation is closely related to the physico-chemical properties of the soil, climatic and geomorphologic conditions, and human activities.

Studies reviewed have shown that badlands have been investigated for their peculiar features and processes in the frame of landscape evolution, spatial and temporal distributions, denudation rates, effects of man's activity, and erosion risk assessment or mitigation, but several aspects of their genesis and evolution are still unclear.

Various factors and geomorphic processes seem to interplay in *calanchi* genesis and evolution, in relation to local microclimatic conditions, geological features and land use changes.

Some parameters, however, are not yet sufficiently considered: grain size, physical and mineralogical characteristics of clay sediments, soil chemistry, and chemical features of surface and underground water.

The results of the presented study show that the erosional mechanism involves morphological and geographic exposure and climatic elements as well as grain size, mineralogy, chemistry and exchangeable processes of soils. They are important characteristics of eroded soil to give a further contribution to the issue of *calanchi* genesis, in an attempt to integrate pre-existing studies with new risk factors.

It's possible to define erosional risk factors as granulometrical, mineralogical from a chemical perspective. These indicators of soil erodibility risk can be applied in different erosional development and can be used to update the current model for erosion prediction in term of soil-erodibility factor K in the RUSLE equation.

Acknowledgement

Erosional and salinization study were supported by fund from Programma Operativo FESR Basilicata 2007/2013, "Pro-Land Project".

The authors gratefully acknowledge comments by Paula Preston on the English revision of this chapter.

Author details

Vito Summa and Maria Luigia Giannossi*

*Address all correspondence to: marialuigia.giannossi@imaa.cnr.it

Institute of Methodologies for Environmental Analysis of the National Research Council of Italy (IMAA-CNR), Tito Scalo (PZ), Italy

References

[1] Alexander D. Difference between 'calanchi' and 'biancane' badlands in Italy. In Badland Geomorphology and Piping, Bryan R, Yair A (eds). Geobooks: Norwich; 1982; p. 71–88.

[2] Bryan RB, Yair A, Hodges WK. Factors controlling the initiation of runoff and piping in Dinosaur Provincial Park badlands, Alberta, Canada. Z. Geomorph. Suppl. Bd. 1978; 29: 151-68.

[3] De Santis F, Giannossi ML, Medici L, Summa V, Tateo F. The influence of physic-chemical material properties on different badland development (Aliano, Southern Italy). Catena. 2012; 81: 172–81

[4] Gerits J, Imeson AC, VersUraten JM, Bryan RB. Rill development and badland regolith properties. Catena Suppl. 1987; 8: 141-60.

[5] Hodges WK, Bryan RB. The influence of material behaviour on runoff initiation in the Dinosaur Badlands, Canada. In: Bryan RB, Yah A. (Eds.), Badland Geomorphology and Piping. Geoabstracts, Norwich. 1982; p. 13-46.

[6] Imeson AC, Verstraten JM. Rills on badland slopes: a physico-chemically controlled phenomenon. Catena Suppl. 1988; 12: 139-50.

[7] Imeson AC, Kwaad FJ, Verstraten JM. The relationship of soil physical and chemical properties to the development of badlands in Morocco. In Badland Geomorphology and Piping, Bryan R, Yair A (eds). Geobooks: Norwich. 1982; p. 47–71.

[8] Summa V, Tateo F, Medici L, Giannossi ML. The role of mineralogy, geochemistry and grain size in badland development in Pisticci (Basilicata, Southern Italy). Earth Surf. Process. Landf. 2007; 32: 980–97.

[9] Torri D, Bryan RB. Micropiping processes and biancane evolution in southeast Tuscany, Italy. Geomorphology. 1997; 20: 219–35.

[10] Battaglia S, Leoni L, Sartori F. Mineralogical and grain size composition of clays developing calanchi and biancane erosional landforms. Geomorphology. 2002; 49: 153–70.

[11] UNCCD. United Nations Convention to combat desertification in countries experiencing serious drought and/or desertification, particularly in Africa. A/AC.241/27, Paris; 1994.

[12] Barrow CJ. Land Degradation: Development and Breakdown of Terrestrial Environments.Cambridge: Cambridge University Press; 1991.

[13] Cooke RU, Doornkamp JC. Geomorphology in environmental management. An introduction. Clarendon press, Oxford; 1974.

[14] EEA-European Environment Agency. SOER 2005 – The European environment – State and outlook 2005. EEA, Copenhagen; 2005.

[15] Clotet N, Gallart F, Balasch J. Medium term erosion rates in a small scarcely vegetated catchment in the Pyrenees. Calena Supplement. 1988; 13:37-47.

[16] Gerits JJP, Imeson AC, Verstraeten JM. Chemical thresholds and erosion in saline and sodic materials. In F. Lopez-Bermudez and J. Thornes (Eds) Estudios Sobre la Geomorfologìa del Sur de Espana. Murcia; 1986: p. 71-4.

[17] Oostwoud Wijdenes DJ, Ergenzinger P. Erosion and sediment transport on steep marly hillslopes, Draix, Haute-Provence, France. An experimental field study. Catena. 1998; 33: 179-200.

[18] Regués D, Pardini G, Gallart F. Regolith behaviour and physical weathering of clayey mudrock as dependent on seasonal weather conditions in a badland area at Vallcebre, Eastern Pyrenees. Catena. 1995; 25:199-212.

[19] Campbell LA. Badlands and badland gullies. In D.S.G. Thomas (Ed.) Arid Zone Geomorphology. Belhaven/Halsted Press, London; 1989: p. 159-83.

[20] Bryan R, Yair A (eds). Badland Geomorphology and Piping. Geobooks, Norwich; 1982.

[21] Kirkby MJ, Atkinson K, Lockwood J. Aspect, vegetation cover and erosion on semi-arid hillslopes. In J. Thornes (Ed.) Vegetation and Erosion. Wiley, Chichester; 1990: p. 25-39.

[22] Solé A, Calvo A, Cerdà A, Làzaro R, Pini R, Barbero J. Influence of micro-relief patterns and plant cover on runoff related processes in badlands from Tabernas (SE Spain). Catena. 1997; 31: 23-38.

[23] Cantòn Y. Efectos hidrologicos y geomorfologicos de la cubierta y proprietades del suelo en paisaje de carcavas. Unpublished PhD thesis, Universidad de Almeria, Spain; 1999.

[24] Alexander DE. Calanchi – accelerated erosion in Italy. Geography. 1980; 65: 95–100.

[25] Farabegoli E, Agostini C. Identification of calanco, a badland landform in the northern Apennines, Italy. Earth Surface Processes and Landforms. 2000; 25: 307–18.

[26] Calzolari C, Ungaro F. Characterization and quantitative spatial analysis of geomorphic features of a badland (biancane) area, central Italy. Catena. 1998; 31: 237–56.

[27] Clarke ML, Rendell HM. The impact of the farming practice of remodelling hillslope topography on badland morphology and soil erosion processes. Catena. 2000; 40: 229–50.

[28] Torri D, Colica A, Rockwell D. Preliminary study of the erosion mechanisms in biancane badland (Tuscany, Italy). Catena. 1994; 23: 281–94.

[29] Gray DH, Sotir RB. Biotechnical Soil Bioengineering Slope Stabilization: A Practical Guide for Erosion Control. John Wiley & Sons, New York, NY; 1996.

[30] Farifteh J, Soeters R. Factors underlying piping in the Basilicata region, southern Italy. Geomorphology. 1999; 26: 239–51.

[31] Farifteh J, Soeters R. Origin of biancane and calanchi in East Aliano, southern Italy. Geomorphology. 2006; 77: 142–52.

[32] Regues D, Guardia R, Gallart F. Geomorphic agents versus vegetation spreading as causes of badland occurrence in a Mediterranean subhumid mountainous area. In: Badlands in changing environments. Catena. Special Issue. 2000; 40 (2): 173–87.

[33] Piccarreta M, Faulkner H, Bentivenga M, Capolongo D. The influence of physico-chemical material properties on erosion processes in the badlands of Basilicata, Southern Italy. Geomorphology. 2006; 81: 235–51.

[34] Piccarreta M, Capolongo D, Boenzi F, Bentivenga M. Implications of decadal changes in precipitation and land use policy to soil erosion in Basilicata, Italy. Catena. 2006; 65: 138–51.

[35] Della Seta M, Del Monte M, Fredi P, Palmieri EL. Direct and indirect evaluation of denudation rates in Central Italy. Catena. 2007; 71: 21-30.

[36] Kasanin-Grubin M, Bryan R. Lithologic properties and weathering response on badland hillslopes. Catena. 2007; 70: 68–78.

[37] Calvo A, Harvey AM, Paya-Serrano J. Processes interactions and badland development in SE Spain. In M. Sala, J.L. Rubio and J.M. Garcia-Ruiz (Eds) Soil Erosion Studies in Spain. Geoforma Ed., Logrono;1991: p. 75-90,

[38] Calvo A, Harvey AM, Paya-Serrano J, Alexander RW. Response of badland surfaces in SE Spain to simulated rainfall. Cuaternario y geomorfologia. 1991; 5: 3-14.

[39] Scheidegger AE, Schumm SA, Fairbridge RW. Badlands. In R. Fairbridge (Ed.) Encyclopedia of Geomorphology. Dowden, Hutchinson and Ross, Inc., USA. 1968: p. 43-8.

[40] Terzaghi K, Peck RE. Soil Mechanics in Engineering Practice, 2nd Edn. John Wiley & Sons, NewYork; 1967.

[41] Taylor RK, Smith TJ. The engineering geology of day minerals: swelling, shrinking and mudrock breakdown. Clay Minerals. 1986; 21: 235-60.

[42] Imeson A. Investigating volumetric changes in clayey soils related to subsurface water movement and piping. Zeitschrift fur Geomorphologie Supplement Band. 1986; 60: 115-30.

[43] Solé A, Josa R, Pardini G, Aringhieri R, Plana F, Gallart F. How mudrock and soil physical properties influence badland formation at Vallcebre Pre-Pyrenees (NE Spain). Catena. 1992; 19: 287-300.

[44] Benito G, Gutierrez M, Rancho C. The influence of physico- chemical properties on erosion processes in badland areas, Ebro basin, N–E Spain. Z. Geomorphol. N. F. 1993; 37: 199– 214.

[45] Gutierrez M, Sancho C, Benito G, Sirvent J, Desir G. Quantitative study of piping processes in badland areas of the Ebro basin, NE Spain. Geomorphology. 1997; 20: 237–53.

[46] Lowery B, Hickey W. J, Arshad M.A, Lal R. Soil water parameters and soil quality. In: Doran J.W, Jones A.J, editors. Methods for assessing soil quality. Madson, WI; 1996: p. 143-55.

[47] Hillel D. Introduction to Soil Physics. Academic Press, Inc. San Diego; 1982.

[48] Arshad MA, Lowery B, Grossman B. Physical tests for monitoring soil quality. In: Doran, J.W., Jones, A.J. (Eds.), Methods for Assessing Soil Quality. SSSA Special Publication No. 49. SSSA, Madison, WI; 1996: p. 123–41.

[49] Rienks SM, Botha GA, Hughes JC. Some physical and chemical properties of sediments exposed in a gully donga in northern KwaZulu-Natal, South Africa and their relationship to the erodibility of the colluvial layers. Catena. 2000; 39: 11–31.

[50] Igwe CA, Ejiofor N. Structural stability of exposed gully wall in Central Eastern Nigeria as affected by soil properties, International Agrophysics. 2005; 19: 215-22.

[51] Cantòn Y, Domingo F, Solé-Benet A, Puigdefábregas J. Hyrological and erosional response of a badlands system in semiarid SE Spain. Journal of Hydrology. 2001; 252: 65–84.

[52] Faulkner H, Spivey D, Alexander R. The role of some site geochemical processes in the development and stabilisation of three badland sites in Almerìa, Southern Spain. Geomorphology. 2000; 35: 87–99.

[53] Faulkner H, Alexander R, Wilson BR. Changes to the dispersive characteristics of soils along an evolutionary slope sequence in the Vera badlands, southeast Spain: implications for site stabilisation. Catena. 2003; 50: 243–54.

[54] Faulkner H, Alexander R, Teeuw R, Zukowskyj P. Variations in soil dispersivity across a gully head displaying shallow sub-surface pipes, and the role of shallow pipes in rill initiation. Earth Surfaces Processes and Landforms. 2004; 29: 1143–60.

[55] Rengasamy P, Greene RSB, Ford GW, Mehanni AH. Identification of dispersive behaviour and the management of red–brown earths. Australian Journal of Soil Research. 1984; 22: 413–31.

[56] Del Prete M, Bentivenga M, Amato M, Basso F, Tacconi P. Badland erosion processes and their interactions with vegatation: a case study from Pisticci, Basilicata, southern Italy. Geografia Fisica e Dinamica Quaternaria. 1997; 20: 147–55.

[57] Rendell HM. Soil erosion and land degradation in southern Italy. In Desertification in Europe, Fantechi R, Margaris NS (eds). Commission of European Communities, Brussels; 1986: p. 184–93.

[58] Robinson DA, Phillips CP. Crust development in relation to vegetation and agricultural practice on erosion-susceptible, dispersive clay soils from central and southern Italy. Soil Till. Res. 2001; 60: 1–9.

[59] Liberti M, Simoniello T, Carone MT, Coppola R, D'Emilio M, Macchiato M. Mapping badland areas using LANDSAT TM/ETM satellite imagery and morphological data. Geomorphology. 2009; 106: 333-43.

[60] APAT. Stato dell'Ambiente: Sezione Geosfera. 2003; 7: 352–61. Available from: http://www.apat.gov.it/site/it-IT/APAT//Pubblicazioni/Annuario_dei_dati_ambientali//Documento/stato_ambiente_2003_10.html#Sommario

[61] Del Prete M, Bentivenga M, Coppola L, Rendell H. Aspetti evolutivi dei reticoli calanchivi a sud di Pisticci. Convegno Nazionale Giovani Ricercatori di Geologia Applicata. Geol. Romana. 1994; 30: 295–306.

[62] Rendell HM. Clay hillslope erosion rates in the Basento Valley, S. Italy. Geografiska Annaler. 1982; 64A: 141–47.

[63] Vittorini S. La degradazione in un campo sperimentale nelle argille plioceniche della Val d'Era (Toscana) e i suoi riflessi morfogenetici. Rivista Geografica Italiana. 1971; 78: 142–69.

[64] Vittorini S. Ruscellamento, deflusso ipodermico ed erosione nelle argille plastiche. Rivista Geografica Italiana. 1979; 86(3): 338–46.

[65] Guerricchio A, Melidoro G. Fenomeni franosi e neotettonici nelle argille grigio azzurre calabriane di Pisticci (Lucania) con saggio di cartografia. Geol. Appl. E Idrogeol. 1979; 14: 105–38.

[66] Boezi F, Palmentola G, Valduga A. Note Illustrative della Carta Geologica d'Italia Foglio 200. Servizio Geologico Italiano; 1971.

[67] Vezzani L. Il bacino plio-pleistocenico di S. Arcangelo (Lucania). Atti dell'Accademia Gioenia di Scienza Naturali in Catania S6. Supplemento di scienze geologiche. 1967; 18: 207–27.

[68] Lentini F. Le Unità Sifilidi della Val d'Agri (Appennino Lucano). Geol. Romana. 1979; 18: 215–24.

[69] Caldara M, Loiacono F, Morlotti E, Pieri P, Sabato L. I depositi Pliopleistocenici della parte nord del Bacino di S. Arcangelo (Appennino Lucano): Caratteri geologici e Paleoambientali. Mem. Soc. Geol. Ital. 1988; 41: 391–410.

[70] Carbone S, Catalano S, Lazzari S, Lentini F, Monaco C. Presentazione della carta geologica del bacino del fiume Agri (Basilicata). Memorie della Società Geologica Italiana. 1991; 41: 109–20.

[71] Pieri P, Sabato L, Loiacono F, Marino M. Il Bacino di piggyback di Sant'Arcangelo: evoluzione tettonico-sedimentaria. Boll. Soc. Geol. It. 1994; 113: 465–81.

[72] Patacca E, Scandone P. Late thrust propagation and sedimentary response in the thrust belt-foredeep system of the Southern Appenines (Pliocene–Pleistocene). In: Vai, G.B., Martini, I.P. (Eds.), Anatomy of an Orogen: The Appenines and Adjacent Mediterranean Basins. Kluwer Academic Publ; 2001: p. 401–40.

[73] Patacca E, Scandone P. Constraints on the interpretation of the CROP-04 seismic line derived from Plio-Pleistocene foredeep and thrust-sheet-top deposits (Southern Appenines, Italy). Boll.Soc.Geol.It. (Ital.J.Geosci.), Spec. Issue. 2007;. 7: 241–56.

[74] Zavala C, Mutti E. Stratigraphy of the Plio-Pleistocene Sant'Arcangelo Basin Basilicata, Italy. Atti Riunione Gruppo Sedimentologia del C.N.R. Catania; 1996: p. 279–82.

[75] Lentini F, Vezzani L. Note illustrative del Foglio 506 S. Arcangelo. V. di 46, I.R.P.I., Cosenza; 1974.

[76] Pieri P, Sabato l, loiacono F. Carta Geologica del Bacino di Sant'Arcangelo (tra il Torrente Sauro e il Fiume Agri). Ed. Paternoster, Matera; 1993.

[77] Verstappen HTh. Geomorphology of the Agri valley, Southern Italy. ITC Journal. 1983; 4: 291–301.

[78] Piccarreta M, Capolongo D, Boenzi F. Trend analysis of precipitation and drought in Basilicata from 1923 to 2000 within a Southern Italy context. International Journal of Climatology. 2004; 24: 907–22.

[79] Piccarreta M. Aspetti Evolutivi della Morfogenesi Calanchiva nelle Argille Pliopleistoceniche della Basilicata, PhD Thesis. Università degli Studi di Bari; 2005.

[80] Piccarreta M, Capolongo D, Bentivenga M, Pennetta L. Influenza delle precipitazioni e dei cicli umido–secco sulla morfogenesi calanchiva in un'area semi-arida della Basilicata (Italia Meridionale). Geografia Fisica e Dinamica del Quaternario, Supplement. 2005; VII: 281–89.

[81] Barahona Fernandez E. Arcillas de ladrilleria de la privincia de Granada: Evaluacion de algunos ensayos de materias primas. Ph.D. thesis, Univ. Granada, Spain; 1974.

[82] Franzini M, Leoni L, Saitta M. Revisione di una metodologia analitica per fluorescenza-X, basata sulla correlazione completa degli effetti di matrice. Rendiconti della Società Italiana di Mineralogia e Petrologia. 1975; 31: 365–78.

[83] Ministero delle Risorse Agricole, Alimentari e Forestali (MiRAAF). etodi Ufficiali di Analisi Chimica del Suolo. Roma; 1992.

[84] Alexander RW, Harvey AM, Calvo A, James PA, Cerda A. Natural stabilisation mechanism on badlands slopes: Tabernas, Almerìa.In Environmental Change in Drylands (Biogeographical and Geomorphological Perspectives), Millington AC, Pye K (eds). Wiley: Chichester; 1994: p. 85–111.

[85] Guerricchio A, Melidoro G. New views on the original of the badlands in the Plio-Pleistocenic clays of Italy. Poc. IV Congr. IAEG; 1982: p. 2.

[86] Pettijohn FJ. Sedimentary Rocks, 3rd edn. Harper and Row, New York; 1975.

[87] APAT. La realizzazione in Italia del progetto Corine Land Cover 2000. APAT, Rapporti. 2005; 36: p. 86.

[88] Gerits J, Imeson AC, Verstraten JM, Bryan RB. Rill development and badland regolith properties. In: Bryan, R.B. (Ed.), Rill Erosion: Processes and Significance: Catena, Suppl. 1997; 8: 141–60.

[89] Sherard JL, Dunningan LP, Decker RS. Identification and nature of dispersive soils. J Geotech. Eng. Div. 1976; 102: 287–301.

[90] Sotelo RR. Identificaciòn de arcillas erodibles dispersivas utilizzando ensayos agronòmicos de suelos. Cience & Tecnica. Comunicaciones Cientificas y Tecnològicas. 1999; 1: 200–4.

[91] Harvey AM. The role of piping in the development of badlands and gully systems in south-east Spain. In: Bryan, R., Yair, A. (Eds.), Badland Geomorphology and Piping. Geobook, Norwich; 1982: p. 317–35.

Chapter 4

Development of Topographic Factor Modeling for Application in Soil Erosion Models

Anna Hoffmann Oliveira,
Mayesse Aparecida da Silva,
Marx Leandro Naves Silva, Nilton Curi,
Gustavo Klinke Neto and
Diego Antonio França de Freitas

Additional information is available at the end of the chapter

1. Introduction

The fast changes in soil use and higher vegetal resource demands favor the triggering of water erosion that needs to have its rates expressed in space and time for the proper adaptation of control practices and resources for agricultural planning. The physical processes of disaggregation, transport and soil deposition that define the erosive process are hydrologically directed and the movement of the water on the soil undergoes the interference of the topography, climate, soil class and land use, so that the studies regarding the theme are based on the intense experimentation of the effects of the variations of these factors on the sediment production. The estimate of the topographic variables, although benefitted by automatic generation and spatial distribution made possible by the Geographical Information Systems (GIS's), is the target of controversy related to the formulation of algorithms for this end, so that its three-dimensional calculation is not a current procedure in the geoprocessing programs [1].

Given the scope of the process, the topographic modeling in the erosion analyses can differ in terms of complexity, processes considered and data required for model use and calibration, which can be empirical, physical and conceptual [2]. In the empirical models most used, such as USLE (Universal Soil Loss Equation) [3] and the revised version of USLE (Revised Universal Soil Loss Equation - RUSLE) [4], the topographic factor is expressed by the association of the steepness and the length of the slope called, respectively, factors S and L. Considering the proper formulation of USLE and its adaptation to the work context in the Digital Elevation

Model (DEM), obviously the advantages associated to DEM derive almost entirely on issues related to the topographic LS factor, because it can be evaluated with the aid of DEM and where the precision of the extracted parameters can become apparent [5].

An aspect that hinders the estimate of appropriate topographic factor (LS) values for applications in GIS and results in high limitation in the use of the USLE and RUSLE erosion models [6, 7] are the dynamics of the erosive process in complex reliefs and hydrographic basins, since USLE was primarily developed for the prediction of the erosion in not very accentuated and uniform slope stretches, in other words, not considering if they are concave, convex [8], or in combination. The limitation of the empirical modeling in the soil loss estimates in complex profiles impelled the development of conceptual models (or semi-empirical), such RUSLE 3D and USPED (Unit Stream Power Erosion and Deposition) [9]. Derived from USLE, these models intend to represent their advancements by adding a physical basis that tries to relate the morphology of the relief and the erosion defining parameters.

In this sense, given the strategic need for generation and diffusion of algorithms for automatic mapping of the topographic variables used in the operationalization of the digital analyses of water erosion, the objective of the present chapter was to conduct a review of the topographic factor development in erosion equations applied in computational geoprocessing systems, with prominence for USLE and RUSLE, addressing the main theories and algorithms used in the digital treatment of the data.

2. Digital Elevation Model (DEM)

In the landscape, the topography determines the behavior of the surface runoff, the phase of the hydrologic cycle that is most directly associated to the water erosion and that requires a rigorous and effective analysis throughout its entire extension, make possible with the use of digital elevation models (DEM). The analyses developed on a DEM allow: to visualize the model in planar geometric projection; generating gray scale images, shaded images and thematic images; calculating fill (embankment) and cut volumes; conducting profile analyses on predetermined trajectories; and generating derivative maps, such as steepness and exposure maps, drainage maps, contour maps and visibility maps. Products of the analyses can even be integrated with other geographical data types aiming at the development of several geoprocessing applications, such as urban and rural planning, agricultural suitability analyses, risk area determination, environmental impact report generation [10], elaboration of digital soil maps, as well as maps of soil attributes such as soil organic matter content [11], among others. Therefore, DEM should faithfully represent the relief allowing to capture the topographic variations presented.

The elaboration and creation of a DEM, indispensable for the representation of a real surface on the computer, can be represented by analytical equations or a network (grid) of points, in a way that transmits the spatial characteristics of the land to the user [12]. Therefore, the information contained before in specific points (vectors) are transformed into a continuous spatial distribution of the relief (raster), enabling new inferences about the local relief. Different methods exist for the interpolation of the data and DEM generation, which are built through regular rectangular grids, such as the Topogrid [13], or triangulated irregular networks (TIN) [14]. For the choice of efficient DEM in the evaluation of the erosive process, an intense preliminary analysis of information, from a hydrologic point of view, is recommended, because the development of the water erosion occurs in response to the manner the water moves through and on the landscape [15].

The geomorphological and hydrological consistency of a DEM is reached when the matrix image faithfully represents the relief features, such as the hydrographic basin watershed, thalwegs and concave and convex elements, and it assures the convergence of the surface runoff for the mapped drainage network. In this sense, several water erosion analysis and modeling works have used the TIN model [16-18], as well as the Topogrid model [19-21] for DEM generation. In a research conducted with the objective of defining the drainage network in a sub-basin [22], the original contour curves (scale 1:10.000) were compared to the curves generated by DEM's of the Topogrid, linear TIN and TIN natural neighbor interpolators. A higher Topogrid hydrological consistency was observed verified in the better continuity of the contour curves and higher drainage area and watershed detailing, resulting in a smaller amount of flat areas and in more detailed drainage pathways. The authors emphasized that all of the generated models have high reliability due to the precise topographic surveys data from which they originated, a fact also observed by [15, 18]. Starting from this same comparison among interpolators, other works [23, 24] made similar observations regarding the behavior and reliability of the models.

The precision of the data collection will influence the quality of the corresponding digital model and the choice of the database to be used in the construction of DEM becomes fundamental. Such data can be derived from contour curves, elevation points, photogrammetric analysis from aerial photography, information collected by stereoscopic satellite images, or radar [25, 26]. The digital database can result from a digitized manual survey obtained through direct readout digital equipment (total stations, topographic GPS), or by remote sensing equipment (radar, laser) in which different DEM generation models can be applied supplying DEM's with varying precision [26]. The spatial distribution and the amount of errors propagated can vary according to the spatial resolution, therefore, the effect of the spatial resolution on supplying useful information for the determination of an appropriate resolution should be investigated [27]. The higher the scale of a map, the higher detail it presents, and to the contrary, the lower the scale in a map, the higher the degree of generalization seen which increases the minimum cartographic area of the land [27]. With better horizontal resolution DEM's, it is possible to include more relief roughness aspects, reducing the length of the slope straight line segments and increasing the accuracy of the L and S factors [18].

The evolution of the geographical information systems and the growing availability of better quality radar images enabled the obtaining of terrain elevation models (DEM) with increasingly better spatial resolutions. However, data surveys considered as having high precision (below 10 meters of resolution) still possess high costs, limiting their use in research. Currently, the most widely used digital database for the generation of DEM's originates from of the digitization of topographic maps or those obtained by remote sensing maintaining their original precision. Because the influence of the pixel size has a significant weight in the analyses derived from DEM's, the choice of the spatial resolution proportional to the scale of the primary data must have certain considerations, among them, the original contour curve scale and the characteristics of the mapped relief. For instance, with a minimum horizontal distance between the curves on the order of 20 m, the spatial resolution of 15 m can be shown appropriate for the detailing of the relief presented in the original base, considering that lower resolutions would tend to generate erroneous information (nonexistent), while higher resolutions would not detail the relief in a satisfactory way [11]. Furthermore, a sufficient spatial resolution cannot only depend on the aimed information and/or the precision used in the collection of this information, but also on the topography. In areas where simple hillsides exist with flat topography, a coarse resolution (> 20 m) may not lead to major errors in hydrographic basins, being able to be used with little uncertainty. In a complex topography, with accentuated slopes, a coarse resolution can result in great uncertainty, and a better resolution could be necessary.

Along those lines, studies have been developed seeking to define the best DEM resolution that is capable of precisely representing the variations of the relief, thus reducing the uncertainties of the erosion prediction models that need topographic modeling. In Slovakia, the S factor derived from a MED with 50 m resolution, obtained from digitized contours of the topographic maps in a 1:50.000 scale, promoted a sufficient level of detail for this type of regional evaluation and the spatial resolution selected reflected the scale of the primary data [28]. To determine the resolution of DEM when the objective is to promote an impartial average global estimate, the global variance of the LS factor can be used [29].

The local variance of a cell measure the average local space variability, while the global variance measures the global variation of the estimates showing a considerable difference of behavior with the increase of the DEM resolution. Thus, in [29] the objective of the research was to evaluate the appropriate DEM resolution for the spatial prediction of the LS factor. A decrease of the global variation of the LS values obtained in the 30 by 100 meter resolution in DEM was observed, followed by a stabilization after 100 m. The local average variance and semivariance of a cell increased with the 30 by 50 m resolution and it decreased after 50 m. The highest local variance and semivariance of a cell was at the 50 m resolution and thus it could be considered appropriate for the necessary detailing of the spatial information (distribution and variability) for the LS factor prediction [29].

A study developed in Thailand analyzed the influence of the spatial resolution on the results of the LS factor [30]. Two DEM resolutions extracted from a SRTM (Shuttle Radar Topography Mission) radar image with 90 m of resolution (original image resolution) and 30 m (resampling of the 90 m resolution) were appraised. The grid size change affected the steepness values, compromising the L and S factor values, since the L factor depends on the grid size and the steepness and the S factor only on the steepness. When affecting the L and S factors, the resolution also affected the sediment transport ratio. The best sediment production estimates were observed in DEM with resolution of 30 m. A fundamental observation is made by the authors of the study, who highlight that the better results of the 30 m resolution compared to the 90 m using the USLE methodology, is probably due to fact this resolution is closer to the 22.4 m slope length, the length used in the derivation of the USLE relationships. Another study

[31] compared SRTM radar images with spatial a resolution of 90 x 90 m and digitized hypsometric curves to determine the topographic factor. The highest detail was obtained by the SRTM images that evidenced lower LS ranges. They concluded that the difference occurred due to the higher detailing in plane areas, where LS is lower (ramp height lower than 40 m, with low steepness).

3. Water erosion modeling

3.1. USLE and RUSLE empirical modeling

The empirical models are the simplest ones, generally possessing less data and lower computational base than the physical and conceptual ones. The empirical models normally have a high aggregation of time and space and are based on analyses of the erosion process using statistic techniques. For this reason, they are particularly useful as the first step to identify the sediment sources. Universal Soil Loss Equation (USLE) and Revised Universal Soil Loss Equation (RUSLE) have been the most used models in the world for predicting erosion processes due to their simplicity and the availability of information.

The topographic factor is the most sensitive parameter of USLE/RUSLE in the soil loss predictions, where a higher relative effect of the steepness factor is observed in a simple analysis of sensitivity. However, an interaction of the steepness (S) and the slope length (L) exists and, LS being used as the "only" parameter in the sensitivity analysis, its influence is even higher on the soil loss than the remaining parameters, including L and S individually [32]. By definition, the slope length (L) is the distance from the point of origin of the surface flow to the point where each slope gradient (S) decreases enough for the beginning of deposition or when the flow comes to concentrate in a defined channel [3]. The soil losses increase with the increase of the slope length and steepness, conditions where the surface flow reaches high-speeds.

The procedure to obtain the slope length was originally manual in these models, which may be adapted to GIS framework. It initially consists of slope length identification through information plans of slope steepness and classified aspects, such as the exposition angle between the slopes and the north. From each rill slope or polygon, the average slope steepness (degrees) and the altitude are calculated [33, 34]. It is possible to calculate the slope length through the following equation [33]:

$$\mathbf{L} = \frac{\mathbf{DH}}{\sin\alpha} \tag{1}$$

Where L is slope length (m); DH is altitude difference (m); and α is average slope steepness (degrees).

The slope angle α corresponds to the inverse of the tangent angle that may be calculated dividing sin α (altitude difference) by cos α (distance between level curves and/or quoted points). The α angle of a surface defined by two points (A and B) is calculated with the horizontal as show in Figure 1 [35].

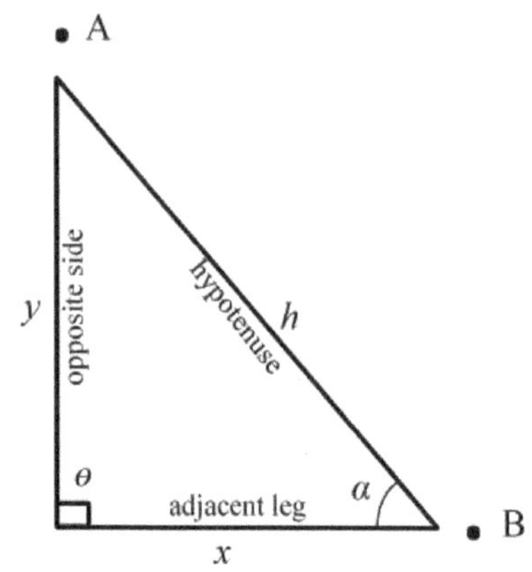

Figure 1. Trigonometric variables in the calculation of the slope.

The slope length (L) and slope steepness (S) factors of the USLE were developed for uniform slopes based on empirical models, which means that they use dependent field measurements. For USLE/RUSLE, they are calculated from the comparison with a ramp length of 22.1 m and 9% slope with the use of a factor m for different steepness classes [3].

The L factor can also be obtained by pixel size in the DEM. If one size of the pixel is considerate as flow length, an equal value is determined for the flown length; therefore it is assumed that the inclination is composed by segments of equal dimensions, consequently with different inclinations, which is not true. However, this approach is considerably feasible if a pixel of suitable dimension is used [36]. A study developed in Thailand evaluated two DEM resolutions and observed better results from 30 m resolution using USLE methodology due to the fact that this one is closer to 22.1 m slope length, which is used for the derivation of model relations [30].

The calculation methodology of LS factors proposed by the USLE was improved in the equation revision, named RUSLE [4], considered more extensive than the previous model. The L factor, in both USLE and RUSLE, is expressed as [3]:

$$\mathbf{L} = \left(\frac{\lambda}{22.1}\right)^{m} \tag{2}$$

Where L is slope length effect on soil loss standardized for 22.1 m length; λ is field slope length (m); and m is slope length exponent.

In the USLE, the m recommend value is from 0.2 to 0.5 for slope levels lower than 1%; 1-3%; 3.5-4.5%; and 5% or more, respectively. Therefore, if a slope gradient is higher than 5%, slope length factor do not change with slope inclination. However, in the RUSLE, m continues to increase with slope inclination (Equation 3). Thus, in the RUSLE, the slope length effect is a function of the erosion ratio of rill to interrill [36].

$$m = \frac{\beta}{(1+\beta)} \tag{3}$$

$$\beta = \frac{\left(\dfrac{\sin\theta}{0.0896}\right)}{\left[3(\sin\theta)^{0.8} + 0.56\right]} \tag{4}$$

Where β is ratio of rill to interrill erosion; and θ is slope angle.

Researchers observed that the m = 0.5 exponent of USLE is better adapted for very accentuate slopes [37]. When the slope increases from 9% to 60%, the *m* exponent increases from 0.5 to 0.71. The slope length exponent, *m*, is 0.7 for a 50% slope with 60 m length and a more moderated ratio of rill and interrill erosion. When the 0.7 factor is used, the RUSLE predicts an addition of 22% of soil loss than the USLE (m = 0.5) through a 60 m of length slope. When the slope is lower than 9%, the USLE will predict a higher soil loss than RUSLE and, being steeper than 9%, the RUSLE will predict a higher soil loss than USLE. The higher difference occurs in much accentuated slopes.

The equation used in the USLE slope factor (S) (Equation 5) was modified to obtain more accurate results in RUSLE model (Equation 6), probably due to changes in the slope factor [34, 38], which depends on the slope angle Θ.

$$S = 65.4\sin^2\theta + 4.56\sin\theta + 0.0654 \tag{5}$$

$$S = 10.8\sin\theta + 0.03,\ \text{for } \theta < 9\% \text{ or } S = 16.8\sin\theta - 0.50 \tag{6}$$

3.1.1. *USLE and RUSLE limitations*

The restrictions of the USLE and RUSLE empirical models frequently occur because neither examines the hydrologic phenomena in their geographical context, using a simplified representation of spatial elements that assumes the hydrographic basin as uniform [39]. Many methods have been developed seeking to include complex slopes, common in a context of hydrographic basins [40]. In a comparison of several manual methods it was concluded that there is no obviously better method [41]. The errors of the empirical models are produced because the water erosion, being a hydrologically driven process, is not evaluated in relation to the surface runoff [42-45]. In the soil loss estimates using the USLE and RUSLE models the surface runoff is not considered in a direct way, though they indirectly consider that the flow transports the eroded sediment and the concentration of sediments depends on the kinetic energy level of the rain, in the sample space of a parcel [44]. Thus, the surface runoff in the empirical models is a primitive factor. This presupposition limits the potential of these models in predicting erosive factor changes, on the scale of basins or drainage systems, which are favored in models based on physical and semi-empirical processes where the surface runoff constitutes a fundamental factor in the water erosion prediction.

For local conservation planning, the LS factor is usually estimated or calculated from length and inclination measurements in the field [6], or even through manual procedures on cartographic bases, making the procedure very difficult and slow due to the difficulty of individualization of each slope [1]. The measurement of the ramp length is made from the evaluated point in relation to the watershed. Besides possessing the difficulty of locating the watershed, this procedure considers the straight line distance until the watershed, concealing the importance of the relief form, because the erosion is affected by the torrent that comes from the whole contribution area. These labor intensive in-field measurements rendered the soil erosion modeling obviously unviable on a regional scale [6] leading to the determination of the ramp length based on the estimate of an average value for hydrographic basins, which is an oversimplification of the true situation [7]. Furthermore, an underestimate of LS values, obtained manually, and consequently also of the erosion risk is observed when compared to the irregular slopes considered in automated models [46].

A second shortcoming of these models is the evaluation of only the erosion, without sediment deposition prediction [47]. When adopting an average rate for an entire slope or hydrographic basin, addressing the erosion using the USLE and RUSLE models does not offer any information as to the sources and sinks of the erosion materials. In spite of the methodology of dividing complex landscapes into series of semi-homogeneous planes used by these models, to provide some consideration as to the convexity and concavity of the inclination, the erosion is only calculated along the flow in a rectilinear manner, without full consideration of the convergence and divergence flow influence [45]. No approach adequately supplies spatially distributed information on the erosion necessary for effective control of the erosion and sediments. Thus, on a hydrographic basin or landscape scale, the spatial distribution of the soil erosion predicted by such models will distort the current conditions and will tend to overestimate the erosion [47, 48]. Some studies mention overestimates in lower soil losses and underestimates for the high losses using the USLE and RUSLE models [44, 49, 50]. As a solution, studies recommend to first identify those portions of the landscape subject to the deposition and to exclude them from the analysis when applying the USLE and RUSLE models [9].

USLE was related to GIS due to the advantages of handling great amounts of spatial data. Until the middle of the 1990's, a great limitation in the use of the USLE and RUSLE erosion models on a regional landscape scale was the difficulty in estimating appropriate LS factor values for applications in GIS [7], since such models evaluate the effects of the topography on the erosion in a two-dimensional way. In that context, the use of models distributed in space came to represent a powerful environmental analysis tool, highlighting soil erosion by water on the hydrographic basin scale.

3.2. Conceptual modeling

The conceptual methods incorporate the impact of different erosive processes through empirical parameters [51] usually obtained through calibration with observed data, such as flow discharge and sediment concentration [52]. Therefore, these models represent the processes within the scale in which they were simulated [53]. It is noteworthy, particularly on a large scale, to mention that deposition patterns and sediment residence time are still little understood in a way that the erosion prediction and the sediment deposition rates on these scales are based, usually, on empirical or semi-empirical studies that are applied in a uniform way throughout the whole area [54].

The semi-empirical LS factor explains the double phenomenon of drainage convergence and furrow [27]. The result of the LS factor thus comes to be equivalent to the traditional LS factor on flat surfaces, but with the advantage of being applicable to slopes with complex geometries [55-57]. When substituting the empirical topographic factor by the semi-empirical one in USLE, the laminar and concentrated flow in complex terrains is considered in the spatial distribution of the erosion, making the estimate more precise.

In the conceptual models the slope length factor is substituted by the upstream contribution area [9, 46, 55, 56] whose modeling conducted in the digital elevation model (DEM) allows to determine the drainage network considering the direction of the surface runoff and the accumulated flow. For each cell, the contribution area upstream is obtained from DEM initially calculating the steepness and aspect maps, building the water flow paths later. The upstream contribution area map is determined from the water flow path lines and the DEM spatial resolution. The precision of the model is related to the uncertainty of the empirical parameters used in the LS factor equation, the accuracy and resolution of DEM and to the methods for derivation of the topographical variables related to LS, such as steepness, aspect and contribution area [27]. In that way, the topographical LS factor can be finally obtained.

Incorporating this concept, an equation modified to compute the LS factor in the form of finite difference in a grid of cells representing a segment of the hillside was derived [46]. Another model, called RUSLE 3D (Revised Universal Soil Loss Equation 3D) presented a simple and continuous form of the LS factor equation considering the impact of the convergent flow [9]. Also considering the contribution area, the USPED model (Unit Stream Power Erosion and Deposition) was developed from the drainage force unit theory [56, 57] for analysis of the erosion and deposition.

3.2.1. Contribution area modeling

The modeling of the contribution area is conducted resorting to DEM, because it contains information that allows to determine the surface runoff network. As such, based on DEM, the flow direction and the accumulated flow and the steepness are determined. The area of contribution of each cell (pixel) of DEM, considering a grid of cells, is its own area plus the area of the upstream neighbors that possess some drained fraction for the pixel in question. The contribution area (A) of a specific grid of cells is calculated from the product of the accumulated flow (χ) and the area of each cell (η) [58]:

$$A = \chi\eta \tag{7}$$

The determination of the accumulated drainage areas (or accumulated flow), which allows the simulation of the hydrographic network, are defined based exclusively on the flow directions. The accumulated flow represents the amount of rain that will drain through each cell, supposing that all of the rain become torrents and there is no interception, evapotranspiration, or loss of underground water. Each pixel receives a value corresponding to the sum of the areas of all of the pixels whose drainage contributed to the analyzed pixel [59].

The flow direction defines the flow path of water, as well as sediments and nutrients, in areas adjacent to the lower altitude points in all of the positions in the hydrographic basin [60]. Independent of the magnitude of the rain event, the flow algorithm in a GIS establishes a one-dimensional flow network connecting each cell with other cells of the hydrographic basin in DEM until the point where the whole surface runoff generated inside the hydrographic basin meets, defined as the mouth [61]. As such, the hydrological relationships are built between different points within a hydrographic basin, topographical continuity being necessary so that functional drainage exists [62].

The estimate of the flow direction is based on the physical principle that the mass of controlled gravity proceed in the direction of the most accentuated slope. The slope is characterized identifying the plane tangent to the topographical surface in the center of the cell. The maximum plane elevation change rate characterizes the inclination gradient, while the correspondent cardinal direction of this larger difference is the aspect [63] (Figure 2).

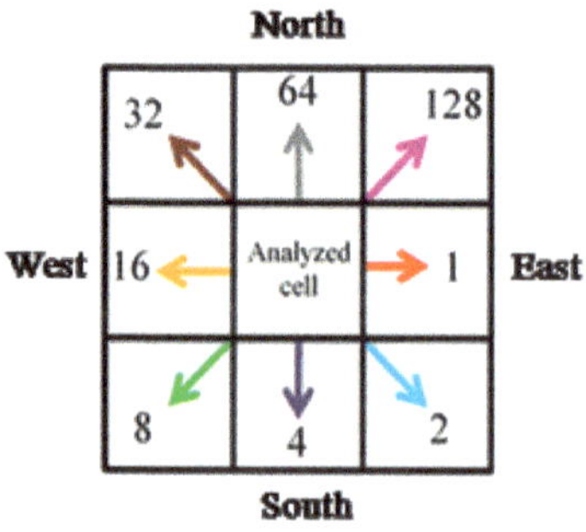

Figure 2. Code of flow direction in analogy to cardinal points generated on the aspect map.

Many algorithms have been developed to consider the contribution area. The flow direction methods, profoundly different, are classified in: concentrative, also called single direction or eight directions, that transfer the whole source pixel matter to the downstream pixel in question [59, 64-68]; and dispersive or multiple direction, that divide the matter of the source cell among several receptors [63, 66, 69-73]. That means that, while those of single flow consider all of the slopes as concave or parallel, the multiples also differentiate the convex ones [46]. In water erosion analysis, the most thoroughly used methods are the concentrative [1, 7, 29, 47, 74-81].

The first and simpler method to specify the flow direction attributes the flow of each pixel to one of their eight neighbors, be it adjacent or diagonal, in the direction of the steepest hillside slope. This method, designated Deterministic 8 Algorithm (D8), is based on the fact that the water can move in 8 possible directions, as demonstrated in Figure 2 (8 flow directions) [64]. The D8 approach has disadvantages arising from the determination of the flow distributed equally in only one of the eight possible directions, separate by 45°, that is expressed in parallel (or convergent) flow patterns in the directions of the cardinal or diagonal points, intermediate values not being possible [60]. In a complex topography, however, the divergent flow frequently can occur causing a significant impact on the delimitation of the basin contribution area [80].

Is suggested to overcome these random flow direction attribution problems for one of the descending neighbors, with the probability proportional to the slope [65]. Other flow direction methods [69, 70] have also been suggested in an attempt to solve the limitations of D8. These attribute a fractional flow to each smaller neighbor, proportional to the slope (or, in the case of the Freeman method, inclination to an exponent) for that neighbor. The multiple flow direction method, here designated by MS (based on multiple slope directions), have the disadvantage that the pixel flow is dispersed for all of the neighboring pixels with lower elevation.

An algorithm was developed using the aspect associated to each pixel to specify the flow directions [71]. The flow is directed as if it was a ball rolling on a plane, liberated from the center of each cell grid. This plane is suited to the elevation of the corners of the pixel, and such corner elevations are estimated by the average of the elevations of the central elevation of the adjacent pixels. This procedure has the advantage of continually specifying the flow direction (an angle between 0 and 2π) without dispersion. Extending the ideas of the previous methodology, a group of elaborated procedures was presented called Demon [66]. Gridded elevation values are used as pixel corners, instead of limiting to the center, and a plane surface is formed for each pixel. The authors recognize the flow as two uniform dimensional origins along the area of the pixel, instead of flow paths drawn from the center of each pixel. The upstream area is evaluated through the construction of detailed flow tubes. It is presupposed that a local plane adjustment for each pixel requires approximation because only three points are necessary to determine a plane. The best adjustment plane, in general, cannot cross the four elevations in the corners, leading to a surface representation discontinuity on the edges of the pixel. The local plane adjustment for specific combinations can lead to inconsistent or deduced flow directions that are a problem in the Lea and Demon methods [63].

Being such, a new procedure to represent the flow directions and calculation of the upstream areas using a grid based DEM was suggested, see reference [63], called infinite D or D∞. The D∞ method calculates the water flow direction according to the steepness of the terrain, distributing the flow proportionally among the neighboring cells. This procedure continually specifies the flow direction (an angle between 0 and 2π) taken in the most accentuated hillside slope, distributing it among the eight facets generate by a 3 x 3 pixel mesh that contains the analyzed pixel in the center. These facets avoid the approximation involved in the plane adjustment and the influence of neighbors with higher altitudes on the upstream water flow

[82]. When the direction does not follow one of the cardinal (0, $\pi/2$, π, $3\pi/2$) or diagonal ($\pi/4$, $3\pi/4$, $5\pi/4$, $7\pi/4$) directions, the accumulated flow is calculated from the flow contribution of a pixel between the two upstream pixels according to the proximity of the flow angle in relation to a right angle for the central pixel. A great advantage of the method D∞ is in considering the form of the divergent surface, in other words, the flow also becomes divergent [83]. Comparing results of the statistical tests and map influence and dependence analysis on the calculation of the upstream area in DEM, a better performance of the D∞ method was observed in relation to the D8, MS and Lea methods, being comparable to Demon, but overcoming its problems of frequent inconsistencies [63].

On defining the drainage directions, it is expected that the resulting drainage network is located within the river channel. Depending on the method used, significant DEM differences in the distribution of the contribution area are obtained. Figure 3 presents the accumulated flow map using the D8 and D∞ methods. As the D8 method routes the whole flow to the cell of higher gradient, rectilinear drainage lines are observed. In turn, the D∞ method, since it considers a proportional distribution among the pixels according to the steepness, does not present the characteristic angular tracings of the flow path restriction.

Using low and high resolution DEM assay data examples, differences are more notable with the increase of the DEM data resolution, especially on the slopes scale [63]. Furthermore, the drainage direction determination methods can produce different results that do not always agree with the reality, mainly when applied in plane areas [60, 84, 85], because they depend on the treatment that each algorithm gives to these regions.

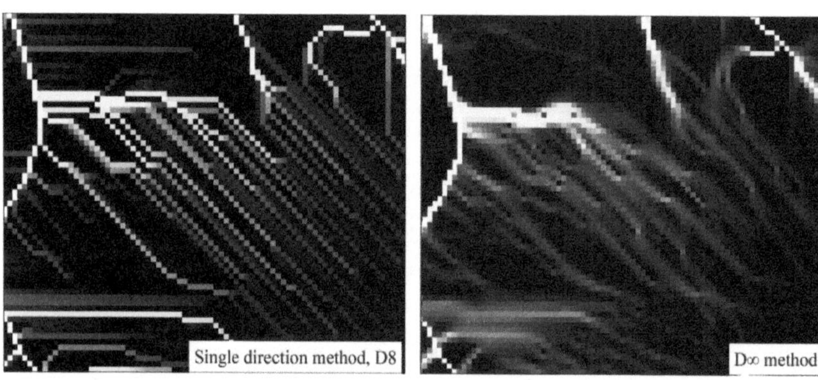

Figure 3. The accumulated flow by the D8 and D-infinity (D∞) methods. (Source: Elaborated by the authors).

Seeking to define a hydrologically consistent digital elevation model and to obtain a sub-basin drainage network, a study [22] tested the D8 [64] and D∞ [63] methods. In the analysis of the mean error between the observed and estimated drainage, the D∞ method provided higher drainage pathway detailing and agreement among the drainage networks. The D8 method provoked errors in the orientation of the drainage network matrix.

The inability of the single direction method (D8) to simulate the flow direction along the inclination of the hillside was also noted [60], which emphasized best performance of the multiple direction method (D∞). The same observations were made in erosive process spatial distribution analysis studies [20, 83, 86-88]. These results were even verified in a study whose purpose was the modeling of the topographic factor [46], which opted for the multiple flow direction method due to its better adjustment in the erosive process analyses.

3.2.2. *Desmet & Govers algorithm*

The great disadvantage observed in the USLE/RUSLE models is the two-dimensional evaluation used to determine the effects of the topography. In these models, the landscape has been generically treated as homogeneous, with plane characteristics. The first research that developed a procedure for the soil loss calculation capable to consider the slope form divided the irregular slopes into a limited number of uniform segments was [89]. Continuing this study, weights were attributed for the slope stretches according to their convexity or concavity [3].

Extending the study of [89], the upstream contribution area concept was introduced for the calculation of the L factor [46] that was applied to the RUSLE LS factor equations [90, 91]. For the calculation of the contribution area, a multiple flow direction algorithm was used [70]. The L factor is expressed according to the equation:

$$L_{i,j} = \frac{\left[\left(A_{i,j-in} + D^2\right)^{m+1} - \left(A_{i,j-in}\right)^{m+1}\right]}{\left[D^{m+2} x_{i,j}^{\ m} (22.13^m)\right]} \tag{8}$$

Where $L_{i,j}$ is the slope length factor of a cell with coordinates (i, j); $A_{i,j-in}$ is the contribution area of a cell with coordinates (i, j) (m²); D = is the cell grid size (m); $x_{i,j}$ is the flow direction value, obtaining the equation $x = sen\alpha + cos\alpha$, where α is the flow direction angle; m is the coefficient that assumes the values: 0.5, if $S \geq 5\%$ (S the steepness degree); 0.4, if $3\% \leq S < 5\%$; 0.3, if $1\% \leq S < 3\%$; and 0.2, if $S < 1\%$.

For the steepness calculation, the following algorithm was employed [92]:

$$G_{ij} = G_x^{\ 2} + G_y^{\ 2} \tag{9}$$

Where G_x and G_y are, respectively, the gradient in the direction x (m/m) and the gradient in the direction y (m/m).

The LS factor for a grid of cells can be thus obtained by the insertion of Li,j and Gij in the LS factor equations of the chosen USLE or RUSLE approach.

This algorithm makes calculations of the steepness, flow direction and the amount of flow that accumulated upstream from a pixel for each pixel [46]. As such, the pixel to pixel topographical factor is calculated along complex slopes. As result, it is possible to define where there is significant distance from the watershed and where there is flow convergence (concave slopes), as well as high steepness, where the LS value tends to be high. In compensation, that value is low in the interfluves (hill tops and plateaus), because the slope length and the steepness are reduced.

The method is not limited to express the sediment transport capacity by the runoff, but it also considers the surface flow, the ramp geometry - if concave or convex - and the erosion way [94]. Comparing the results obtained in the automatic method [46] and manual one [3], a research work [94] verified similar LS values in areas of low steepness. However, in the case of complex slopes, in more sloping areas, the LS values generated by the Desmet & Govers algorithm [46] were significantly superior. That can probably be explained by the assimilation of the convergence and by the respective flow accumulation of this method, which does not occur with the Wischmeier & Smith method [3].

Refining the method of Desmet & Govers, another study [80] incorporated an infinite flow direction model (D∞) and additional methods to isolate the slope length factor. Such method can be applied in situations where a complex topography can influence the surface runoff path and where the excessively long slope lengths, calculated from DEM, can be in need of new landscape detailing. The authors validated the method by the comparison of the statistical distribution of the LS values in GIS with the LS distribution, calculated from field observation data, providing support for applicability of the GIS method to obtain spatial heterogeneity and LS factor magnitude. The evaluation in GIS presented statistical distributions of the LS factor values very similar to those described in the field data, supplying strong support for the use of GIS based methods to represent the spatial heterogeneity and LS factor magnitude for the first time.

3.2.3. *RUSLE–3D*

From Equation 8, derived by Desmet & Govers, an LS factor equation was generated that is used by the RUSLE 3D model [9]. The model includes irregular hillsides integrating a wide spectrum of hillside convexities and concavities and it incorporates the contribution area *A* for the determination of the LS factor:

$$G_{ij} = G_x^{\ 2} + G_y^{\ 2} \tag{10}$$

Where $A_{(r)}$ is the upstream contribution area (m^2); $\beta_{(r)}$ is the slope inclination angle (degrees); and m and n are flow type dependent parameters.

Typical m values are 0.4-0.6 and for n 1-1.3. The exponents for the runoff and slope terms in the soil detachment and sediment transport equations reflect the interaction among different flow, detachment and soil transport types. Figure 4 shows the spatial pattern of the topographic potential by RUSLE-3D with different values for the exponent m. For laminar flow (m = 0.1), the detachment and transport of sediments increase relatively little with the amount of water. This type of flow is typical for areas with good plant covering, but also for a severely compacted soil, where the compacting prevents the detachment and formation of furrows. The value of the exponent *m* of this flow is low, represented by the contribution area [51].

In case the area presents both flow types, usually due to the spatial variability of soil use and properties, an *m* value=0.4 balances the impact of the surface laminar and turbulent flow and it supplies average satisfactory results [51] (Figure 4). When furrow and gully erosion in degraded soils vulnerable to the formation of deep furrows prevails, there are high water flow turbulence conditions and higher water impact, reflecting in a high exponent (m=0.6). The dense vegetation impedes the creation of furrows and maintains a dispersed water flow, while in situations of bare soil, the detachment caused by the flow turbulence increase leads to the formation of furrows [51].

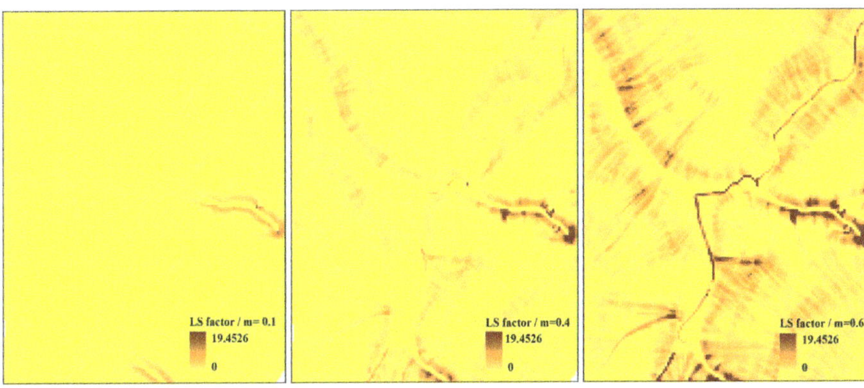

Figure 4. LS factor of RUSLE-3D with m value of = 0.1 for laminar flow; m = 0.4 for laminar and concentrated flow impact; and m = 0.6 for high flow impact in the erosion pattern. (Source: Elaborated by the authors).

The exponent of the spatial variable based on the covering can vary seeking to increase the negative impact of the disturbed areas and to reduce the impact in vegetated areas [51]. For instance, for forest m=0.2, for pasture m=0.4, for degraded pasture m=0.5 and for degraded areas m=0.6. In a study of these coefficient calibrations for two hydrographic sub-basins with forest vegetation, pastures and native field (prairie) in Mexico [95], the estimated value for m was 0.49. In the evaluation of the water erosion in forest systems [17] the value adopted for the coefficient was 0.4. In a hydrographic sub-basin with prevalence of native forest in Australia, the coefficient m=0.6 was considered more representative [96]. It was observed that the m value is low (m=0.1) when sediment detachment and transport increase relatively little with the amount of water. Thus, the geometric properties of the topography (slope, curvatures) play a more important role in the evolution of the soil detachment and erosion/deposition pattern than the water flow pattern [95].

The RUSLE model supplies an exponent m expressed in function of the slope angle that reflects the predominating dispersed flow in plane hillsides of gentle slope, while the flow in accentuated slopes is more turbulent. However, in the RUSLE-3D model this exponent supplies satisfactory results only for short segments. The formula for m based on the slope can result in values of 0.8 or higher in longer slopes. For slopes with hundreds of meters in length or for the concentrated flow this exponent predicts extremely high erosion rates in RUSLE-3D due to the contribution area [51]. As such, the RUSLE equation for the variable m was developed for the slope length and traditional field applications, and therefore it is not recommended for use as contribution area without evaluating the results through field measurements.

Considering this question, a study conducted the calibration of the m and n parameters through the comparison of the result of each soil loss estimated by the different coefficients, with the soil losses obtained in sample portions inserted in an analyzed sub-basin [48]. Joint analysis the values that presented good performance in relation to the mean error and mean difference for the soil loss estimates were determined. The obtained results, m=0.5 and n=1.0, correspond to those mentioned in [51] (m=0.4 and n=1.0) for areas with high spatial variability of use and soil properties under which both flow types, laminar and concentrated, occur. In fact, the forest use is predominant in the sub-basin and so the laminar flow is favored. At the same time, the high variability of soils and their respective properties also influenced by the relief can generate concentrated flows.

3.2.4. USPED

As the RUSLE-3D model, the Unit Stream Power Erosion and Deposition (USPED) [9] model is derived from USLE and represents its modifications or improvements. The model was developed considering the limitation of the empirical models when estimating the soil loss for convergent and divergent terrain in large areas allied to the Geographic | Information System (GIS). Proposing an adaptation of the contribution area variable the LS topographic factor was derived using the drainage force unit theory to describe the erosive process associated to the laminar and furrow flow in steep hillsides starting from a DEM [55-57].

An advantage of USPED is the fact that it predicts the spatial distribution of the erosion, as well as the deposition rates under conditions of uniform surface flow and high precipitation. Thus, this model can be applied in complex terrains where the erosion is limited by the capacity of the runoff to transport sediment. The topographic index represents the change in the transport capacity of the flow direction, being positive for areas with topographic potential for deposition and negative for areas with erosion potential. The contribution area is used as the representation of the water flow in a place or grid of cells. In USPED, the LS factor equation is [9]:

$$LS = (m+1)\left[A_{(r)} / 22.13 \right]^m \left[\sin\beta_{(r)} / 0.09 \right]^n \quad (11)$$

Where A is the contribution area (m^2); θ is the slope angle; and *m* and *n* are constants that depend on the flow and soil property types. For situations where the furrow erosion dominates, these parameters are usually established as m = 1.6 and n = 1.3; where the laminar erosion prevails, m = n = 1.0 is considered [57, 97].

In this model, the water erosion in a DEM cell is dependent on the surface runoff in this cell that in turn depends on the upstream drainage area. When substituting the slope length the upstream contribution area generates the erosion network calculated as the convergence of the sediment flow and the deposition network obtained by the alteration in the sediment transport capacity.

Due to USPED computing divergence of the sediment flow, the impact of the exponents is more complex when compared to RUSLE 3D [51]. In USPED the water flow exponent controls the ratio between the erosion extension and deposition, reflecting the fact that the turbulent flow can transport sediments and the impact of the concentrated erosion will be wider than if the flow was dispersed throughout the vegetation. Figure 5 shows the spatial pattern of topographic potential by USPED with different values for the exponent m. For m = 1, a case of dispersed laminar flow and deposition along the hillside. With m = 1.4 we have the case of the influence of both flow types, laminar and in furrows, on the erosion and deposition, with the deposition beginning in the lower third of the hillside and gullying beginning in headwater areas. For m=1.6, furrows and concentrated flows prevail beginning with great force in the headwater areas and turning the erosion even longer and wider with potential for gullying. In this situation, the extension of the deposition areas is even more reduced.

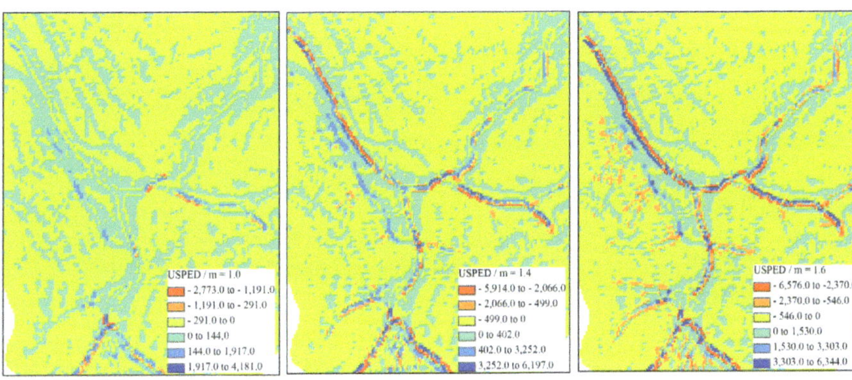

Figure 5. Spatial pattern of topographic potential for erosion and deposition by USPED with m=1.0; m=1.4; and m=1.6. (Source: Elaborated by the authors).

The variation of the LS factor coefficients of USPED interferes in the soil loss estimates by varying with the relief forms, plant covering and erosive processes. For this reason, the value of the exponents has been documented and established for different climates and areas. For the United States, the value suggested for n in [98] varies from 0.3 to 2.0. In laminar erosion situations the coefficient n=1.0 prevails and where the erosion in furrows is dominant, n=1.3. For a hydrographic basin of forest and agricultural use in Italy, the values adopted were m=n=1 [99], as well as in [95] after the calibration of these parameters. In the identification of the mining impact in an agricultural area of India, the coefficients m=1.6 and n=1.3 were adopted [100], while in Poland, they opted for m=1.4 and n=1.2 for two hydrographic basins with the presence of intense erosive processes [101]. In a sub-basin of forest use located in Brazil, the coefficients m=n=1.0 were determined [48].

As the conceptual models reflect the physical processes that govern the system describing them with empirical relationships, various quantitative evaluation studies of erosion risk in hydrographic basins have opted for the association of the USLE model with an LS factor that reflects the expected surface drainage according to the topography, in order to reach soil loss estimates closer to reality [1, 55, 56, 77, 79, 102-104]. The analysis of the erosion risk in a sub-basin was conducted evaluating the performance of four topographic factor models (USLE, RUSLE, RUSLE 3D and USPED) in the USLE model [48]. The USPED (0.1286 ton ha^{-1}) and RUSLE 3D (0.0668 ton ha^{-1}) models did not present statistical differences in relation to the field losses (0.1354 ton ha^{-1}) and they generated a water erosion distribution meditated by the accumulated flow, while the LS factors of RUSLE (2.74 ton ha^{-1}) and USLE (3.65 ton ha^{-1}) overestimated the soil losses [48]. The model considered most efficient in the modeling of the erosion was USPED. This model represented the erosive process in a broad manner when estimating potential erosion and deposition areas, thus allowing to more precisely define the priority areas for conservationist practices under different management sceneries, agreeing with other studies [51, 99]. The advantages of USPED related to the possibility of predicting the spatial distribution of the erosion as well as the deposition rates were stood out in [105], after its comparison with the LS factor of RUSLE 3D. In this context, several research works are opting for the use of the USPED model [9, 45, 51, 98-100, 106, 107].

4. Models in real environmental scenarios

Application these topographic models in the real environmental scenarios have permitted more accuracy and faster estimate of soil erosion in different regions, reliefs and land uses than manual methods. This way, studies have tried to figure out better results applying different LS factor in the water erosion models to estimate and understand this process on watersheds.

LS RUSLE was utilized by [38] in the USLE model to estimate water erosion distribution caused by forest ecosystems in a small watershed and generating soil loss prediction maps according different land use situations. This same LS factor was used in USLE model by [108] allowing identification of the water erosion potential in a watershed forested with eucalyptus and by [109] to estimate of the sediment delivery ratio in a watershed upstream from the hydroelectricity plants.

[110] applied USPED to identify the influence of changing land use on erosion and sedimentation in different land use situations in watershed. [48] applied USLE, RUSLE 3D and USPED in a small watershed for predicting water erosion by eucalyptus plantation founding best results for USPED model. USPED and RUSLE were also applied for assessing the impact of soil erosion/deposition on the archaeological surface at the archaeological site in Greece and USPED presented better results [111]. Using USPED, [112] presented a modeling approach to implement the support practices factor using geographic systems information where data are

unavailable and they concluded that USPED with adequate support practices permit to reduce erosion process.

Thus, considering the erosion problems in the world and data available efforts have been done to improve erosion models.

5. Final considerations

The analysis and obtaining of the topographic factors conducted in the digital environment has become a fundamental piece in erosion model progress, because they address the systematic analyses from specific Geographical Information Systems (GIS's) tools, as well as allow the empirical processing of the data through adaptations of analogical techniques, thus maintaining researcher interpretation. The analysis of the topography in GI enables the analyses of the landscape on a large scale, considers the effects of the topographical complexity more fully in the soil erosion, makes the data processing easier and faster and reduces the relative cost. It is stood out that the reliability of these estimates is directly related to the precision of the topographic surveys used for the derivation of the digital elevation model (DEM).

Evolution of LS factor driven by advanced technology allows the application of different topographic models in the USLE/RUSLE equations modernizing, and improving the estimate of those models. In semi-empirical algorithms, where the contribution area constitutes the central concept, advantages include application in slopes of complex geometries, representation of the surface runoff paths and incorporation of the convergent and divergent flow impact by calibration of empirical coefficients that allow to indicate the qualitative and quantitative effects of the changes in the land use without demanding large spatial and temporal databases. Such models, besides determining the erosion areas on a hydrographic basin level, in the case of the USPED model, even allows to determine the deposition areas, including the erosive process to its full extent.

Author details

Anna Hoffmann Oliveira[1*], Mayesse Aparecida da Silva[1,2], Marx Leandro Naves Silva[1], Nilton Curi[1], Gustavo Klinke Neto[3] and Diego Antonio França de Freitas[1]

*Address all correspondence to: anna.ufla@gmail.com

1 Soil Science Department, Federal University of Lavras, Lavras, MG, Brazil

2 CAPES and CNPq (Visiting scholar) scholarships, Brazil

3 Vitaramae Environmental Consulting Ltd., Lavras, MG, Brazil

References

[1] Bloise, G.L.F.; Carvalho Júnior, O.A.; Reatto, A.; Guimarães, R.F.; Martins, E.S.; Carvalho, A.P.F. Avaliação da suscetibilidade natural à erosão dos solos da Bacia do Olaria – DF. Planaltina: Embrapa Cerrados, 2001. 33 p. (*Boletim de pesquisa e desenvolvimento)*

[2] Merritt, W.S.; Letcher, R.A.; Jakeman, A.J. A review of erosion and sediment transport models. *Environmental Modelling & Software*, Camberra, v. 18, p. 761-799, 2003.

[3] Wischmeier, W. H.; D. D. Smith. *Predicting rainfall erosion losses*; a guide to consevation planning. Washington: USDA, 1978. Departament of Agriculture. 58 p. (Agriculture Handbook n. 537).

[4] Renard K.G.; Foster G.A.; Weesies D.A.; Mccool D.K.; Yoder, D.C. Predicting soil erosion by water: a guide to conservation planning with the revised universal soil loss equation (RUSLE). 1997. *Agriculture handbook* No. 703. USDA, Washington, DC.

[5] Ferrero, V. O. *Hidrologia computacional y modelos digitales del terreno: teoría, práctica y filosofía de una nueva forma de análisis hidrológico.* [S.l.: s.n.], 2004. 364 p. Disponível em: <http://www.gabrielortiz.com/descargas/Hidrologia_Computacional_MDT_SIG.pdf>. Acesso em: 2 dez. 2010.

[6] Hickey, R. Slope angle and slope length solutions for GIS. *Cartography,* 2000. vol. 29, n. 1, pp. 1-8.

[7] Van Remortel, R.D.; Maichle, R.W.; Hickey, R.J. Computing the LS factor for the RevisedUniversal Soil Loss Equation through array-based slope processing of digital elevation data using a C++ executable. *Computers & Geosciences*, 2004: 30, 1043–1053.

[8] Bertoni, J.; Lombardi Neto, F. *Conservação do solo*. São Paulo: Ícone, 2005. 355p.

[9] Mitasova, H.; Hofierka, J.; Zlocha, M.; Iverson, L.R. Modelling topographic potential for erosion and deposition using GIS. *International Journal Geographical Information System, 1996*. 10: 629– 641.

[10] Felgueiras, C. A.; "Análises sobre Modelos Digitais de Terreno em Ambiente de Sistemas de Informação Geográfica". *VIII Simpósio Latinoamericano de Percepción Remota y Sistemas de Información Espacial.* Sesión Poster. Mérida, Venezuela, 2 a 7 de Novembro de 1997.

[11] Oliveira, A.H.; Silva, M.A. Da; Silva, M.L.N.; Avanzi, J.C.; Curi, N.; Lima, G.C.; Pereira, P.H. Caracterização ambiental e predição dos teores de matéria orgânica do solo na Sub-Bacia do Salto, Extrema, MG. *Semina: Ciências Agrárias*, 2012a. 33: (1)143-154.

[12] Camara, G.; Souza, R.C.M.; Freitas, U.M.; Garrido, J. Spring: Integrating remote sensing and GIS by object-oriented data modeling. *Computers & Graphics,* 20: (3) 395-403, 1996.

[13] Hutchinson, M. F. A new procedure for gridding elevation and stream line data with automatic removal of spurious pits. *Journal of Hydrology*, 1989. 106: (3-4)211-232.

[14] Câmara, G.; Davis. C.; Monteiro, A.M.; D'alge, J.C. *Introdução à Ciência da Geoinformação.* São José dos Campos, INPE, 2001.

[15] Chagas, C. S.; Fernandes Filho, E.I.; Rocha, M.F.; Carvalho Júnior, W. De; Souza Neto, N.C. Avaliação de modelos digitais de elevação para aplicação em um mapeamento digital de solos. *Revista Brasileira de Engenharia Agrícola e Ambiental*, 2010. 14: (2)218-226.

[16] Ferraz, S. F. B.; Marson, J.C.; Fontana, C.R.; Lima, W.P. Uso de indicadores hidrológicos para classificação de trechos de estradas florestais quanto ao escoamento superficial. *Scientia Forestalis*, 2007. (75)39-4.

[17] Ferreira, A.G.; Gonçalves, A.C.; Dias, S.S. Avaliação da Sustentabilidade dos Sistemas Florestais em Função da Erosão. *Silva Lusitana*, 2008. 55 - 67.

[18] Liu H.; Fohrer, N.; Hörmann, G.; Kiesel. J. Suitability of *S* factor algorithms for soil loss estimation at gently sloped landscapes. *Catena*, 2009. 77: 248–255.

[19] Freitas, L.F. De; Carvalho Júnior, O.A. De; Guimarães, R.F.; Gomes, R.A.T.; Martins, E.S.; Gomes-Loebmann, D. Determinação do potencial de erosão a partir da utilização da EUPS na bacia do Rio Preto. *Espaço & Geografia*, 2007. 10: (2)431:452.

[20] Mata, C.L.; Carvalho Júnior, O.A. De; Carvalho, A.P.F. De; Gomes, R.A.T.; Martins, E.S.; Guimarães, R.F. Avaliação multitemporal da susceptibilidade erosiva na bacia do rio Urucuia (MG) por meio da Equação Universal de Perda de Solos. *Revista Brasileira de Geomorfologia*, 2007. 8: (2)57-71.

[21] Bilaşco, Ş.; Horvath, C.; Cocean, P.; Sorocovschi, V.; Oncu, M. Implementation of the USLE model using GIS techniques. Case study the Someşean plateau. *Carpathian Journal of Earth and Environmental Sciences*, 2009. 4: (2)123 – 132.

[22] Oliveira, A.H.; Silva, M.L.N.; Curi, N.; Klinke Neto, G.; Silva, M.A. da; Araújo, E.F. Consistência hidrológica de modelos de elevação digital (MED) para avaliação da erosão hídrica na Sub-bacia hidrográfica do horto florestal Terra Dura, Eldorado do Sul, RS. *Revista Brasileira de Ciência do Solo*, 2012b. 36: (4)1259-1267.

[23] Zeilhofer, P. Modelação de relevo e obtenção de parâmetros fisiográficos na Bacia do Rio Cuiabá. *Revista Brasileira de Recursos Hídricos*, 2001. 6: (3)95-109.

[24] Redivo, A.L.; Guimarães, R.F.; Ramos, V.M.; Carvalho Júnior; Martins, E.S. *Comparação entre diferentes interpoladores na delimitação de bacias hidrográficas*. Documentos/Embrapa Cerrados, 2002. 20p.

[25] Aronoff, S. *Geographic information systems: A managment perspective*. WDL Publications, Otawa, 1989. 294 p.

[26] Hutchinson, M.F.; Gallant, J.C. Digital elevation models and representation of terrain shape. In: WILSON, J.P.; GALLANT, J.C. (ed.). *Terrain analysis: Principles and applications*. New York: John Wiley & Sons, 2000. p.29-50.

[27] Gertner, G.; Wang, G.; Fang, S.; Anderson, A.B. Effect and uncertainty of digital elevation model spatial resolutions on predicting the topographical factor for soil loss estimation. *Journal of Soil and Water Conservation*, 2002. 57: (3)164-174.

[28] Cebecauer, T.; Hofierka, J. The consequences of land-cover changes on soil erosion distribution in Slovakia. *Geomorphology*, 2008. 98: 187–198.

[29] Wang, G.; Gertner, G.; Parysow, P.; Anderson, A. Spatial prediction and uncertainty assessment of topographic factor for revised universal soil loss equation using digital elevation models. *Journal of Photogrammetry & Remote Sensing*, 2001. 56: 65–80.

[30] Bhattarai, R.; Dutta, D. Estimation of Soil Erosion and Sediment Yield Using GIS at Catchment Scale. *Water Resources Management*, 2007. 21: 1635–1647.

[31] Fornelos L.F.; Neves S.M.A.S. Uso de modelos digitais de elevação (mde) gerados a partir de imagens de radar interferométrico (SRTM) na estimativa de perdas de solo. *Revista Brasileira de Cartografia, 2007*. 59(01).

[32] Truman, C.C.; Wauchope, R.D.; Sumner, H.R.; Davis, J.G.; Gascho, G.J., Hook, J.E.; Chandler, L.D.; Johnson, A.W. Slope length effects on runoff and sediment delivery. *Journal of Soil and Water Conservation*, 2001. 56: (3)249.

[33] Rocha, J.V.; Lombardi Neto, F.; Bacellar, A.A.A. Metodologia para determinação do fator comprimento de rampa *(L)* para a Equação Universal de Perdas de Solo. *Caderno de Informações Georreferenciadas* (CIG), 1997. 1: (2).

[34] Silva, A.M.; Mello, C.R.; Curi, N.; Oliveira, P.M. Simulação da variabilidade espacial da erosão hídrica em uma sub-bacia hidrográfica de Latossolos no sul de Minas Gerais. *Revista Brasileira de Ciência do Solo*, 2008. 32: 2125-2134.

[35] Bateira, C. Cálculo e cartografia automática dos declives: Novas tecnologias versus lelhos problemas. *Revista da Faculdade de Letras – Geografia*. Porto / Portugal. v. XII/XIII, , 1996/7, pp. 125-143.

[36] Ismail, J.; Ravichandran, S. Using Remote Sensing and GIS. *Water Resources Management*, 2008. 22: 83-102.

[37] Liu, B.Y.; Nearing, M.A.; Shi, P.J.; Jia, Z.W. Slope length effects on soil loss for steep slopes. In: Stott, D.E.; Mohtar, R.H., Steinhardt, G.C. *Sustaining the Global Farm*, 2001. 784-788.

[38] Silva, M.A. Da; Silva, M.L.N.; Curi, N. L.; Norton, D.; Avanzi, J.C.; Oliveira, A.H.; Lima, G.C. (2010) - Water erosion modeling in a watershed under forest cultivation through the USLE model. In: *Proceedings 19th World Congress of Soil Science*. Brisbane, Australia, IUSS, p. 173-176.

[39] Tucci, C. E. M. *Modelos hidrológicos*. Porto Alegre: Ed. Universidade, UFRGS, 1998. 669 p.

[40] Wilson, J.P. Estimating the topographic factor in the universal soil loss equation for watersheds. *Journal of Soil and Water Conservation*, 1986. 41: 179-184.

[41] Griffin, M.L.; Beasley, D.B.; Fletcher, J.J.; Foster, G.R. Estimating soil loss on topographically nonuniform field and farm units. *Journal of Soil and Water Conservation*, 1988. 43: 326-331.

[42] Kandel, D.D., Western, A.W., Grayson, R.B.; Turral, H.N. Process parameterization and temporal scaling in surface runoff and erosion modelling. *Hydrological Processes*, 2004. 18: 1423-1446.

[43] Kinnell, P.I.A. Slope length factor for applying the USLE-M to erosion in grid cells. *Soil Tillage Research*, 2001. 58: 11–17.

[44] Kinnell, P.I.A. Why the universal soil loss equation and the revised version of it do not predict event erosion well. *Hydrological Processes*, 2005. 19: 851–854.

[45] Warren, S.D.; Mitasova, H.; Hohmann, M.G.; Landsberger, S.; Iskander, F.Y.; Ruzycki, T.S.; Senseman, G.M. Validation of a 3-D enhancement of the Universal Soil Loss Equation for prediction of soil erosion and sediment deposition. *Catena*, 2005. 64: 281-296.

[46] Desmet, P.J.J.; Govers, G. A GIS procedure for automatically calculating the USLE LS factor on topographically complex landscape units. *Journal of Soil and Water Conservation*, 1996. 51: 427–433.

[47] Van Remortel, R., Hamilton, M.; Hickey, R. Estimating the LS factor for RUSLE through iterative slope length processing of digital elevation data. *Cartography*, 2001. 30: (1)27-35.

[48] Oliveira, A.H. *Erosão hídrica e seus componentes na sub-bacia hidrográfica do horto florestal Terra Dura, Eldorado do Sul, RS.* PhD thesis. Universidade Federal de Lavras, 2011.

[49] Risse, L.M.; Nearing, M.A.; Nicks, A.D., Laften, J.M. Error assessment in the universal soil loss equation. *Soil Science Society of America Journal*, 1993. 57: 825–833.

[50] Rapp, J.F.; Lopes, V.L.; Renard, K.G. Comparing soil erosion estimates from Rusle and Usle on natural runoff plots. In: Aschough II, J.C., Flanagan D.C. (eds). *Proceedings of International Symposium Soil Erosion Research for the 21st Century*, Jan. 3–5, Honolulu, HI. American Society Agricultural Engineers: St Joseph, MI; p. 24–27. 2001.

[51] Mitasova, H.; Mitas, L.; Brown, W.M.; Johnston, D. *Terrain modeling and Soil Erosion Simulation: applications for Ft. Hood Report for USA CERL.* University of Illinois, Urbana-Champaign, IL. 2001.

[52] Zhou, Q.; Liu, X. Error assessment of grid-based flow routing algorithms used in hydrological models. *International Journal of Geographic Information Science*, 2002. 16: (8)819-842.

[53] Arnold, J.G. *SWAT: Soil and Water Assessment Tool / User's Manual*, USDA-ARS. 1996.

[54] Lu, H., Moran, C., Prosser, I.; Sivapalan, M. Modelling sediment delivery ratio based on physical principles. In: Pahl-Wostl, C.; Schmidt, S.; Jakeman, T. (Editors), *Interna-*

tional Congress: "Complexity and Integrated Resources Management". Osnabrueck: International Environmental Modelling and Software Society, p. 600, 2004.

[55] Moore I.D.; Burch, G.J. Modeling erosion and deposition. Topographic effects. *Transactions of the America, Science Agricultural Engineering*, 1986. 29: 1624-1640.

[56] Moore, I.D., Burch, G.J. Physical basis of the length-slope factor in the universal soil loss equation. *Soil Science Society of America Journal*, 1986. 50: 1294-1298.

[57] Moore, I.D.; Wilson, J.P. Length-slope factors for Revised Universal Soil Loss Equation (RUSLE): simplified method of estimation. *Journal of Soil and Water Conservation*, 1992. 47: (5)423-428.

[58] Moore, I.D., Turner, A.K., Wilson, J.P., Jenson, S. K.; Band, L. E. Gis and land surface subsurface process. In: Goodchild, M.F.; Bradley, O. *Environmental Modeling with GIS*. Parks & Louis T. Steyaert (eds), 1993. pp. 196-230.

[59] Jenson, S. K.; Domingue, J. O. Extracting topographic structure from digital elevation data for geographic information system analysis. *Photogrammetric Engineering and Remote Sensing*, 1988. 54: (11)1593-1600.

[60] Nardi, F.; Grimaldi, S.; Santini, M.; Petroselli, A.; Ubertini, L. Hydrogeomorphic properties of simulated drainage patterns using digital elevation models: the flat area issue. *Hydrological Sciences Journal*, 2008. 53: (6)1176-1193.

[61] Maidment, D. R., Oliveira, F., Calver, A., Eatherall, A. and Fraczek, W. Unit Hydrograph Derived From A Spatially Distributed Velocity Field. *Hydrological Processes*, 1996. 10: 833-844.

[62] Rennó, C.D.; Nobre, A.D.; Cuartas, L.A.; Soares, J.V.; Hodnett, M.G.; Tomasella, J.; Waterloo, M.J.H. A new terrain descriptor using SRTM-DEM: Mapping terra-firme rainforest environments in Amazonia. *Remote Sensing of Environment*, 2008. 112: 3469-3481.

[63] Tarboton, D. G. A new method for the determination of flow directions and upslope areas in the grid digital elevation models. *Water Resources Research*, 1997. 33: (2)309- 319.

[64] O'callaghan, J.F.; Mark, D.M. The extraction of drainage networks from digital elevation data. *Computer Vision Graphics Image Processing*, 1984. 28: 323–344.

[65] Fairfield, J.; Leymarie, P. Drainage networks from grid digital elevation models. *Water Resources Research*, 1991. 27: (5) 709-717.

[66] Costa-Cabral, M.C.; Burges, S. J. Digital elevation model networks (DEMON): A model of flow over hillslopes for computation of contributing and dispersal areas. *Water Resources Research*, 1994. 30: (6)1681-1692.

[67] Garbrecht, J.; Martz, L.W. The assignment of drainage direction over flat surfaces in raster digital elevation models. *Journal of Hydrology*, 1997. 193:204-213.

[68] Orlandini, S.; Moretti, G.; Franchini, M.; Aldighieri, B.; Testa, B. Path-based methods for the determination of nondispersive drainage directions in grid-based digital elevation models. *Water Resources Research*, 2003. 39: (6)1144.

[69] Freeman, T.G. Calculating catchment area with divergent flow based on a regular grid. *Computers & Geosciences*, 1991. 17: 413–422.

[70] Quinn, P.; Beven, K.; Chevallier, P.; Planchon, O. The prediction of hillslope flow paths for distributed hydrological modeling using digital terrain models. *Hydrological Processes*, 1991. 5: 59-79.

[71] Lea, N.L. An aspect driven kinematic routing algorithm. In: *Overland Flow: Hydraulics and Erosion Mechanics*, Eds. PARSONS, A.J.; Abrahams, A.D. Chapman & Hall, New York, USA. 1992.

[72] Lindsay, J.B. A physically based model for calculating contributing area on hillslopes and along valley bottoms. *Water Resources Research*, 2003. 39: (12)1332.

[73] Seibert, J., Mcglynn, B.L. A new triangular multiple flow direction algorithm for computing upslope areas from gridded digital elevation models. *Water Resources Research*, 2007. 43.

[74] Rieke-Zapp, D.H.; Nearing, M.A. Slope Shape Effects on Erosion: A Laboratory Study. *Soil Science Society American Journal*, 2005. 69: 1463–1471.

[75] Mendes, C.A.B.; Ordoñez, J.E.S.; Grehs, S.A. Arcabouço de modelo hidrológico de escala continental utilizando-se a topologia da rede de drenagem simulada: aplicação na Bacia Hidrográfica Amazônica. *Geografia*, 2006. 15: (2)21-49.

[76] Patriche, C.V.; Capatana, V.C.; Stoica, D.L. Aspects regarding soil erosion spatial modeling using the USLE / RUSLE within GIS. *Geographia Technica*, 2006. 2: 87-97.

[77] Efe, R.; Ekinci, D.; Cürebal Erosion analysis of Sahin Creek watershed (NW of Turkey) using GIS based on RUSLE (3d) method. *Journal of Applied Sciences*, 2008. 8: (1)49-58.

[78] Ferreira, A.G.; Gonçalves, A.C.; Dias, S.S. Avaliação da Sustentabilidade dos Sistemas Florestais em Função da Erosão. *Silva Lusitana*, 2008. (n. especial) 55 - 67.

[79] Ozcan, A.U.; Erpul, G.; Basaran, M.; Erdogan, H.E. Use of USLE/GIS technology integrated with geostatistics to assess soil erosion risk in different lan uses of Indagi Mountain Pass-Çankiri, Turkey. *Environmental Geology*, 2008. 53: 1731-1741.

[80] Winchell, M.F.; Jackson, S.H.; Wadley, A.M.; Srinivasan, R. Extension and validation of a geographic information system-based method for calculating the Revised Universal Soil Loss Equation length-slope factor for erosion risk assessments in large watersheds. *Journal of Soil and Water Conservation*, 2008. 63: (3)105-111.

[81] Zhang, Q.; Wang, L.; Wu, Fa-qi. GIS-Based Assessment of Soil Erosion at Nihe Gou Catchment. *Agricultural Sciences in China*, 2008. 7: (6)746-753.

[82] Tarboton, D.G.; Mohammed, I.N. *Terrain analysis using digital elevation models*. TauDEM, version 5.0. Software. 2010.

[83] Bogaart, P.W.; Troch, P.A. Curvature distribution within hillslopes and catchments and its effect on the hydrological response. *Hydrology and Earth System Sciences*, 2006. 10: 925–936.

[84] Ramos, V.M.; Guimarães, R.F.; Redivo, A.L.; Carvalho Júnior, O.A. De; Fernandese, N.F.; Gomes, R.A.T. Avaliação de metodologias de determinação do cálculo de áreas de contribuição. *Revista Brasileira de Geomorfologia*, 2003. 4: (2)41-49.

[85] Buarque, D.C.; Fan, F.M.; Paz, A.R. Da; Collischonn, W. Comparação de Métodos para Definir Direções de Escoamento a partir de Modelos Digitais de Elevação. *Revista Brasileira de Recursos Hídricos*, 2009. 14: (2)91-103.

[86] Günter, A.; Seibert, J.; Uhlenbrook, S. Modeling spatial patterns of saturated areas: An evaluation of different terrain indices. *Water Resources Research*, 2004. 40: 5114.

[87] Pan, F.; Peters-Lidard, C.D.; Sale, M.J.; King, A.W. A comparison of geographical information systemsbased algorithms for computing the TOPMODEL topographic index. *Water Resources Research*, 2004. 40: 6303, 2004.

[88] Erskine, R.H.T.R.G.; Ramirez, J.A.; Macdonald, L.H. Comparison of grid-based algorithms for computing upslope contributing area. *Water Resources Research*, 2006. 42:9416.

[89] Foster, G.R.; W.H., Wischmeier. Evaluating irregular slopes for soil loss prediction. *Transactions of ASAE*, 1974. 17: 305-309.

[90] Mccool, D.K., Brown, L.C.; Foster, G.R.. Revised slope steepness factor for the Universal Soil Loss Equation. 1987. *Transactions of the ASAE*, vol. 30, pp. 1387-1396.

[91] Mccool, D.K., Foster, G.R.; Mutchler, C.K.; Meyer, L.D. Revised slope length factor for the Universal Soil Loss Equation. 1989. *Transactions of the ASAE*, vol. 32, pp. 1571-1576.

[92] Zevenbergen, L.W.; Thorne, C.R. Quantitative analysis of land surface topography. *Earth Surface Processes and Lnadforms*, 1987. 12: 475.

[93] Carvalho Junior, O.A.; Guimarães, R.F. Implementação em ambiente computacional e análise de emprego da área de contribuição no cálculo do fator topográfico (*LS*) da USLE. In: Simpósio Nacional de Controle da Erosão, 7. *Anais...* Goiânia, 2001. (CD-ROM).

[94] Silva, V.C. Cálculo automático do fator topográfico (*LS*) da EUPS, na Bacia do Rio Paracatu. *Pesquisa Agropecuária Tropical*, 2003. 33: (1)29-34.

[95] Suárez, M.C.G. *Metodologia de cálculo del factor topográfico, LS, integrado em los modelos RUSLE y USPED. Aplicación em Arroio Del Lugar, Guadalajara (España)*. 2008. Universidad Politécnica de Madrid, Madri. 391 p. (Tese de Doutorado)

[96] Simms, A.D.; Woodroffe, C.D.; Jones, B.G. Application of RUSLE for Erosion Management in a Coastal Catchment, Southern NSW. In: *Proceedings of MODSIM 2003*. Townsville, Queensland, Australia, p. 678-683, 2003.

[97] Foster, G.R. Comment on "Length-slope factors for the Revised Universal Soil Loss Equation: simplified method of estimation". *Journal of Soil and Water Conservation*. 1994. 49: 171–173.

[98] Pricope, N.G. Assessment of Spatial Patterns of Sediment Transport and Delivery for Soil and Water Conservation Programs. *Journal of Spatial Hydrology*, 2009. 9: (1)21-46.

[99] Pistocchi, A.; Cassani, G.; Zani, O. Use of the USPED model for mapping soil erosion and managing best land conservation practices. In: Rizzoli, A. E.; Jakeman, A. J. (eds.). *Integrated Assessment and Decision Support, Proceedings of the First Biennial Meeting of the International Environmental Modelling and Software Society*, v. 1, 2002. pp. 163-168. Disponível em: <iEMSs, 2002. http://www.iemss.org/iemss2002/proceedings/>.

[100] Kandrika, S.; Dwivedi, R.S. Assessment of the impact of mining on agricultural land using erosion-deposition model and space borne multispectral data. *Journal of Spatial Hydrology*, 2003. 3: (2).

[101] Drzewiecki, W.; Mularz, S. Model USPED jako narzędzie prognozowania efektów erozji i depozycji materiału glebowego. In: *Annals of Geomatics (Associação Polonesa de Informação Espacial)*, 2005. 3: (2)52–54.

[102] Andrade, A.C.; Leal, L.R.; Carvalho Júnior, O.A.; Martins, E.S.; Reatto, A. *Estudo dos processos erosivos na Bacia do Rio Grande (BA) como subsídio ao planejamento agroecológico*. Planaltina, DF: Embrapa Cerrados, 2002. 26 p. Boletim técnico n. 63.

[103] Erdogan, E.H.; Erpul, G.; Bayramin, İ. Use of USLE/GIS Methodology for Predicting Soil Loss in a Semiarid Agricultural Watershed. *Environmental Monitoring and Assessment*, 2007. 131: 153–161.

[104] Jain, M.; Kothyari, U.C. Estimation of soil erosion and sediment yield using GIS. *Hydrological Science Journal*, 2000. 45: (5)771-786.

[105] Saavedra, C.P.; Mannaerts, C.M. Erosion estimation in an Andean catchment combining coarse and fine resolution satellite imagery. In: *Proceedings of the 31st International Symposium on Remote Sensing of Environment: global monitoring for sustainability and security*. Saint Petersburg, 20-24 June, 2005. 4 p.

[106] Liu, J.; Liu, S.; Tieszen, L.; Chen, M. Estimating soil erosion using the USPED model and consecutive remotely sensed land cover observations. *Proceedings of the 2007 summer computer simulation conference*, San Diego, 2007.

[107] Alimohammadi, A.; Sheshangosht, S.; Soltani, M.J. Evaluation of relations between DEM-Based USPED Model Output and Satellite-based spectral indices. In: Conference Proceedings of Map India 2006. *Anais...* Disponível em: <http://www.gisdevelopment.net/proceedings/mapindia/2006/index.htm>.

[108] Avanzi, J. C. *Soil properties, condition and soil losses for south and east Brazilian forest areas.* PhD thesis. Universidade Federal de Lavras, 2009.

[109] Beskow S.; Mello, C. R.; Norton, L. D.; Curi, N.; Viola, M. R.; Avanzi, J. C. Soil erosion prediction in the Grande River Basin, Brazil using distributed modeling. *Catena*, 2009. 79: 49-59.http://www.sciencedirect.com/science/article/B6VCG-4WKJ5JC-1/2/fb138c9c89ca86d834b52c3124987424 (accessed 19 October 2012).

[110] Leh, M.; Bajwa, S.; Chaubey, I. Impact of land use change on erosion risk: an integrated Remote sensing, geographic information system and modeling methodology. *Land Degradation & Development*, 2011. https://engineering.purdue.edu/ecohydrology/Pubs/2011%20LDD%20Leh-Bajwa-Chaubey.pdf. (accessed 19 October 2012).

[111] Gouma, M.; Wijngaarden, G. J. V.; Soetens, S. Assessing the effects of geomorp hological processe son archaeologic al densities: a GIS case stud y on Zakynth os Island, Greece. *Journal of Archaeological Science*, 2011. 38: 2714-2725

[112] Pelacani, S.; Märker, M.; Rodolfi, G. Simulation of soil erosion and deposition in a changing land use: A modelling approach to implement the support practice factor. *Geomorphology*, 2008. 99: 329 – 340

Chapter 5

Integrated Indicators for the Estimation of Vulnerability to Land Degradation

Vito Imbrenda, Mariagrazia D'Emilio,
Maria Lanfredi, Tiziana Simoniello,
Maria Ragosta and Maria Macchiato

Additional information is available at the end of the chapter

1. Introduction

The setting up of sustainable development strategies, able to balance the opposite demands of economic growth and environmental protection, is one of the fundamental challenges for the international community. Our developing world is experiencing growing pressures on its land, water, and food production systems and the role of the human society in determining change within the Earth environment is becoming ever more central [1]. In this context, preserving the land productivity is a prior goal, especially in those areas, such as drylands, which are particularly fragile from an ecological point of view.

One of the most serious problem threatening these areas is land degradation, which is defined as the (persistent) reduction of biological and economic productivity [2] or, equivalently, as the reduction in the capacity of the land to provide ecosystem goods and services and to assure its functions [3,4]. Land degradation is due to a mix of predisposing factors (thin soil horizons, low soil organic matter, sparse vegetation cover, etc.) frequently accentuated by human mismanagement and periodic drought.

As a crucial component of terrestrial ecosystems, soil plays a prominent role in triggering or exacerbating land degradation. The combined action of climatic factors (aridity, extreme events, rainfall erosivity) and human pressure (overgrazing, deforestation, intensification of agriculture, tourism development, see e.g., [5]) can result in a general soil degradation and in some cases in a irretrievable loss of lands suitable for agricultural/grazing/forest use [6].

In particular, as far as the anthropic pressure is concerned, the demographic boom and the economic growth have caused a rapid and unplanned change of land use patterns [7-9] as a consequence of the conversion of natural and semi-natural areas in areas often managed through intensive farming techniques. These mainly consist in the use of a considerable amount of external inputs (frequent use of fertilizers, pesticides and genetically modified organisms, see [10-12]) and in a set of unsuitable management practices (too deep ploughing, large irrigation schemes, monoculture, etc., [13]). It is evident that the progressive intensification of agricultural practices can accelerate soil degradation phenomena especially in those areas marked by poor soil qualities [14]. In fact, cropping and grazing cause land degradation more than non-agricultural uses of soil [15].

According to the European Commission, six soil degradation processes (water, wind and tillage erosion, loss of soil organic carbon, compaction, salinization and alkalinization, contamination, and decline in biodiversity) were identified as induced or worsened by bad agricultural practices [13].

Also livestock husbandry can represent a potential degradation driver when a high number of head of cattle is strongly concentrated in limited areas, as it often occurs in Southern Europe (overgrazed land, e.g., [16]).

Furthermore, degradation phenomena affect land surface processes and particularly vegetation covers which play a decisive role in the surface energy exchanges and water balance [17,18]. Therefore vegetation assessment is crucial for evaluating land degradation vulnerability, particularly in areas that are still productive. Stressed vegetation, characterized by a decrease of photosynthetic activity and/or patch fragmentation processes, can have negative repercussions on the other biophysical components (soil and climate, [19]). This is particularly true for Mediterranean landscapes, often marked by a gradual reduction of biological productivity (e.g., [20, 21]), low resilience of vegetation [7,9] and abrupt modifications due to wildfires [22,23] and land use/land cover changes [24,25].

On the whole, today, a quarter of world population is threatened by the effects of degradation phenomena [26], which affect nearly 84% of agricultural lands [26]. Then it is clear the reason why land degradation is listed among the most important socio-environmental issues having direct and indirect effects on food security, climate change at local scale, eco-refugees and wars linked to the exploitation of natural resources [28-30].

The need to halt and prevent soil/land degradation has urged the international scientific community to improve the knowledge on causes and consequences of the interest phenomena and identify efficient monitoring tools. These have to help policy makers in developing effective conservation/rehabilitation measures adapted to each involved area. In particular, scientists must provide efficient tools for the early detection of sensitive areas by classifying them in different levels of land degradation vulnerability [8]. At this aim many different methodologies have been used to study land degradation (field measurements, visual interpretation, social enquiries, mathematical models, remote sensing, environmental indicators, etc.), including the use of simple models based on indicators that synthesize information on the state and tendency of complex processes [31].

In particular, in the context of the Mediterranean basin the most used methodology is the indicator-based Environmentally Sensitive Areas (ESA) model developed within the MEDA-

LUS project [32]. This combines information concerning the biophysical component (climate, soil and vegetation) and the anthropic one to detect areas prone to degradation and defines, at the same time, relative values of vulnerability. The standard scheme of the ESA model is not free from faults consisting in too little detailed guidance on the choice and the distribution in vulnerability classes of anthropic indicators, lack of dynamical information on the vegetation component and lack of an objective weighting system based on statistical analysis for the used indicators [33,34]. Nevertheless, the ESA model is the most frequently applied in the Mediterranean basin enabling comparability with other similar studies. This is due to the immediacy of the adopted approach in dealing with land degradation and the consequent easy and rapid interpretation of the produced cartography. Moreover, the flexibility of the model, allowing inclusion/exclusion of variables, is particularly suitable to match local biophysical and socio-economic peculiarities of each examined area [35].

In this chapter we approach the assessment of the vulnerability to land degradation of a typical Mediterranean environment using a modified version of the ESA model. This approach combines analyses of the socio-economic component with analyses of the vegetation trends.

According to the standard ESA strategy, different indicators representing the impact of agricultural and grazing activities are used. The main feature of these indicators is that they are census-based and consequently suitable only for the analysis at municipal scale. Therefore we have also elaborated a mechanization index (proxy for soil compaction induced by agricultural machineries) that uses land cover and morphological data [36], enabling high spatial resolution and faster rate of update.

The indicators related to the anthropic impact are integrated into an overall Land Management Index (LMI) and in each area it is possible to enhance the main contributing factors to highlight the prevailing forces that drive human-induced degradation processes.

In order to include vegetation in the vulnerability map we analyze satellite vegetation index NDVI (Normalized Difference Vegetation Index) which is recognized as ideal tool for monitoring long term trends of degradation phenomena and assessing different values of severity of the concerned processes [37,38].

The final result of our analyses is an integrated vulnerability map of the investigated region, accounting for management and vegetation factors, which allows us to identify priority sites where restoration/rehabilitation interventions are urgent.

The adopted procedure can be easily applied to geographic contexts characterized by high complexity in terms of land cover type and economic vocation (intensive agriculture, grazing, industrial activities) thus enabling an early detection of the areas most vulnerable to land degradation.

2. Study area

The Basilicata region covers an area of about 10000 km^2 in the core of Southern Italy (Fig. 1). This is recognized as a region at potential risk of land degradation by several studies [39-41].

In this area, as in all the Southern Italy, vulnerability to land degradation results from the co-occurrence of some specific bioclimatic features (uneven reliefs with steep slopes, highly erodible soils, wide climate variability, recurrent drought) and from an improper land use (urbanization intensive farming, industrial pollution). For example, inappropriate agricultural practices may significantly contribute to land degradation, determining a strongly impact on the economic value of the lands [42].

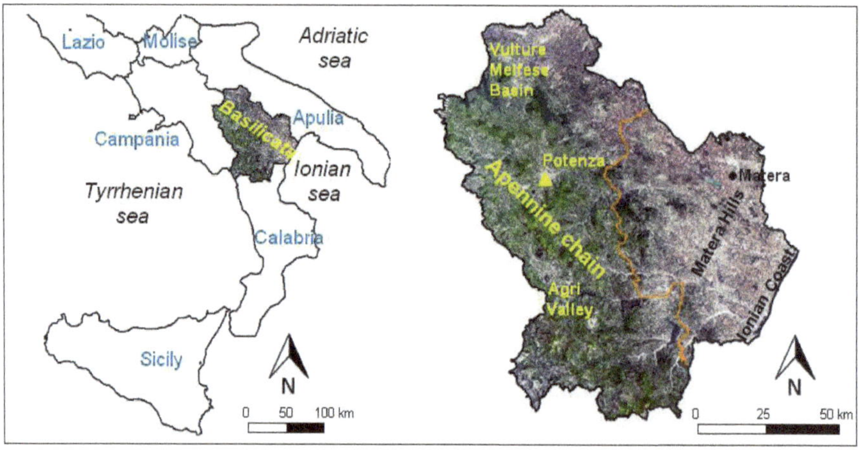

Figure 1. Location of the study area within Southern Italy and its main placenames

From a geographic point of view, Basilicata is a mountain region, including only a small percentage of lowland (less than 10% of the total surface) in the Ionian coastal area.

In the study area, soils often show a high susceptibility to degradation due to different causes. In the Ionian coastal area (Metaponto plain) we find soils affected by salinization phenomena caused both by coastline regression and by an incorrect agro-forestry management [43,44]; in the Central-Eastern hills, soils show singular geo-mineralogical composition, irregular morphology and are exposed to strong climatic fluctuations shaping the badlands (see e.g., [45,46]).

Vegetation is highly heterogeneous according to the different orography: dense and widespread vegetation in the central area, occupied by the Apennine chain, where broad-leaved forests, maquis and pastures are dominant; sparse vegetation and bare soils in the Eastern part of the region. On the Ionian coast several irrigation schemes enable a diversified agriculture including different cultivation types: orchards, permanent crops and arable lands. These last are also prevalent in the Northern zone, near to the Apulia region.

The Basilicata region is not univocally classified in a single climatic zone. Along the coasts climate is typically Mediterranean (rainy and mild autumns-winters, hot and dry summers) while the mountain areas are characterized by cold winters and by abundant precipitations; finally, inland areas, (Melfi industrial area, Basento valley and Agri valley), are characterized by very warm summers and mild winters with annual rainfall lower than 600 mm. In these areas, the period 1994-2003 has shown a significant decrease of the average annual and winter precipitation compared with the precipitation observed from 1916 to 1980s [47] thus evidencing an increase of dryness also in the wettest periods of the year.

The specific geomorphological characteristics of this region and a limited infrastructure network determine the concentration of industrial districts in small dedicated areas (Melfi area, Basento valley and Agri valley area). At now the tertiary is the prevalent economic sector. In the agriculture sector, though farms and cultivated lands decreased in the last decade (-31.9% and -4.7 respectively, [48]), the number of employees is still very high (about one fifth of the total employees, [49]).

Intensive and often inadequate farming practices have worsened degradation phenomena under way especially where climatic conditions are particularly unfavorable (e.g. badlands, [50]); mountainous areas have experienced a remarkable dynamism in the zootechnical sector, with a net increase in the number of head of cattle and in the size of farms.

3. Data

3.1. Satellite data

In order to evaluate the state of vegetation cover and its variations we used a vegetation index time series (2000-2010) acquired by the MODIS (Moderate Resolution Imaging Spectroradiometer) sensor. We analyzed NDVI (Normalized Difference Vegetation Index) values available at full spatial resolution (250m) as 16-day composite from the MODIS dataset by NASA LP DAAC (Land Processes Distributed Active Archive Center). Among different vegetation indices available in literature, NDVI is one of the best-known and best-working indices, and is recognized as a suitable proxy for vegetation activity. It is defined as the ratio [51,52]:

$$NDVI = \frac{NIR - RED}{NIR + RED} \tag{1}$$

where RED is the reflectance in the red band of the sensor and NIR is the reflectance in the near infrared band. NDVI takes values between -1 and 1; negative values indicate water and thick clouds, very low positive values correspond to barren areas (mainly rock, sand) or snow cover, whereas high positive values correspond to vigorous and healthy vegetation cover (Fig. 2).

The choice of MODIS sensor has been determined by its peculiar characteristics. High temporal resolution (2 images per day), moderate spatial resolution (250m), and the availability of a time series since 2000 make it suitable for monitoring vegetation variability at the national/regional scale. Furthermore, MODIS data are widely used to analyze vegetation conditions in the context of land degradation studies [53-56].

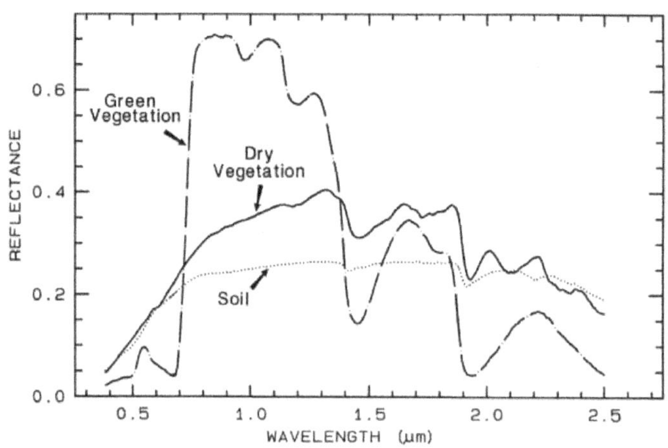

Figure 2. Spectral reflectance of natural surfaces (see http://bluemarble.ch/wordpress/2003/01/07/)

3.2. Census data

In order to estimate anthropic pressure indicators we extracted information from census database. The main source has been the Agricultural Census carried out by ISTAT (Italian National Institute of Statistics) for the years 1990 and 2000 (latest available census). Data are provided by municipality (i.e., the minimum administrative level) for the Basilicata region.

In particular, we gathered data on:

- Utilized Agricultural Area (UAA, years 1990 and 2000);

- Permanent grass and Pasture areas (PP, year 2000);
- Number of heads of cattle (bovines, buffalos, sheep, goats and equines, year 2000).

3.3. Ancillary data

For the elaboration of the Mechanization Level Index (MLI), we used the following ancillary data:

- level-3 Corine Land Cover (CLC) 2000 map (Fig. 3), downloaded from the High Institute for Environment Protection and Research (ISPRA - former APAT, see http://www.eea.europa.eu/data-and-maps/data/corine-land-cover-2000-clc2000-seamless-vector-database-4);
- number of machinery passes per cultivation type (source ENAMA – Italian National Agency of Agricultural Mechanization);
- 20m resolution DEM (Digital Elevation Model, Fig. 4) of the Basilicata provided by the Basin Authority of the Region.

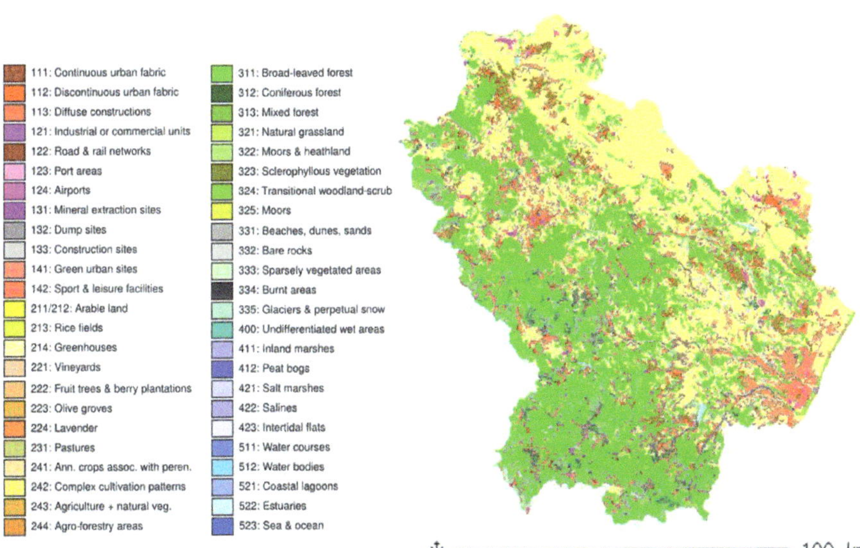

Figure 3. CLC map for Basilicata region

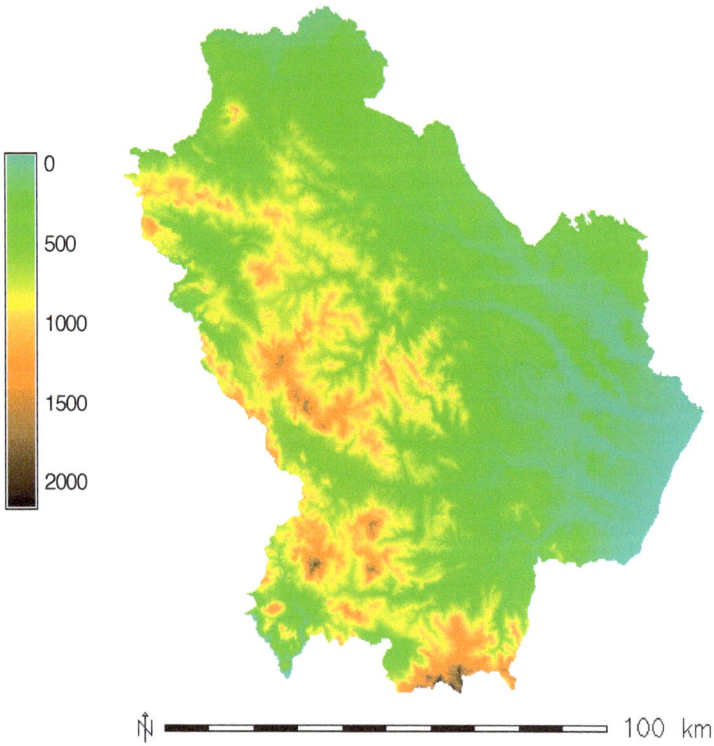

Figure 4. Digital Elevation Model for Basilicata region

4. Methodological procedure

4.1. Estimation of the vulnerability due to anthropic factors

In the last years, despite scientists have paid much attention to anthropogenic factors as potential land degradation drivers [57,34], the socio-economic component still remains difficult to explore. The main problems are related to the qualitative character, the strong spatial aggregation, and the infrequent update of the information [58]. Our approach takes into account the so called "agricultural impact" hypothesis [59] as potential explanation for the most part of the land degradation processes, by focusing on crop intensification/land abandonment and overgrazing in Southern Italy. Among the indicators already adopted in similar studies [60-62], we selected the following ones: variation of cultivated surfaces, percentage of permanent grass and pasture on the total agricultural area, grazing intensity and mechanization level. The first three indicators are based on census data, the last is calculated combining information on land cover and the other ancillary data.

According to the ESA model, in order to make the used indicators comparable, we classified them in a common range of vulnerability levels starting from 1 (the lowest vulnerability to land degradation) up to 2 (the highest vulnerability to land degradation).

4.1.1. Census based indicators

The first indicator calculates the percentage variation of the cultivated surfaces (UAA_VAR) referred to a time horizon of ten years, as follows:

$$UAA\_VAR = \left| \frac{UAA_{t_2} - UAA_{t_1}}{UAA_{t_1}} \cdot 100 \right| \quad (2)$$

where UAA_{t_1} and UAA_{t_2} are the Utilized Agricultural Area (arable land, permanent grassland, permanent crops and other agricultural land such as kitchen gardens, see http://epp.eurostat.ec.europa.eu/statistics_explained/index.php/Category:Agriculture_glossary) at the start and at the end of the investigated period (t_1=1990 and t_2=2000 in this study). The absolute value makes this indicator a good proxy both for agricultural intensification and land abandonment (Table 1).

Vulnerability class	UAA_VAR Values	
	Decreases	Increases
(2) high	< - 50	> 50
medium - high	-50 : -20	20 : 50
medium	-20 : -10	10 : 20
medium - low	-10 : -5	5 : 10
(1) low	-5 : 5	-5 : 5

Table 1. Distribution of vulnerability classes for the index of agricultural area variations (UAA_VAR)

In fact, both these processes are considered potential land degradation drivers: the increase in cultivated surfaces means a reduction in natural lands and requires additional inputs (water resources, fertilizers, tilling, etc.) that strongly impact on the environment; on the other hand, the decrease in cultivated areas is associated to the abandonment of marginal lands (lack of maintenance of drainage network, terracing, etc.) causing acceleration of degradation [63,64], or urbanization/industrialization phenomena with consequent soil sealing and pollution.

The second indicator estimates the percentage of Permanent grass and Pasture surfaces (*Sur_PP*) with respect to the total Utilized Agricultural Area (UAA) according to this formula:

$$PP\_UAA = \frac{Sur\_PP}{UAA} \cdot 100 \tag{3}$$

The rationale behind this indicator is the basic assumption that grass and pasture can be considered low-impact covers because they do not require considerable amount of external input (fertilizers, herbicides, mechanization and irrigation scheme), accomplishing an important protection function against erosional processes [61]. Therefore, the higher the indicator value, the lower the vulnerability level (Table 2).

Vulnerability class	PP_UAA Values
(2) high	< 5
medium - high	5 : 10
medium	10 : 30
medium - low	30 : 50
(1) low	50 : 100

Table 2. Distribution of vulnerability classes for the percentage of permanent grass and pasture on the Utilized Agricultural Area (PP_UAA)

The third indicator is used to estimate the Grazing Intensity (GI), by evaluating the amount of Adult Bovine Unit (ABU) on the total area of permanent grass and pasture (expressed in hectares), as follows:

$$GI = \frac{ABU}{Sur\_PP} \tag{4}$$

where ABU is computed accounting for the unit number of various livestock types (referred to the 2000 year), homogenizing them to the size of adult bovine [60]:

$$\text{ABU} = \text{n.bovines} + \text{n. buffalos} + \text{n. equines} + \frac{\text{n. goats}}{10} + \frac{\text{n. sheep}}{10} \tag{5}$$

Overgrazing remains a typical driver of degradation in many areas of Southern Italy, resulting from the inappropriate practice of grazing too many livestock for too long periods exceeding the productive capacity of the considered areas. Livestock hooves remove vegetation cover, exposing soil to be washed away and reducing its capacity of water storage, previously facilitated by vegetation [65]. As additional effects, soil compaction arises and runoff increases. On this basis, the highest vulnerability scores are associated to the highest values of the indicator (Table 3).

Vulnerability class	GI Values
(2) high	> 100
medium - high	30 : 100
medium	10 : 30
medium - low	3 : 10
(1) low	0 : 3

Table 3. Distribution of vulnerability classes for grazing intensity (GI)

4.1.2. Land cover based indicator

The index of mechanization level is a proxy for soil compaction due to heavy equipments used in agriculture. Multiple passes of machinery on the same lanes facilitate the formation of a compacted layer of soil (ploughsole) with a severe deterioration of many soil properties, such as porosity, hydraulic conductivity and root penetration [66-68]. The plant roots often spread out horizontally exhibiting stunted growth because of the insufficient access to soil water and nutrients [69]. Altogether, mechanization can increase risk of runoff [70], flood events and loss of nutrients by leaching [71].

The mechanization level index adopted in this work follows a new formulation based on land cover and morphological data [36], so as to obtain information more flexible for resolution, update frequency, and quality compared to census data, which are normally used to calculate this indicator [72,73]. Our indicator estimates soil compaction due to heavy vehicle traffic by taking into account the variable number of passes for each cultivation type (extracted from the land cover map and ancillary information) and the different impact on soil produced by using tyres or tracks (evaluated thanks to morphological data).

As a first step, starting from level-3 CLC we separated cultivable from natural or anthropized classes. Then we associated an average number of passes, obtained from the aggregation of ENAMA data (Table 4), for each agricultural CLC class.

Cultivation type and corresponding CLC2000 level3 code	Number of average passes
Arable land (cereals, legumes, crops, vegetables, etc.) - 2.1.1/2.1.2	7,5
Permanent crops (vineyards, fruit trees, olive groves) - 2.2.1/2.2.2/2.2.3	7
Pastures - 2.3.1	3
Annual crops associated with permanent crops - 2.4.1	5
Complex cultivation patterns - 2.4.2	4
Land principally occupied by agriculture, with natural areas - 2.4.3	3
Agroforestry areas - 2.4.4	1
Other classes	0

Table 4. Number of average passes for CLC2000 class, obtained aggregating ENAMA data for cultivation type.

In order to take into account the different equipments of the agricultural machinery, consisting in tyres or tracks, we applied a threshold (20%) on the slope map derived from the 20m resolution DEM since land on steep slope can be managed only by tracked vehicles, whereas tyres are adopted in all the other cases. Soil compaction induced by tracks is limited to the topsoil, that can be rather easily restored, whereas tyres mostly damage subsoil layers that are more difficult to restore [74,75]. Neglecting such a variable means to estimate equal vulnerability levels in very different conditions of soil tillage. According to this evaluation, we introduced a correction factor (f) associating a lower vulnerability to areas where tracked vehicles are used (f =1) with respect to those managed with tyred vehicles (f =1.5). The final formulation of the index (MLI) is the following:

$$MLI = N_p \cdot f \tag{6}$$

where Np is the number of average passes for each CLC class, and f represents the correction factor accounting for track or tyre use. The indicator was classified within the ESA range (1-2) to provide values comparable with the values of other land management indicators (Table 5).

Vulnerability class	MLI Values
(2) high	>9
medium - high	7 : 9
medium	5 : 7
medium - low	3 : 5
(1) low	<3

Table 5. Distribution of vulnerability classes for mechanization level indicator at pixel scale (MLI)

4.1.3. Land management index

The overall land management index (LMI) is calculated for each pixel as the geometric mean of the scores of the four indicators previously described:

$$LMI = (MLI \times UAA\_VAR \times PP\_UAA \times GI)^{1/4} \tag{7}$$

4.2. Estimation of the vulnerability due to vegetation component

The ESA model is devised to assess only the structural (potential) vulnerability to land degradation, which is connected, in the specific case of vegetation, to the different sensitivity of the different land cover classes. Nevertheless, it is frequent to detect areas showing similar vulnerability levels from a structural point of view and exhibiting, on the contrary, very different actual signs of degradation. In addition, vegetation conditions change in time and this temporal evolution can be very interesting for singling out degradation processes. Thus, moving from the assumption that land degradation should not be regarded as something static but as a dynamic process [76], multitemporal investigations using satellite time series can be profitably used for estimating not only the current state of vegetation but also the changes occurred over time. At this aim, in this chapter, we used NDVI_PV, already adopted by APAT [77], as a reliable indicator to carry out a multitemporal analysis of the vegetation activity [78].

4.2.1. NDVI_PV indicator

NDVI_PV provides the spatial variability of the changes in the study area at the satellite resolution and is based on the estimation of NDVI interannual variations compared with the starting conditions. It is calculated as follows:

$$NDVI - PV = \frac{\left[\dfrac{Y\sum_{i=1}^{Y} MVC_{p,i} \cdot y_i - \sum_{i=1}^{Y} MVC_{p,i} \sum_{i=1}^{Y} y_i}{Y\sum_{i=1}^{Y} y_i^2 - \left(\sum_{i=1}^{Y} y_i\right)^2}\right] Y}{MVC_{p,in}} \tag{8}$$

where Y = the number of years (11 in this work); y_i = given year; $MVC_{p,i}$ = Maximum Value Composite for the given pixel and year i; $MVC_{p,in}$ = Maximum Value Composite for the given pixel at the first year of the investigated time series.

The normalization to the initial value reported in the formula takes into consideration that the vulnerability of an area is strongly linked to the starting value and to the type of vegetation cover corresponding to different typical values of NDVI. This aspect is particularly important, because the same change (trend magnitudo and direction) has a different weight if the examined cover is a densely or sparsely vegetated. Therefore, the percentage variation rather than the absolute values allows for better estimating degradation levels. This indicator is able to enhance increase/decrease of vegetation activity and to identify slow variations, long-term processes (e.g., decline of forest areas), and sudden changes (e.g., fire events).

Finally, the NDVI_PV indicator has been classified within the ESA range 1-2 (Table 6).

Vulnerability class	NDVI_PV values
high	<-20
medium- high	-10 : -20
medium	-5 : -10
medium -low	0 : -5
low	>0

Table 6. Distribution of vulnerability classes for NDVI_PV indicator.

4.3. Integration of the anthropic and vegetation components

In order to take into account the information provided by the evaluation of the anthropogenic and vegetation components (LMI and NDVI_PV), we integrated them through the geometric mean. We defined a modified index based on the ESA final index [32]:

$$ESA_{mod} = (NDVI\_PV \times LMI)^{1/2} \quad (9)$$

4.4. Main contributing factor

Once defined the different vulnerability levels of a composite index, it is possible to identify spatial patterns of the main contributing factor (MCF) so as to point out the prevalent driving forces acting at pixel scale on the ongoing degradation processes. This is strategic to address ad hoc measures of conservation/mitigation/rehabilitation towards the specific involved factors. In GIS environment such an analysis is carried out by means of a simple maximizing algorithm applied on the comparable layers (rasters) representing each land management indicator:

$$OUTPUT = MAX(RASTER\,1, RASTER\,2, RASTER\,3, \ldots RASTER\,N) \quad (10)$$

The output raster shows the spatial dominance of one factor with respect to the other ones.

5. Results

5.1. Analysis of the land management indicators

Among the anthropic indicators, the highest vulnerability values were found for the UAA_VAR indicator (Fig. 5). Most of the vulnerable municipalities seem to be equally distributed in the study area, confirming that the abandonment of marginal lands (especially in inland areas), and the agriculture intensification (in lowlands and along the Ionian coast) represent important human-induced causes of degradation for Basilicata region [79-81].

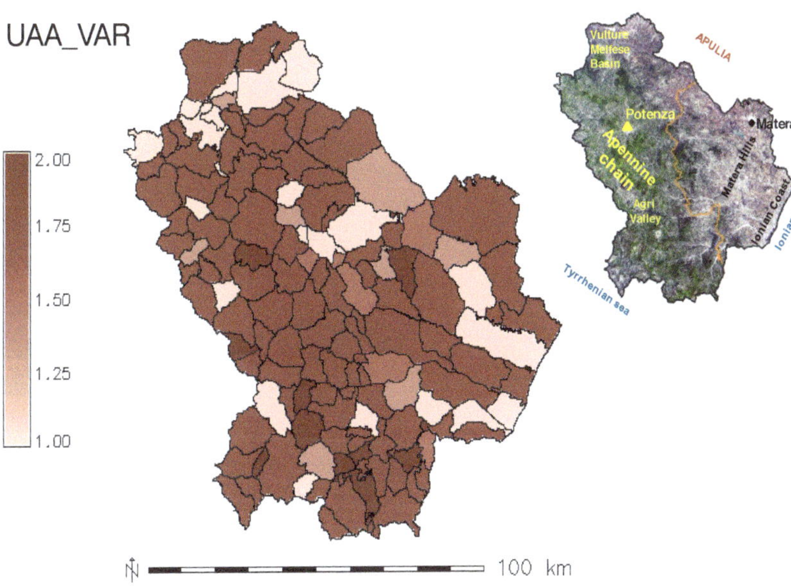

Figure 5. Classification of UAA_VAR in vulnerability classes. In the upper right corner it is shown the geographical reference map

As far as PP_UAA is concerned (Fig. 6), this is an important vulnerability factor only for a limited number of municipalities. In these areas, UAA is prevalently devoted to intensive farming activities (permanent crops, arable lands and heterogeneous agricultural areas) rather than to less-impacting practices that are normally carried out in grass, pasture and agroforestry areas; conversely, the Apennine and sub-Apennine zones show medium-low or low values of vulnerability, because the municipal UAA encompasses a fairly considerable proportion of grass and pasture (see http://censagr.istat.it/basilicata.pdf).

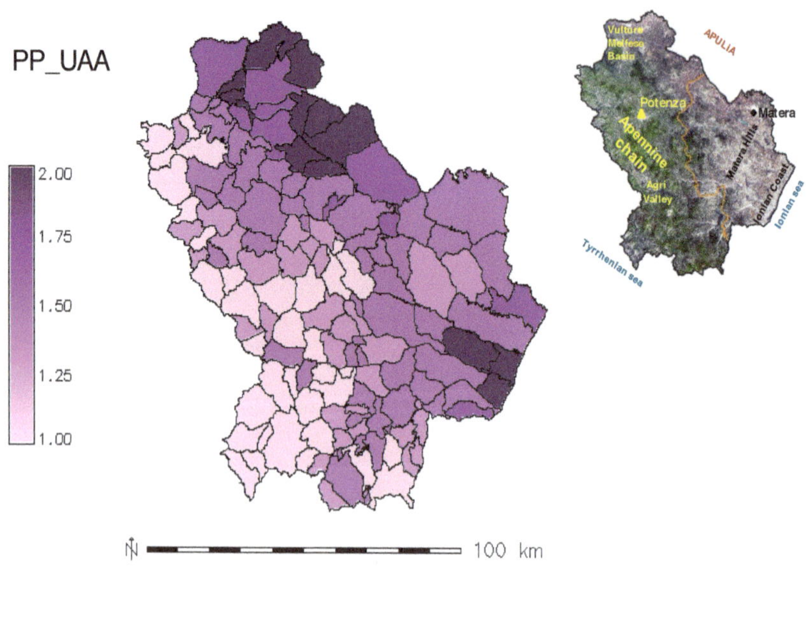

Figure 6. Classification of PP_UAA in vulnerability classes. In the upper right corner it is shown the geographical reference map

The vulnerability map of Grazing Intensity (GI - Fig. 7) reveals at a glance that the least impacting degradation factor in Basilicata region is overgrazing, because we found high vulnerability values only in a very few municipalities, whereas the rest of the examined areas shows prevalently low vulnerability values.

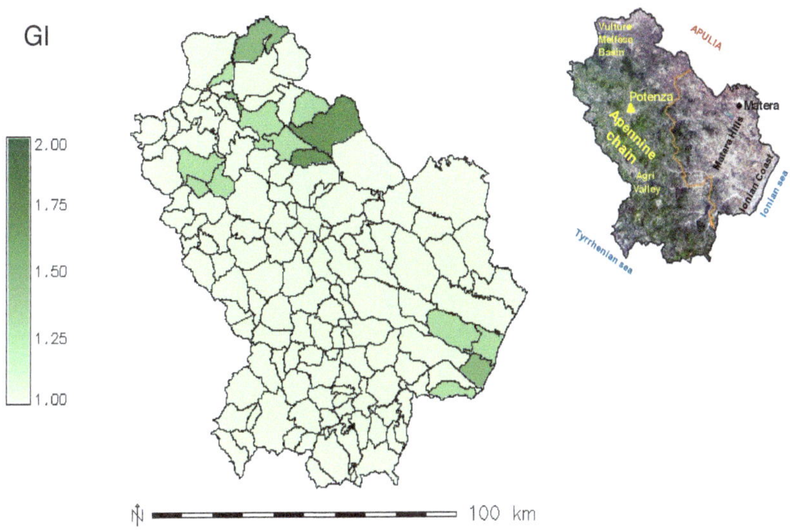

Figure 7. Classification of GI in vulnerability classes. In the upper right corner it is shown the geographical reference map

This agrees with the indications inferred from the previous indicators: even though livestock husbandry is a well-established economic platform comprising a large number of small to medium size enterprises in Basilicata (also in mountainous areas), the fairly even abundance of pastures and grasses allows to graze without exceeding the regeneration capacity of vegetation. As illustrated in Fig. 8, the mechanization level indicator (MLI), which is displayed with the spatial resolution of the pixel (20m as the original DEM), allows a quick discrimination of different vulnerability values also inside the municipal areas.

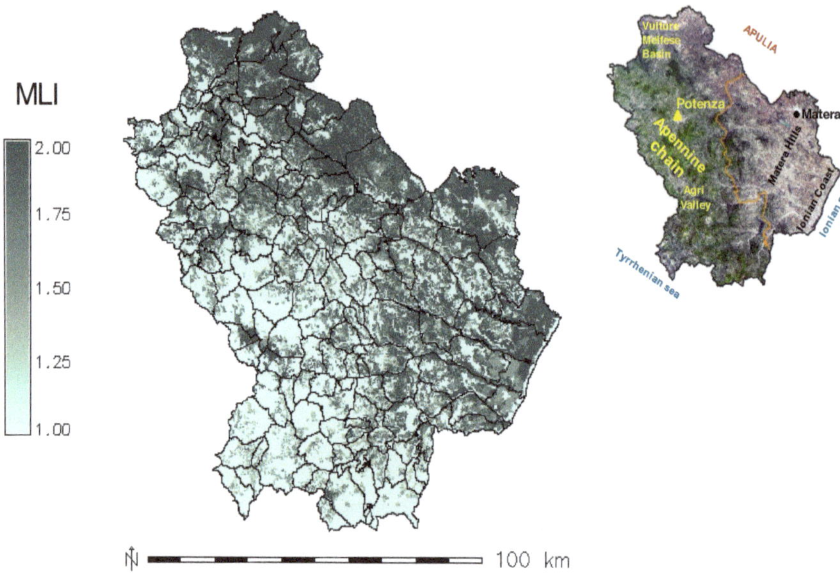

Figure 8. Classification of MLI in vulnerability classes. In the upper right corner it is shown the geographical reference map

This is a first improvement with respect to previous analyses made at the municipal level, enabling a better identification of the local critical aspects in terms of induced environmental impacts. In particular, the arrangement of the vulnerable areas reflects the agricultural productivity patterns of Basilicata, providing a picture of the actual conditions of the investigated region which is more realistic of that provided by census-based indicators [82].

We found high and medium-high vulnerability for areas located in lowlands (wide stripe in the Northeastern part of the region) and along the coast as well as in a large part of the hilly landscape (e.g., medium and low hills surrounding the city of Matera), which is particularly devoted to (intensive) farming practices; low vulnerability levels are found instead in mountain areas, less suitable to be exploited for agricultural purposes.

Finally, the Land Management Index (LMI), exhibiting the same resolution of the MLI indicator, is shown in Fig. 9. It is evident that the most severe management problems related to agriculture/grazing activities are concentrated in the cluster in the Northeastern part of the region and in some of the coastal areas along the Ionian sea. The rest of the seaboards are characterized by medium/medium-high levels of vulnerability as well as hilly areas in the Matera province and some areas surrounding the city of Potenza. The management state for the Western side of the region, dominated by natural areas, is quite satisfactory, even if there are patches having medium vulnerability values (Vulture-Melfese and Agri valley).

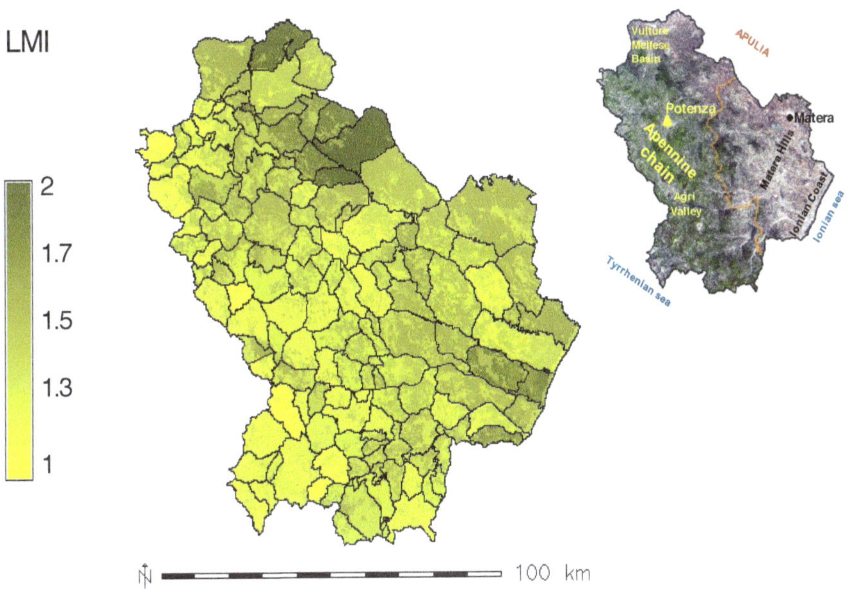

Figure 9. Classification of LMI in vulnerability classes. In the upper right corner it is shown the geographical reference map

5.2. Spatial pattern of Main Contributing Factors (MCF) related to anthropic pressure

We performed a preliminary analysis consisting in area-weighted average calculations of the adopted indicators (see radar chart, Fig. 10). According to our results, UAA_VAR shows the highest average value (1,57). Also MLI and PP_UAA are not negligible (respectively 1,45 and 1,42) whereas the role of GI seems to be nonessential (1,05).

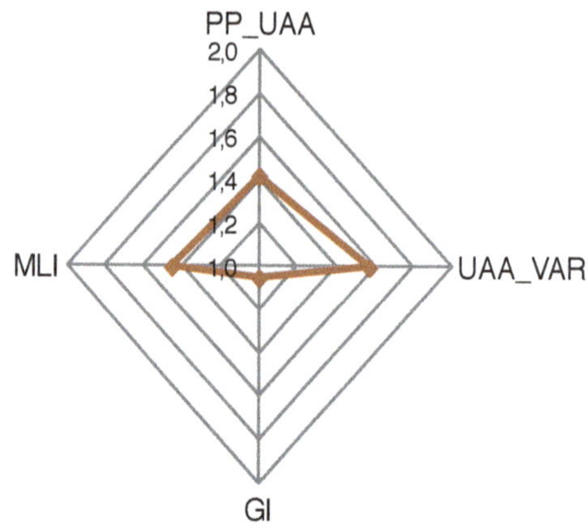

Figure 10. Radar chart showing the comparison among the area-weighted average values of land management indicators for the whole investigated region

In order to investigate the role of each indicator we applied the MCF algorithm (see section 4.4) at the pixel scale. It should be remarked that (see Fig. 11) 70% of the regional surface shows a unique MFC, while the remaining part of the investigated areas is characterized by two (about 24% of the total surface), three indicators (about 4% of the total surface), or no prevailing indicator (about 2% of the total surface). In the last case all the four indicators reach the maximum vulnerability value.

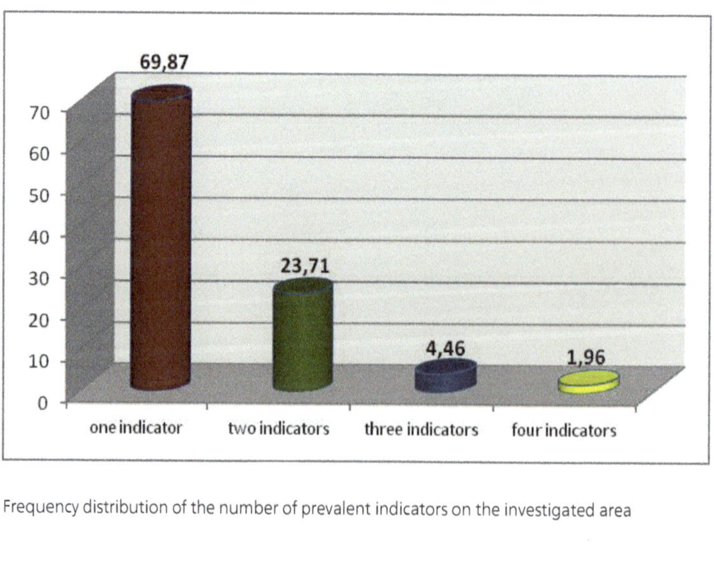

Figure 11. Frequency distribution of the number of prevalent indicators on the investigated area

The analysis of the areas in which just one indicator is dominant (Fig. 12) brings out the importance of the UAA_VAR as the most significant driver of degradation (about 58% of the considered area). In these areas the degradation mainly comes from the decrease in cultivated surfaces.

Apart from the appreciable contribution of the mechanization indicator (MLI, about 29% of the examined area), neither the scarce presence of grass and pasture (PP_UAA, about 13% of the examined area) nor the overgrazing (GI, no area involved) contribute meaningfully to degradation.

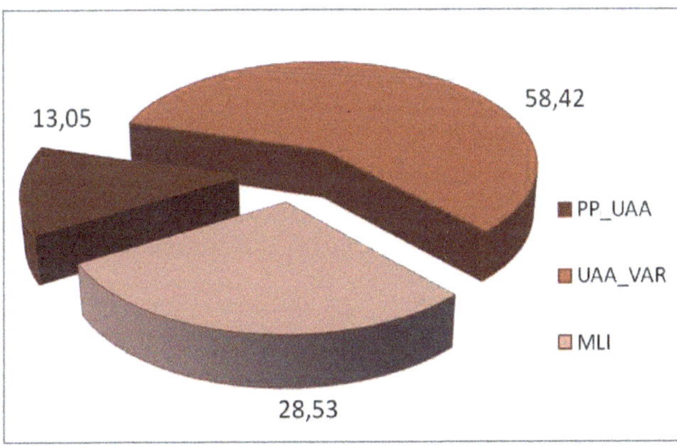

Figure 12. Frequency distribution of the prevalent indicators (areas having just one indicator prevalent)

The analysis of the pixels having two dominant indicators (Fig. 13) shows a large prevalence of the synergy between MLI and PP_SAU (about 75%). On the contrary, the variation of cultivated lands (UAA_VAR) jointly with PP_UAA or MLI (respectively about 17% and 9% of analyzed areas) seems not to be particularly diffused as a degradation driver. Owing to the negligible role of grazing, areas exhibiting simultaneously three dominant indicators are always characterized by the values of MLI, PP_UAA, and UAA_VAR.

On the whole, the analysis aimed at identifying the MCF for the anthropic component indicates that UAA_VAR plays the main role in inducing degradation followed by excessive mechanization (MLI), whereas PP_UAA and particularly GI seem not to play an important role in promoting environmental degradation. This last result is due to the positive effects generated by the widespread presence of grass and pasture, also in non mountainous areas. These covers represent a mainstay of the local agricultural structure enabling a sustainable management because, on the one hand, they counterbalance the man-induced impact caused by intensive agricultural practices (resulting in lower values of the PP_UAA indicator), on the other, they allow a suitable form of grazing (resulting in very low values of the GI indicator).

The spatial patterns of the MCF (Fig. 14) show two opposite paths in the Basilicata region: marginalization of inland rural areas and further intensification of low-sustainable agriculture in lowland areas.

The first phenomenon, arising from complex socio-economic dynamics, involves the inland districts located in the core of the region (prevalently near Potenza town) that were mainly devoted to poor agricultural practices in the recent past. Today, these areas experience depopulation (for further details see http://www.istat.it/it/basilicata) as a consequence of the present economic crisis generating low profitability of agricultural products. This reduction in profit margin, in turn, can be accelerated by natural factors such as growing aridity and natural disasters (flood, landslide, fire, etc.) which induce an increase in agricultural management costs (e.g., irrigation, agrochemicals products, land rehabilitation, etc.) exacerbating land abandonment and culminating in a downward spiral of land degradation [83]. This fact, supported by provisional data of the Sixth National Agricultural Census (indicating a reduction of farm and cultivated areas, see section 2), stresses one of the most critical aspect of the local economic-productive system having serious repercussions on environmental quality and promoting social imbalances between marginal and more populated areas [84]. However, in this case, regional/national policies should be undertaken to strengthen infrastructural facilities and promote the redevelopment of marginal lands.

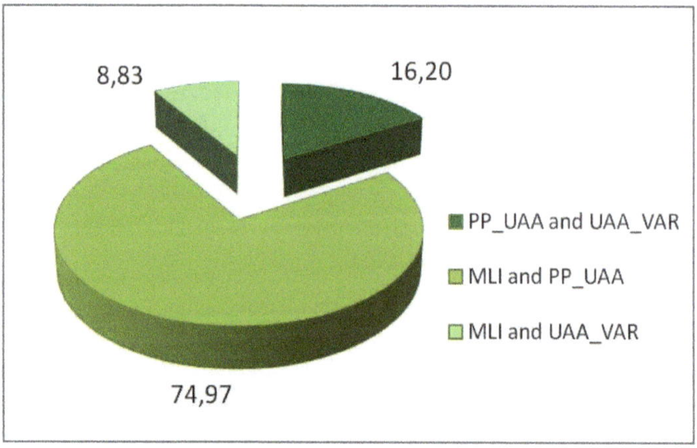

Figure 13. Frequency distribution of the prevalent indicators (areas having two indicators prevalent)

The second phenomenon focuses on the long-term sustainability of intensive farming. Especially in areas where the natural conditions are optimal (e.g., slope) and technologies and infrastructure are easily available, we notice a tendency to increase agricultural production. This occurs at the expense of future land fertility, because enlarging cultivated areas, increasing the use of mechanization and fertilizers and overexploiting water resources contributes to exacerbate land degradation processes. In these places, we observe the reverse problem affecting marginal areas and thus appropriate strategies are required to locally encourage farmers towards sustainable soil management practices and technical skill improvement.

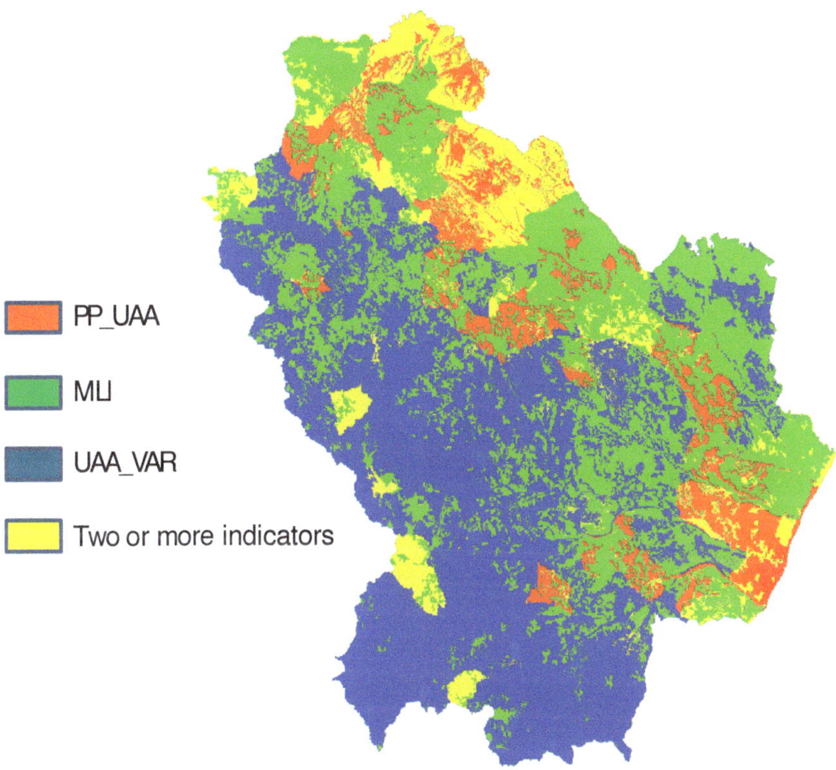

Figure 14. Map of the Main Contributing Factor (MCF) computed for the anthropic component

5.3. Analysis of trends in photosynthetic activity (NDVI_PV)

In Fig. 15 absolute values of NDVI_PV are displayed. Positive values of the indicator (generally fairly high) are visible especially in areas located south of Matera city and they mainly are estimated for permanent crops (fruit trees and olive groves) and, in some cases, for arable lands. Areas mostly characterized by dense vegetation (coniferous and broad-leaved forests) reveal stability or a slight increase in photosynthetic activity. Negative values are detected in correspondence with arable lands (the narrow stripe bordering Apulia region) and industrial districts (geographically concentrated in Tito Scalo, near Potenza and in S. Nicola di Melfi at the northern of Basilicata, where we find one of the most recent FIAT plant, see Fig. 15).

By aggregating the NDVI_PV values in 7 ranges (see Fig. 16) we observe a considerable coverage of stable areas (more than 50%) and a limited extent of areas characterized by low negative values (10%). Areas affected by a strong decrease in vegetation activity are only 1% of the investigated territory; on the contrary, areas marked by positive trends (slight and appreciable increases in photosynthetic activity) altogether amount to 30% of the examined surfaces.

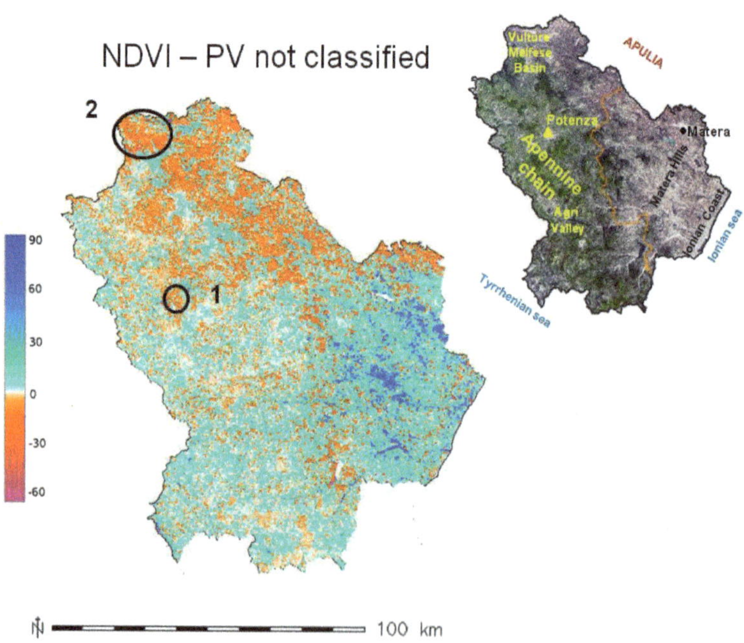

Figure 15. Map of the indicator NDVI_PV (not classified). Areas within the circles 1 and 2 belong to the Tito Scalo and San Nicola di Melfi locations respectively. In the upper right corner it is shown the geographical reference map

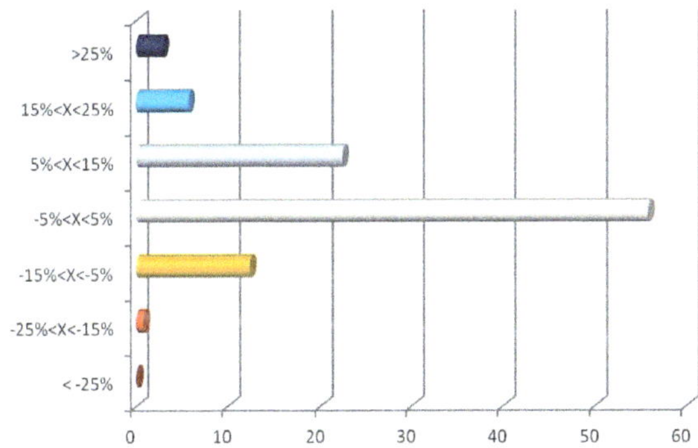

Figure 16. Frequency distribution of the prevalent indicators (values in abscissa represent the percentage of areas included in the given ranges)

By classifying the obtained values of NDVI_PV in the ESA range (1-2), we can extract some further information: highly vulnerable areas (medium-high and high) reach about 5% of the Basilicata surface; there are few medium vulnerability areas (about 10%), whereas the extent of areas with medium-low/low vulnerability is very significant (about 85%, see Fig. 17).

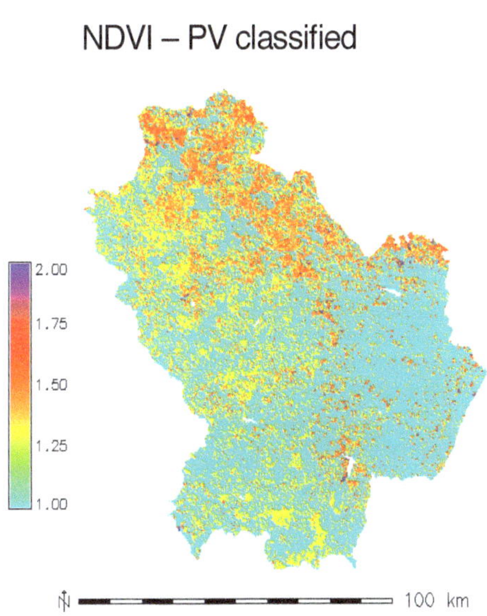

Figure 17. Map of the indicator NDVI_PV, classified in the ESA range (1-2)

5.4. Analysis of the integrated vulnerability map (ESA_{mod})

As we can see from the map in Fig. 18, the combined analysis of the anthropic component and the vegetation one, does not show a particularly critical picture of the Basilicata region. The most vulnerable areas (ESA_{mod}>1.5) are located, as expected, in the Northeastern sector of the region, including the agriculture-oriented lands bordering Apulia region, a part of the Ionian coast and some areas belonging to the hilly zone in the surrounding of Matera city. More densely vegetated areas, but also a large part of grasses, pastures and semi-natural areas, where the anthropic influence is clearly lower, seem to show good health conditions and thus a rather negligible vulnerability.

As established by the ESA methodology, the arrangement of the examined areas in different risk classes points out that about 23% of the region is included in the critical areas (ESA_{mod} > 1.38) and nearly the 30% in the fragile (1.23<ESA_{mod}<1.37); the rest of the investigated territory is characterized by potential or non-threatened areas (ESA_{mod}<1.22; 50% of the regional surface) according with results from independent studies [85]. The composite picture emerging from all these investigations suggests that for areas falling within the first two categories (critical and fragile) several measures should be put in place to prevent more severe degradation processes by promoting mitigation/restoration actions. As for the third category (potential and non-threatened areas), a periodic monitoring can be a great (and sometimes cost-effective) solution.

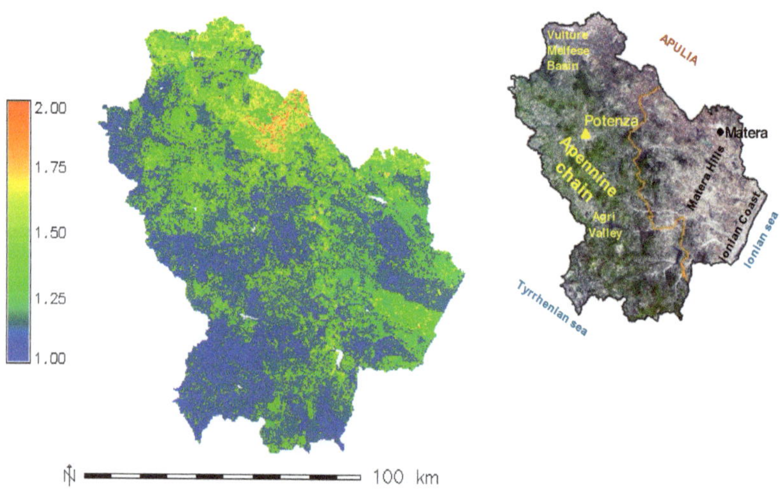

Figure 18. Map depicting the integration of the analyzed components (ESA_{mod}). In the upper right corner it is shown the geographical reference map

Finally, Fig.19 shows the ESA_{mod} map segmented according to four different levels of influence of MLI and NDVI_PV.

The extent of areas having both the anthropogenic component (LMI) and the biophysical one (NDVI_PV) not exceeding the value of 1.4 (vulnerability threshold) is very considerable (blue pixels). These pixels are principally concentrated in the Western side of the region and belong to various type of land cover including mainly forested and seminatural areas and some human-influenced covers such as arable lands. These last dominate, instead, in two of the four classes: areas showing both negative vegetation trends and inappropriate land management (red pixels), and areas affected by substantial decreases of photosynthetic activity (yellow pixels) but where management is quite satisfactory. Finally, a lot of permanent crops occupy largely those areas experiencing positive trends of vegetation activity but unsuitable agricultural practices (green pixels).

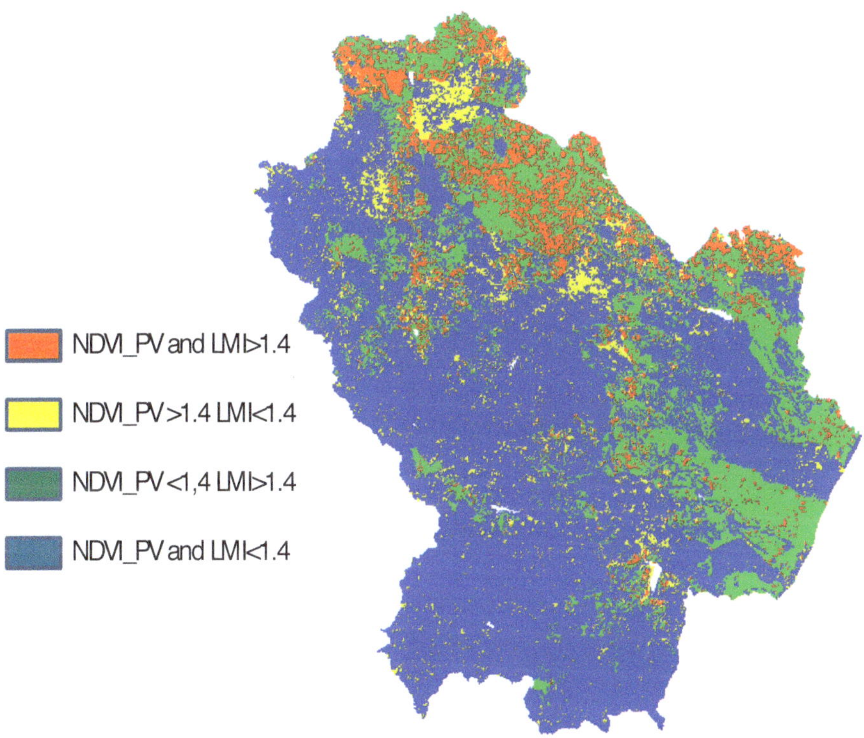

Figure 19. Zones of influence resulting from the partition of the ESA_{mod} map

6. Conclusions

In order to estimate the vulnerability to land degradation of a typical Mediterranean region (Basilicata) we have jointly considered the impact of the anthropic component and the vegetation conditions, using socio-economic indicators related to agriculture/grazing activities and analyzing trends of photosynthetic activity. As regards anthropic pressure we have used census-based indicators (UAA_VAR, PP_UAA and GI computed at municipal scale) and the mechanization indicator (MLI) based on land cover map and morphological information (DEM). Thanks to its formulation, the new indicator we elaborated is independent from census data, enabling a faster rate of update and providing a better discrimination of the vulnerability values because the adopted spatial resolution is connected to the used land cover map or DEM in state of the municipal level. It allows friendly exportability to different monitoring scales, which can be obtained by selecting the most opportune land cover map, and high adaptability, thanks to the possibility of selecting the number of classes for the satellite data classification.

We have combined all the socio-economic indicators to define the Land Management Index (LMI) and have carried out an analysis aimed at identifying the dominant factors driving human-induced degradation processes.

In order to estimate trends of vegetation activity we have calculated the NDVI_PV indicator using a time series (2000-2010) of the MODIS sensor observations. This indicator is able to compute interannual variations of NDVI compared with the starting conditions, so that it is possible to detect also slow variations and long-term processes of increase/decrease of the photosynthetic activity in the analyzed period.

The final map of the ESA_{mod} index, taking into account the vulnerability due to the anthropic and vegetation components, depicts a very complex picture characterized by a wide range of vulnerability values and by many combinations of degradation causes.

The adopted procedure, which integrates remote sensing data (synoptic view, multi-temporal availability) and socio-economic indicators, is a valuable tool for estimating vulnerability to land degradation in large anthropized areas, which are highly complex in terms of land cover type and economic vocation (intensive agriculture, grazing, industrial activities).

Our methodology allows the early detection of the most vulnerable areas and the identification of the local prevailing stress factors, providing key information for the setting up of sustainable development strategies.

Acknowledgments

Our activity was carried out in the framework of "Assessment methodologies for controlling land degradation processes and impacts on the environment" (Programma Operativo FESR Basilicata 2007-2013).

Author details

Vito Imbrenda[1*], Mariagrazia D'Emilio[1], Maria Lanfredi[1], Tiziana Simoniello[1], Maria Ragosta[2] and Maria Macchiato[3]

*Address all correspondence to: vito.imbrenda@imaa.cnr.it

1 National Research Council of Italy, Institute of Methodologies for the Environmental Analysis, CNR- IMAA, Tito Scalo (Pz), University of Basilicata, Dep. Environmental Engineering and Physics - DIFA, Potenza, DSF-University of Naples Federico II, Naples, Italy

2 University of Basilicata, Dep. Environmental Engineering and Physics - DIFA, Potenza, Italy

3 DSF-University of Naples Federico II, Naple, Italy

References

[1] Steffen W, Crutzen PJ, McNeill JR. The Anthropocene: Are Humans Now Overwhelming the Great Forces of Nature. AMBIO: A Journal of the Human Environment 2007;36(8) 614–21.

[2] UN/FAO. Report on Land Degradation Assessment in Drylands – LADA, 2003.

[3] Adeel Z, Safriel U, Neimeijer D, White R. Ecosystems and human well-being: desertification synthesis. Washington: World Resources Institute; 2005.

[4] Nachtergaele F, Petri M, Biancalani R, Van Lynden G, Van Velthuizen H. Global Land Degradation Information System (GLADIS). Beta Version. An Information database for Land Degradation Assessment at Global Level. LADA, 2010. Technical Report No.:17.

[5] Puigdefábregas J. Ecological impacts of global change on drylands and their implications for desertification. Land Degradation & Development 1998;9(5) 393–406.

[6] Costantini EAC, Bocci M, L'Abate G, Fais A, Leone G, Loj G, Magini S, Napoli R, Nino P, Urbano F. Mapping the state and risk of desertification in Italy by means of remote sensing, soil gis and the EPIC model. Methodology validation in the Sardinia island, Italy. International Symposium: Evaluation and Monitoring of Desertification. Synthetic Activities for the Contribution to UNCCD, February 2, 2004. Tsukuba, Ibaraki (Japan), NIES publication. 2004

[7] Lanfredi M, Simoniello T, Macchiato M. Temporal persistence in vegetation cover changes observed from satellite: Development of an estimation procedure in the test site of the Mediterranean Italy. Remote Sensing of Environment 2004;93(4) 565–76.

[8] Hill J, Stellmes M, Udelhoven T, Roder A, Sommer S. Mediterranean desertification and land degradation: Mapping related land use change syndromes based on satellite observations. Global and Planetary Change 2008;64(3-4) 146–57.

[9] Simoniello T, Lanfredi M, Liberti M, Coppola R, Macchiato M. Estimation of vegetation cover resilience from satellite time series. Hydrology and earth system sciences 2008;12(4) 1053–64.

[10] [10]Botta GF, Rivero D, Tourn M, Melcon FB, Pozzolo O, Nardon G, et al. Soil compaction produced by tractor with radial and cross-ply tyres in two tillage regimes. Soil and Tillage Research. 2008;101(1-2) 44–51.

[11] [11]Jefferies RL, Maron JL. The embarrassment of riches: atmospheric deposition of nitrogen and community and ecosystem processes. Trends in Ecology & Evolution 1997;12(2) 74–78.

[12] James C. Global status of commercialized transgenic crops (1999). ISAAA Brief 2000;17 1–78.

[13] Gay SH, Louwagie G, Sammeth F, Ratinger T, Marechal B, Prosperi P, Rusco E, Terres J, van der Velde M, et al. (Eds) Louwagie G, Gay SH, Burrell A. 2009. Final report on the project, Sustainable Agriculture and Soil Conservation (SoCo). Luxembourg, European Commission.

[14] UNEP. United Nations Environment Programme. Climate Change Science Compendium 2009. Nairobi: EarthPrint, 2009.

[15] FAO/UN. Land degradation assessment in drylands – LADA. Methodology and Results, Final draft, Rome, 2011.

[16] Yassoglou N J. Land, desertification vulnerability and management in Mediterranean landscapes. In: Balabanis P, Peter D, Ghazi A, Tzogas M. (eds.) Mediterranean desertification: Research results and policy implications. Luxembourg: European Commission – Directorate General Research, EUR 19303, 1999. p 87–113.

[17] Pitman AJ. The evolution of, and revolution in, land surface schemes designed for climate models. International Journal of Climatology 2003;23(5) 479–510.

[18] Betts A K, Ball J H, Beljaars A C M, Miller M J, Viterbo P A. The land surface–atmosphere interaction: A review based on observational and global modeling perspective. Journal of Geophysical Research 1996;101(D3) 7209–25.

[19] Balling R C Jr, Klopatek J M, Hilderbrandt M L, Moritz C K, Watts C J. Impacts of Land Degradation on Historical Temperature Records from the Sonoran Desert. Climatic Change 1998;40(3-4) 669–681.

[20] Lanfredi M, Lasaponara R, Simoniello T, Cuomo V, Macchiato M. Multiresolution spatial characterization of land degradation phenomena in Southern Italy from 1985

to 1999 using NOAA-AVHRR NDVI data. Geophysical Research Letters 2003;30(2), 41-44.

[21] del Barrio G, Puigdefabregas J, Sanjuan ME, Stellmes M, Ruiz A. Assessment and monitoring of land condition in the Iberian Peninsula, 1989–2000. Remote Sensing of Environment 2010;114(8) 1817–32.

[22] Keeley JE, Rundel PW. Fire and the Miocene expansion of C4 grasslands. Ecology Letters. 2005;8(7) 683–90.

[23] Shakesby RA, Wallbrink PJ, Doerr SH, English PM, Chafer CJ, Humphreys GS, et al. Distinctiveness of wildfire effects on soil erosion in south-east Australian eucalypt forests assessed in a global context. Forest Ecology and Management 2007;238(1–3) 347–64.

[24] Van Eetvelde V, Antrop M. Analyzing structural and functional changes of traditional landscapes-two examples from Southern France. Landscape and Urban Planning 2004;67(1–4) 79–95.

[25] Wood R, Handley J. Landscape Dynamics and the Management of Change. Landscape Research. 2001;26(1) 45–54.

[26] FAO, 2008. Land degradation on the rise – One fourth of the world's population affected says new study. http://www.fao.org/newsroom/en/news/2008/1000874/index.html. (accessed 30 October 2010).

[27] UNEP. United Nations Environment Programme. UNEP Year Book: New Science and Developments in our Changing Environment. Nairobi: EarthPrint, 2009.

[28] Eswaran H, Lal R, Reich PF. Land degradation: An overview. In: Bridges EM, Hannam ID, Oldeman LR, DeVries WTP, Scherr SJ, Sombatpanit S. (eds.) Response to Land Degradation. Enfield: Science Publishers Inc; 2001. p 20-35.

[29] Arribas A, Gallardo C, Gaertner MA, Castro M. Sensitivity of the Iberian Peninsula climate to a land degradation. Climate Dynamics 2003;20(5) 477–89.

[30] Myers N. Environmental refugees: a growing phenomenon of the 21st century. Philosophical Transactions of the Royal Society of London. Series B: Biological Sciences 2002;357(1420) 609 –13.

[31] Rubio JL, Bochet E. Desertification indicators as diagnosis criteria for desertification risk assessment in Europe. Journal of Arid Environments 1998;39(2) 113–20.

[32] Kosmas C, Kirkby M, Geeson N. Manual on: Key indicators of desertification and mapping environmentally sensitive areas to desertification. European Commission, Energy, Environment and Sustainable Development, EUR 18882; 1999.

[33] ARPAS - Agenzia Regionale per la Protezione dell'Ambiente della Sardegna. Progetto pilota di lotta alla desertificazione nelle cinque regioni italiane maggiormente a rischio: Regione Sardegna, 2009. Final report.

[34] Salvati L, Zitti M. Assessing the impact of ecological and economic factors on land degradation vulnerability through multiway analysis. Ecological Indicators 2009;9(2) 357–63.

[35] Contador JFL, Schnabel S, Gutiérrez AG, Fernández MP. Mapping sensitivity to land degradation in Extremadura. SW Spain. Land Degradation & Development 2009;20(2) 129–44.

[36] Imbrenda V, Simoniello T, Carone T, Coppola R, D'Emilio M, Lanfredi M, Liberti M, Macchiato M. A new index for the evaluation of land management in the framework of land degradation assessment. In: Maktav D. (ed.) Remote Sensing for a Changing Europe: proceedings of the 28th EARSeL Symposium and Workshops, 2-7 June 2008, Istanbul, p 535-542.

[37] Hostert P, Röder A, Hill J. Coupling spectral unmixing and trend analysis for monitoring of long-term vegetation dynamics in Mediterranean rangelands. Remote Sensing of Environment 2003;87(2–3) 183–97.

[38] Geymen A, Baz I. Monitoring urban growth and detecting land-cover changes on the Istanbul metropolitan area. Environmental Monitoring and Assessment 2008;136(1-3) 449–59.

[39] APAT-CNLSD La vulnerabilità alla desertificazione in Italia: raccolta, analisi, confronto e verifica delle procedure cartografiche di mappatura e degli indicatori a scala nazionale e locale. Manuali e linee guida, 2006. (Collection and analysis of land degradation maps in Italy) ©APAT, CRA- UCEA, ISBN 88-448-02010-4.

[40] Costantini EAC, Urbano F, Aramini G, Barbetti R, Bellino F, Bocci M, et al. Rationale and methods for compiling an atlas of desertification in Italy. Land Degradation & Development 2009;20(3) 261–76.

[41] Basso B, De Simone L, Cammarano D, Martin EC, Margiotta S, Grace PR, et al. Evaluating Responses to Land Degradation Mitigation Measures in Southern Italy. International Journal of Environmental Research 2012;6(2) 367–80.

[42] Salvati L, Carlucci M. Estimating land degradation risk for agriculture in Italy using an indirect approach. Ecological Economics 2010;69(3) 511–18.

[43] Ferrara A, Xiloyannis C. Monitoraggio ed azioni di lotta alla desertificazione nella regione mediterranea europea. DesertNet2 2008, Final Report.

[44] Satriani A, Loperte A, Imbrenda V, Lapenna V. Geoelectrical Surveys for Characterization of the Coastal Saltwater Intrusion in Metapontum Forest Reserve (Southern Italy). International Journal of Geophysics 2012, Article ID 238478, 8 pages. doi: 10.1155/2012/238478.

[45] Summa V, Tateo F, Medici L, Giannossi ML. The role of mineralogy, geochemostry and grain size in badland development in Pisticci (Basilicata, Southern Italy). Earth Surface Processes and Landforms 2007;32(7) 980–97.

[46] De Santis F, Giannossi ML, Medici L, Summa V, Tateo F. Impact of physico-chemical soil properties on erosion features in the Aliano area (Southern Italy). Catena 2010;81(2) 172–81.

[47] Caloiero T, Coscarelli R, Ferrari E, Mancini M. Precipitation change in Southern Italy linked to global scale oscillation indexes. Natural Hazards and Earth System Sciences 2011;11(6) 1683–94.

[48] ISTAT, 2011. Sesto Censimento generale dell'agricoltura - Risultati provvisori. http://www3.istat.it/salastampa/comunicati/non_calendario/20110705_00/ (accessed 20 June 2012).

[49] SVIMEZ, 2011. Rapporto SVIMEZ 2011 sull'economia del mezzogiorno – Introduzione e sintesi. http://www.svimez.it/ (accessed 20 June 2012).

[50] Piccarreta M, Capolongo D, Boenzi F, Bentivenga M. Implications of decadal changes in precipitation and land use policy to soil erosion in Basilicata, Italy. Catena 2006;65(2) 138–51.

[51] Rouse JW, Haas RH, Schell JA, Deering DW, Harlan JC. Monitoring the vernal advancements and retrogradation (green-wave effect) of nature vegetation. NASA/GSFC 1974, Final Report.

[52] Herrmann SM, Anyamba A, Tucker CJ. Recent trends in vegetation dynamics in the African Sahel and their relationship to climate. Global Environmental Change 2005;15(4) 394–404.

[53] Beck PSA, Atzberger C, Hogda KA, Johansen B, Skidmore AK. Improved monitoring of vegetation dynamics at very high latitudes: A new method using MODIS NDVI. Remote Sensing of Environment 2006 15;100(3) 321–34.

[54] Lin ML, Chen CW, Wang QB, Cao Y, Shih JY, Lee Y T, et al. Fuzzy model-based assessment and monitoring of desertification using MODIS satellite imagery. Engineering Computations 2009;26(7-8) 745–60.

[55] Jacquin A, Sheeren D, Lacombe JP. Vegetation cover degradation assessment in Madagascar savanna based on trend analysis of MODIS NDVI time series. International Journal of Applied Earth Observation and Geoinformation 2010;12 S3–S10.

[56] Veron SR, Paruelo JM. Desertification alters the response of vegetation to changes in precipitation. Journal of Applied Ecology 2010;47(6) 1233–41.

[57] Montanarella L. Trends in land degradation in Europe. In: Sivakumar MVK, Ndiangui N. (eds.) Climate and Land Degradation. Berlin: Springer; 2007. p. 83–104.

[58] Salvati L, Zitti M. Multivariate analysis of socio-economic indicators as a measure of sensitivity to land degradation in the ESA model. International Journal of Ecological Economics and Statistics 2009b;15 93-102.

[59] Wilson GA, Juntti M. Unravelling desertification: policies and actor networks in Southern Europe. Wageningen: Wageningen Academic Publishers; 2005.

[60] Motroni A, Canu S, Bianco G, Loj G. Environmentally sensitive areas to desertification in Sardinia. Servizio Agrometeorologico Regionale per la Sardegna 2004, Technical Report.

[61] Genesio L, Magno R, Capecchi V, Crisci A, Bottai L, Ferrari R, Angeli L, Gardin L. CNR - Ibimet 2004, Final Report of Desert-Net project – Pilot Action WP10.

[62] Salvati L, Zitti M. The Environmental «Risky» Region: Identifying Land Degradation Processes Through Integration of Socio-Economic and Ecological Indicators in a Multivariate Regionalization Model. Environmental Management 2009;44(5) 888–98.

[63] Kosmas C, Danalatos NG, Gerontidis S. The effect of land parameters on vegetation performance and degree of erosion under Mediterranean conditions. Catena 2000;40(1) 3–17.

[64] Zalidis G, Stamatiadis S, Takavakoglou V, Eskridge K, Misopolinos N. Impacts of agricultural practices on soil and water quality in the Mediterranean region and proposed assessment methodology. Agriculture Ecosystems & Environment 2002;88(2) 137–46.

[65] Zucca C, Canu A, Della Peruta R. Effects of land use and landscape on spatial distribution and morphological features of gullies in an agropastoral area in Sardinia (Italy). Catena 2006;68(2-3) 87–95.

[66] Rete rurale nazionale, 2010. Rapporto di applicazione della condizionalità in Italia. http://www.reterurale.it/flex/cm/pages/ServeBLOB.php/L/IT/IDPagina/3984 (accessed 20 June 2012).

[67] Soane B, van Ouwerkerk C. Soil Compaction in Crop Production. Developments in Agricultual Engineering, 11. Amsterdam: Elsevier; 1994.

[68] Raper RL. Agricultural traffic impacts on soil. Journal of Terramechanics. 2005;42(3-4) 259–80.

[69] Wolkowski RP, Lowery B. Soil compaction: Causes, Concerns, and Cures. Madison: Cooperative Extension Publishing; 2008.

[70] Fleige H, Horn R. Field experiments on the effect of soil compaction on soil properties, runoff, interflow and erosion. In: Horn R, Van den Akker JJH, Arvidsson J. (eds.) Advances in Geoecology. Reiskirchen: Catena; 2000. p 258-268.

[71] Van-Camp L, Bujarrabal B, Gentile AR, Jones RJA, Montanarella L, Olazabal C, Selvaradjou SK.. Reports of the Technical Working Groups Established under the Thematic Strategy for Soil Protection EUR 21319 EN/6; 2004. Luxembourg, Office for Official Publications of the European Communities.

[72] APAT (formely ANPA) 2000. Sviluppo di indicatori per il suolo e i siti contaminati, RTI CTN_SSC 1/2000, Italian National Environmental Protection Agency.

[73] APAT (formely ANPA) 2003. In: Annuario dei Dati Ambientali - APAT, sezione: Geosfera, Italian National Environmental Protection Agency, p 278-279.

[74] Pagliai M, Marsili A, Servadio P, Vignozzi N, Pellegrini S. Changes in some physical properties of a clay soil in Central Italy following the passage of rubber tracked and wheeled tractors of medium power. Soil & Tillage Research 2003;73(1-2) 119–29.

[75] Ansorge D, Godwin RJ. The effect of tyres and a rubber track at high axle loads on soil compaction, Part 1: Single axle-studies. Biosystems Engineering 2007;98(1) 115–26.

[76] Ibanez J, Valderrama JM, Puigdefabregas J. Assessing desertification risk using system stability condition analysis. Ecological Modelling 2008;213(2) 180–90.

[77] APAT (formely ANPA) 2001. Rapporto sullo Stato dell'Ambiente: Popolamento degli indicatori ambientali SINAnet. Italian National Environmental Protection Agency, Vol. 5, p 90-91.

[78] Simoniello T, Lasaponara R, Macchiato M, Cuomo V. An environmental change indicator based on satellite time series: NDVI-PV. Conference proceedings, June 2-5, 2003. Gent (Belgium) 23rd EARSeL Annual Symposium; 2003.

[79] Nunes AN, Coelho COA, de Almeida AC, Figueiredo A. Soil erosion and hydrological response to land abandonment in a central inland area of Portugal. Land Degradation & Development. 2010;21(3) 260–73.

[80] Salvati L, Mancini A, Bajocco S, Gemmiti R, Carlucci M. Socioeconomic development and vulnerability to land degradation in Italy. Regional Environmental Change 2011;11(4) 767–77.

[81] Weissteiner CJ, Boschetti M, Bottcher K, Carrara P, Bordogna G, Brivio PA. Spatial explicit assessment of rural land abandonment in the Mediterranean area. Global and Planetary Change. 2011 Nov;79(1-2):20–36.

[82] Imbrenda V, D'Emilio M, Lanfredi M, Ragosta M, Simoniello T. Indicators of Land Degradation Vulnerability Due to Anthropic Factors: Tools for an Efficient Planning. In: Borruso G, Bertazzon S, Favretto A, Murgante B, Torre CM. (eds.) Geographic Information Analysis for Sustainable Development and Economic Planning: New Technologies. Hershey: IGI Global; p 87-101.

[83] Salvati L, Bajocco S. Land sensitivity to desertification across Italy Past, present, and future. Applied Geography 2011;31(1) 223–31.

[84] Salvati L, Zitti M.. Territorial disparities, natural re source distribution, and land degradation: a case study in southern Europe. Geojournal 2007;70(2-3) 185–94.

[85] Costantini EAC, Urbano F, Bonati G, Nino P, Fais A. Atlante nazionale delle aree a rischio di desertificazione. Roma: INEA; 2007.

Chapter 6

Potassium in Soils of Glacial Origin

Mariusz Fotyma, Piotr Ochal and Jan Łabętowicz

Additional information is available at the end of the chapter

1. Introduction

1.1. Soils in Poland

Poland occupies a territory of 312,7 km2 with 99,7% of its area lying in the Baltic Sea catchment. The average altitude is 173 m and about 90% of the territory is situated below 300m above sea level, hence there is the predominance of lowlands. The majority of the country's territory is drained by two big rivers Vistula (55,7%) and Oder (33,9%) and few small Coastal rivers discharging directly to the Sea (9,3%). Vistula is exclusively the Polish river with springs in Carpathian Mountains in the South, flowing across the middle of the country and discharging the water to Baltic Proper, while river Oder borders the territories of Poland and Germany (Figure 1).

Considering the area and the population, Poland constitute an average sized (38 million inhabitants) country according to European standards. The rural area comprise around 190 km2 i.e. about 60% of the country's territory. From those area around 160 km2 is dedicated to agricultural activities and the rest to rural infrastructure. Poland is a country with the highest ratio of agricultural land compared to other European countries. According to European standards the country is quite densely afforested, with about 30% of forests land. A substantial territory is submerged under the lakes and rivers, including world well known Mazurian and Pomeranian regions. The majority of soils in the country are of glacial origin. The first so-called Narwian glaciation, covering the small area only occurred already in early Pleistocene. The main glaciation, South Poland and Middle Poland occurred in proper Pleistocene era and were split into five sub-periods (named after rivers: Nidian, Sanian 1, Sanian 2, Odranian and Wartanian). The youngest so-called Vistulian glaciation followed in late Pleistocene, dated back to 100 thousands of years (Figures 2, 3).

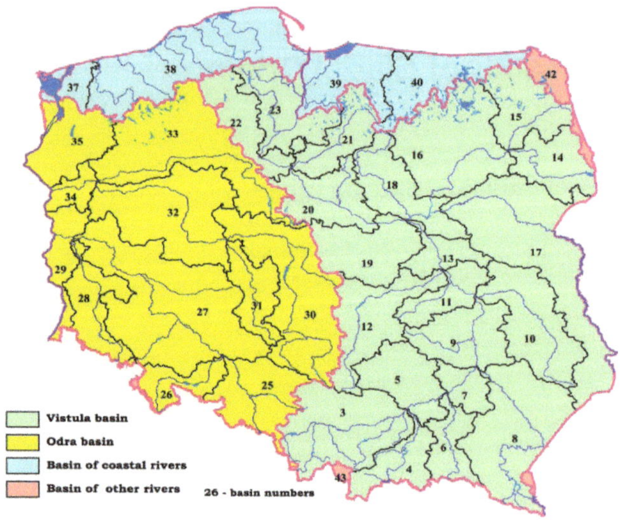

Figure 1. Drainage areas of the Vistula, Oder and Coastal rivers

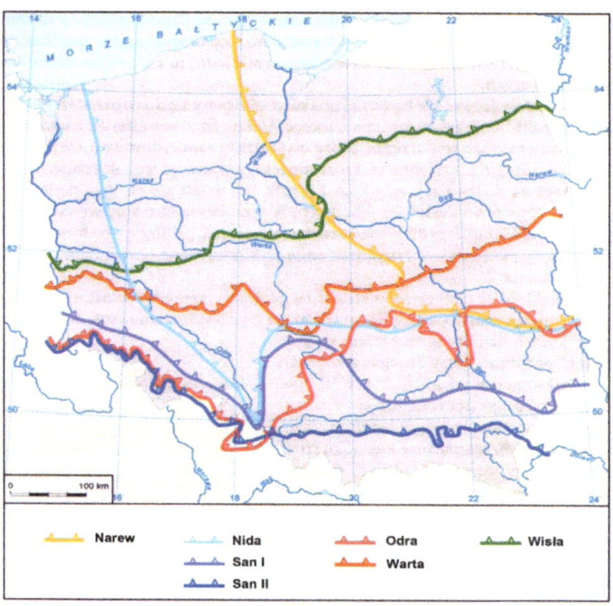

Figure 2. The borders of sequential glaciations in Poland [1]

Crossing the country from the north (Baltic Sea) to the south (Carpathian and Sudety Mountains) the following zones of the soils can be distinguished [2]:

- poorly sorted clays and sands of moraines (Cambisols, Stagno-gleyic Luvisols, Cambic Arenosols)
- fluvi-glacial sands (Podzols)
- broad, flat periglacial zone (Luvisols, Podzols, Arenosols, Fluvisols, Gleysols,Histosols)
- old eroded mountains and hills covered by loess and glacial deposits (Luvisols, Phaezems,Rendzinas)
- mountaineous zone

Pleistocene era												
		South Poland glaciation						Middle Poland glaciation				
Narew glaciation	Early interglaciation	Nida glaciation	Interglaciation 1	San 1 glaciation	Interglaciation 2	San 2 glaciation	Great intergalaciation	Oder glaciation	Interglaciation 3	Warta glaciation	Late interglaciation	Vistula glaciation
Thousands of years B.C.												
1000	850	760	660	630	560	530	430	330	230	210	130	110

Table 1. Sequence of glaciations in Poland [1]

According to geological history above 70% of Polish mineral soils have been formed from Pleistocene boulder clay and sand, strongly bathed and sorted by glacial waters. About 28% of soils are formed from loose and slightly loamy sand as well as from gravel. Soils of loess and alluvial origin cover a very small area of agricultural land (Table 1). It is not surprising that the ratio of soils of low fertility and productivity is estimated at 40% of the area.

For agricultural purposes, the soils in Poland are classified according to their productivity into soil suitability complexes and according to their buffer capacity (with respect to water and nutrients) into soil categories. Both classifications partly overlap. Productivity complexes, altogether thirteen on arable soils and three on grassland, are distinguished on the base of the soil parent's rocks, climate, soil water properties and the position in relief. This classification focuses on the possibility of growing different crops and on potential crop yield. On the very high quality complexes, numbered 1, 2 and 10, all crops can be grown satisfactorily, on the high ones, numbered 3, 4, 8 and 11 practically all crops can be grown as well but yields are significantly lower. On medium quality soil complex numbered 5 triticale, rye, and maize may be grown. Low quality soil complexes, numbered 6, 9 and 12 are suitable for rye, oats and potato and the very low quality ones, numbered 7 and 13 for rye and lupine only. The very high quality and high quality soil suitability complexes cover roughly 50% of arable soils, medium quality complex about 16% and the low and very low quality ones about 34% of arable land. Soil classification into categories is based on the content of fine fraction, i.e. soil particles below 0,02 in diameter. In this study four soil categories have been distinguished: very light up to 10% fine fraction, light 11-20% fine fraction, medium 21-35% fine fraction and heavy above 35% fine fraction. The very light soils cover 23,4% of arable area, light soils 36,1% of area, medium 29,4% and the heavy ones 11,1% of area only. For the purposes of this paper soil classification into such categories will be used extensively. Other factors limiting the fertility and productivity of Polish soils are the low content of soil organic matter SOM and strong acidification of most of the soils. According to the newest survey 7,6% of arable soils show low content of SOM (below 1%), 47,1% soils, medium content (1,1-2,0% SOM), 29,3% high content (2,1-3% SOM) and only 3% of soils, very high content of organic matter (above 3,0%) [3]. The summary of the last four-years period of agrochemical soils monitoring program reveals that 20,2% of soils are very acid (pH_{KCl} below 4,5), 29,4% acid (pH 4,5-5,5), 28% slightly acid (pH5,5-6,5), 14,7% neutral (pH 6,5-7,2) and 7,7% alkaline (pH above 7,2) [4]. There is a strong relationship between soil pH and soil's category. Very light and light soils are simultaneously very acid and acid while medium and heavy soils are much less acidified.

Parent rocks of soils	Total area thousands ha	% share in relation to	
		Total area	Agricultural land
Light loam	2562	15,8	18,8
Medium and heavy loam	2062	10,4	14,2
Loess	1396	3,3	4,8
Alluvial	788	4,7	5,8
Medium sand	2476	10,2	12,4
Sand and slightly loamy sand	4262	34,6	24,8
Very fine sandy soil	739	4,2	4,6
Rendzina	235	1,1	1,6
Massive rocks	599	6,1	3,9
Peat and muck	1414	8,5	9,6
Gravel	88,4	0,9	0,5

Table 2. Parent rocks of soils in Poland [3]

2. Soil potassium–conceptual–functional approach

Potassium is the seventh most abundant element in the Earth's crust representing on average 2,8% of total elemental composition. The average concentration of potassium in mineral soil is 1,4%, ranging between 0,01-3,7% [5] and, hence it is the fourth or fifth the most abundant element. The enrichment ratio, i.e. the relation between K in soil and K in Earth's crust, for potassium is 0,67 and is similar to the one for another important plant nutrient, phosphorus. In plant nutrition potassium is the second, after nitrogen element appearing in the highest concentration. The most important, potassium bearing minerals are primary aluminosilicates (feldspars, biotite and micas-muscovite) and secondary aluminosilicates called phyllosilicates. The last group of minerals is formed either through weathering the primary minerals or through the precipitation of secondary silicates [5]. Phyllosilicates and associated clay minerals play the most important role in the soil processes governing the availability of potassium for plants. Secondary clay minerals make up the greatest part of clay fraction of the soil characterized by particle less then 0,002 mm and, hence soils rich in clay fraction are generally abundant in potassium.

Researches on potassium in agriculture, concerning soil are as a rule based on conceptual-functional approach [6, 7, 8]. In this approach, four forms or pools of potassium have been distinguished (Figure 3). The potassium in soil solution plays the pivotal role in plant nutrition. This pool represents only up to 5% of the average plant demand for this element and not more than 0,1–0,2% of the total content of potassium in the upper soil horizon. The exchangeable or readily available potassium refers to the form of this element reversibly adsorbed on the edges of soil secondary minerals making up about 1-2% of the total potassium. Non-exchangeable or fixed potassium pool is slowly available for plants but represents a considerable 1–10% share of the total amount of this element. This form of potassium is adsorbed on wedge sites of secondary clay minerals, showing selectivity for potassium ions. The remaining pool called structural or lattice potassium is held in the structure of the primary soil minerals, feldspars and micas and becomes available for plants throughout very slowly running weathering processes. A conceptual – functional approach is based on the recognition of soil mineralogy and potassium transformation processes occurring in the soil. The availability of potassium for plants is, in this approach recognized as well.

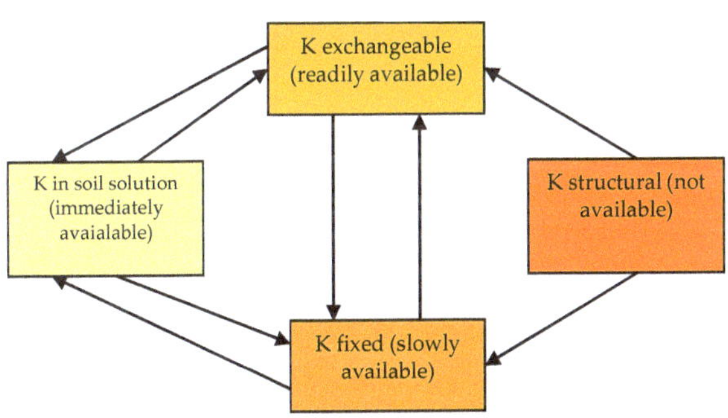

Figure 3. Conceptual–functional approach, forms (pools) of potassium in soil [8]

In the soil's solution, potassium occurs as a monovalent cation K+ in the hydrated form. The diameter of this cation, 0,66 nm is smaller than the diameter of sodium Na+, 0,72 nm. The hydration number of potassium ion, i.e. the number of water particles surrounding K+, being

10,5 is smaller as this number for sodium ion, 16,6. The K+ charge density, as the relation of its charge to the radius, is 0,0075 C•pm-1. Potassium does not form any chelates, ion pairs and/or ion complexes in soil solution. The concentration of potassium in the soil solution is in the range 0,1–1 mmol•L-1 K+ i.e. up to 40 mg K•dm-3. In Poland, in the previous research not referred in this paper, the average concentration of potassium in 136 soil samples of glacial origin was 30 mg K·L-1 and in 80% of samples it did not exceed 50 mg K·L-1 [10]. Due to very high solubility of potassium compounds applied in fertilizers, potassium ions are almost instantly removed from the soils solution by adsorption and plant uptake processes and the solution is very rarely saturated with potassium ions. In the upper 30 cm soil layer, saturated to field water capacity the content of potassium in soil solution is in the range 5–25 kg K·ha-1. This content is far below the potassium requirements of high-yielding crops. However, this pool is rather quickly replenished from the other pools, mainly from the exchangeable potassium pool.

The exchangeable potassium is held at the non-specific adsorption sites of clay minerals and at the phenolic and carboxylic groups of soil, organic matter. Potassium ion is adsorbed non-specifically in the hydrated form. These non-specific adsorption sites occur in planar "p" and edge "e" position of clay minerals and amorphous iron and aluminum hydroxides (Figure 4) [11,13,14].

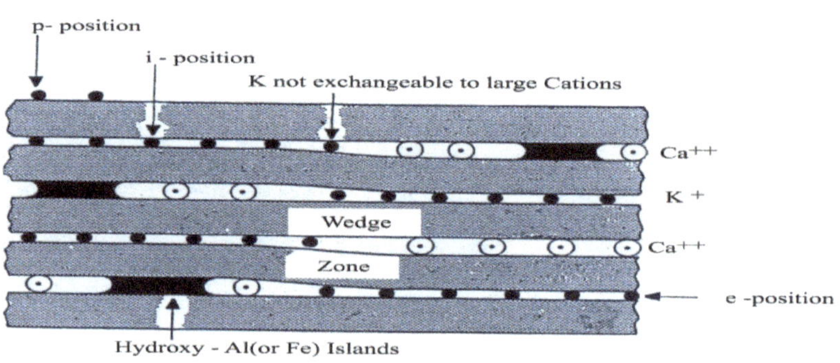

Figure 4. Model of 1:2 type clay mineral (illite) showing the positions of K^+ adsorption [11].

Clay minerals have a sheet-like structure and are built of Si tetrahedral layer and Al octahedral layer(s) in either 1:1 or 1:2 arrangement. Clay minerals are charged negatively due to the isomorphous substitution of ions pairs having the same size and coordination number, but differing in valences, i.e. Al3+ - Si4+ in tetrahedral layer and/or Mg2+-Al3+ in tetrahedral layer. The arising negative charges of clay minerals are balanced by exchangeable cations, e.g. potassium held at planar or edge positions. The 1:1 clays have very little isomorphous substitution and hence low negative charge and cation exchange capacity. Negative charges occur also at broken edges of mineral's crystals at the surfaces of amorphous and oxide minerals and soil organic matter [5,16]. The structure and properties of clay minerals and soil organic matter SOM are presented in Table 3. In Poland the prevailing soil – layered minerals are hydrated micas and illite [Długosz et al. 2005], non-expanding minerals of low cation exchange capacity.

Clay mineral	**Isomorphous substitution**	**Cation exchange capacity cmol(+) kg^{-1}**
Kaolinite 1: 1	None	1 – 15
Montmorillonite 1: 2 (expanding)	Mg for Al, Al for Si	80 – 150
Vermiculite (limited expanding)	Al for Si	140 – 200
Hydrous mica (illite) (non expanding)	Al for Si, Mg for Al	40 – 70
Amorphous oxides		2 – 4
Soil organic matter		200

Table 3. Structure and properties of clay minerals and SOM [12, 13].

The soil solution as well as the soil adsorption complex contains not only potassium but other cations as well. The total amount of cations for a certain element is partitioned amongst the soil solution and the fraction of the element adsorbed as a complex. The prevailing cation in well managed agricultural soils is calcium. To describe the relations between potassium and calcium several kinetic equations of cation exchange have been developed [5, 14]. The best-known and the most often used is Gapon equation [13]:

$$K_G = [\text{K-soil}]n(Ca^{2+})^{1/2}/[Ca_{1/2}\text{soil}]n(K^+)$$

Where: K_G is equilibrium constant often called selectivity coefficient, [K-soil] and [$Ca_{1/2}$ soil] are expressed in mol·kg^{-1} and $(Ca^{2+})^{1/2}$ and (K^+) in mol·L^{-1}

The value of K_G varies between the pair of exchanging cations and on the binding site at the clay mineral. As lower the value of K_G as easier is potassium exchanged for calcium cation. According to [11] the K_G for p-position is 2,21 $(mM \cdot L^{-1})^{-1/2}$, for e-position is 102 $(mM \cdot L^{-1})^{-1/2}$ and for i-position is infinite (Figure 4). The binding selectivity of kaolinite type 1:1 minerals and for the organic matter are close to this at p-position. Potassium adsorbed in p-position and e-position at clay minerals and on the soil organic matter is, therefore, easily exchangeable and available for plants and adsorbed in i-position is practically fixed and hence slowly available only.

Non-exchangeable or fixed potassium is held between tetrahedral layers of clay minerals like hydrous micas and vermiculite (Figure 4). Potassium ion in non-hydrated form has almost the same size as oxygen ion and fits perfectly into the spaces between sheets of 2:1 types of clay minerals [15].The binding forces between K+ and mineral's surfaces are greater than between these cations themselves. It results in the partial collapse of clay structure and "entrapping" potassium ions at wedge position between tetrahedral layers. The ions of the similar size as NH_4^+, but not much larger Ca^{2+} and Mg^{2+}, fit into these positions and can replace (exchange) entrapped K^+.

Structural or lattice potassium is covalently bonded with crystal structure of potassium rich primary silicates minerals like micas (biotite, muscovite) showing layered structure and feldspars (orthoclase, microcline) showing framework structure. Micas contain 8,7–9,8% of potassium. These minerals are found in coarser fractions of soils, silt, sand and here included potassium is freed in the long-term weathering processes (Figure 5). This pool of potassium is practically unavailable for plants, at least in the perspective of crop rotation. The typical weathering process runs from muscovite through hydro muscovite to illite and mixed clay minerals kaolinite-illit or illit-montmorollinite. In these processes potassium ion is substituted by hydroksyanion H_3O^+ and, hence each next product contains less potassium [16]. Clay minerals can be formed as well as a secondary ones as a result of synthesis from the end-products of weathering the feldspars.

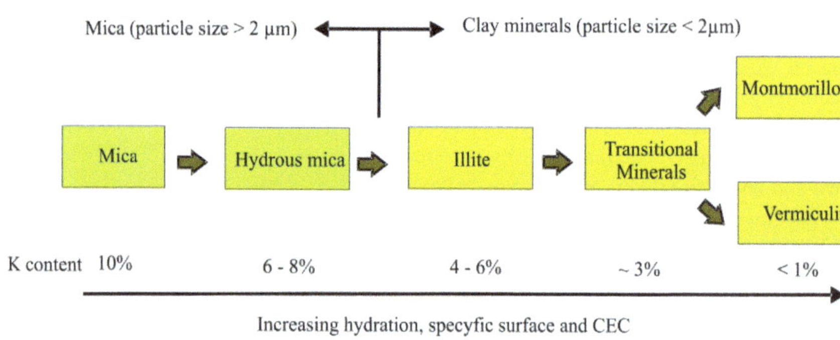

Figure 5. Model of freeing potassium in the sequence of weathering processes the primary soil minerals [17].

3. Methods and materials

3.1. Operational–analytical approach

Conceptual – functional approach is theoretically oriented, and its focus is on the chemistry of soil potassium and principles of potassium availability for crops. From agricultural point of view more interesting is, however analytical approach focused on distinguishing by laboratory methods the potassium forms and linking these forms with crop's potassium requirements. In this approach, five analytical forms of potassium have been distinguished : water soluble, available, reserve, nominal total and total (Table 4).

Potassium form	Brief description of method	Remarks
Water soluble K_{H2O}	Extraction with water at soil/water ratio 1: 5	Close to soil solution, form immediately available for plants
Available (exchangeable) K_{ex}	Extraction with ammonium acetate buffered to pH 7,0	Classical method of estimation , form recognized as available
Available K_{DL}	Extraction with calcium lactate buffered with HCl to pH 3,5	Official method in Poland and Latvia
Available K_{CAL}	Extraction with calcium acetate and calcium lactate, buffered with acetate acid to pH 3,7	Official method in Austria and Germany
Available K_{AL}	Ammonium lactate, buffered with acetate acid to pH 3,7	Official method in Lithuania, Slovenia and Hungary
Available K_{Meh}	CH_3COOH,0,25NH_4NO_3, NH_4F, HNO_3, EDTA	Mehlih-3 method, official in Estonia, Czech Rebublic, Slovakia
Reserve K_{res}	Extraction with boiling mol·dm^{-3} HNO_3	Close to fixed, form slowly available for plants
Nominal total K_{sem}	Extraction with hot HCl and HNO_3 (Aqua Regia)	Very slowly available for crops
Total K_{tot}	Fluorescence atomic spectrometry	Total content of potassium, unavailable for crops
Soil texture	Except the soils from KALIFERT project soil texture was evaluated by the "finger" method directly in the field. In Kalifert, the full particle composition of soil was analyzed quantitatively by laser method.	

Table 4. Analytical form of potassium in the soils of glacial origin, mg K·kg^{-1} soil [18, 19]

Analytical forms of potassium, however, based on the conceptual pools differ from them quite substantially. The amount of potassium extracted with water is higher than the amount in soil solution. For this reason, it is expressed in mg K·kg^{-1} soil and not in mg K·L^{-1}. Available potassium includes the water soluble one and a part of exchangeable form of this element. Otherwise the relations between exchangeable and available forms are statistically very close. Reserve potassium has an analytical meaning only, although conceptually reflects the potassium pool entrapped in layers of non-expanding clay minerals. In the analytical approach two forms of "total" potassium is distinguished. Nominal total K_{sem} is extracted with boiling Aqua Regia and, apart from previously mentioned three forms includes a part of structural potassium of soil's primary and secondary minerals. Total potassium can be measured either by fusion with alkali or by roentgen spectrometry, and includes all, already distinguished forms of this element and the whole potassium in soil minerals and organic substance. The analysis for nominal potassium is much easier than for the total one and, therefore it is often applied in research on soil potassium, which sometimes is a source of misunderstanding. Operational - analytical approach was applied in the own investigation on potassium in soils of Poland, executed in three research projects funded by Polish Ministry of Science and Higher Education. These projects have been running in the years 2002-2005 (1st project), 2006 – 2008 (2^{ed} project) and 2009-2012 (3rd project). The 1st project was focused on so called available forms of potassium and the relation between water soluble potassium K_{H2O} and available one K_{DL} and the 2^{ed} on different forms of potassium in Polish soils. The 3rd project is still running in scope of international collaboration with 10 countries belonging to MOEL group (Mttelosteuropaische Lander). This collaboration was established already in 1998 for exchanging information and standardizing methods of soil fertility investigation and fertilizer recommendations [23]. The main aim of this project including not only the soils of glacial origin from Estonia, partly Germany, Latvia, Lithuania and Poland but also soils from Austria, Czech Republic, Hungary, Slovakia and Slovenia was to follow long-term changes in the soil's potassium forms depending on fertilization. Another aim was to compare the methods of estimation the content of so called available potassium in the soils (see table 4) in different countries.

The results of all three projects have been partly published [8,18,19,21] but rather in a reports-like way. In this paper the cross-synthesis of all projects, subordinated to the specific problems is presented. In operational-analytical approach, the content of potassium forms is linked with the soil texture and not with the mineralogical composition of soil. Mineralogical soil analysis is very cumbersome, and depends considerably on the method of extraction the clay minerals. Otherwise it is well known that the content of potassium increases from the coarse sand to the clay soil's fractions. According to Brogowski et al. [20] in soil composed of 38% of sand, 52% silt and 10% of clay the share of total potassium in the fraction of sand was merely 13% in the fraction of silt 64% and in the fraction of clay 23%. The conclusion is that total potassium is under-proportionally located in sand fraction, almost proportionally in silt and over-proportionally in clay fraction.

4. Results

4.1. Water soluble K_{H2O}–versus available K_{DL} potassium

In the years 2006 and 2007 each of the 17 Agrochemical Laboratories operating in Poland collected about 1.600 representative soil samples and analyzed them for soil texture, soil pH and the content of available potassium K_{DL}. In the same soil samples, the content of water soluble K_{H2O} potassium was estimated in the laboratory of the Institute of Soil Science and Plant Cultivation at Puławy using method described in Table 4. In the population of almost 24.000 soil samples the content of both potassium forms was significantly differentiated depending on soil texture (soil category) and soil pH. The mean values of the K_{H2O} and K_{DL} contents in the classes of soil texture and soil reaction are presented in tables 5 and 6. Due to the far from normal distribution of the data, medians were presented along with average values.

Soil category	No of samples	K_{DL} mg K·kg^{-1} soil		K_{H2O} mg K·kg^{-1} soil		% K_{H2O} in K_{DL}	
		average	median	average	median	average	median
very light	2044	86,3	78,0	27,5	24,2	33,3	32,1
light	10290	115,2	108,7	32,1	28,6	28,7	27,4
medium	9492	130,7	124,5	28,2	24,5	21,9	20,8
heavy	1832	135,2	120,8	25,2	21,1	19,2	17,8
all soils	23658	120,4	112,0	29,6	26,1	25,6	24,2

Table 5. The content of potassium forms depending on the soil category [19]

Soil reaction,pH	No of samples	K_{DL} mg K·kg^{-1} soil		K_{H2O} mg K·kg^{-1} soil		% K_{H2O} in K_{DL}	
		average	median	average	median	average	median
very acid	4012	87,9	75,5	21,2	18,4	26,9	25,4
acid	6934	117,8	111,2	27,9	25,2	25,0	23,5
slightly acid	6488	132,7	126,2	32,2	29,8	25,0	23,9
neutral	3668	137,3	128,2	34,7	31,1	25,4	24,4
alkaline	2554	123,6	115,4	33,0	29,1	26,9	25,5

Table 6. The content of potassium forms depending on soil pH [19]

The content of available potassium increases in the direction from the very light and light soils to the medium and heavy soils and from the very acid soils to the neutral ones. Alkaline soils contain less available potassium than the slightly acid and neutral ones. The content of water soluble potassium is admittedly the lowest in the very light soils, but soils of remaining categories show rather similar content of this potassium form. The content of water soluble potassium depends more on the soil's acidity and increases significantly in the direction from the very acid to alkaline soils. In the whole population of soil samples, the mean share of water soluble potassium makes about a quarter of the available form of this element. This share decreases significantly in the direction from the very light to heavy soils. Due to the opposing tendency in changing the content of available (increase) and water soluble (decrease) content of potassium in conformity with soil pH, the share of K_{H2O} in K_{DL} is practically independent of this soil characteristic. Generally, the heavier soils are more abundant in available forms of potassium, but this element is less accessible for the crops in comparison with the coarse-textured soils. Increasing soil pH influences, however, positively both the content of available and water-soluble forms of potassium. The latter is easily accessible for the crops. These statements are very important in Poland due to the prevalence the very light and light soils and the high soil acidity. Between the content of available potassium and potassium soluble in water exists a significant correlation and this relation was quantified by linear regression models. Two approaches were applied here: multiplicative regression analysis for each soil category separately and model of comparison the linear regressions including four soils categories jointly. The calculations were performed using two adequate procedures, a multiplicative regression model and comparison of regression lines model, offered by statistical package Statgraphic 5+. The regression equations for multiplicative model are presented below in frame.

Very light soils: $K_{DL} = 4{,}7689 \cdot K_{H2O}$^0,8698 $R^2 = 0{,}74$

Light soils : $K_{DL} = 6{,}9677 \cdot K_{H2O}$^0,8068 $R^2 = 0{,}71$

Medium soils: $K_{DL} = 12.5141 \cdot K_{H2O}$^0,7023 $R^2 = 0{,}65$

Heavy soils: $K_{DL} = 10{,}8869 \cdot K_{H2O}$^0,7788 $R^2 = 0{,}70$

On the base of these equations the content of available potassium K_{DL} can be calculated for a given range of potassium soluble in water K_{H2O} contents, separately for soil categories. Such calculation has been performed in a new approach to calibrate the content of available potassium K_{DL} in soils of glacial origin in Poland (Table 7). For a better visualization of the dependence on the relation $K_{DL}K_{H2O}$ the comparison of regression lines model offered by Statgraphic 5+ package has been applied (Figure 6) as well.

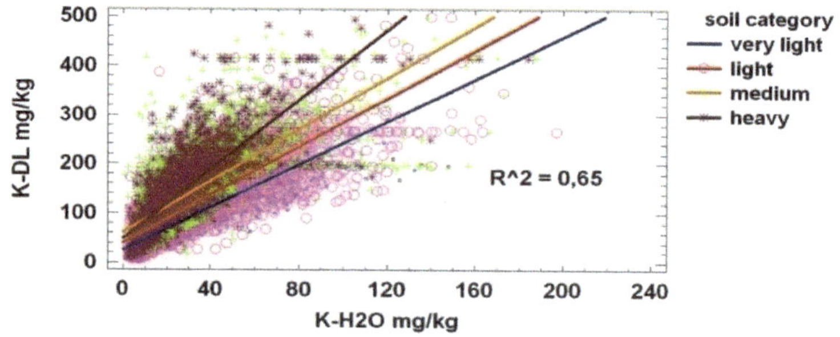

Figure 6. Relation between available K_{DL} and water soluble K_{H2O} potassium depending on soil category (comparison of regression lines)

From this model the conclusion is easily drown, that as heavier is a soil as steeper is the slope of the regression line corresponding to this relation. In practice it means that by the same content of available potassium K_{DL} the concentration of potassium ions, hence its accessibility for plants is much higher in the very light against the heavy soils. This model, however gives a transparent picture, is limited to straight regression lines only and therefore its predictability (one correlation coefficient only) is lower than the already presented multiplicative regression model (in frame). By interpretation the already presented data, one must remember that the number of soil's samples and therefore its representation was the highest for light and medium soils as well as for soils characterized by acid and slightly acid reaction. Data for the very light and very acid soils (Tables 5, 6) were less representative.

Since 1985, the five classes of available potassium content K_{DL} (along with phosphorus) are implemented in Poland. The ranges of potassium content in such classes are originating from the previous partition into three classes and have not been changed ever since. Grounding on the large number of soil samples (24.000) in which the content of water soluble and available soil potassium has been measured simultaneously the modification of official classes has been suggested. The idea underlying this suggestion was to link the content of available potassium K_{DL} with the content of water soluble K_{H2O} one, which is directly accessible to plant roots, according to the following procedure. The whole range of data concerning K_{H2O},for each soil category separately, was split into pentiles, around the median value. For the lower and upper limits of each pentile the corresponding K_{DL} have been calculated using the regression equations included above in the frame. Such calculated ranges of K_{DL} values are proposed as the new classes of available potassium content in soils of glacial origin in Poland. The details of this idea have been presented in separate publication [19]. For the sake of further consideration the official and proposed critical values of available potassium content are, however, presented in Table 7.

Soil category	K_{DL}mgK·kg^{-1} soil, official classes					K_{DL}mgK·kg^{-1} soil, proposed classes				
	Very low	Low	Medium*	High	Very high	Very low	Low	Medium	High	Very high
very light	< 20	21-62	63-103	104-145	>146	< 50	51-70	71-90	91-120	> 121
light	< 41	41-83	84-124	125-166	> 167	< 65	66-90	91-115	92-140	>141
medium	< 62	63-104	105-166	167-207	>208	<85	86-110	87-140	141-170	>171
heavy	<83	84-124	125-207	208-249	>250	< 90	91-120	92-155	156-200	>201

*threshold range focused on in the fertilizer recommendation system in Poland

Table 7. The official and proposed [19] classes of available potassium in soils of Poland

From the comparison of the official and proposed classes of available potassium content appears that official system undervalues potassium contents in very light and light soils as well as in the very low and low classes of K_{DL}. Overvalued are, indeed values for high and very high values of potassium content, independently of the soil texture. There is resemblance of both classifications in the range of medium potassium content and, with the exception of heavy soils quite similar are medians for proposed and the official systems. The author's proposition is focused on improving the officially accepted and being in use for over 25 years system of soil classification for available potassium content. In this proposition, the special position has the medium class for K_{DL}. Medium class is proposed as the threshold range for potassium content, which should be accomplished and kept on in the sustainable system of fertilization. However this proposition needs the further development, particularly in the very light and heavy soils categories for which the number of samples was not fully representative.

4.2. The content of available potassium K_{DL} in soils of Poland

In Poland, 17 Agrochemical Laboratories subjected to Ministry of Agriculture and Rural Development and covering with their activity the whole country's territory, are operating. The main task of these Laboratories is soil and plant testing and launching fertilizer recommendations. Every year several thousand soil samples are being analyzed for the soil pH and the content of available forms of potassium, phosphorus and magnesium. The results are making available for farmers, but they are collected in the data bank as well. Every fourth year the data are subjected to statistical analysis and published with the aim to monitoring the soil fertility status in Poland. The synthesis of the data concerning the content of available potassium, including the results of over 950.000 soil samples analyzed over the years 2004–2008 is presented in the following tables [4]. The synthesis reveals that the content of available potassium K_{DL} depends on the soil category and soil pH (Table 8) which is in accordance with the data already presented for 24.000 soils samples.

Soil category	mg K·kg^{-1} soil		Soil acidity	mg K·kg^{-1} soil	
	average	median		average	median
very light (37.170)*	91,2	76,0	very acid	88,9	74,7
light (376.602)	126	112	acid	118	105
medium (413.098)	159	143	slightly acid	138	126
heavy (130.681)	195	183	neutral	144	129
total (957.551)	148	130	alkaline	134	120

*number of soil samples

Table 8. The content of available potassium in soils of Poland [4]

The numeric data have been categorized and presented in the tables of contingency (Tables 9 and 10), which better visualize the link between potassium content and soil categories, and/or soil pH. Categorization of samples is based on the official classes of potassium content (Table 7)

Soil category	% samples in the classes of available potassium content				
	very low	low	medium	high	very high
very light	7,06	42,97	28,84	12,29	8,84
light	12,20	31,19	27,66	15,78	13,18
medium	17,25	24,45	31,53	12,86	13,91
heavy	15,88	20,02	41,46	9,67	12,97
total	14,68	27,21	31,26	13,55	13,30

Table 9. The ratio of soils samples in different classes of potassium content depending on soil category [4]

As follows from Table 9 one-half of the area of the very light soils is very poor in available potassium, while 20% of these soils show high and very high content of this element. Most of the medium and heavy soils contains medium to very high content of potassium.

Soil acidity	% samples in the classes of available potassium content					total
	very low	low	medium	high	very high	
very acid	25,85	37,08	24,37	7,49	5,21	20,21
acid	13,97	28,76	32,47	13,47	11,33	29,39
slightly acid	9,96	22,35	34,30	16,56	16,83	28,00
neutral	10,48	21,69	32,52	15,83	19,48	14,74
alkaline	13,30	23,66	31,22	14,44	17,38	7,66

Table 10. The ratio of soils samples in different classes of potassium content depending on soil acidity [4]

The data in Table 10 prove that very acid and acid soils are simultaneously poor in available potassium, while over 60 % of neutral and alkaline soils contain medium to very high K content. In the whole Poland around 42 % of soils shows the very low and low content of available potassium. On this area the recommended rates of potassium fertilizers exceed the of-take of this element with the crop yields. The surplus of potassium would contribute to increase the content of available nutrient and should bring the soil in the future to the medium class. The content of available potassium is substantially differentiated among the country's regions (Figure 7). In eight regions (administrative units), the ratio of soils poor in potassium is over 40% and only in three regions it does not exceed 30%. These differences are partly grounded on pedological soil origin but also on different management practices in agriculture. It concerns, mainly, different among regions consumption of potassium fertilizers and limestone, which influence the soil acidity and indirectly the content of available potassium forms. This problem lays, however outside the scope of the paper.

4.3. Comparison different method for estimation the available form of potassium

In the Central-Eastern European countries (see MOEL group in Materials) different methods are applied for estimation the available potassium in soil. In scope of the MOEL co-operation, the research project has been launched on comparison the methods used in ten countries (see last chapter). Altogether, 132 soil samples have been collected from long-term field experiments in the treatments without and with potassium fertilization and two soil layers, 0-25 and 25-50 cm. The samples were analyzed for the content of exchangeable potassium K_{ex}, which was included as a reference method, and available potassium by methods presented in Table 4. Considering very different texture of individual soils the variability of data was big enough for making the reliable comparison of the applied methods (Table 11).

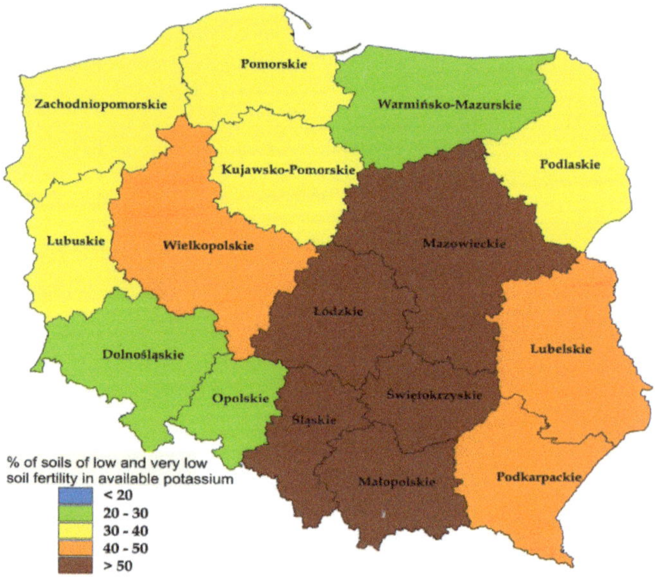

Figure 7. The percent of soils poor in potassium in regions of Poland [4]

Potassium form mg K•kg⁻¹soil	light soils(46)*		medium soil(26)		heavy soils(24)		very heavy soils(36)		total (132)	
	average	std**	average	std**	average	std**	average	std**	average	Std²
K_{ex}	72,2	37,1	122	75,0	135	72,5	168	91,1	120	78,6
K_{DL}	75,5	35,4	119	83,9	112	69,1	116	78,2	102	66,7
K_{CAL}	54,6	32,9	90,0	61,8	83,3	47,9	93,3	60,0	77,3	52,5
K_{AL}	128	55,6	186	111	197	95,6	238	127	182	106
K_{Meh}	88,6	40,1	137	74,2	144	64,9	176	88,6	132	75,3

*number of samples, **standard deviation

Table 11. The content of available potassium by different methods [21]

The average amounts of potassium extracted by different methods were in the following decreasing order: $K_{AL} > K_{ex} > K_{Meh} > K_{DL} > K_{CAL}$. Exchangeable potassium K_{ex} is a form (pool) defined conceptually. Therefore, it may be presumed that the method of available potassium determination is as better as its results correlate closer with this theoretically grounded pool (Table 12).

Test	K_{ex}	K_{DL} (X)	K_{CAL} (X)	K_{AL} (X)	K_{Meh} (X)
K_{Meh}	0,98	0,92	0,93	0,97	-
K_{AL}	0,97	0,94	0,95	-	K_{AL}=-0,69+1,38X
K_{CAL}	0,91	0,97	-	K_{CAL}=-8,14+0,47X	K_{CAL}=-8,64+0,65X
K_{DL}	0,90	-	K_{DL}=5,96+1,24X	K_{DL}=-6,32+0,59X	K_{DL}=-6,36+0,82X
K_{ex}	-	K_{ex}=11,27+1,07X	K_{exm}=14,17+1,36X	K_{ex}=-10,2+0,71X	K_{ex}=-15,9+1,03X

Table 12. Relationships between pairs of examined soil tests of available potassium

The closest correlation has been found between exchangeable potassium K_{ex} and the Mehlich 3 method K_{Meh}. Besides, the slope of the regression line between these two methods is close to one, which means that both lines run parallel. However, using Mehlich 3 method somewhat higher amounts of potassium are extracted. Similar proportionality has been also found between K_{ex} and K_{DL}, although the correlation coefficient between these two potassium forms is weaker than between K_{ex} and K_{Meh}. The highest amounts of potassium are extracted using K_{AL} method, and it seems that with this method, not only exchangeable potassium but a part of reserve form K_{res} is estimated.

4.4. The content and proportion of different potassium forms in Polish soils

In the years 2002–2004, in the areas of the oldest glaciation periods (Sanian I and Sanian II, see Figure 2) seven hundreds soils samples representing medium and heavy soils have been collected. The samples, from the soil layers 0-25 cm and 25-50 cm were analyzed for soil pH, and the content of different forms of potassium by methods described in Table 3. Factor which the most influence the content of all potassium forms was soil categories, i.e. the content of soil's particles less then 0,02 mm (Table 13).

Soil category	No of samples	K_{H2O}		K_{ex}		K_{res}		K_{sem}	
		0-25 cm	25-50 cm	0-25 cm	25-50 cm	0-25 cm	25-50 cm	0-25 cm	25-50 cm
v.light	32	17	8,87	59,1	31,1	156	97	234	226
light	177	24,3	14,9	91,3	56,0	254	195	503	440
medium	253	22,0	13,8	112	72,6	382	204	872	789
heavy	238	21,6	10,1	166	104	655	470	2372	2256

Table 13. The content of potassium forms in two soil layers (medians) mg K·kg^{-1} soil [18,19,22].

The content of all forms of potassium, except the water soluble one increased significantly from the very light to the heavy soils. The water-soluble potassium K_{H2O} makes 0,7-5%, exchangeable potassium K_{ex} 7-18% and reserve potassium K_{res} 23-45% of the nominal total potassium. In these ranges, the ratio of available or slowly available potassium forms in the nominal total potassium was the highest in the very light and the lowest in heavy soils.Therefore the coarse-textured soils although contain considerably less potassium, the greater part of it appears in the form immediately available and readily available for crops.The relative content of potassium in the subsoil against the plow soil layer was 56%, 60%, 66% and 87% for water soluble, exchangeable, reserve and nominal total potassium forms respectively. It means that the plow soil's layer (0-25 cm) is enriched, particularly in the potassium forms available for crops.

4.5. Long–term changes in potassium forms

In 2010 in 10 countries belonging to MOEL group (see Materials) at least one or more long-term fertilizers experiments, including treatments with NPK and without NP potassium fertilization has been selected. In 2010 soil samples from these treatments and from two soil layers 0-25(30) cm and 25(30)-50(60cm) were collected and prepared for analysis. All soils from Estonia,Latvia, Lithuania and Poland were of glacial origin and have been characterized by coarse texture as light ones. These soils fall in the topic of the paper and other soils, of much heavier structure make a frame of reference. Soil analysis has been performed by methods presented in Table 4. It is worth to emphasize that in this particular project, the soil texture was analyzed by laser method and the total content (not only seemingly total) of potassium has been estimated by the fluorescence atomic spectrometry. The content of all potassium forms are included in Table 14.

Potassium pool mgK•kg^{-1} soil	Soil category*				Soil level cm		Fertilization	
	light**	medium	heavy	v.heavy	0-25	25-50	NP	NPK
samples	46	26	24	36	66	66	61	71
K_{H2O}	14,6	17,0	14,8	12,3	16,8	12,0	11,8	17,0
K_{ex}	72,2	122	135	168	135	104	100	137
K_{res}	339	525	762	964	651	595	618	628
K_{sem}	1624	2903	4719	5892	3513	3692	3552	3640
K_{total}	14051	14675	15360	15877	14852	14968	14798	15009

* light < 20% of silt, medium 20-35% silt, heavy 35-50% silt, very heavy > 50% silt, silt particles less then 0,02 mm. ** soils of glacial origin friom Estonia, Latvia, Lithuania and Poland

Table 14. The content of potassium forms depending on soil texture (category) [8]

The factor which most strongly determinates the content of all pools of potassium, except the immediately available $K_{H2O,}$ is a soil category, i.e. the percent of silt. The differences in the content of K_{ex}, K_{res} and K_{sem} between the light and very heavy soils are two and three-folds and in the content of total potassium K_{tot} reaches almost 15%. The content of immediately available potassium K_{H2O} is practically independent of soil category and is generally higher in light and medium, than in heavy and very heavy soils. The content of two pools of potassium (K_{H2O} and K_{ex}), accessible for plants was higher in the upper soil layer in comparison to the subsoil while slowly available K_{res} and structural pools (K_{sem} and K_{tot}) seem to be more uniformly distributed in the soil profile 0-50 cm. Long term application of potassium fertilizers showed a positive effect on the content of K_{H2O} and K_{ex} pools of this element. However, the reserve K_{res} and total potassium K_{tot} contents in the soil have hardly changed in spite of long-term soil mining from this element in the control treatment. There is a strong correlation between immediately available K_{H2O} and readily available K_{ex} potassium and the weaker between K_{H2O} and K_{res}. No correlation exists between immediately available and total (K_{sem}, K_{tot}) potassium. Readily available potassium K_{ex} correlated the strongest with slowly available one K_{res} and significantly with the pools of total potassium. The strongest estimated correlation was between the slowly available and total potassium and, between both operationally distinguished pools of total potassium. The ratio of potassium forms has been calculated against the pool of total potassium K_{tot} (Figure 8) and/or the pool of nominal total potassium K_{sem} (Figure 9). Both approaches provide inconsistent results, when related to the soil texture.

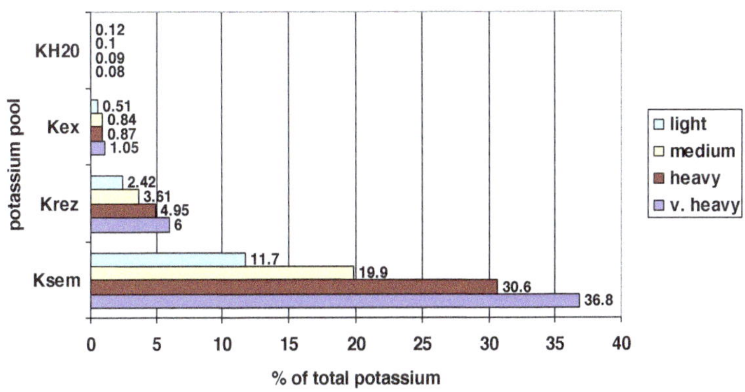

Figure 8. The ratio of potassium pools in the content of total potassium K_{tot}.

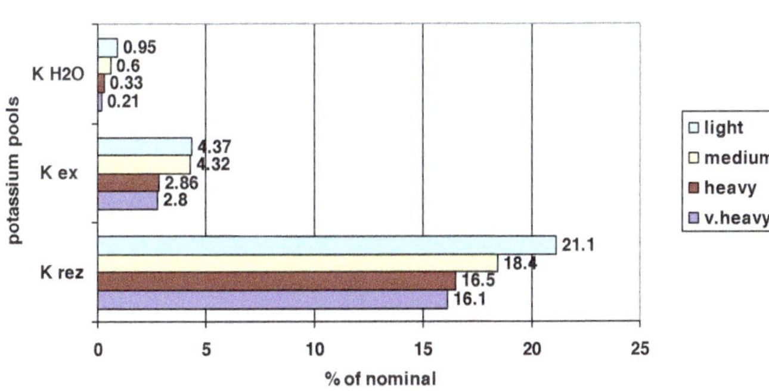

Figure 9. The ratio of potassium pools in the content of nominal total potassium K_{sem}

The relative content of all potassium forms (except the water soluble one) calculated against the total potassium K_{tot} increased systematically from the very light to very heavy soils. However, the ratio of all potassium forms calculated against the nominal total potassium, decreased in the same direction. This discrepancy has not yet deserve special attention in the available literature and needs further research. The relative content of potassium forms in light soils (mainly of glacial origin), against the nominal total potassium is very close to those found in the research conducted in scope of the 1st project in the years 2002-2005 [18]. From this finding, it can be concluded and recommended that by relating the content of different potassium forms to the total potassium a clear distinction should be made between the nominal total K_{sem} and total K_{tot} forms. Such differentiation is particularly relevant, due to analytical problems in many research conducted on soil potassium, for which only nominal total form of this element is estimated.

It has been already mentioned that in this research, the content of clay (soil particles less than 0,002 mm) and silt (soil particles less than 0,02 mm) was quantitatively estimated by a diffraction laser method. It was therefore, possible to calculate the degree of clay and silt saturation by different forms of potassium (Table 15).

Potassium pool	mg K per 1% of clay		mg K per 1% of fine fraction**	
	without potassium NP	with potassium NPK	without potassium NP	with potassium NPK
Exchangeable K_{ex}	1,15 (0,69)*	2,03 (1,53)	0,30 (0,15)	0,50 (0,30)
Reserve K_{rez}	7,25 (3,45)	9,01 (4,88)	1,88 (0,74)	2,06 (0,76)
nominal total K_{sem}	37,7 (12,3)	47,1 (23)	10,1 (2,98)	11,6 (3,79)
Total K_{tot}	208 (123)	253 (163)	56 (30)	60,0 (32,6)

*In parenthesis the standard deviation, ** particles <0,02 mm

Table 15. Saturation of clay and fine fraction with different potassium forms, depending on the long-term fertilization with this element [8]

Results from Table 15 indicate, that the arable soils in Central-Eastern Europe, properly fertilized contain about 2 mg of readily available K_{ex} and about 9 mg of slowly available K_{res} potassium per 1% of clay and respectively 0,5 mg of readily available and about 2 mg slowly available potassium per 1% of fine fraction. Long–term soil mining from potassium resulted in diminishing the saturation of clay and silt with readily available potassium K_{ex} by about 70% and the saturation with slowly available potassium K_{rez} by about 20%.

According to the results achieved in this project, comparison of potassium content in different forms at the beginning of experiments, with the amount taken up by crops in this period of time was also preliminary considered (Table 16). This problem is beyond the scope of this paper and, besides might deserve the further research.

K pool	Light soils			Medium soils			Heavy soils			Very heavy soils		
	0-25	25-50	off*take	0-25	25-50	off take	0-25	25-50	off take	0-25	25-50	off take
K_{H2O}	75	57		86	66		79	44		69	37	
Kex	390	259		653	483		694	464		797	614	
Krez	1368	1236	2476	2152	1938	2871	3150	2580	3193	3999	3445	3374
Ksem	6253	7115		10992	11126		17973	18404		22742	22512	
Ktot	53028	53470		55064	55560		58120	58184		60220	59910	

*sum of the potassium removed from the soil by all crops grown in experiments

Table 16. The amount of potassium at the beginning of experiments, in the soil layers 0-25 cm and 25-50 cm, against its off take with crop yields in the NP treatment, kg K•ha^{-1}.[8]

Independently, of soil texture the amount of exchangeable potassium K_{ex} in the first 0-50 cm of the soils profile 0-50 cm does not meet the long-term plant's requirements for this element. Therefore, this pool needs to be constantly replenished from the pool of slowly available potassium K_{rez}. In the light soil even this pool in the soil profile 0-50 cm hardly matches the long-term plant requirements and had to be supplemented from the structural pool of potassium K_{sem}. In the medium soil, the amount of slowly available potassium K_{rez} in the upper soil level is too small against the plant requirements and only the amount in the soil profile 0-50 cm suffices to fulfill these requirements. In the heavy and very heavy soils, the upper soil level, 0-25 cm contains enough slowly available potassium to cover the long-term plant's requirements for this element. The amount of available and slowly available potassium was higher in the upper soil level in comparison to the subsoil. This regularity does not concern the nominal total K_{sem} and total K_{tot} potassium pools, which practically do not differ between the soil levels. It leads to the conclusion that upper level of arable soils is enriched with plant accessible potassium

5. Summary and conclusions

Most of the soils in Poland are of glacial origin and are formed from poorly sorted clays and sands of moraines and fluvial-glacial sands. These soils are deeply leached down, poor in bases and, therefore, acid. Almost 60% of the soils belong to very light and light categories, i.e. contain less than 20% of fine fraction and over 50% of soils are very acid and acid. The content of organic matter is generally low and very low even at the limit established in Poland, which is below 2,0% of SOM. In the paper, two approaches to potassium in the soil are presented. The first approach, conceptual – functional, is based upon the literature data, and it focuses on the defined potassium pools and physicochemical processes governing the potassium dislocation between these pools. The second approach, operational-analytical, concerns the potassium forms isolated from the soil samples using chemical methods. The second approach has been applied in the own research carried on in scope of three scientific projects in the years 2002 – 2012. The soils of glacial origin in Poland have been characterized as regards the content of water soluble K_{H2O}, available (in different extracts, mainly calcium lactate K_{DL}), reserve (in boiling nitrate acid K_{res}), nominal total (in Aqua Regia) and total (Roentgen spectrometer) K_{tot} potassium forms. The content of all forms of potassium, except water soluble one depended significantly on soil granulometric composition and soils pH and increases from the very light to heavy, and from very acid to neutral soils. Properly managed soils contain 2 mg K_{ex} and 9 mg of K_{res} per 1 % of clay fraction (<0,002 mm). The new calibration figures for available potassium K_{DL} by linking the K_{DL} content to water soluble potassium K_{H2O} is proposed. Applying these figures would promote more sustainable potassium fertilizer's management. The soils of glacial origin in Poland are generally poor in available potassium and the ratio of soils showing very low and low potassium (K_{DL}) content exceeds 40 %. The water-soluble potassium (K_{H2O}) makes 0,7-5%, exchangeable potassium (K_{ex}) 7-18% and reserve potassium (K_{res}) 23-45% of the nominal total potassium. In the light soils, the ratio of nominal total to total potassium is slightly over 10 %. In long-term experiments with potassium fertilization, the amounts of potassium in different forms were compared with the uptake of this element by crops grown in control (without K fertilization)treatments. Independently, of soil texture the amount of exchangeable potassium in the soil profile 0-50 cm does not suffice to cover the crop's requirements for this element. Therefore, this pool needs to be constantly replenished from the pool of slowly available potassium K_{rez} and even from the pool of nominal total potassium. The problem of potassium fertilization and its influence on the forms of potassium in the soil is beyond the scope of this paper and, besides might deserve the further research.

Acknowledgements

The experimental part of this paper is a summary of three research projects funded by the Ministry of Science and Higher Education in Poland : Nr 3 PO6R 087 23 "Nawkal 2002 – 2005, No 2 PO6R 48 30 "Kalpol" 2006-2008 and NN 310 204437 "Kalifert" 2009 – 2012. Author expresses sincere thanks to all colleagues from the MOEL group for providing the soil samples in scope of the last project.

Author details

Mariusz Fotyma[1], Piotr Ochal[1] and Jan Łabętowicz[2]

*Address all correspondence to: fot@iung.pulawy.pl

1 Institute of Soil Science and Plant Cultivation, Puławy, Poland

2 Life Science University, Warsaw, Poland

References

[1] Harasimiuk, K. Rodzoś, Geografia Polski. UMCS, Lublin ; (2004).

[2] Soil Atlas of EuropeEuropean Soil Bureau Network ;(2005).

[3] Terelak, H, Krasowicz, S, & Stuczynski, T. srodowisko rolnicze Polski i racjonalne użytkowanie rolniczej przestrzeni produkcyjnej. In: Igras J., Pastuszak M. (ed.) Contribution of Polish agriculture to emission of nitrogen and phosphorus compounds to the Baltic Sea : Puławy; (2012). , 70.

[4] Ochal, P. Wykorzystanie syntetycznego wskaźnika do oceny agrochemicznego stanu gleb w Polsce. PhD, manuscripts: Puławy ;(2012).

[5] Essington, M. E. Soil and water chemistry, an integrated approach. London :RC Press; (2004).

[6] Romheld, V, & Kirkby, E. A. (2010). Research on potassium in agriculture; needs and prospects. Plant Soil.2010; , 335, 155-180.

[7] Syers J,K.Soil and plant potassium in agriculture. IFS Proceedings 1998; No. 411.

[8] Fotyma, M. The consequences of soil mining from potassium stock. Nawozy i Nawozenie- Fertilizers and Fertilization. (2011). , 43, 5-20.

[9] Epstein, E, & Bloom, A. J. (2005). Mineral nutrition of plants. Principles and perspectives. Second edition; 2005. Sinauer Associates, Inc, Publishers, Sunderland, Massachusetts.

[10] Labetowicz, J. Skład chemiczny roztworu glebowego w zróżnicowanych warunkach glebowych i nawozowych; (1995). Fundacja "Rozwój SGGW" Warszawa.

[11] Mengel, K, & Kirkby, E.A. . Principles of plant nutrition third edition ;1982. International Potash Institute, Bern.

[12] Barber, S. A. Soil nutrient bioavailability, a mechanistic approach, second edition; (1995). John Wiley & Sons,Inc.

[13] Kim, H. Tan. Principles of soil chemistry second edition ;(1993). Marcel Dekker Inco.

[14] Sparks, D. L. Environmental soil chemistry, second editions ;(2003). Academic Press.

[15] Troeh,F.R., Thompson,L. M, Soils and soil fertility, sixth edition;.Blackwell Publishing

[16] Pracz, J. Podstawy mineralogii ;(2003). SGGW, Warszawa.

[17] Schroeder, D. Soils- facts and concepts ;(1984). International Potash Institute,Bern.

[18] Fotyma, M. (2007). Content of potassium in different forms in the soils of southeast Poland. Polish J. Soil Sci; 2007. XL., 1, 19-32.

[19] Fotyma, M. (2009). Forms and tests of available potassium in soils. Nawozy i Nawozenie- Fertilizers and Fertilization; 2009. , 34, 9-24.

[20] Brogowski, Z, Uziak, S, & Komornicki, T. . Distribution of potassium in granulometric soils fractions. Polish Journal of Soil Science; 2009. XLII/1: 30-40.

[21] Fotyma, M. (2011). Testy glebowe potasu łatwo dostępnego dla roślin. Nawozy i Nawożenie- Fertilizers and Fertilization ;2011. , 44, 5-16.

[22] Grzebisz, W, & Fotyma, M. Recommendations and use of potassium fertilizers in Central-Eastern Europe. The International Fertilizer Society. Proceedings;(2007). (621)

[23] Loch, J. years of MOEL consultative meetings in retrospection. Nawozy i Nawozenie- Fertilizers and Fertilization;(2009). , 37, 7-16.

Chapter 7

Soil Contamination with Heavy Metals and Petroleum Derivates: Impact on Edaphic Fauna and Remediation Strategies

Raphael Bastão de Souza,
Thiago Guilherme Maziviero,
Cintya Aparecida Christofoletti,
Tamaris Gimenez Pinheiro and
Carmem Silvia Fontanetti

Additional information is available at the end of the chapter

1. Introduction

Soil is characterized as a complex and dynamic system. It is constituted by several layers that differ in relation to the physical, chemical, mineralogical and biological nature, which are influenced by the climate and activities of the living organisms. Besides contributing to the maintenance of all forms of life that occur in the terrestrial surface, soil plays an important role in protecting the groundwater acting as a collector filter of organic and inorganic residues, helping in sequestering possible toxic compounds [1].

During the last decades of the twentieth century there was an awareness of the importance of the soil as an environmental component and recognition of the need to maintain or improve its capacity to allow it to perform its various functions. At the same time there was a confirmation that the soil is not an inexhaustible resource and, if used improperly or poorly managed, its characteristics can be lost in a short period of time, with limited opportunities for regeneration [2].

However, the final disposal of potentially toxic residues in the soil has become a practical and inexpensive alternative and can cause alterations in the arthropod community [3, 4]. These species can present individual biological alterations (physiological, morphological and behavioural), which can be extrapolated to field studies in order to analyze ecological aspects, such as population dynamics and richness of diversity in the contaminated areas. Therefore, the gathering of biological studies, both laboratorial and field, combined with chemical analysis of the contaminants, provides a real scenario of the effects that the toxic substances can cause in the ecosystem.

Among the substances released in the soil it can be highlighted the petroleum derivatives and heavy metals [5]. In soils contaminated with petroleum and derivatives, some contaminants stand out compared to others, such as benzene, toluene, ethylbenzene and xylenes, known as BTEX, polycyclic aromatic hydrocarbons (PAH) and total petroleum hydrocarbons (TPH) [6, 7]. Pollution by heavy metals is derived from the anthropogenic activity, mainly associated to the industrial process and natural sources, such as volcanic eruptions [8].

Although researches involving soil quality are facing an important technologic challenge with several actions being taken in order to assess, correct and reduce the risks of contaminants in the soil, standardized monitoring combined with remediation strategies are still needed [5].

Thus, several researches aiming to remediate the effects of the soil contaminants have been carried out worldwide. Remediation of a contaminated area involves the application of one or more techniques aiming to remove or contain harmful substances in order to allow the reuse of the area with acceptable risk limits for human and environmental health. For this purpose, an ideal remediation process must remove all the contaminants of the soil or, at least, reduce the percentage of contamination of the environment to acceptable limits; should also avoid the migration of contaminants to other areas.

For the remediation of soils contaminated with petroleum and heavy metals, several physical, chemical and biological techniques have been developed for the removal or degradation *in situ* or *ex situ* of the pollutant [6, 9]. In this context, the chapter aims to provide a thorough revision of techniques for the removal or degradation of the pollutants as well as a discussion on the implementation of such techniques for the development of remediation strategies and policies.

2. Dynamic of pollutants in the soil

Geosphere, or terrestrial layer, is that part of the earth on which the human beings live and extract the maximum of its resources. Erstwhile it was believed that the earth had unlimited capacity to absorb the impacts of humankind. Currently, the geosphere is considered very fragile and vulnerable to injuries originating from anthropogenic activities. According to Manahan [10] the definition of pollutant can be described as the increase in the concentration of a certain substance to higher levels than that they occur naturally, arising from an external source, generally related to the human activity.

There is great difficulty in predicting the behaviour of a xenobiotic in the soil, since its composition is totally complex and heterogeneous. Therefore, the knowledge of the physico-chemical characteristics of the contaminant compounds and the environment is fundamental to predict its dynamic [11].

It should be noted that several soils have the capacity to assimilate and neutralize such pollutants, since chemical and biochemical phenomena are capable of attenuating the harmful nature of the pollutants. These phenomena include processes of oxi-reduction, hydrolysis, acid-base reactions, precipitation, adsorption and biochemical degradation. Some hazardous organic chemical products can be degraded to innocuous products on the soil and the heavy metals can be sorbed, immobilized or mineralized. In general, a lot of care should be taken in the elimination of the residues, rejects and other potentially hazardous materials to the soil, particularly where there is the possibility of contaminating the existing water.

When the contaminant reaches the soil, either on purpose or accidentally, it suffers the action of geochemical and biological phenomena and is distributed by the subsurface in the vaporized, residual or adsorbed phases, free phase and dissolved phase. The distribution of such phases will depend on their physico-chemical characteristics and also on the type of the soil [12]. Thus, the mobility of the contaminants and, consequently, their toxicity are directly related to the capacity of the soil in maintaining them retained in their solid phase, making them unavailable to be absorbed by plants, eroded and/or leachate [13]. Among the factors that determine the binding of contaminants to the soil there is the available surface area of the particles (m^2/g). Moreover, the electrical charges of the particles of the soil matrix also influence in the adsorption of the contaminants to the environment. It is noteworthy that in relation to their physico-chemical properties the contaminants are classified as Dense Non-Aqueous Phase Liquid (DNAPL), when the substance is more dense than the water and Light Non-Aqueous Phase Liquid (LNAPL), when it is less dense [14].

The main processes of interaction between the organic compounds or metals and the environment are the retention by adsorption, absorption or precipitation; biotic and abiotic transformations and transport by volatilization, leaching or runoff [15]. There are compounds highly resistant to degradation that can interact strongly in a reversible or irreversible way with the colloidal components of the soil. This process is called sorption, both for adsorption and absorption. Adsorption is characterized as an interfacial process while absorption differs for involving the penetration of the compound in the particles of the soil and can be accumulated inside the absorber system [11].

In general, the dynamic of the contaminants in the soil can be modelled by three mechanisms of mass transference, namely: advection, dispersion and attenuation.

a. *Advection* – it consists in the mechanism where the contaminants coincidentally follow the flow vectors and keep a direct relationship with the speed of percolation in the soil. It is the mechanism responsible for the formation and mobilization of the free phase of hydrocarbons.

b. *Dispersion* – Consists in the mechanism responsible for the decrease in the concentration of the contaminants in the fluid percolation and that can occur by two processes: hydrodynamic dispersion and molecular diffusion. Hydrodynamic dispersion occurs by the flow restriction in the pores of the soil that generates the reduction in the percolation velocity of the more viscous components while the molecular diffusion is, intrinsically, a phenomenon of dilution of the more soluble compounds, and is the main formation process of the dissolved phase, responsible for the greater mobility of the contaminants. In the case of emulsions, such as hydrocarbons, the dispersion can occur in a more complex mechanism, due to the phenomena of hysteresis (delay) of the entrainment of the contaminants, especially in the saturation fronts and capillary fringe. This process is associated to the formation of the adsorbed phase and also by the production of a fraction of emulsions that can compose the dissolved phase.

c. *Attenuation* – Consists in the reduction of contaminants transported by advection or dilution by chemical or physico-chemical reactions. Chemical attenuation is the more intense in soils with higher cation exchange capacity and acts reducing compounds in the free and adsorbed phase. Also in the list of reactions there are the bioconversion reactions, in which a part of the hydrocarbons is transformed or totally oxidized in organic acids. Chemical attenuation is more intense in the region with higher availability of oxygen.

Physico-chemical attenuation is responsible for the formation of the adsorbed phase and consists in the imprisonment of the contaminants that adhere to the grains of the soil, especially to the grumes of clay with higher activity. However, associated with the mechanisms of chemical attenuation, it is responsible for the formation of the dissolved phase (facilitated by the reduction of pH) [16].

3. Contamination of soil and its effects on the edaphic fauna

Soil ecosystem harbours an enormous biodiversity and is increasingly being recognized that this diversity is essential for the maintenance of the function of other ecosystems [17], since the activities of the invertebrates have significative effects in its organization and structure, dynamics of the organic matter and in the growth of plants [18]. Despite this importance, soil has become a practical and cheap alternative for the final disposal of several toxic residues, resulting in negative consequences [4].

Contaminants can be resistant to the decomposition processes and, therefore, can be accumulated in the soil [19]. Invertebrates easily become exposed to such contaminants, which can affect their ecological function [20] and influence indirectly the ecosystem and alter the ratio predator/prey and affect the complex food chain [21]. In order to evaluate the ecological effects of this contamination it is developed tests that aim to quantify the abundance, mortality and reproduction of the organisms exposed [20].

In this sense, the following topic will address the effects caused in the edaphic fauna due to the contamination of the soil by heavy metals and petroleum derivatives.

3.1. Petroleum and its derivatives

According to Leblond [22], it is expected the production of 95 million barrels of petroleum per day in order to meet the growing worldwide demand of this resource. Crude petroleum is a complex mixture constituted, mainly, by hydrocarbons, organic sulphur compounds, nitrogen and oxygen [23]. Although about 80% of the total production of crude petroleum is generated from terrestrial fields, few studies about its impact on the soil are available [24].

Studies on the toxicity of petroleum have shown that some species present higher sensitivity to these contaminants. Survival of earthworms (*Eisenia andrei* and *E. fetida*) and enchytraeids (*Enchytraeus crypticus*) can be reduced in soil containing crude petroleum [25, 26], while the abundance of Isopoda and Hymenoptera in areas contaminated with complex mixtures derived from refineries can be higher in relation to uncontaminated areas [27].

Among the petroleum derivatives, the Polycyclic Aromatic Hydrocarbons (PAH) have a prominent role. Chemically, they are aromatic compounds formed by two or more benzene rings, constituted exclusively by atoms of carbon and hydrogen, arranged in a linear, angular or grouped form [28], and are residues of combustion, petroleum refinery and other industrial processes of high temperature [29]. There are thousands of these substances in the environment, each one differing in the number and position of the aromatic ring [30], but only 16 substances cause environmental concern: acenaphthene, acenaphthylene, anthracene, benzo(a)anthracene, benzo(a)pyrene, benzo(b)fluoranthene, benzo(ghi)perylene, benzo(k)fluoranthene, chrysene, dibenzo(a, h)anthracene, phenanthrene, fuoranteno, fuoreno, indeno(1,2,3-cd)pyrene, naphthalene and pyrene [31].

Although van Brummelen et al. [19] asserted that the exposure of invertebrates to PAH accumulated in the soil can affect the ecological function of these organisms, little is known about their effects [32, 33]. However, it is known that terrestrial invertebrates do not have the ability to metabolize aromatic compounds, with exception of some species that have microorganisms associated to the intestine [34], which implies in a broader problem, since it generates the bioaccumulation in the organism, enhancing the possibility of contaminating their predators via the food chain [35].

A small review performed by Souza et al. [7] discusses the main ecotoxicological assays that can be applied in soils contaminated by petroleum hydrocarbons. In this review, the authors affirm that bioassays with invertebrates have been efficient, thanks to the important role that these animals play in the ecological processes of the soil, such as cycling and decomposition.

Studies using earthworms as bioindicator organisms of contamination of the soil by PAH showed that the impact in these organisms is limited. Both the survival and reproduction rates were not altered and the concentrations of these substances in the individuals were low, suggesting low absorption by them [36]. Schaub and Achazi [apud 36] observed that PAH did not influence the survival and growth of the earthworm *E. fetida* in the concentration of 100.8 mg/kg, but the reproduction was affected in the concentration of 1.008 mg/kg. Chrysene did not alter the survival of *E. fetida* in a study carried out by Bowmer [37].

The non-toxicity of PAH for earthworms can be explained by the fact that there is a mutual interference between them [38, 39]. Earthworms are responsible for assisting the elimination of the PAH in the soil by improving the natural conditions of biodegradation, contributing to the increase of its oxygenation due to the intimate contact of the microorganisms present in their intestine with the soil [36].

In relation to Collembola, Sverdrup et al. [33] affirm that they are more sensitive to PAH when compared to other organisms, such as earthworms, being, therefore, good models of toxicity for this class of contaminants. To reach this conclusion, the authors tested 16 different PAH, from which eight affected the reproduction and survival rates of Collembola. Eom et al. [40] corroborate the fact that Collembola are more sensitive to PAH. Isopoda did not show to be more sensitive to contamination by PAH. The species *Oniscus asellus* presented a small alteration in the abundance after exposure to benzo(a)anthracene and no effect after exposure to benzo(a)pyrene. The species *Porcellio scaber* did not present alteration for any of the two substances [19].

PAH can also act indirectly on organisms and cause alterations in the populations, since the increase in the density of the soil due to their presence and their hydrophobic properties decrease the inhabitable space within the pores of the soil. Moreover, PAH can also act as fungicides, eliminating the source of food of some organisms [41].

Due to the reduced number of studies there is not a base to predict alterations in the community of invertebrates caused by contaminations of PAH. Studies with this focus does not seem to be a promising tool to assess the risks of this substance and the use of more sensitive biochemical markers (concentration of metabolites, damages in the DNA) are better strategies for this purpose [3].

In this sense, besides the traditional tests with Annelida and Collembola, studies with other terrestrial invertebrates have been developed to assess the quality of soils [42]. Diplopoda also make part of the edaphic fauna and are continuously exposed to the contaminants present in the soil. In these animals, histopathological markers have been applied [43-47].

Tissular alterations in the midgut and perivisceral fat body of the diplopod *Rhinocricus padbergi* were studied by Souza and Fontanetti [46] and Souza et al. [42], after exposure of these animals to a landfarming soil. According to the authors, the chemical analysis showed the presence of high concentrations of compounds such as PAH and metals, the authors also inferred that the histological and physiological alterations observed can be an attempt of defence of the animals exposed to this residue, in an attempt to eliminate and/or neutralize the assimilation of toxic residues [42].

3.2. Heavy metals

As a consequence of the technological development and global population growth, the agricultural and industrial activities have intensified, leading to a considerable increase of metals in the different compartments of the environment. Unlike organic pollutants, the toxicity of metals is intrinsic to their atomic structure and they cannot be transmuted/mineralized to a total innocuous form [48].

Pollution by heavy metals in terrestrial ecosystems has been recognized as a serious environmental concern, due to their non-biodegradability and tendency to accumulate in plants and animal tissues [49]. The extreme sensitivity of the macrofauna to the conditions of the soil make them potential indicators of the disturbance occurred in this environment [50]. For studies of this nature, the most used organisms are nematodes, earthworms, Collembola [apud 51], as well as molluscs [49] and ants, despite the last two ones be quite resistant to this type of contamination [52].

Among the most common responses of these organisms to the contamination by heavy metals it can be highlighted the decrease in the diversity of species due to changes in the composition of the community that eliminate the most sensitive species [53, 54] and promote the tolerance of opportunistic species [55] (Syrek et al., 2006) or invasive species [56].

Despite the increase in the abundance, richness and/or uniformity is not commonly found, some studies reported these types of alterations with the increase of pollution by heavy metals. It can be cited the studies developed by Russell and Alberti [57] that observed that Protura present tolerance to heavy metals, since this group was the only one found in sites highly contaminated; by Nahmani and Lavelle [51] that also found that the abundance of some groups of arthropods, such as larvae of beetles of the subfamily Hoplinae and family Staphylinidae, was positively correlated with pollution by heavy metals; by Migliorini et al. [58] that verified increase in the abundance of Protura, Diplura and Collembola with increase in the pollution by metals; by Grzés [52] that presented clear evidence of the increase of the diversity of species of ants with the increase in the pollution by metals.

Although no direct explanation for these patterns has been proposed, some of the authors point out the importance of the interactions between fauna and soil, mainly related to decrease of predation and competition between the edaphic organisms [52]. For this reason, considering all the macrofauna communities as indicator seems logical, since they have a wider range of adaptive mechanisms than a single taxonomic group [51, 59].

Pollution by metals can still influence directly the communities by the alteration of the abiotic conditions such as temperature and humidity. If the pollution decreases the density of the vegetation, the temperature of the environment will increase and this will facilitate the increase in the diversity of thermophilic organisms [52]. According to the same author, pollution by metals can favour species with affinity to humidity by reducing the microbial activity, allowing an accumulation of organic matter [52].

4. Types of treatment of contaminated soils

Geochemical and biological processes that determine the mobilization and transformation of the compounds in the soil involve countless variables, making the remediation process a complex task. Thus, for the remediation be satisfactory and complies the environmental legislation, it is necessary to know the treatment technologies available, their advantages and disadvantages (table 1), cost-benefit relationships, applicability regarding the hydrogeology of the place and the nature of the contaminant [60].

TECHNOLOGY	ADVANTAGES	DISADVANTAGES
Solidification/stabilization	• Simple design • Cost-effective • Large soil volume can be treated • Very recommended for metals	• Does not promote the treatment of the contaminant, promotes only immobilization • Short-lived • Dependent on the soil characteristics and homogeneity of the mixture • Process hindered by the depth of the contaminant
Advanced Oxidative Processes (AOP)	• Cost-effective • Mineralization capacity • Recommended for soils with high permeability • Different reagents may be employed	• Mass transfer of the adsorbed phase to the aqueous phase • Risk of aquifer contamination by not recovered solvent • Limitations for large-scale application (*ex-situ* treatment) • The use of strong acids causes destruction of the basic structure of the soil
Advanced Oxidative Processes (AOP)	• *In situ* treatment • Cost-effective • Rapid process • Little or no waste is generated	• Lower efficiency for insoluble compounds • Susceptible to changes in pH • May be harmful to soil microorganisms
Thermal desorption or extraction with supercritical CO_2	• High efficiency for volatile compounds • Soil aeration can facilitate the bioremediation process • Rapid process • Low environmental impact	• Low efficiency for soils with low permeability • Not recommended in saturated areas • Treatment of the realeased vapors is required
Incineration	• High efficiency • Rapid process • Compounds mineralization • May be used where other processes are not effective	• High cost • Release of secondary compounds to the atmosphere • Periodic and rigorous monitoring are riquered • *In situ* treatment is not possible
Adsorption with clay	• Cost-effective • Simple design • Can be combined with other techniques	• Soil exchange is required • Limited by buffer capacity of the soil • Selectivity for specific ions
Electrokinetic	• High efficiency • *In situ* treatment	• Treatment time depends on the distance between the electrodes • pH change in areas near the electrode

TECHNOLOGY	ADVANTAGES	DISADVANTAGES
	• May be combined with bioremedition techniques	• Cost depends on the contaminant concentration and soil • Lower efficiency in soils with low permeability

Table 1. Advantages and disadvantages in the use of different techniques in the remediation of soils contaminated by petroleum and heavy metals

According to Andrade et al. [6], the technique to be used depend on some factors, such as: physical, chemical and biological conditions of the contaminated site, concentration of the contaminants and time needed for the degradation/removal of the target compounds, according to the technique to be employed.

The main processes of interaction between the hydrocarbons or metals and the environment are retentions (adsorption, absorption or precipitation); biotic and abiotic transformations, transport by volatilization, leaching or runoff [15]. There are compounds highly resistant to degradation that can interact strongly in a reversible or irreversible way with the colloidal components of the soil. This process is called sorption, both for adsorption and absorption. Adsorption is characterized as an interfacial process while absorption differs for involving the penetration of the compound in the particles of the soil and can be accumulated inside the absorber system [11].

Since 1993, information of the Environmental Protection Agency (EPA) was considered to indicate the need for innovative technologies, such as remediation, to replace conventional processes. New technologies have as objective the treatment of organic compounds, however, few alternatives are available for the removal of metals in the soil, particularly *in situ*.

Among the existing remediation processes it can be highlighted the technologies of immobilization, destruction of the contaminants and separation. Immobilization technology consists in the creation of physical barriers to avoid the migration of the contaminants, such as processes of solidification/stabilization (encapsulation of the contaminants). The processes of destruction are based, mainly, on the use of high temperatures and chemical methods, such as incineration, chemical reduction, chemical oxidation, photolysis and bioremediation; and the separation consists in retaining, isolating or extracting the contaminants to a phase of easier management or to a more concentrated phase, reducing the volume of the material to be remediated or disposed, such as processes of thermal desorption, washing the soil, extraction by solvent and supercritical extraction [61, 62].

4.1. Solidification/stabilization

The process of solidification/stabilization, also known as immobilization, modifies the physico-chemical characteristics of the residue to contain the contaminants. Metals are commonly remediated by solidification *ex situ* by encapsulation and sometimes complexation.

The encapsulation technology has become an important alternative treatment for the disposal of hazardous residues in landfills and control of contaminated areas, since it provides an improvement of the physical and toxicological characteristics of the residue and/or soil, facilitating its management in a safe and effective form. Moreover, the cost of the encapsulation has been considered low in relation to other treatment techniques, fact that has stimulated the development of this technology in the last years. However, there is an increasing interest in more durable and safer solutions [63].

The frequently used agents for encapsulation are Portland cement and lime. In physical terms, the cement presents response in a smaller interval of time than lime, since its curing takes place in less time. Chemically, both act to alkalinize the environment, increasing the pH of the compound, decreasing the solubility of the contaminants, since it is known that the solubility is dependent on the pH [64]. Physically, it occurs the cementing of the particles, causing a decrease in the mobility of the contaminant within the soil. Therefore, the reduction in the mobility of the contaminant can be enhanced by the alkalinisation of the environment and also by the cementing effect of the particles.

After application of the encapsulation technique, some assays become necessary for analyzing the effectiveness of the method, which consist in chemical and physical analysis of the treated compound. Chemical analysis are performed based on leaching assays and chemical extraction. Physically, it is performed analysis of compressing, resistance to simple compression, permeability, durability, among others [63].

Another solidification technique involves the vitrification by the passage of an electric current between electrodes. This process results in the retention of solids and incorporation of metals in the vitrified method. This technology is being commercially evaluated and presents very promising results. Vitrification has been used for capturing mercury and other volatile metals such as lead and arsenic [65].

4.2. Washing and extraction by solvent and chemical oxidation

One technique of separation of organics in soils very used is the extraction by organic solvents. In these cases, the organic contaminant is extracted from the contaminated site and later destined to the destruction treatment. The process occurs by washing the soil using adequate solvents for each type of contaminant, such as detergents for oils or petroleum and chelators for metals. It has the disadvantage of being a process that requires specific machinery, demands specialized staff and, at the end of the process, generates great quantities of contaminated liquid residues, which must be adequately treated and disposed posteriorly [11].

Chemical oxidation or *In Situ Chemical Oxidation* (ISCO) has shown to be a promising technique for the remediation of soils contaminated by organic compounds [66]. This technique is based on the application of strong oxidant agents to degrade the organic. It has been applied both *in situ* and *ex situ*, and its application in the field is more appropriate.

ISCO also has its limitations, especially regarding the reactivity of the agent with the contaminant and mass transference between the adsorbed and aqueous phases, where generally occurs the oxidation reaction [67]. The most used agents in ISCO processes are ozone (O_3), hydrogen peroxide (H_2O_2) and potassium permanganate ($KMnO_4$). Each one has advantages and disadvantages and the application depends on the environment to be treated and the contaminant to be degraded [68].

4.3. Advanced Oxidative Processes (AOP)

The most effective processes in the destruction of organic pollutants are known as advanced oxidative processes (AOP). AOP are characterized by the generation of hydroxyl radicals (HO*), which presents high potential pattern of oxidation, superior to those of other oxidant species, such as O_3, H_2O_2 and chloride (Cl_2) [69], capable of reacting with practically all classes of organic and inorganic compounds. These processes are emerging as a promising alternative for the treatment of matrices contaminated with highly toxic and recalcitrant substances, leading them to total mineralization or formation of more biodegradable intermediates [70, 71].

Although there are more economical processes, not always the time needed to achieve the expected results allow their use, thus, AOP can be used when these limits of time and other logistics become hierarchically more important.

Fenton system is one of the most known advanced oxidative processes and consists in the combination of hydrogen peroxide and ferrous ions to form hydroxyl radicals. The oxidizing power of Fenton's reagent (H_2O_2/Fe^{2+}) is attributed to the hydroxyl radicals resulted from the catalytic decomposition of hydrogen peroxide in acid medium, whose general reaction is represented by:

$$Fe^{2+} + H_2O_2 \text{->} Fe^{3+} + OH^- + {}^*OH$$

Hydroxyl radicals generated oxidize the organic compounds of the environment, generating intermediates that are attacked again by other hydroxyl radicals and can reach the complete mineralization (CO_2 and H_2O). This system has been widely studied in the oxidation of organic compounds of high toxicity.

The reagents that compose the Fenton system present advantages over the others because they are compounds relatively inexpensive and non-toxic, besides the reaction occurs at room temperature and pressure. It is known that the hydroxyl radical oxidize effectively organic compounds in aqueous phase, including the polychlorinated biphenyls (PCB) [72].

The efficiency of the chemical oxidation in soil is influenced, mainly, by factors such as concentration of iron, concentration of peroxide, presence of other organic compounds competitive by hydroxyl and pH [62]. Moreover, some researchers have observed a strong increase in the oxidant power of the Fenton reagent when combined with radiation UV or UV-visible, called Photo-Fenton. This technique has shown to be an extremely promising alternative, especially on tropical countries, like Brazil, where the incidence of sunlight is high practically during the entire year, configuring an important source of energy, hitherto unexplored [11].

4.4. Thermal desorption or extraction with supercritical CO_2

The most applied system of thermal desorption is the process of injecting steam water in the soil with a system of pumps and vacuum, i.e., it is installed in the area to be remediated a series of pipes from which will be injected in the soil steam water and other suction pipes. The steam at high temperature drags the contaminants, extracting them from the soil, which are then sucked by vacuum sites and sent to filters or condensers to receive appropriate treatment [66].

Extraction of compounds using supercritical fluid consists in making the extraction of the contaminants by passing a gas at high pressure (400 bar) and high temperature (150ºC) through the contaminated soil. In general, CO_2 is the fluid chosen due to its low toxicity and environmental acceptability and this extraction has shown to be very efficient for compounds with high solubility in CO_2, such as PAH, PCB, dioxins and organochlorine pesticides [11].

In the United States, an area with more than 170 tons of soil contaminated with benzene, arsenic, chromium and PAH was remediated using the process of thermal desorption [73].

4.5. Incineration

The use of heat to destroy toxic compounds is a very old practice. Incineration has been used for centuries to destroy or diminish the volume of domestic or agricultural residues that are unnecessary or undesirable. However, during the combustion process occurs the formation of undesirable by-products, such as dioxins and furans, highly toxic and carcinogenic. To avoid the formation of such compounds it is necessary to have strict control over the combustion conditions [74].

To remediate soils contaminated with PAH, this process is one of the most efficient and used, despite the high cost due to the need of soil excavation, transport and treatment with heat [75]. Although the treatments with high temperatures are effective in the treatment of organic residues, a serious problem occurs when the residue has metals, since a fraction of them will volatilize during the treatment and, after the gas cooling, they will condense on particles of metal [76].

4.6. Adsorption with clay

Clays have structures in layers of lamellae that consist on sheets of silicon oxide alternating with sheets of aluminium oxide. The sheets of silicon oxide are arranged in tetrahedra in which each atom of silicon is surrounded by four atoms of oxygen, some variations can present geometrical structure in form of octahedrons. Many clays contain large quantities of sodium, potassium, magnesium, calcium and iron and other metals. Clays can attract cations such as Ca_2^+, Mg_2^+, K^+, Na^+ and $NH4^+$, retaining them between their lamellar structure in order to avoid leaching by water, but maintain then available in the soil as nutrients for the plants [10].

Thus, heavy metals and other charged species are strongly attracted and adsorbed in the clay surfaces. Heavy metals have different sorption characteristics and the mechanisms depend on the adsorbents. The sorption mechanisms include complexation of the surface (adsorption) and ion exchange. Adsorbents show difference in the sequence of selectivity for different metals. One example is lead when compared to other metals, since it is highly attracted and adsorbed by several types of clay.

4.7. Electrokinetic

Electrokinetic remediation, also called electrokinetic processing of the soil, electromigration, electrokinetic decontamination or electrocorrection, can be used to extract metals and some types of organic residues, such as PAH, of saturated or unsaturated soils, sludges and sediments [66]. This technique consists on the application of a direct current of low intensity between the electrodes located in the soil. The materials used for the construction of the electrodes can be graphite, stainless steel and platinum. Electrolysis of the water (in the disperse electrolyte) produces ions H^+ in the anodes and ions OH^- in the cathodes, generating a localized change of pH, which leads to the desorption of the contaminated ions.

Some variations of this technique involve the direct extraction of metallic ions already in the metal form and the others involve the extraction of metallic ions using a posterior process of ion exchange resins. Electrokinetic remediation can be also used to delay or prevent the migration and/or diffusion of the contaminants, directing them to specific sites and diverting them from the freatic sheets.

Currently, the application of electrokinetic process has been considered promising, especially for the remediation of low permeability contaminated soils, where the electric field generated mobilizes electrically charged species, particles and ions in the soil by the processes of electromigration, electrophoresis and eletroosmosis [66]. For the migration process in the electrodes, the contaminants can be removed by reduction in the cathode, precipitation, pumping next to the electrode, or in a more complex form with ion exchange resins.

However, the electrokinetic process is limited by the solubility of the contaminant and by desorption of the contaminants in the surface of the soil. Heavy metals in their metallic state are not being sufficiently dissolved and separated from the samples of soil. The process is also not efficient when the concentration of the ions to be removed is low and the concentration of diverse ions is high. Moreover, factors such as heterogeneity and anomalies in the local surface (boulder, large quantities of iron or iron oxides, large rocks and gravel or materials such as shells) can reduce the efficiency of removal.

The cost of remediating soils contaminated by metals, using the electrokinetic technique is strongly influenced by the soil conductivity, since the consumption of energy is directly related to the conductivity of the soil between the electrodes. The electrokinetic treatment of the soils with high ion conductivity may not be feasible due to the high cost [63].

Another method that uses the electrokinetic technology is the electroacoustic decontamination of the soil. This technology combines the eletrokinetic with the sonic vibration. The properties of the liquid contaminant in the soil can be altered in order to increase the rate of the contaminant removal by the application of a mechanical vibratory energy in the form of sonic or ultra-sonic energy. The elecroacoustic technology is technically feasible for the removal of inorganic species from the soil with clay (and partially effective for the removal of hydrocarbons) [63].

4.8. Bioremediation

According to Yeung et al. [77], biological processes are gaining increasing importance in the treatment of soils. To meet the challenges presented by environmental pollution, the objective of bioremediation (along with prevention and physical and chemical methods for remediation) is reduce the quantity and availability of hazardous chemical compounds and convert them into useful products and/or less innocuous [48]. However, biological processes, when compared to the conventional physical and chemical processes, are safer, less costly and less aggressive to the environment [78].

Bioremediation process can be defined as the use of microorganisms, such as bacteria, fungi, yeasts and algae or their enzymes to treat polluted areas or "return" them to their original condition [48, 79, 80]. In general, bioremediation is based on the biochemical degradation of contaminants [6, 81], resulting in the transformation in metabolites or their mineralization [78].

The types of treatment involved in the remediation process can be of two types: *ex situ*, in which there is an excavation and removal of the contaminated soil to another place and the *in situ*, where the treatment is performed in the local. The *in situ* bioremediation is the most worldwide used type of process regarding the place of treatment [6].

Briefly, the main techniques involved in the bioremediation process are:

a. *Bioattenuation* (natural process) - used to described the passive remediation of the soil, which involves several natural processes, such as biodegradation, volatilization, dispersion, dilution and adsorption of the contaminants, promoted in the sub-surface by native microorganisms [6, 80].

b. *Biostimulation* (or accelerated natural attenuation) - consists in the addition of nutrients and/or descompacting agents in the contaminated soil, increasing the population of endogenous or native microorganisms [42].

c. *Biomagnification* (or bioaugmentation) - characterized by the increase of the native microbiota by the inoculation of exogenous microorganisms (allochthonous) [6, 82, 83]. In this case, according to the literature, generally, the used microorganisms are bacteria, philamentous fungi and yeasts.

d. *Landfarming* - is an *ex situ* remediation technique, based on the placement of the contaminated soil in layers with at maximum 40 cm of thickness and their processing with agricultural machines [84].

e. *Biopiles* - is an *ex situ* technology of bioremediation, which involves the stacking of contaminated soils, which stimulates the aerobic microbial activity, accelerating the degradation of the pollutant by aeration, addition of nutrients and correction of humidity.

f. *Composting* - technology that involves the addition of organic structuring agents in the contaminated soil/compounds, increasing the porosity and airflow in them. Such agents still serve as easy access source of carbon to the biomass growth. The energy released during the degradation of the organic matter result in temperature increase, which facilitates the action of different microbiological phases: mesophilic, thermophilic, cooling and maturation [85].

g. *Phytoremediation* - technique that uses plants as decontamination agent. Involves several mechanisms such as phytoextraction, phytostabilization, rhizofiltration, phytodegradation, phytostimulation, phytovolatilization, vegetative strains, artificial ponds and hydraulic barriers [86].

4.8.1. Bioremediation of sites contaminated by petroleum derivatives and heavy metals

Contaminations of soils with petroleum hydrocarbons have become a worldwide problem in the mid 80's [77]. The contamination sources by these compounds are related with exploration, production, storage, transport, distribution and final disposal of petroleum and their derivatives.

In the biological treatment of soils contaminated by petroleum, microorganisms, being bacteria the most studied, use hydrocarbons, major components of petroleum, as source of carbon and alternative energy in the production of biomass. This process involves the transformation of hydrocarbons into smaller unities and later incorporation as cellular material (biotransformation) or conversion to carbon dioxide (mineralization), resulting in the reduction of the concentration of the petroleum hydrocarbons [87].

There are, in the scientific literature, a considerable number of studies on bioremediation of soils contaminated by PAH, using different remediation methodologies such as treatment of the solid phase, landfarming/composting, phytoremediation, biostimulation among several others [85].

In the landfarming process, petroleum derivatives are removed by volatilization, biodegradation and absorption. The more volatile products, such as gasoline, are removed by volatilization during the aeration process and a small portion is degraded by the microorganism respiration. Derivatives such as diesel and kerosene have less volatile constituents than gasoline and, therefore, the biodegradation is more significative than volatilization. The heavier compounds, such as lubricating oil, are not volatile, suffering only biodegradation [88].

Composting has obtained success in the bioremediation of petroleum derivatives using different compounds, such as mushrooms [89], soot residues [90], green residues [91, 92], maple leaves and alfalfa [93] and horse manure [94]. Plants, by phytoremediation, have shown positive results in the degradation of PAH, since it stimulates the growth and microbial activity in the rhizosphere (interface soil/root) [95].

Besides the individuals use of these processes, it is possible to combine more than one technology in the bioremediaton of contaminated soils. According to Straube et al. [96], microorganisms naturally present in the soil that degrade PAH can have their degradation capacity limited due to several environmental factors, such as low solubility and low bioavailability of PAH and limitation of nitrogen or other nutrient. Thus, it is possible to combine landfarming with biostimulation and bioaugmentation to increase the efficiency of the technique. In the bioremediation process by biopile it can be also employed procedures such as aeration, bioaugmentation, biostimulation and composting in order to increase the efficiency of the remediation of petroleum hydrocarbons [97].

Mohan et al. [85] and Megharaj et al. [80] presented a review on the main techniques used in the bioremediation of soils contaminated by organic pollutants. The different strategies of the bioremediation process have specific advantages and disadvantages (table 2), which, according to the same authors, need to be considered in several situations, since there are many factors that limit the efficiency of the microbial degradation of organic pollutants: bioavailability of the pollutant, low temperatures, anaerobic conditions, low levels of nutrients and co-substrates, presence of toxic substances and physiological potential of microorganisms.

TECHNOLOGY	ADVANTAGES	DISADVANTAGES
Landfarming	• Simple desing and implementation • Cost-effective • Large soil volumes can be treated • Favourable public opinion • Complete destruction of waste material	• Large treatment area is required • Risk of human pollutant exposure • Limited to removal of biodegradable pollutants
Phytoremediation	• Cost-effective • Easy to implement and operate • Environment-friendly • Favourable public opinion • Reduced pollutant exposure	• Slower than other methods • Soil properties, toxicity level and climate should allow plant growth • Limitations for large-scale application
Bioaugmentation	• Cost-effective • Increase the bioavailability of pollutants • Short treatment times	• Laboratory strains of microorganisms rarely grow in contaminated soil • The use of genetically modified organisms does not have public acceptance • Recent and under development • Possible environmental risk by introducing non-indigenous microorganisms
Biostimulation	• Improve the degradation potential of the inhabiting microbial population	• Dependent on the indigenous organisms

TECHNOLOGY	ADVANTAGES	DISADVANTAGES
		• Inorganic nutrients that are injected may precipitate metals, swell clays, change redox potentials and conductivity
Bioattenuation	• Efficient and continuous process • Equipment are not required • Less impact on the environment	• Preliminary studies are required • Slow and unpredictable process • Periodic and rigorous monitoring are riquered • May be costly
Biopiles	• Rapid process	• *Ex situ* treatment • High cost • Other sites may be contaminated
Composting	• Cost-effective • Simple design • High efficiency	• Depending on the compound employed, there may be a small increase in contamination, pH, solubility and mobility of toxic elements • Poor public reception due to odour and insects

Table 2. Advantages and disadvantages in the use of different techniques in the bioremediation of soils contaminated by petroleum and heavy metals

To illustrate the difficulty and success/failure of the bioremediation of soils contaminated by petroleum and its derivatives, there are some studies performed in different parts of the world, which use different techniques of bioremediation. Bento et al. [98] assessed the efficiency of the natural attenuation, biostimulation and bioaugmentation in the degradation of TPH (Total Petroleum Hydrocarbons) in soils contaminated by diesel, in samples from California and Hong Kong. After 12 weeks of incubation, the authors observed that the three techniques employed show different effects in the degradation of light fractions (C_{12}-C_{23}) and heavy fractions (C_{23}-C_{40}) of TPH in the soil samples. However, the authors noted that the number of microorganisms that degrade diesel and the heterotrophic population were not influenced by the treatments, suggesting, therefore, that detailed studies on the characterization of the site are needed before deciding the adequate bioremediation method.

Haderlein et al. [93] studied the effects of composting or simple addition of manure in the soil, during the mineralization of pyrene and benzo[a]pyrene. It was reported that composting and addition of manure had no effect on the mineralization of benzo[a]pyrene. In contrast, the mineralization rate of pyrene increased dramatically with the amount of time that the soil was composted (more than 60% of mineralization after 20 days).

Bioremediation of metals face major obstacles in relation to the bioremediation of organic compounds, since metals introduced in the environment cannot be degraded. They persist indefinitely and can cause pollution of water, air and soil, and the main strategies in the control of their contamination are the reduction of their bioavailability, mobility and toxicity [99].

The oxidation state, solubility and association of metals with other organic and inorganic molecules can vary, however, the microorganisms, as well higher organisms can play an important role in the bioremediation of the concentration of metals, so that they become less available and less hazardous [48].

Among the main methods involved in the remediation of environments contaminated by heavy metals it is included the phytoremediation [99] and the use of microorganisms [100].

In this context, phytoremediation of heavy metals present in the soil, also called phytoextraction, is the technique that uses the capacity of the plants to absorb the metals [101]. As a general rule, metals bioavailable for absorption by plants include Cd, Ni, Zn, As, Se and Cu. Metals moderately bioavailable are Co, Mn and Fe; while the least are Pb, Cr and U [apud 101]).

Phytoremediation process can be divided into three types: phytoextraction, phytostabilization and rhizofiltration. Phytoextraction uses species of hyperaccumulator plants to transport metals to soil and concentrate them into the roots or buds, which will be later collected; in the phytostabilization the plants are used to limit the mobility and bioavailability of metals in the soil by sorption, precipitation, complexation or reduction of the valences of metals; rhizofiltration uses roots of plants in order to absorb, concentrate and precipitate metals from residual waters, which can include soil leachates [102].

Microorganisms, frequently used in the bioremediation of organic pollutants, can be also used in the bioremediation of soils contaminated with metals by biosorption (process in which metals are absorbed and/or complexed in live or dead biomass), alterations in the redox state (transformations catalyzed by enzymes) [103, 104]), by biosurfactants [105], bioleaching (immobilization of metals by excretion of organic acids or methylation reactions), biomineralization (immobilization of metals by the formation of insoluble sulphides or polymeric complexes) and intracellular accumulation [apud 100].

Since metals cannot be biodegraded in CO_2 and water, microorganisms can only modify their speciation, converting them into non-toxic forms [105]. In order to ensure the efficiency of the bioremediation process, the microorganisms added in the contaminated site must have, besides enzymes of biodegradation, resistance to the metal target [101].

5. Current policies for soil remediation

Bredehoeft [106] suggested that the problem of the remediation of toxic substances would be present in the society for a long time and taking into consideration the policies and expenses of the period with the issue, it would exist until mid twenty first century. Fifteen years after this statement, management of contaminated soils and waters still continue to be a current environmental issue due to the great number of areas around the world that face this problem [107].

In the countries member of the EEA (European Environment Agency), according to estimates performed in 2007, about 250,000 areas need remediation. Potentially toxic activities occurred in about 3 million areas, which are under study to determine the need for remediation and, if this tendency continues, the number of areas requiring remediation will increase in 50% until 2025. In these countries, approximately 35% of the costs with remediation were public [108].

In the USA, the report of USEPA [109] state that, despite much has already been done in the last decades of the last century, a considerable amount of work is still needed. According to the report, about 300,000 areas will still need remediation in the next three decades. The estimate cost for the remediation of these areas is around 209 billion dollars, funded by the responsible for the contamination, private or public entity.

According to Fernandes et al. [110], the resources needed, both human and economic, to overcome the challenges in the implementation of remediation programs can be great. The resources destined to this purpose will not be the same in different countries. Some countries are more prepared to deal with the costs of the remediation programs in relation to others, since they have appropriate mechanisms (technical and economic) to implement projects on a large scale. In the countries where this is not possible, the existence of contaminated areas should be a livelong problem.

New remediation technologies are under development in the physical and/or chemical areas, however most of them are still in the initial phase of elaboration [111]. However, the trend of emerging technologies are focused in methods in which the contaminants can be destroyed or carefully removed with low risk of secondary contamination [112]. The methodologies traditionally used, physical and chemical, simply transfer the contaminants, creating other sources of contamination and not eliminating the problem [113].

According to Koenigsberg et al. [114] there is the intention to use tools of molecular biology of microorganisms in contaminated areas, which can and must influence the conception and management of bioremediation engineer and open new paradigms so that the closure of a contaminated site does not occur.

Due to this, in the last years, the bioremediation methodologies have a significative portion of the remediation market [112]. According to Singh et al. [115], bioremediation entered in a new era with the use of genetically modified bacteria, however its use is still limited due to the fact that environmental factors can interfere in the process, making the results unpredictable. A study performed by Liu et al. [116] using genetically modified bacteria showed that its use is a promising strategy in the bioremediation process of environments contaminated by arsenic. Other technologies of bioremediation in development include the use of protein engineering, metabolic engineering, transcriptome and proteomics [117, 118].

For the development of tests in field using genetically modified bacteria, the major obstacle is the environmental concern and political restrictions for the use of these organisms [113].

As most of the researches on this theme are still basic, there is a growing need for regulatory and cost protocols and, thus, transform this potential technology into reality [119].

6. Conclusion

Contamination of the soil by petroleum and heavy metals has shown to be one of the major environmental problems that the governments and researchers must solve in the next decades. Several studies available in the literature warn about the negative effects of these substances in the living organisms, mainly in terrestrial invertebrates, since they are in direct contact with the contamination. In order to avoid that this problem become more serious, several remediation technologies have been elaborated and improved. Physical and chemical techniques are very used, however bioremediation, as it is ecologically correct, has gained great prominence, both in the remediation of petroleum and heavy metals.

Soil remediation standards are based on the protection of human health and on the protection of the ecosystem. Critical values for concentration in the soil are calculated based on human toxicology and others based on ecotoxicology. The most critical value is retained as soil remediation standard. The methodology for site specific risk assessment is based on the approach followed to derive soil remediation standards. A generic approach is followed for the derivation of soil remediation standards, while for site-specific risk assessments certain parameters, such as soil properties, can be evaluated.

Author details

Raphael Bastão de Souza, Thiago Guilherme Maziviero, Cintya Aparecida Christofoletti, Tamaris Gimenez Pinheiro and Carmem Silvia Fontanetti*

*Address all correspondence to: fontanet@rc.unesp.br

Departament of Biology, São Paulo State University, Rio Claro, Brazil

References

[1] Sousa A, Pereira R, Antunes SC, Cachada A, Pereira E, Duarte AC, Gonçalves F. Validation of avoidance assays for the screening assessment of soils under different anthropogenic disturbances. Ecotoxicology and Environmental Safety 2008;71 661-670.

[2] Nortcliff S. Standardisation of soil quality attributes. Agriculture, Ecosystems and Environment 2002;88 161–168.

[3] Van Straalen. The use of soil invertebrates in ecological surveys of contaminated soils. Developments in Soil Science. 2004;29 159-195.

[4] Fontanetti CS, Nogarol LR, Souza RB, Perez DG, Maziviero GT. Bioindicators and biomarkers in the assessment of soil toxicity. In: Pascucci S. (ed.) Soil Contamination. Rijeka: InTech; 2011. p.143-168.

[5] Pascucci S., editor. Soil Contamination. Rijeka: InTech; 2011.

[6] Andrade JÁ, Augusto F, Jardim ICSF. Biorremediação de solos contaminados por petróleo e seus derivados. Eclética química 2010;35(3) 17-43.

[7] Souza TS, Christofoletti CA, Fontanetti CS. Ecotoxicological assays applied in soils contaminated by petroleum hydrocarbons. In: Visser J. (ed.) Ecotoxicology around the globe. New York: Nova Science Publishers; 2011. p247-26.

[8] Aras S, Aydin SS, Körpe DA, Dönmez Ç. Comparative genotoxicity analysis of heavy metal contamination in higher plants. In: Begun G (ed.) Ecotoxicology. Rijeka: InTech; 2012. p.107-124.

[9] Mulligan CN, Yong RN, Gibbs BF. Remediation technologies for metal-contaminated soils and groundwater: an evaluation. Engineering Geology 2001;60 193-207.

[10] Manahan SE., editor. Introduction to Chemistry: Fundaments of Environmental Chemistry. Boca Raton; 2001.

[11] Higarashi MM. Processos oxidativos avançados aplicados à remediação de solos brasileiros contaminados com pesticidas. PhD thesis. Universidade Estadual de Campinas; 1999.

[12] Silva DN, Zagatto PJP, Guardani R, Nascimento CAO. Remediação de Solos Contaminados com Linear Alquil Benzenos Usando Reagentes de Fenton. COBEQ, 2004.

[13] McBride MB., editor. Environmental chemistry of soils. New York: University Press; 1994.

[14] Otten A, Alphenaar A, Pijls C, Spuij F, Wit H. In Situ Soil Remediation. Kluwer Academic Publishers; 1997.

[15] Lavonreti A. Comportamento de herbicidas no meio ambiente. In: Workshop sobre bidegradação - Embrapa, Jaguariuna, São Paulo. p. 81-83; 1996.

[16] Azambuja E, Costa AFU, Nanni AS. O emprego da prospecção geofísica na avaliação de plumas de contaminação por hidrocarbonetos. In: XXVII Congresso Interamericano de Engenharia Sanitária e Ambiental. ABES, Porto Alegre, Rio Grande do Sul; 2000.

[17] Copley J. Ecology goes underground. Nature 2000;406 452-454.

[18] Lavelle P. Diversity of soil fauna and ecosystem function. Biology International 1996; 33 3-16.

[19] Van Brummelen TC, Verweij RA, Wedzinga SA, Van Gestel CAM. Polycyclic aromatic hydrocarbons in earthworms and isopods from contaminated forest soils. Chemosphere 1996;32(2) 315-341.

[20] Smith R, Pollard SJT, Weeks JM, Nathanail CP. Assessing significant harm to terrestrial ecosystems from contaminated land. Soil Use and Management 2006;21 527–540.

[21] Edwards CA. Assessing the effects of environmental pollutants on soil organisms, communities, processes and ecosystems. European Journal of Soil Biology 2002;38 225–231.

[22] Leblond D. (2008). Total world oil output to reach 95 million b/d by 2020. Oil Gas J 2008;106 32–33.

[23] Ko JY, Day JH. A review of ecological impacts of oil and gas development on coastal ecosystems in the Mississipi Delta 2004;47 597-623.

[24] Wang Y, Chen H, Chen H, Wu J. Influences of chronic contamination of oil field exploitation on soil nematode communities at the Yellow River Delta of China. Front. Biol. China 2009;4(3) 376-383.

[25] Filimonova ZV, Pokarzhevskii AD. Enchytraeid Enchytraeus crypticus as a Test Organism for Crude Oil Contamination of Soil. Bull. Environ. Contam. Toxicol. 2000;65 407-414.

[26] Hanna SHS, Weaver RW. Earthworm survival in oil contaminated soil. Plant and Soil 2002;240 127–132.

[27] Faulkner BC, Lochmiller RL. Increased Abundance of Terrestrial Isopod Populations in Terrestrial Ecosystems Contaminated with Petrochemical Wastes. Arch. Environ. Contam. Toxicol. 2000;39 86–90.

[28] Netto ADP, Dias JCM, Arbilla G, Oliveira LF, Barek J. 2000. Avaliação da contaminação humana por Hidrocarbonetos Policíclicos Aromáticos e seus derivados nitratos: Uma revisão metodológica. Química Nova 2000;23(6) 765-773.

[29] Bispo A, Jourdain MJ, Jauzein M. Toxicity and genotoxicity of industrial soils polluted by polycyclic aromatic hydrocarbons (PAHs). Organic Geochemistry 1999;30 947-952.

[30] Eisler R. Polycyclic aromatic hydrocarbon hazards to fish, wildlife, and invertebrates: a synoptic review. U.S. Fish and Wildlife Service Patuxent - Wildlife Research Center 1987.

[31] United State Environment Protection Agency - Quality criteria for water 1986. USEPA 440/5-86-001. Washington, DC. 1986.

[32] Erstfeld KM, Snow-Ashbrook J. Effects of chronic low-level PAH contamination on soil invertebrate communities. Chemosphere 1999;39(12) 2117-2139.

[33] Sverdrup LE, Kelley AE, Krogh PH, Nielsen T, Jensen J, Scott-Fordsmand JJ, Stenersen J. Effects of eight polycyclic aromatic compounds on the survival and reproduction of the springtail Folsomia fimetaria L. (Collembola, isotomidae).Environ. Toxicol. Chem. 2001;20(6) 1332-1338.

[34] Kaplan DL, Harteinstein R. Studies on monooxygenases and dioxygenases in soil macroinvertebrates and bacterial isolates from the gut of the terrestrial isopod, Oniscus asellus L. Comp. Biochem. Physiol. 1978;60 47-50.

[35] Faber JH, Heijmans JSM. Polycyclic Aromatic Hidrocarbons in Soil Detritivores: specific accumulation and effects. In:Van Straalen NM (ed.) Bioindicator Systems for Soil Pollution. Dordrecht: Kluwer Academic Publishers; 1996. p.31-44.

[36] Eijsackers H, van Gestel CAM, de Jonge S, Muijs B, Slijkerman DME. Polycyclic aromatic hydrocarbon-polluted dredged peat sediments and earthworms: a mutual interference. Ecotoxicology 2001;10 35-50.

[37] Bowmer CT, Roza P, Henzen L, Degeling C. The development of chronic toxicological tests for PAH contaminated soils using the earthworm Eisenia fetida and the springtail Folsomia candida. TNO report IMW-R 92/387, The Netherlands, 1992.

[38] Hund K, Traunsperger W. Ecotox-evaluation strategy for soil bioremediation exemplified for a PAH-contaminated site. Chemosphere 1994;29 371-390.

[39] Potter CL, Glaser JA, Chang LW, Meier JR, Dosani MA, Hermann RF. Degradation of polynuclear aromatic hydrocarbons under bench-scale compost conditions. Environ. Sci. Technol. 1999;33, 1717-1725.

[40] Eom IM, Rast C, Veber AM, Vasseus P. Ecotoxicity of a polycyclic aromatic hydrocarbon (PAH)-contaminated soil. Ecotoxicology and Environmental Safety 2007;67 190-205.

[41] Blakely JK, Neher DA, Spongberg AL. Soil invertebrate and microbial communities, and decomposition as indicators of polycyclic aromatic hydrocarbon contamination. Applied Soil Ecology 2002;21 71-88.

[42] Souza TS; De Angelis DF, Fontanetti CS. Histological and histochemical analysis of the fat body of Rhinocricus padbergi (Diplopoda) exposed to contaminated industrial soil. Water, Air and Soil Pollution 2011; 221 235-244.

[43] Triebskorn R, Köhler HR, Zanh T, Vogt G, Ludwing M, Rumpf S, Kratzmann M, Alberti G, Storch V. Invertebrate cells as targets for Hazardous substances. Zeitschfirt fuer Angewandte Zoologie 1991;78 277-287.

[44] Godoy JAP, Fontanetti CS. Diplopods as bioindicators of soils: analysis of midgut of individuals maintained in substract containing sewage sludge. Water, Air and Soil Pollution 2010;210 389-398.

[45] Nogarol LR, Fontanetti CS. Acute and subchronic exposure of diplopods to substrate containing sewage mud: Tissular responses of the midgut. Micron 2010;41 239-246.

[46] Souza TS, Fontanetti, CS. Morphological biomarkers in the Rhinocricus padbergi midgut exposed to contaminated soil. Ecotoxicology and Environmental Safety 2011; 74 10-18.

[47] Souza RB, Fontanetti CS. Alterations in the fat body cells of Rhinocricus padbergi (Diplopoda) resulting from exposure to substrate containing sewage sludge. Microscopy and Microanalysis 2012;18 317-323.

[48] Bonaventura C, Johnson FM. Healthy environments for healthy people: bioremediation today and tomorrow. Environmental Health Perspectives 1997;105 5-20.

[49] Otitoloju AA, Ajikobi DO, Egonmwan RI. Histopathology and Bioaccumulation of Heavy Metals (Cu & Pb) in the Giant land snail, Archachatina marginata (Swainson). The Open Environmental Pollution & Toxicology Journal 2009;1 79-88.

[50] Paoletti MG., editor. Invertebrate biodiversity as bioindicators of sustainable landscapes, practical use of invertebrates to assess sustainable land use. Amsterdam: Elsevier Science BV; 1999.

[51] Nahamani J, Lavelle P. Effects of heavy metal pollution on soil macrofauna in a grassland of Northern France. European Journal of Soil Biology 2002;38 297- 300.

[52] Grzés, I. M. Ant species richness and evenness increase along a metal pollution gradient in the Bolesław zinc smelter area. Pedobiologia 2009;53 65 – 73.

[53] Del Val C, Barea JM, Azon-Aguilar C. Diversity of arbuscular mycorrhizal fungus populations in heavy- metal-contaminated soils. Applied and Environmental Microbiology 1999;65 718–723.

[54] Beyrem H, Mahmoudi E, Essid N, Hedfi A, Boufahja F, Aissa P. Individual and combined effects of cadmium and diesel on a nematode community in a laboratory microcosm experiment. Ecotoxicology and Environmental Safety 2007;68 412–418.

[55] Syrek D, Weiner WM, Wojtylak M, Olszowska GY, Kwapis Z. Species abundance distribution of collembolan communities in forest soils polluted with heavy metals. Applied Soil Ecology 2006;31 239–250.

[56] Piola RF, Johnston EL. Pollution reduces native diversity and increases invader dominance in marine hard-substrate communities. Diversity Distributions 2008;14 329–342.

[57] Russell DJ, Alberti G. Effectsoflong-term, geogenic heavy metal contamination on soil organic matter and microarthropod communities, in particular Collembola. Applied Soil Ecology 1998;9 483–488.

[58] Migliorini M, Pigino G, Bianchi N, Bernini F, Leonzio C. The effects of heavy metal contamination on the soil arthropod community of a shooting range. Environmental Pollution 2004;129 331–340.

[59] Koehler HH. Soil animals and bioindication, Bioindicator systems for soil pollution. In: Van Straalen, DAKNM, Krivolutsky DA. (eds.) Dordrecht; 1996.

[60] Nobre MM, Nobre RCM. Remediação de solos - Técnicas alternativas melhoram desempenho. Revista Química e Derivado 2003;417.

[61] Koustas RN, Fischer D. Review of separation technologies for treating pesticide-contaminated soil. Journal of the Air & Waste Management Association 1998;48 434-440.

[62] Policarpo N A. Tratamento de solos contaminados com bifenilas policloradas (PCBs). PhD thesis. Univesridade de São Paulo; 2008.

[63] Pedrazzoli CD. Remediação eletrocinética de chumbo em resíduos industriais. PhD thesis. Universidade Federal do Paraná; 2004.

[64] Heineck KS, ConsolI NC, Cruz RC, Knop A. Encapsulamento de um solo contaminado por óleo diesel. Revista Geotecnia 2007;110 3-17.

[65] United State Environmental Protection Agency. Engineering Bulletin: Technology Alternatives for the Remediation of Soils Contaminated with Arsenic, Cadmium, Chromium, Mercury, and Lead. USEPA - Office of Emergency and Remedial Response. 1997.

[66] United State Environmental Protection Agency. Engineering Bulletin: Technology Alternatives for the Remediation of Soils Contaminated with Arsenic, Cadmium, Chromium, Mercury, and Lead. USEPA - Office of Emergency and Remedial Response. 1997.

[67] Watts RJ. In situ application of modified fenton's reagent spells success for BETX, MBTE site cleanup. Hazardous Waste Consulant 1998;3.

[68] United State Environmental Protection Agency. In Situ Chemical Oxidation. USEPA 2005.

[69] Mattos IL, Shiraishi KA, Braz AD, Fernandes JR. Peróxido de hidrogênio: importância e determinação. Química Nova 2003;26(3) 273-380.

[70] Miller CM, Valentine RL, Roehl ME, Alvarez PJJ. Chemical and microbiological assessment of pendimethalin-contaminated soil after treatment with fenton's reagent. Water Research 1996;20(11) 2579-2586.

[71] Muszkat L, Bir L, Feigelson L. Solar photocatalytic mineralization of pesticides in polluted waters. Journal of Photochemistry and Photobiology A: Chemistry 1995;87 85-88.

[72] Yeh CK, Wu H, Chen T. Chemical oxidation of chlorinated non-aqueous phase liquid by hydrogen peroxide in natural sand systems. Journal of Hazardous Materials B. 2003;96 29-51.

[73] United State Environmental Protection Agency - FRTR (Federal Remediation Technologies Roundtable): Abstracts of Remediation Case Studies and Technology Assessment – fact sheet. USEPA 2002;6.

[74] Gullet TBK, Lemieux PM. Role of combustion and sorbent parameters in prevention of polychlorinated dibenzo-p-dioxin and polychlorinated dibenzofuran formation during waste combustion. Environmental Science Technology 1994;29(1) 107-118.

[75] Henner P, Schiavon M, Morel JL, Lichtfouse E. Polycyclic aromatic hydrocarbons (PAH) occurrence and remediation methods. Analusis Magazine 1997;25(9-10) 1-8.

[76] Ho TC, Chu HW, Hopper JR. Metal volatilization and separation during incineration. Waste Management 1993;13 455-466.

[77] Yeung PY, Johnson RL, Xu JG. Biodegradation of petroleum hydrocarbons in soil as affected by heating and forced aeration. Journal of Environmental Quality 1997;26 1511–1516.

[78] Molina-Barahona L, Rodrýguez-Vázquez R, Hernández-Velasco M, Vega-Jarquýn C, Zapata-Pérez O, Mendoza-Cantú A, Albores A. Diesel removal from contaminated soils by biostimulation and supplementation with crop residues. Applied Soil Ecology 2004;27 165–175.

[79] Glazer AN, Nikaido H., editors. Microbial Biotechnology – Fundamentals applied of microbiology. New York: WH Freeman and Company Editors; 1995.

[80] Megharaj M, Ramakrishnan B, Venkateswarlu K, Sethunathan N, Naidu R. Bioremediation approaches for organic pollutants: a critical perspective. Environmental International 2011;37 1362-1375.

[81] Bernoth L, Firth I, Mcallister PE, Rhodes S. Biotechnologies For Remediation and Pollution Control in the Mining Industry. Minerals & Metallurgical Processing 2000;17(2) 105-111.

[82] Sarkar D, Ferguson M, Datta R, Birnbaum S. Bioremediation of petroleum hydrocarbons in contaminated soils: Comparison of biosolids addition, carbon supplementation, and monitored natural attenuation. Environmental Pollution 2005;136 187-195.

[83] Hamdi H, Benzarti S, Manusadzianas L, Aoyama I, Jedidid N. Soil bioaugmentation and biostimulation effects on PAH dissipation and soil ecotoxicity under controlled conditions. Soil Biology and Biochemistry 2007;39 1926–1935.

[84] Berger TM. Biorremediação de solos contaminados com hidrocarbonetos totais de petróleo – enfoque na aplicação do processo Terraferm. PhD Thesis. Universidade Federal do Rio Grande do Sul; 2005.

[85] Mohan SV, Kisa T, Ohkuma T, Kanaly R.A, Shimizy Y. Bioremediation technologies for treatment of PAH-contaminated soil soil and strategies to enhance process efficiency. Rev Environ Sci Biotechnol. 2006;5 347-374.

[86] Parrish ZD, Banks MK, Schwab AP. Effectiveness of Phytoremediation as a Secondary Treatment for Polycyclic Aromatic Hydrocarbons (PAHs) in Composted Soil. International Journal of Phytoremediation 2004;6(2) 119–137.

[87] Martins A, Dinardi AL, Formagi VM, Lopes TA, Barros RM, Coneglian CMR, Brito NN, Sobrinho GD, Tonso S, Pelegrini R. Biorremediação. III Fórum de Estudos Contábeis, Faculdades Integradas Claretianas, Rio Claro, SP. www.ceset.unicamp.br/lte/artigos/3fec2401 (accessed 20 June 2012).

[88] Maila MP, Cloete TE. The use of biological activities to monitor the removal of fuel contaminants – perspective for monitoring hydrocarbon contamination: a review. International Biodeterioration and Biodegradation 2005;55 1-8.

[89] Lau KL, Tsang YY, Chiu SW. Use of spent mushroom compost to bioremediate PAH-contaminated samples. Chemosphere 2003;52 1539–1546.

[90] Moretto LM, SIlvestri S, Ugo P, Zorzi G, Abbondanzi F, Baiocchi C, Iacondini A. Polycyclic aromatic hydrocarbons degradation by composting in a soot-contaminated alkaline soil. Journal of Hazardous Materials 2005;126(1–3) 141–148.

[91] Van Gestel K, Mergaert J, Swings J, Coosemans J, Ryckeboer J. Bioremediation of diesel oil-contaminated soil by composting with biowaste. Environmental Pollution 2003;125 361–368.

[92] Antizar-Ladislao B, Lopez-Real J, Beck AJ. Laboratory studies of the remediation of polycyclic aromatic hydrocarbon contaminated soil by in-vessel composting. Waste Management 2005;25(3) 281–289.

[93] Haderlein A, Legros R, Ramsay BA. Pyrene mineralization capacity increases with compost maturity. Biodegradation 2005;17 293-303.

[94] Kirchmann, H, Ewnetu W. Biodegradation of petroleum-based oil wastes through composting. Biodegradation 1998;9 151–156.

[95] Parrish ZD, White JC, Isleyen M, Gent PNM, Iannucci-Berger W, Eitzer BD, Kelsey JW, Mattina MI. Accumulation of weathered polycyclic aromatic hydrocarbons (PAHs) by plant and earthworm species. Chemosphere 2006; 64 609-618.

[96] Straube WL, Nestler CC, Hansen LD, Ringleberg D, Pritchard PH, Jones-Meehan J. Remediation of polyaromatic hydrocarbons (PHAs) trough landfarming with biostimulation and bioaugmentation. Acta Biotechnologica 2003;23 179-196.

[97] Jorgensen KS, Puustinen J, Suortti AM. Bioremediation of petroleum hydrocarbon-contaminated soil by composting in biopiles. Environmental Pollution 2000;107 245-254.

[98] Bento FM, Camargo FAO, Okeke BC, Frankenberger WT. Comparative bioremediation of soils contaminated with diesel oil by natural attenuation, biostimulation and bioaugmentation. Bioresource Technology 2005;96 1049-1055.

[99] Mohammed AS, Kapri A, Goel R. Heavy Metal Pollution: Source, Impact, and Remedies. Earth and Environmental Science 2011;10 1-28.

[100] Abioye P. Biological remediation of hydrocarbon and heavy metals contaminated soil. In: Pascucci S (ed.) Soil Contamination. Rijeka: InTech; 201. P.127-142.

[101] Garbisu C, Alkorta I. Basic concepts on heavy metal soil bioremediation. The European Journal of Mineral Processing and Environmentak Protection 2003;3(1) 58-66.

[102] Kumar N, Dushenkov V, Motto H, Raskin I. Phytoextraction: the use of plants to remove heavy metals from soils. Environ. Sci. Technol. 1995;29 1232-1238.

[103] Lovley DR, Coates JD. Bioremediation of metal contamination. Environmental Biotechonology 1997;8 285-289.

[104] Congeevaram S, Dhanarani S, Park J, Dexilin M, Thamaraiselvi K. Biosorption of chromium and nickel by heavy metal resistant fungal and bacterial isolates. Journal of Hazardous Materials 2007;146 270–277.

[105] Singh P, Cameotra SS. Enhancement of metal bioremediation by use of microbial surfactants. Biochemical and Biophysical Research Communications 2004;319 291–297.

[106] Bredehoeft JD. Hazardous waste remediation: a 21st century problem. Ground Water Monitoring & Remediation 1994; 95-100.

[107] Lemming G. Environmental assessment of contaminated site remediation in a life cycle perspective. PhD thesis. Technical University of Denmark; 2010.

[108] Europen Environmental Agency. EEA: Progress in management of contaminated sites (CSI 015). http://www.eea.europa.eu/data-and-maps/indicators/progress-in-management-of-contaminated-sites/progress-in-management-of-contaminated-1 (accessed 19 July 2012).

[109] United States Environmental Protection Agency – USEPA. Cleaning up the nation's waste sites: markets and technology trends. Office of Solid Waste and Emergency Response 2004.

[110] Fernandes HM, Recio MS, Forsstrom H, Carson PM. International cooperation and support in environmental remediation e is there any room for improvement? Journal of Environmental Radioactivity 2011;DOI 10.1016/j.jenvrad.2011.06.011

[111] Olexsey RA, Parker RA. Current and future in situ treatment techniques for the remediation of hazardous substances in soil, sediments, and groundwater. Soil and Water Pollution Monitoring, Protection and Remediation 2006; 3–23.

[112] Lynch LJ, Moffat AJM. Bioremediation – prospects for the future application of innovative applied biological research. Annals of Applied Biology 2005;146 217–221

[113] Shukla KP, Singh NK, Sharma S. Bioremediation: Developments, Current Practices and Perspectives. Genetic Engineering and Biotechnology Journal 2010; 1-20.

[114] Koenigsberg SS, Hazen TC, Peacock AD. Environmental Biotechnology: A Bioremediation Perspective. Environmental Biotechnology 2005; DOI: 10.1002/rem.20057

[115] Singh S, Mulchandani A, Chen W. Highly Selective and Rapid Arsenic Removal by Metabolically Engineered Escherichia coli Cells Expressing Fucus vesiculosus Metallothionein. Applied and Environmental Microbiology 2008;74(9) 2924–2927.

[116] Liu L, Zhang F, Chen J, Sun G. Arsenic removal from contaminated soil via biovolatilization by genetically engineered bacteria under laboratory conditions. Journal of Environmental Sciences 2011;23(9) 1544–1550.

[117] Alcade M, Ferrer M, Plou FJ, Ballesteros A. Environmental biocatalysis: from remediation with enzymes to novel green processes. Trends in Biotechnology 2006;24 281-287.

[118] Woods TK. Molecular approaches in bioremediation. Current Opinion in Biotechnology. 2008;19 572-578.

[119] Singh JS, Abhilash PC, Singh HB, Singh RP, Singh DP. Genetically engineered bacteria: An emerging tool for environmental remediation and future research perspectives. Gene 2011;480 1–9.

Section 2

Soil Organic Matter Dynamics

Chapter 8

Stability of Organic Matter in Anthropic Soils: A Spectroscopic Approach

M.C. Hernandez-Soriano, A. Sevilla-Perea,
B. Kerré and M.D. Mingorance

Additional information is available at the end of the chapter

1. Introduction

1.1. Stability of soil organic matter

The soil organic matter (SOM) plays an essential role in soil biogeochemical processes (Bot and Benites, 2005). Thus, a productive and healthy soil must present a balance among SOM protection and soil biological functioning (Wander, 2004). However, the prediction of organic matter dynamics in soil is hampered by the complexity of SOM distribution and chemical composition (Foereid et al., 2012). The integration of organic inputs in the physicochemically defined organic pools in soil (Six et al., 2002) and their effect on native organic matter has been described to vary with land use, soil physicochemical properties (Strong et al., 2004; Denef and Six, 2005); and composition of the organic inputs (Kimetu and Lehmann, 2010).

The term soil organic matter refers to all organic substances in the soil: plant and animal residues, substances synthesized through microbial and chemical reactions and biomass of soil micro-organisms. The processes responsible for the stabilization of SOM constitute an essential component of global biogeochemical cycles (Lehmann et al., 2007). Overall, the chemical composition of the organic matter (OM) and the interactions with other soil components such as the mineral phase largely drive the mechanisms for SOM stabilization (Baldock and Skjemstad, 2000), which can be summarized as: (1) biochemical stabilization, (2) physical stabilization and (3) chemical stabilization (Six et al., 2002; von Lützow et al., 2006). The extent of protection offered by each mechanism (Fig. 1) depends on the chemical and physical properties of the mineral matrix and the morphology and chemical structure of the organic matter (Six et al., 2002). Thus, each mineral matrix presents a unique and finite capacity to stabilize organic matter (Baldock and Skjemstad, 2000).

The *physical stabilization* is the preferential location of OM in the soil structure which results in lower access to OM by soil micro-organisms. Thus, integration of OM in soil aggregates reduces the availability of OM for microbial transformation (Six et al., 2002).

The *biochemical stabilization* is a selective enrichment of organic compounds, and refers to the inherent recalcitrance of specific organic molecules against degradation by microorganisms and enzymes. Thus, compounds like lignin, lipids and polyphenols will remain more stable in the soil matrix compared to more labile compounds like polysaccharides and proteins (Six et al., 2002; Kögel-Knabner et al., 2008).

The *chemical stabilization* involves all intermolecular interactions between organic and inorganic substances leading to a decrease in availability of the organic substrate due to surface condensation and changes in conformation, i.e., sorption to soil minerals and precipitation. The chemical stabilization of SOM results mainly from the interaction of SOM with minerals and metal ions (Fig. 1). These interactions include organo-mineral associations such as complexation of organic substances with polyvalent cation bridges, weak hydrophobic interactions (Van der Waals and H-binding) and sorption of SOM to soil minerals (von Lützow et al., 2006; Jastrow et al., 2007). Therefore, some authors have pointed clay fraction as an inhibitor of SOM decomposition (Kleber et al., 2007). For instance, Merckx et al. (1985) described that the stabilization of C and N in soils is positively correlated to the content of clay and silt. Moreover, other authors have indicated that the specific type of clay present in the soil, i.e. clay mineralogy, is most relevant for the capability of a particular soil to stabilize OM (Sollins et al., 1996; Denef and Six, 2005). Consequently, it might be adequate to evaluate specific surface and surface reactivity of soil minerals as predictors of OM stabilization rather than clay content (Baldock and Skjemstad, 2000).

According to Kogel-Knabner et al. (2008), the protection of OM against decomposition by the described mechanisms decreases in the order: chemically protected > physically protected > biochemically protected > non-protected (Fig. 1).

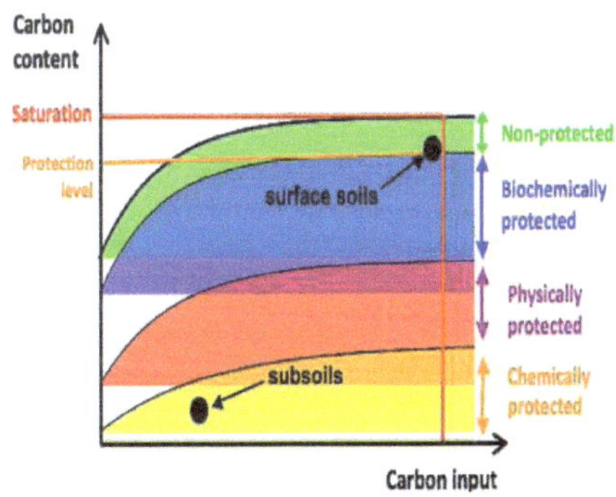

Figure 1. Protection of organic matter (OM) in soil. Chemical protection refers to the interactions of OM with minerals; physical protection renders OM poorly accessible to microbes and enzymes; and biochemical protection results from the differential degradability of organic structures. Saturation is thus defined as the theoretical C storage capacity of a soil (adapted from Six et al., 2002).

The complexation of OM with mineral surfaces occurs mostly through ligand exchange, which is an organo-mineral association between OH groups on mineral surfaces and ionized phenolic OH and carboxylic groups of the OM (Korshin et al., 1997). This interaction is particularly relevant in acidic soils with minerals presenting protonated OH groups and is reverted with increasing pH values in soils (von Lützow et al., 2006).

Another important mechanism for organo-mineral association is the formation of polyvalent cation bridges. The presence of negatively charged acidic functional groups (COO-) in organic molecules results in repulsion from negatively charged surfaces of clay minerals. However, polyvalent cations function as a cation bridge between those two negatively charged sites (von Lützow et al., 2006). In alkaline and neutral soils, the most abundant multivalent cations are Ca2+ an Mg2+ while Fe3+ and Al3+ predominate in acid soils, and present higher binding strength. Besides, the role of cation bridges in the stabilization of SOM is particularly relevant for soils with a predominance of 2:1 clays such as smectite and illite (Jastrow et al., 2007).

Adsorption processes also contribute to protect OM against biological degradation (Balesdent et al., 2000). Although feldspars and quartz are the most common minerals in soils, their specific surfaces are rather low (approximately 0.1 m^2/g), while clay particles provide a significant surface area (specific surface >10 m^2/g) for the adsorption of OM (Jastrow et al., 2007). Therefore, soils with a high content of clays may provide higher SOM protection than sandy soils, resulting in higher total contents of organic matter (Blanco-Canqui and Lal, 2004). Multivalent cations also contribute to the stabilization of OM by inducing flocculation. Clay particles saturated with multivalent cations remain in a flocculated state, which reduces the exposure of adsorbed organic materials on the clay surface. Thus, flocculation and condensation of organo-mineral complexes can effectively isolate and protect OM from decomposition (Baldock and Skjemstad, 2000). This mechanism has been mainly described for soils with low-charge clays (Jastrow et al., 2007).

Kleber et al. (2007) proposed a conceptual model for organo-mineral interactions in soils, partly based on the self-organizing molecular structure of SOM. These authors suggest that SOM is adsorbed on the mineral surfaces in three discrete zones. Thus, in a contact zone, amphiphilic organic fragments accumulate mainly on charged surfaces through electrostatic interactions, thereby directing hydrophobic parts outwards toward the aqueous solutions. This organization results in a membrane-like bilayer with a hydrophobic core. In the outer region, denominated as kinetic zone, further accumulation of organic fragments likely occurs, and the process is assumed to be mediated by the presence of multivalent cations. However, further research is still necessary to advance our understanding on the dynamics and structure of organo-mineral interactions to validate this zonal concept.

Although abundant research has been previously conducted, the described mechanisms responsible for the reduction of OM decomposition rates, such as sorption of OM to minerals, are not yet well understood. Different stabilization mechanisms may act simultaneously and those proposed in the literature remain speculative and poorly supported by data, mainly due to methodological constraints (von Lützow et al., 2006). Therefore, novel research strategies attempt a better understanding of the mechanisms of OM stabilization in soil by studying the molecular composition of SOM in specific soil fractions. The implementation of spectroscopic techniques (Cory and McKnight, 2005; Lehmann et al., 2010) provides novel and promising methodological strategies to undertake the challenge of characterizing SOM composition and spatial distribution in soil.

1.2. Stability of organic matter in artificial soils

The production of artificial soils or technosols (WRB, 2006) aims to recover landscapes or increase soil productivity. Overall, the addition of soils with organic materials aims to benefit soil quality or increase crop yields by regulating nutrient supply and improving soil structure (Wagner et al., 2007). Addition of single or composite organic wastes is expected to have a positive effect in initiating soil aggregation of structurally degraded topsoils (Wagner et al., 2007 and references therein). For instance, composted or stabilized municipal sewage sludge is frequently applied to soil as organic amendment for restoration purposes. Otherwise, abandoned Fe mine tailings provide a source of Fe-rich mud (FeM), that constitutes an environmental challenge for its adequate disposal, being currently stored in large open-air ponds. A suitable approach might be to use such FeM as a substrate to obtain artificial soils. The Fe oxides and hydroxides present in the FeM might provide a suitable surface for adsorption of organic compounds and may favor the formation of organo-mineral associations. which may result in a pool of chemically stabilized OM. Chemical stabilization by organo-mineral associations is a main mechanisms leading to soil aggregation (Kögel-Knabner et al., 2008) and such strategy might result in pools of carbon with long residence time in soil (Macias and Camps Arbestain, 2010).

The objective of this study was to evaluate the chemical composition of OM in artificial soils obtained from organic wastes combined with the FeM at different ratios. The analysis of OM composition at a molecular level and the characterization of the spatial distribution among different pools by Fourier transform infrared spectroscopy (FTIR) can be directly related to SOM stability, soil respiration and OM decomposition rates. Previous research has demonstrated that intensities of distinct peaks obtained by FTIR analysis can be a measure of decomposition of organic carbon in soil (Haberhauer et al., 1998).

Otherwise, dissolved organic matter (DOM) constitutes a highly available carbon source for microorganisms while playing a fundamental role in the mobilisation of organic compounds (Kalbitz et al., 2003). Variations in the composition of the OM present in this pool are an essential component to the knowledge on SOM dynamics (Kalbitz et al., 2000). Therefore, this pool of OM deserves particular attention in our attempt to characterize the evolution of SOM pools. The spectrofluorometric analysis of the soil solution extracted from the different scenarios assayed provides a fingerprinting of the composition of DOM (Cory and McKnight, 2005). Thus, excitation-emission matrix spectroscopy provides the sensitivity to examine subtle changes in DOM fluorescence and provide a valuable insight into variations on the DOM pool composition.

2. Materials and methods

2.1. Studied area

The studied area is an Fe mine dump in Southeast Spain (Alquife, Granada) planned to be used for residential, leisure and agro-industrial activities. The soil is a degraded technosol with high infiltration rate under a continental Mediterranean climate.

2.2. Technosols

A collection of five technosols (Table 1) was obtained by combining a composted mixture (19.4% organic carbon) of sewage sludge from wastewater treatment and olive pruning (SVC) with FeM (44% Fe oxides) and/or a biodiesel byproduct (DRS) with a high concentration of glycerol, following saturation and incubation at 28º C for 30 d. The mineral waste was originated in milling activities carried out in the mine site. A second collection of five technosols was obtained by controlled acidification (H_2SO_4 5M) of mixtures as such described for the first set of technosols (Table 1).

Technosol	SVC	FeM	DRS	%OC
T1	100%	-	-	19.4
T2	90%	-	10%	21.0
T3	90%	10%	-	17.5
T4	85%	5%	10%	19
T5	70%	30%	-	13.6

Table 1. Composition of technosols.

A second batch of technosols (TS) was prepared with equivalent composition but SVC and FeM were saturated and incubated for 1 week before mixing. For TS3 and TS4, FeM was acidified with H_2SO_4 5N during the preconditioning step.

Samples of the different technosols were collected after 2, 9, 20 and 30 d of incubation and stored at -20º C for posterior analysis.

2.3. Carbon mineralization rates

To determine carbon mineralization upon application of the technosols in the mine dump, subsamples of soil from two different plots in the dump (AL7 and AL14) were amended at 2% with the different technosols and placed in air tight incubation jars with a volume of 300 mL and moisture content adjusted to field capacity. The lids of the incubation jars were fitted with three-way valves to allow sampling the air from the headspace. The jars were stored in an incubation room at 25º C for circa 120 d. Headspace in the jars was periodically sampled with 60 mL syringes and the CO_2 concentration measured with an infrared gas analyzer (LI-COR; Li-820). The amount of carbon respired was calculated using the ideal gas equation and expressed as percentage of carbon respired relative to the total carbon content in the amended soil.

2.4. FTIR-microscopy

Microaggregates-like structures (100-200 μm) were isolated and collected from the different technosols after 2 and 30 d of incubation and analyzed with a Fourier transform infrared spectrophotometer (Varian 620-IR IR microscope) coupled to a microscope (FTIR-microscope) using a KBr splitter and a liquid nitrogen cooled Focal Plane Array detector for spectrochemical imaging and a CCD camera. The spectra were recorded for the microaggregates-like structures in the mid-infrared range (4000–800 cm^{-1}) by combining 32 scans at a resolution of 1 cm^{-1}. The spectra were recorded in absorbance units. Peak area integration and analysis of the spectral features distribution in the microaggregates were performed using the software Agilent Resolutions Pro. Spectra and image analysis presented were obtained as the average of 5 spectra.

2.5. Water soluble organic matter

The fraction of water soluble organic matter (WSOM) was obtained from the technosols sampled at different incubation times, through centrifugation (10 min at 3000 g) using the 'double chamber' method (Bufflap and Allen, 1995). After centrifugation, the soil solution samples were immediately filtered through a 0.45-μm filter. The solutions were analyzed for dissolved organic carbon (DOC) using a TOC-analyser (Analytical Sciences Thermalox). The UV-absorbance was measured with a UV-VIS spectrophotometer (Perkin-Elmer, Lambda 20, quartz cells).

Variation in the ratio of absorbance to DOC was used to characterize the quality of DOM, through the specific UV absorbance at 254 and 340 nm (Tipping et al., 2009).

2.6. Spectrofluorometry

The soil solution samples were diluted such that the absorbance at 254 nm was less than 0.2 prior to the collection of fluorescence spectra (Miller et al., 2010). Fluorescence excitation-emission spectra were obtained for the pore water solutions using a JY HORIBA Fluorolog-3 spectrofluorometer with an excitation range set from 240 to 400 nm and an emission range set from 300 to 500 nm in 2 nm increments. Instrumental parameters were excitation and emission slits, 5 nm; response time, 8 s; and scan speed, 1200 nm min^{-1}. Spectra were analyzed using the software FluorEssence.

2.7. Adsorption of gallic acid on Fe mud

Sorption isotherms were carried out using a batch equilibration method, with 5 g of FeM and 20 mL of an aqueous solution of gallic acid (GA, Sigma Aldrich) at concentrations ranging 5–50 mM. The samples were mechanically shaken end-over-end in a thermostatic chamber at 20 ± 1 °C for 24 h. The samples were centrifuged at 3500 rpm and 15 °C for 15 min. The isotherms were run in duplicate. A GA solution without addition of FeM was used as control, to account for possible degradation during the batch process.

The difference between initial concentration of GA and the concentration of GA remaining in solution after reaching equilibrium was attributed to sorption of GA on FeM. The sorption equilibrium partition coefficient Kd (L kg^{-1}) was calculated as $K_d = X / C_e$, where X is the concentration of GA in the FeM (mg kg^{-1}) and Ce is the concentration of GA in the solution at equilibrium (mg L^{-1}). The adsorption experiment was described by the empirical Freundlich equation ($X = K_f C_e^n$), where K_f is the Freundlich adsorption coefficient (L kg^{-1}) and n a constant which depends on the adsorbate, the adsorbent and the temperature.

3. Results

3.1. Carbon mineralization rates

Results obtained from the carbon mineralization assays are summarized in Figures 2-4, which describe cumulative respiration determined for the application of the different technosols to samples of soil collected from the mine dump.

Soil addition with the first batch of technosols (Table 1) increased carbon mineralization rates for all the technosols applied (Fig. 2) compared to control soil (C). For technosols produced solely from SVC and Fe mud, a higher ratio SVC:FeM (T3: SVC:FeM, 90:10) resulted in lower CO_2 production than T5 (SVC:FeM 70:30), regardless the OC content.

The preconditioning step significantly affected carbon mineralization rate. Addition of technosols TS2 and TS4 (saturated) to the dump soil resulted in lower mineralization rates than application of non-preconditioned technosols (Fig. 3). Thus, for technosols obtained solely from SVC/FeM, saturation of wastes before mixing (TS5) significantly decreased the mineralization of OC added to the soil when FeM was acidified during the preconditioning (TS3, Fig. 4).

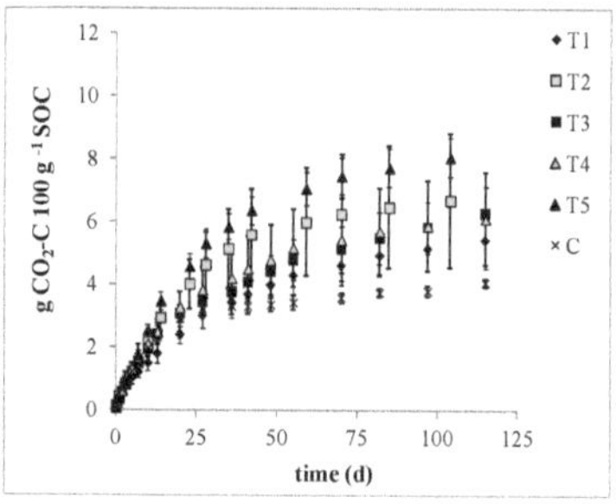

Figure 2. Cumulative CO_2-C respired for T1-T5 and control soil.

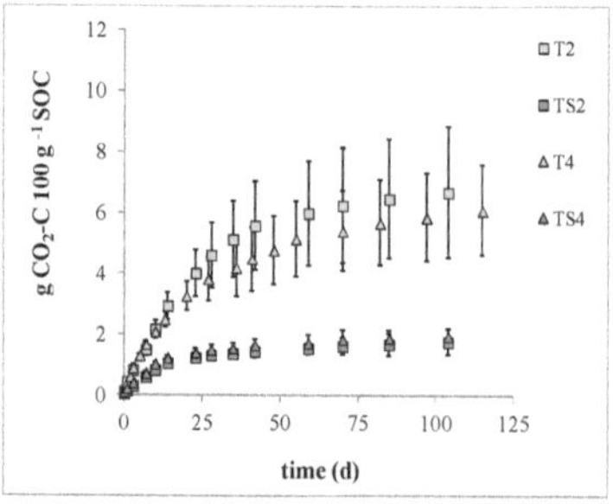

Figure 3. Cumulative CO_2-C respired for T2, T4, TS2 and TS4.

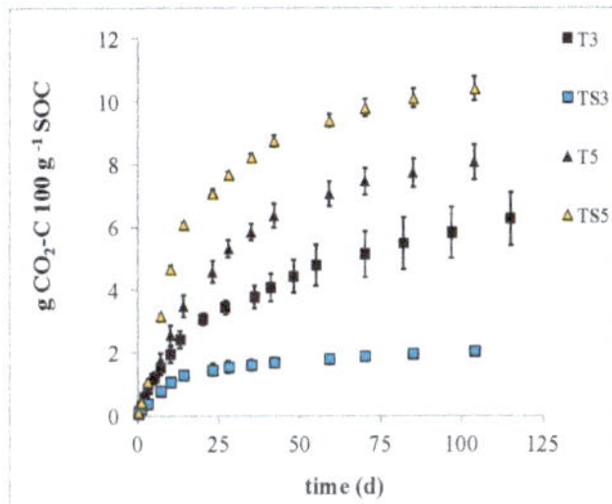

Figure 4. Cumulative CO_2-C respired for T3, T5, TS3 and TS5.

3.2. Dissolved organic carbon: concentration and composition

Total organic carbon content and UV-absorbance analysis performed for the water soluble organic carbon collected from the technosols indicated that DOC from technosols prepared with SVC+FeM (90/10) was highly humified, as suggest by SUVA and extinction coefficients values at 254 and 340 nm (Fig. 5 and 6).

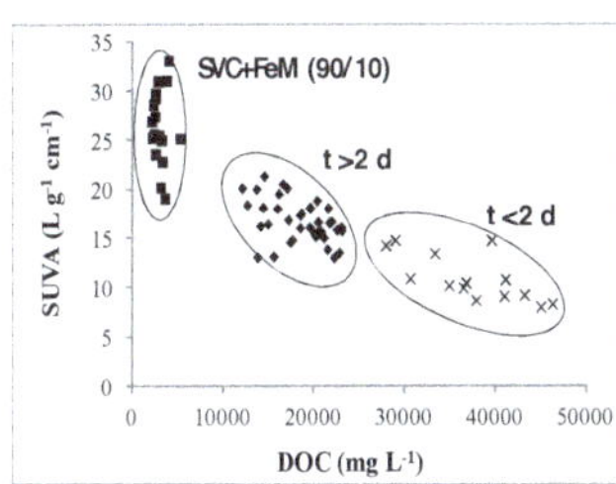

Figure 5. Relationship SUVA-DOC. Squares, technosol prepared with SVC+FeM (90/10); diamonds, rest of technosols after incubation time > 2 d; X, rest of technosols after 2 incubation days.

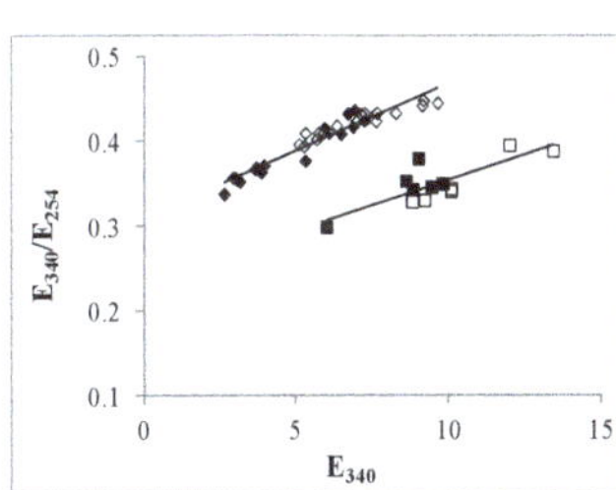

Figure 6. Relationship ratio of extinction coefficients at 340 and 254 nm and the extinction coefficient at 340. Full squares, T3, T9 (t=2d); empty squares, T3, T9 (t=30 d); full diamond, T1-T2, T4-T8, T10-T12 (t=2 d); empty diamond, T1-T2, T4-T8, T10-T12 (t=30 d).

Table 2 summarizes the values determined for the different fluorescence indexes derived from the spectrofluorometric analysis. Overall, higher humification index values (HIX) were derived for increasing concentration of FeM in a particular technosol, which confirms higher humification of the water soluble organic carbon. The freshness index values (β:α) confirmed the expected predominance of recently derived DOC. Besides, the high fluorescence index values determined (FI) indicated that DOC was originated from intense microbial activity.

Overall, preconditioning the materials prior to the production of the technosols resulted in higher values of the HIX, likely due to an increase in organo-metal interactions, while changes on the β:α and FI values were neglectable. Variation of the fluorescence indexes over time was not conclusive.

Technosol	Incubation (d)	HIX	β:α	FI
T1	2	5.25	0.68	2.17
T1	30	5.72	0.66	2.09
T3	2	7.25	0.60	1.99
T3	30	7.16	0.58	1.88
T4	2	4.62	0.7	2.18
T4	30	6.14	0.67	2.12
TS1	2	4.91	0.74	2.19
TS1	30	5.51	0.67	2.25
TS3	2	6.85	0.60	1.96
TS3	30	7.13	0.62	1.91
TS4	2	6.57	0.58	2.21
TS4	30	7.46	0.58	1.82

HIX: Humification index. β:α: freshness index. FI: Fluorescence index.

Table 2. Spectroscopic analysis.

Otherwise, results from the excitation emission matrixes (Ex/Em) collected for T1 and T3 (Fig. 7) indicated a substantial increase in the fraction of UV (Ex/Em 260/400-460) and visible (Ex/Em 320-360/400-460) humic-like organic matter for T3 compared to T1, which suggest the presence of a pool of highly stable, low degradation rate OM. Moreover, the strong increase in fluorescence intensity suggests that the added OM might complex metal ions in solution, which can result in a protective effect for DOM against rapid mineralization. The attenuation of the signal at Ex/Em 320-360/400-460 over time suggest the precipitation of organo-metal complexes in the solid phase.

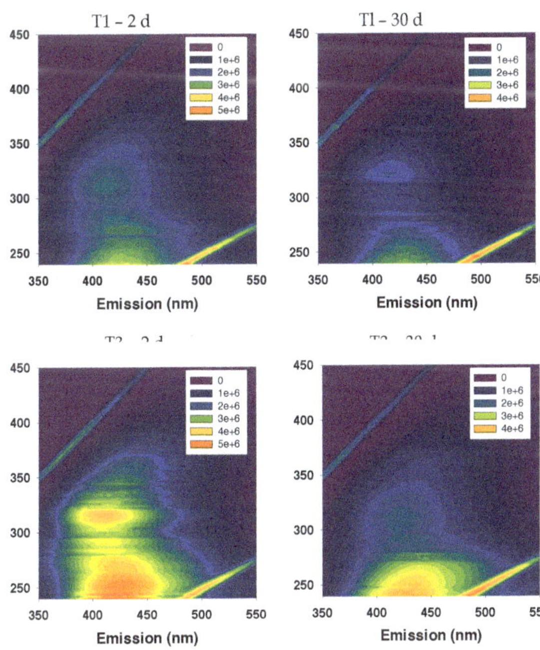

Figure 7. Excitation-Emission matrixes for technosols T1 and T3 after 2 and 30 d of incubation.

3.3. FTIR-microscopy

Spectra obtained for microaggregate-like structures (Fig. 8) showed a consistent absence of aliphatic-C (2900 cm^{-1}); the presence of aromatic compounds-C assigned to signals at 1400-1500 cm^{-1} and at 1600 cm^{-1}; and aromatic overtones at 1790, 1865 and 1998 (T3, Fig. 8), according to previous literature (Demyan et al., 2012). Polysaccharide-C were identified in the fingerprint region (between 800 and 1200 cm^{-1}) while peak at 3620 cm^{-1} are related to the presence of clay like compounds (Lehmann et al., 2007). Additionally, a peak at 3700 cm^{-1} was obtained for the analysis of T5, which might also be related to clay-like compounds.

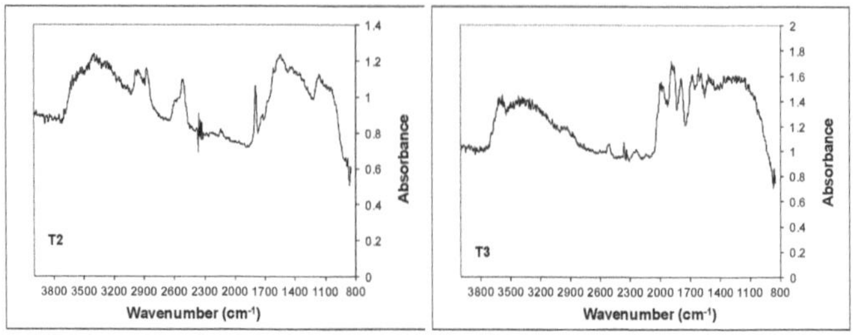

Figure 8. FTIR spectra collected for technosols T2 and T3.

Analyzing the distribution of such spectral features in soil microaggregates revealed polysaccharides homogeneously dispersed on the surface of the microaggregates, as depicted for T1 (Fig. 9) and T3 (Fig. 10). Otherwise, distribution analysis for T1 and T3 (Fig. 9 and 10) suggested the presence of cores of aromatic compounds (Wan et al., 2007).

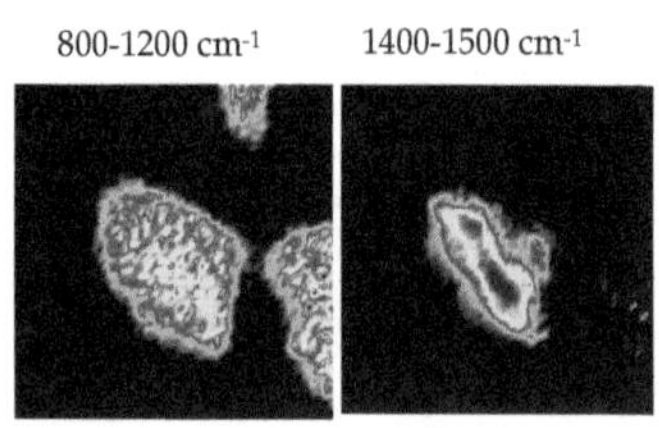

Figure 9. Distribution of polysaccharides (800-1200 cm^{-1}) and aromatic compounds (1400-1500 cm^{-1}) in a microaggregate from T1 (without FeM).

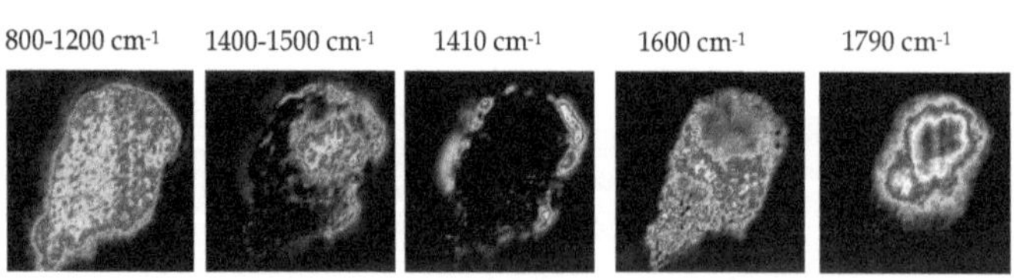

Figure 10. Distribution of chemical compounds in a microaggregate from T3 (SVC plus 10% FeM) obtained from FTIR spectra. Polysacharides at 800-1200 cm^{-1}, aromatic compounds at 1400-1500, 1600, 1410 and 1790 $cm^{-1}$and organo-mineral associations at 1410 cm^{-1}.

The distribution of clay-like compounds in the microaggregates isolated from T1, T3 and T5 indicates an increasing presence of such compounds for higher concentrations of FeM in the technosols (Fig. 11).

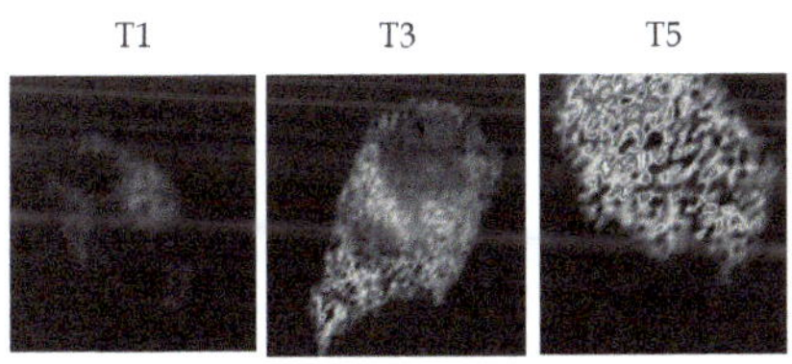

Figure 11. Distribution analysis for the signal recorded at 3620 cm^{-1} in microaggregates from the three technosols.

Thus, a weak signal was detected for T1 that might be related to oxidized compounds that overlap with the signal corresponding to clay-like compounds, while a strong signal was obtained for T3 and T5, consistently with the different ratios of FeM in their composition. Moreover, the distribution of clay-like compounds obtained for T3 (Fig. 11) overlaps with the distribution depicted for aromatic compounds (1600 cm^{-1}, Fig. 10), which confirms the presence of organo-mineral associations.

3.4. Adsorption of gallic acid on Fe mud

Results from batch adsorption assays confirmed the capability of FeM to adsorb gallic acid (300 mmol kg^{-1}), probably through interaction of the carboxyl and phenolic groups with the Fe oxide surface, as determimed by the decrease in the signals at 220 and 270 nm for increasing concentrations of FeM in solution (Fig. 12). The adsorption constant derived, K_d=231.5 L kg^{-1}, indicates high adsorption of the acid in the FeM n=0.1, weak sorption of the second layer.

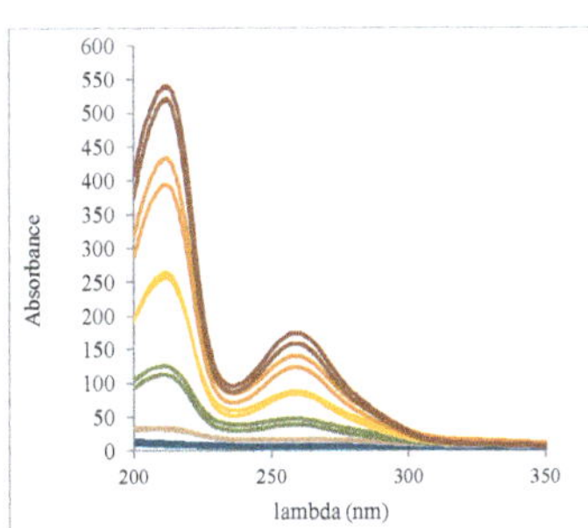

Figure 12. Gallic acid solution UV spectra. K_d=231.5 L kg^{-1}, high adsorption of the acid in the FeM n=0.1, weak sorption of the second layer.

4. Discussion

Fresh organic inputs applied to a harsh environment such as a mine dump in the arid Mediterranean climate can be expected to be rapidly mineralized, as previously described by Novara et al. (2012). The evolution of DOM in a particular soil largely determines the SOM stability and protective capacity of the soil. Hence, the aromaticity and fluorescent properties of DOM provide an adequate characterization of this pool of OM, which allow adequately predicting carbon mineralization rates and protective capacity for a given soil scenario (Murphy et al., 2010).

The extent and rate of DOM biodegradation and humification were in agreement with previous studies (Kalbitz et al., 2003). The increase in DOM with low aromaticity upon addition of such amendments might enhance the microbial activity in the soil, but poor beneficial effects can be expected in the long term due to the short residence time of the OM added. The application of composite amendments including a source of metal ions might contribute to a longer permanence of OM in the soil (Kaiser and Kalbitz, 2012), which will largely benefit soil quality in the long term. Thus, application of composites with a low percentage of FeM resulted in higher humification indexes. Moreover, soil preconditioning by acidification and incubation under saturated conditions promoted the formation of organo-metal complexes, which resulted in lower mineralization rates (Kaiser and Kalbitz, 2012).

Otherwise, the presence of FeM in the artificial soils provides a pool of Fe oxides and Fe and Al hydroxides that presents a clay-like behaviour (Lehmann et al., 2007). This was confirmed by peak position at 3620 cm^{-1} in spectra collected for T3 (Fig. 10-11). The slight signal recorded for T1 can be attributed to hydroxyl groups in the SVC. Therefore, the incorporation of FeM in the production of technosols results in organo-mineral associations, due to complexation of Fe with phenol/carboxyl groups, which contributes to the protection and stabilization of fresh inputs of organic carbon. Complexation of fulvic acids with Fe oxides surfaces has been linked to the occurrence of a band around 1410 cm^{-1} (Gu et al., 1994). For technosols from SVC+FeM, distribution analysis suggests that such complexes locate in the edges of the microaggregate as depicted for T3 in Fig. 10. However, the protection is limited by the low adsorption capacity of this FeM, which can be attributed to a negatively charged surface (pH=8.5). Organic compounds present in the fresh inputs such as gallic acid (pH=3.5) might induce short-term acidification on the Fe oxide surface, which could allow the adsorption of carboxyl and phenolic groups (Ni et al., 2011). However, increasing the ratio of FeM in the technosol might counteract such effect, resulting in lower protection effect, which explains the higher mineralization rate determined for such technosols.

Polysaccharides were ubiquitous on the microaggregates analyzed and homogeneously dispersed on the surface of the microaggregates, which is consistent with an increase in the microbial activity due to the addition of fresh inputs of organic carbon (Six et al., 2004). The presence of cores of aromatic compounds in the microaggregates-like structures analyzed was consistent with hypothesis previously established in the literature for the formation of aggregates (Six et al., 2002; Six et al., 2004; Wan et al., 2007).

Overall, the results obtained with this study have demonstrated that succesful production of technosols as organic amendments to ameliorate soil quality might highly benefit of the incorporation of a mineral substrate at an optimized ratio. The spectroscopic characterization of DOM and soil aggregates provides a low cost, effective analysis to determine the effect of a particular amendment in soil structure and OM stability.

5. Conclusions

The production of technosols from low-cost wastes provides a suitable strategy for wastes disposal while providing a valuable resource for plant sustaining in soil. Application of composite amendments to degraded soils might constitute a highly effective approach for increasing soil health and productivity and a suitable alternative to conventional strategies based on single organic amendments. Moreover, such composites constitute a balanced soil amendment, which involves a compromise between enhancement of soil biological activity and the establishment of pools of stabilized organic matter.

The approach presented provides general guidances for designing optimized mixtures of C-rich organic materials through characterization of the DOM pool, soil aggregates, and the potential of the composites for chemical stabilization of OM. Thus, spectrophotometric fingerprinting of DOM and molecular characterization of OM in soil aggregates have been demonstrated to provide soil quality benchmarks to develop technosols tailored for an specific environmental scenario.

Rehabilitation plans can be designed according to soil-plant requirements as well as safe and effectively cost means for disposal wastes. Fe-enriched amendments might constitute an essential component for technosols, playing a key role in the chemical stabilization of organic matter in soil.

Overall, an optimized combination of mineral and organic wastes may result in a pool of chemically stabilized organic matter. The proposed technosols present a significant potential to create a sink of C while providing an inexpensive in-situ strategy for wastes disposal.

Acknowledgements

MCHS thanks the KU Leuven (Belgium) for a postdoctoral fellowship (PDMK/10/080). The research reported was partially supported by a Hercules project (2011-2012, KU Leuven, Belgium) and Junta de Andalucía-P08-RNM3526 (Spain).

Author details

M.C. Hernandez-Soriano[1], A. Sevilla-Perea[2], B. Kerré[1] and M.D. Mingorance[2]

1 Division of Soil and Water Management, KU Leuven, Belgium

2 Instituto Andaluz de Ciencias de la Tierra, University of Granada - CSIC, Spain, Spain

References

[1] Baldock, J. A, & Skjemstad, J. O. (2000). Role of the soil matrix and minerals in protecting natural organic materials against biological attack. Organic Geochemistry , 31, 697-710.

[2] Balesdent, J, Chenu, C, & Balabane, M. (2000). Relationship of soil organic matter dynamics to physical protection and tillage. Soil and Tillage Research , 53, 215-230.

[3] Blanco-canqui, H, & Lal, R. (2004). Mechanisms of Carbon Sequestration in Soil Aggregates. Critical Reviews in Plant Sciences , 23, 481-504.

[4] Bot, A, & Benites, J. (2005). The importance of soil organic matter: the key to drought resistant soil, sustained food and production.. FAO, Rome.

[5] Bufflap, S. E, & Allen, H. E. (1995). Comparison of pore-water sampling techniques for trace metals. Water Res. , 29, 2051-2054.

[6] Cory, R. M, & Mcknight, D. M. (2005). Fluorescence Spectroscopy Reveals Ubiquitous Presence of Oxidized and Reduced Quinones in Dissolved Organic Matter. Environmental Science & Technology , 39, 8142-8149.

[7] Demyan, M. S, Rasche, F, Schulz, E, Breulmann, M, Müller, T, & Cadisch, G. (2012). Use of specific peaks obtained by diffuse reflectance Fourier transform mid-infrared spectroscopy to study the composition of organic matter in a Haplic Chernozem. European Journal of Soil Science , 63, 189-199.

[8] Denef, K, & Six, J. (2005). Clay mineralogy determines the importance of biological versus abiotic processes for macroaggregate formation and stabilization. European Journal of Soil Science , 56, 469-479.

[9] Foereid, B, Bellamy, P. H, Holden, A, & Kirk, G. J. D. (2012). On the initialization of soil carbon models and its effects on model predictions for England and Wales. European Journal of Soil Science , 63, 32-41.

[10] Gu, B, Schmitt, J, Chen, Z, Liang, L, & Mccarthy, J. F. (1994). Adsorption and desorption of natural organic matter on iron oxide: mechanisms and models. Environmental Science & Technology , 28, 38-46.

[11] Haberhauer, G, Rafferty, B, Strebl, F, & Gerzabek, M. H. (1998). Comparison of the composition of forest soil litter derived from three different sites at various decompositional stages using FTIR spectroscopy. Geoderma , 83, 331-342.

[12] Jastrow, J, Amonette, J, & Bailey, V. (2007). Mechanisms controlling soil carbon turnover and their potential application for enhancing carbon sequestration. Climatic Change , 80, 5-23.

[13] Kaiser, K, & Kalbitz, K. (2012). Cycling downwards- dissolved organic matter in soils. Soil Biology & Biochemistry , 52, 29-32.

[14] Kalbitz, K, Schmerwitz, J, Schwesig, D, & Matzner, E. (2003). Biodegradation of soil-derived dissolved organic matter as related to its properties. Geoderma , 113, 273-291.

[15] Kalbitz, K, Solinger, S, Park, J, Michalzik, H, & Matzner, B. E., (2000). Controls on the dynamics of dissolved organic matter in soils: a review. Soil Science , 165, 277-304.

[16] Kimetu, J, & Lehmann, J. (2010). Stability and stabilisation of biochar and green manure in soil with different organic carbon contents. Aust. J. Soil Res. , 48, 577-585.

[17] Kleber, M, Sollins, P, & Sutton, R. (2007). A conceptual model of organo-mineral interactions in soils: self-assembly of organic molecular fragments into zonal structures on mineral surfaces. Biogeochemistry , 85, 9-24.

[18] Kögel-knabner, I, Guggenberger, G, Kleber, M, Kandeler, E, Kalbitz, K, Scheu, S, Eusterhues, K, & Leinweber, P. (2008). Organo-mineral associations in temperate soils: Integrating biology, mineralogy, and organic matter chemistry. Journal of Plant Nutrition and Soil Science , 171, 61-82.

[19] Korshin, G. V, Benjamin, M. M, & Sletten, R. S. (1997). Adsorption of natural organic matter (NOM) on iron oxide: Effects on NOM composition and formation of organo-halide compounds during chlorination. Water Research , 31, 1643-1650.

[20] Lehmann, J, Kinyangi, J, & Solomon, D. (2007). Organic matter stabilization in soil microaggregates: implications from spatial heterogeneity of organic carbon contents and carbon forms. Biogeochemistry , 85, 45-57.

[21] Lehmann, J, Solomon, D, Balwant, S, & Markus, G. (2010). Organic Carbon Chemistry in Soils Observed by Synchrotron-Based Spectroscopy. Developments in Soil Science. Elsevier, , 289-312.

[22] Macias, F. Camps Arbestain, M., (2010). Soil carbon sequestration in a changing global environment. Mitigation and Adaptation Strategies for Global Change , 15, 511-529.

[23] Merckx, R. den Hartog, A., van Veen, J.A., (1985). Turnover of root-derived material and related microbial biomass formation in soils of different texture. Soil Biology and Biochemistry , 17, 565-569.

[24] Miller, M. P, Simone, B. E, Mcknight, D. M, Cory, R. M, Williams, M. W, & Boyer, E. W. (2010). New light on a dark subject: comment. Aquatic Sciences , 72, 269-275.

[25] Murphy, K. R, Butler, K. D, Spencer, R. G. M, Stedmon, C. A, Boehme, J. R, & Aiken, G. R. (2010). Measurement of Dissolved Organic Matter Fluorescence in Aquatic Environments: An Interlaboratory Comparison. Environmental Science & Technology , 44, 9405-9412.

[26] Ni, J, Pignatello, J. J, & Xing, B. (2011). Adsorption of aromatic carboxylate ions to black carbon (biochar) is accompanied by proton exchange with water. Environmental Science & Technology , 45, 9240-9248.

[27] Novara, A. La Mantia, T., Barbera, V., Gristina, L., (2012). Paired-site approach for studying soil organic carbon dynamics in a Mediterranean semiarid environment. CATENA , 89, 1-7.

[28] Six, J, Bossuyt, H, Degryze, S, & Denef, K. (2004). A history of research on the link between (micro)aggregates, soil biota, and soil organic matter dynamics. Soil & Tillage Research , 79, 7-31.

[29] Six, J, Conant, R. T, Paul, E. A, & Paustian, K. (2002). Stabilization mechanisms of soil organic matter: Implications for C-saturation of soils. Plant and Soil , 241, 155-176.

[30] Sollins, P, Homann, P, & Caldwell, B. A. (1996). Stabilization and destabilization of soil organic matter: mechanisms and controls. Geoderma , 74, 65-105.

[31] Strong, D. T, De Wever, H, Merckx, R, & Recous, S. (2004). Spatial location of carbon decomposition in the soil pore system. European Journal of Soil Science , 55, 739-750.

[32] Tipping, E, Corbishley, H. T, Koprivnjak, J. F, Lapworth, D. J, Miller, M. P, Vincent, C. D, & Hamilton-taylor, J. (2009). Quantification of natural DOM from UV absorption at two wavelengths. Environmental Chemistry , 6, 472-476.

[33] Von Lützow, M, Kögel-knabner, I, Ekschmitt, K, Matzner, E, Guggenberger, G, Marschner, B, & Flessa, H. (2006). Stabilization of organic matter in temperate soils: mechanisms and their relevance under different soil conditions- a review. European Journal of Soil Science , 57, 426-445.

[34] Wagner, S, Cattle, S. R, & Scholten, T. (2007). Soil-aggregate formation as influenced by clay content and organic-matter amendment. Journal of Plant Nutrition and Soil Science , 170, 173-180.

[35] Wan, J, Tyliszczak, T, & Tokunaga, T. K. (2007). Organic carbon distribution, speciation, and elemental correlations within soil micro aggregates: Applications of STXM and NEXAFS spectroscopy. Geochimica Et Cosmochimica Acta , 71, 5439-5449.

[36] Wander, M. (2004). Soil organic matter fractions and their relevance to soil function In: Magdoff, F.a.R.R.W. (Ed.). Soil Organic Matter in Sustainable Agriculture. CRC Press LLC, Upper Saddle River, NJ.

[37] Wrb, I. W. G. (2006). World reference base for soil resources In: FAO (Ed.). World Soil Resources Reports. FAO, Rome.

Chapter 9

Soil Organic Matter Stability as Affected by Land Management in Steppe Ecosystems

Elmira Saljnikov, Dragan Cakmak and
Saule Rahimgalieva

Additional information is available at the end of the chapter

1. Introduction

1.1. Soil organic matter status

Soil organic matter (SOM) is most reactive and powerful factor in the formation of soil and in its fertility. Formation of soil and accumulation of organic matter are a function of interactions between biological factors and parent rocks under certain hydrothermal conditions and are one of the sections of a continuous chain of the trophic bounds between different life forms, serving as a first and a last section at the same time. The later is because SOM contain the main nitrogen stock, nearly the half of phosphorus, significant part of sulphur and other macro- and micronutrients for sustaining life and productivity of plants. Although soil organic matter comprise only five percent of total soil structure it has been a major research topic throughout the history of soil science, which is generally regarded to have been ongoing for approximately a century [1, 2].

Discovering the role and fate of soil organic matter has been a great challenge for the scientists. There are many argues about definitions of SOM among soil scientist. One of the most dynamic definitions of the SOM was given by [3]: the amount of organic carbon contained in a particular soil is a function of the balance between the rate of deposition of plant residues in or on soil and the rate of mineralization of the residue carbon by soil biota. In fact organic matter in soil always is in a very dynamic state, where transformations of bio-products occur constantly. The mechanisms through which soil organic C can be biologically stabilized depend on the decomposition of the soil mineral phase and the chemical structure of the organic residues added to the soil.

Climate is the most powerful factor that determines the array of plant species at any given location, the quantity of plant material produced, and the intensity of microbial activity in the soil. Climate influences soil organic carbon (SOC) content primarily through the effects of temperature, moisture, and solar radiation. Related studies found that amounts of SOC were positively correlated with precipitation and, at a given level of precipitation, negatively correlated with temperature [4, 5]. Climatic influences on biologically active fractions of SOM are not well understood. Therefore, one of the focuses in this study was investigation of the dynamics of labile SOM under the different hydrothermal conditions of steppe ecosystems.

Another powerful factor determining SOM reserves is plant biomass inputs and outputs. In agricultural systems, where soil and plant residues are often intensively manipulated, human impact on decomposition is especially pronounced [6]. Management practices like tillage, selection of crops and cropping sequences, and fertilization can alter decomposition rates by their effects on soil moisture, soil temperature, aeration, composition and placement of residues. Many studies confirm that under the similar climatic condition, carbon and nitrogen retention in soil is influenced by crop management systems, such as *crop rotation* [7, 8], *tillage* [5, 9], *residue management* [10] and fertilization and fertility [7, 10, 11]. This Chapter will discuss an impact of different land management practices on the labile (biologically active) pool of soil organic matter.

1.2. Decomposition

Decomposition is the progressive break down of organic, ultimately into inorganic constituents. The decomposition process is mediated mainly by soil microorganisms, which derive energy and nutrients from decomposing substrate. Plant litter decomposes very rapidly and although the carbon from plant litter represents only a small fraction of C in soil, about half of the CO_2 output from soil, globally, comes from decomposition of the annual litter fall [12]. Decomposition is central to the biogeochemical cycles in terrestrial, aquatic and atmospheric systems. It releases nutrients and energy associated in organic materials and feeds them back into local and global cycles, thereby affecting land, and air and water quality (Fig. 1).

Three interrelated factors regulate decomposition: the quality of the residue, the physical-chemical environment in which decomposition occurs and the type of organisms in the decomposer community. All organic carbon in soils can serve as potentially suitable as substrate. Vegetation can influence SOC levels as a result of the amount, placement and biodegradability of plant residues returned to the soil. The fate of surface deposited residues depends on the activity of soil microorganisms and fauna and their ability to mix these residues into surface mineral horizons. Microorganisms are the major contributors to soil respiration and are responsible for 80-95% of the mineralization of carbon. Humans can affect decomposition by altering some of these factors, especially in agricultural systems. The current understanding of decomposition processes, learned from field and laboratory studies, is embodied in simulation models, e.g., the first-order kinetic model [110].

One of the effects of global warming is accelerated decomposition of soil organic matter, thereby releasing CO_2 to the atmosphere, which will further enhance the warming trend

[104]. The United Nations Framework Convention on Climate change (Kyoto Protocol of 1997), allows organic carbon stored in arable soils to be included in calculations of net carbon emissions. By altering organic matter production, litter quality, and belowground C allocation, however, changes in vegetation type can influence microbial decomposition [105] and root respiration and therefore soil respiration rates [80]. As a result of global climate change and alterations in land use many ecosystems are currently experiencing concurrent changes in the abiotic and biotic controls on soil respiration. Given the large quantity of CO_2 that soils respire annually and the role CO_2 plays in greenhouse warming, an understanding of SR response to climate change and alterations in vegetation resulting from land use is critical.

1.3. Labile pool of soil organic matter

Labile carbon is the fraction of soil organic carbon with most rapid turnover times and its oxidation drives the flux of CO_2 between soils and atmosphere. Labile organic matter pools are fine indicators of soil quality that influence soil function in specific ways and that are much more sensitive to changes in soil management practice [e.g., 13]. The biggest and main source for labile organic matter is a 'light' fraction organic matter (or particulate OM, or macroorganic matter; [8, 14-17] that consists of partially decomposed plant litter. This 'light' organic matter acts as a substrate for soil microbial activity, a short-term reservoir of nutrients, a food source for soil fauna and loci for formation of water stable macroaggregates.

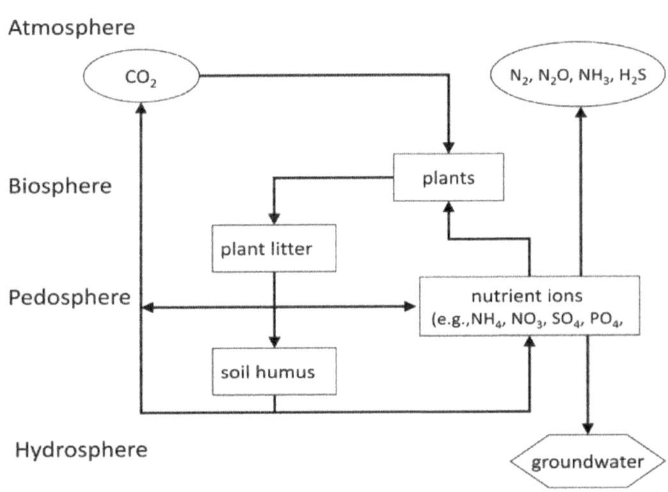

Figure 1. Conceptual model of soil organic matter decomposition (modified from [106])

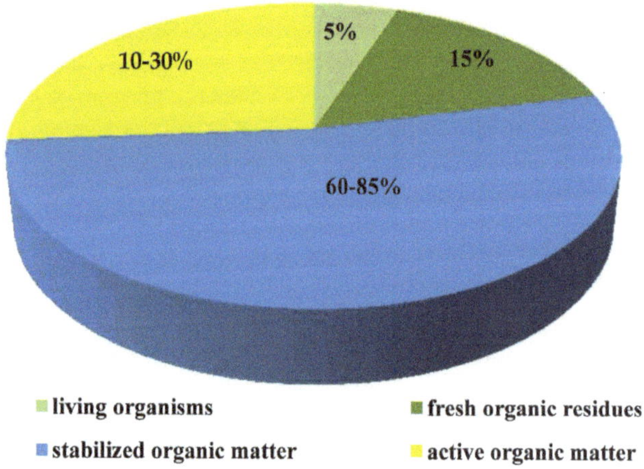

Figure 2. Composition and distribution of fractions of soil organic matter

The biological determination of labile SOM is the carbon decomposed by microorganisms during the microbial growth. This biological definition of labile SOM includes two aspects: labile soil organic carbon chemically and physically assessable; the organic carbon that is chemically decomposable but physically un-assessable due to protection by clay minerals is not considered as a labile organic carbon. Generally, soil organic matter is divided into stable (70-96%), active (2-30%) and plant litter (0-20%) fractions (Fig.2). The active fraction mainly consists of microbial biomass and their metabolites, the organic substrate in different stages of decomposition and non-humic substances, with turnover time from 0.8 to 5 years. The stabilized or passive fraction of SOM is passive, chemically and physically protected matters. The physically protected OM has turnover time from 20 to 50 years; the chemically protected –from 800 to 1200 years.

The 10 to 30% active fraction is responsible for the support of soil microorganisms. This fraction is most sensitive to soil management practices. Although labile OM comprises a small part of total SOM, it is the main source for nutrients and energy for microorganisms and plants, and main source for carbon dioxide flux from soil. The roles of stable and labile SOM differ. An active fraction mainly influences the activity of microorganisms, the stability of macroaggregates, filtration speed, and the speed of nutrient mineralization. Whilst, the stable fraction influences mainly water-holding capacity, soil cation exchangeable capacity and soil microaggregation.

Fresh plant litter decomposes very quickly and the decomposition usually occurs not as a single step, but as a cascade. Fresh material, usually plant residue, undergo hydrolysis and redox reactions and then converted into altered forms. The transformed organic material, so called 'light' fraction (LF), in turn, is susceptible to further decomposition. A small part of LF is utilized for microbial synthesis, which after death contribute back to LF. Greatest part of LF is subjected to further mineralization resulting in mineral products, which is of direct practical interest of soil scientists from agronomical and ecological point of view, because as mentioned above, about half of the CO_2 output from soil, globally, comes from decomposition of the annual litter fall.

Thus, transformations of SOM are generally concentrated within labile pool. The end products of organic matter mineralization (e.g., CO_2, NO_3, NH_4) can give us valuable information about ability of a given soil to supply plants with nutrients and/or ability to stabilize soil organic matter.

Nitrogen (N) is generally the most common growth-limiting nutrient in agricultural production systems Nitrogen taken up by crops is derived from a number of sources, particularly from fertilizer, biological N fixation and mineralization of N from soil organic matter, crop residues, and manures [18]. Large amounts of mineralizable N can accumulate under grassland with the result that crops grown immediately after cultivation of long-term grass may derive much of their N from mineralization. In contrast, soils that have been intensively cropped often mineralize little N, leaving crops heavily dependent on fertilizer nitrogen. This chapter will present the impact of different fertilization experiments on soil labile OM.

Labile fractions of SOM neither have been fully described nor successfully isolated [19, 20]. However, procedurally defined fractions such as carbon and nitrogen mineralized under controlled conditions and "light" fraction organic carbon proved to be good indicators of subtle changes in SOM, because they affects the nutrient dynamics within single growing season, the organic matter content under contrasting management regimes, and C sequestration over extended periods of time. Organic matter quality may also be characterized by estimates of kinetically defined pools obtained by fitting the simulation models to data of carbon and nitrogen mineralization [21, 22]. Although soil labile organic carbon is constituted of amino acids, simple carbohydrates, a fraction of microbial biomass, and other simple organic compounds, a clear chemical or physical definition of soil labile organic carbon is difficult if not impossible. We here present a biological definition of soil labile organic carbon as microbial degradable carbon associated with microbial growth. This biological definition includes two aspects: soil labile organic carbon is both chemically degradable and physically accessible by soil microbes. Organic carbon that is chemically degradable but physically inaccessible by microbes due to clay mineral protection is not regarded here as soil labile organic carbon [13].

Mollisols soils are the most fertile and productive soils and therefore they are often overexploited for agricultural needs. The area under *Mollisols* in Kazakhstan occupies 25.3 million ha and in Ukraine 60.4 million ha. During the Soviet period the political aim was a rapid increase in grain production that was achieved by indiscriminate plowing of as large area of virgin lands as possible. However, such intensive cultivation of these soils resulted in drastic decrease in its humus content. In this Chapter four types of *Mollisols*: *Hupludolls, Argiudolls, Calciustolls* and *Haplustolls* studied for characterization of the "light fraction organic matter" for the scenarios considered and estimation of relationship between C/N ratio and mineralization rates are presented.

2. Materials and methods

2.1. Description of study sites

Four experimental sites from Eurasian steppes were examined for soil organic matter fraction. They are: Kharkov (dry forest-steppe, east Ukraine), Uman (moist forest-steppe, central Ukraine), Kherson (dry steppe, south Ukraine) and Astana (dry steppe, northern Kazakhstan,). The sites are located in different soil-ecological zones and differ in the amount of precipitation, temperature, soil type and vegetation

2.2. Soil sampling and analysis.

Topsoil samples (0-20-cm) were collected in spring-summer. Three sub-samples for chemical and five sub-samples for biological analysis were taken from each sampling point. The soil samples were air-dried followed by grinding, and were passed through a 2-mm sieve for chemical analysis. Samples for biological analysis were stored at fresh condition at 4° C before analysis. The dried soil samples were analyzed for total N concentration using a full automatic analyzer (Shimazu NC-800-13N). Organic C was determined by dichromate oxidation method [23].

2.3. Labile carbon (Potentially Mineralizable Carbon, PMC)

The rate of disappearance of plant residues can be described using a kinetic model. First-order kinetic model is usually used to characterize decomposition of plant residues, assuming that the annual input of plant residues is independent of the rate of their decomposition. Using first-order kinetics to describe decomposition implies that the metabolic potential of the soil microbial biomass exceeds the substrate supply.

Carbon mineralization was determined using laboratory incubation techniques via measuring soil respiration. The fresh soils were brought to 50% of WHC followed by incubation in a square-plastic jar (500-ml) at 30°C for 70 days. The evolved CO_2 was trapped in an alkali solution (10-ml of 1M NaOH) that was replaced every 14 days, and cumulative CO_2 was measured by titration with 0.5*M*HCl. The amount of mineralizable carbon was estimated from the rate of CO_2-C evolved during 70 days of incubation using nonlinear regression according to the following equation [24]: $Cmin=C_0(1-e^{-kt})$,where, *Cmin* is an experimental value of mineralized C (mg kg^{-1} soil) at time *t* (days) that was plotted to fit the equation, C_0 is potentially mineralizable carbon (PMC) (mg kg^{-1} soil) that was calculated after fitting the curve, and *k* is a nonlinear mineralization constant, i.e. the fraction mineralized per day (d^{-1}) [25].

2.4. Labile nitrogen (Potentially Mineralizable Nitrogen, PMN)

Potentially mineralizable N is a measure of the active fraction of soil organic N, which is chiefly responsible for the release of mineral N through microbial action. Mineralizable N is composed of a heterogeneous array of organic substrates including microbial biomass, residues of recent crops, and humus. Despite a continuing research effort [26, 27], chemical tests that are selective for the mineralizable portion of soil N are not available and incubation assays remain the preferred way of estimating mineralizable N.

Mineralized N was determined after incubation of soils for 2-, 4-, 6-, 8-, 10-weeks and analyzed for nitrate and ammonium N content by colorimetric method following extraction with 2*N*KCl solution. Nitrate N was analyzed after reduction of NO_3ions to NO_2 by passing the extraction through cadmium column. Ammonium N was analyzed by salicylate nitroprusside method [18]. The amount of mineralizable N (N_0) was obtained after fitting the data of mineralized N (Nmin) every 14 days to the first order kinetic model [25]: $Nmin=N_0*(1-e^{-kt})$, where, *N*min is an experimental value of mineralized N at a given time (*t*) that was plotted to fit the equation, N_0 is a potentially mineralizable nitrogen (PMN) that was calculated after fitting the curve, and *k* is nonlinear mineralization constant.

2.5. Microbial biomass

Soil microbial biomass measurements have been used in studies of soil organic matter dynamics and nutrient cycling in a variety of terrestrial ecosystems. They provide a measure of the quantity of living microbial biomass present in the soil, and in arable soils account for ~1%–5% of the total soil organic matter [28, 29]. Measurements of the carbon (C) and nitrogen (N) contained in the soil microbial biomass provide a basis for studies of the formation and turnover of soil organic matter, as the microbial biomass is one of the key definable fractions [30]. The data can be used for assessing changes in soil organic matter caused by soil management [31] and tillage practices [32], for assessing the impact of management on soil strength and porosity, soil structure and aggregate stability [33], and for assessing soil N fertility status [21].

Soil microbial biomass was determined by chloroform fumigation-extraction technique as described by [34]. For each sample, four sub-samples of field-moist soil were placed in flasks, moistened to field capacity and conditioned for 3 days at 25°C. Two sub-samples were fumigated with chloroform in a vacuum chamber for 5 days at 25°C and the other two sub-samples (controls) were incubated without fumigation at the same temperature. All samples were extracted with 0.5 m of K_2SO_4 at ratio 5:1. Microbial biomass C was measured by dissolved organic carbon analyzer (TOC-5000) and microbial biomass N was determined by colorimetric method. The microbial biomass C and N were calculated using an equation relating the increased release of C and N as a result of $CHCl_3$ fumigation and a factor representing the fraction of biomass C and N extracted by $K2SO_4$ [35].

2.6. "Light" fraction organic matter (LF)

"Light" fraction organic matter (LFOM) was separated by density separation using reagent-grade NaI solution adjusted to 1.8 g cm^{-3} [36]. 10g of soil was suspended in 40 ml of NaI solution (sp.gr. = 1,7) and the soil dispersed for 30 seconds using a Virtis homogenizer. After centrifugation, the floating material, i.e., the 'light' fraction was transferred directly to a vacuum filtration unit. The LFOM was then washed (three aliquots of 10 ml 0.01M $CaCl_2$ followed by three aliquots of distilled water), dried at 70°C for 15h and weighed. The residue was re-suspended and the procedure was repeated to ensure complete collection of the LF. The composite LF was finely ground and analyzed for N and C concentrations.

2.7. Statistics

Descriptive statistical analyses were performed using SYSTAT-8 software [24]. Variability among treatments in each region was within the range of variability among the regions for all the cases. Sigma Plot 8 software [25] was applied for modeling C mineralization pattern and mineralization rate constant.

3. Impact of soil management practices on content of total organic C and N in soil

3.1. Soil total C and N in fertilization experiment

Mean annual mineralization of humus depends upon many factors. However, in case of unified soil and climatic conditions the limiting factor of soil organic matter mineralization becomes the cultivated plant and the technology of crop cultivation. Time, depth, frequency and intensity of cultivation are directly related to the amount of humus mineralization [37, 38].

The experiment with application of different dozes of mineral and organic fertilizers was conducted on *Mollisols*, in Uman (Table 1). The results of the study confirm the role of manure in contribution to both stable and labile soil organic matter. The content of soil organic carbon was not increased after 36 years application of mineral fertilizer in most of the treatments, compared to the control, while application of high rates of manure (O) alone maintained the higher accumulation of soil organic carbon (Table 2).

Manure contains humic acids [39], which directly contributes to the soil humic acids and favors humification processes [40, 41]. As this experiment has been performing since 1964, the long-term input of high rates of manure contributed to SOM via direct inputs of humic acids into the soil, showing the higher soil organic C than in other treatments. Content of total N in the treatments was not statistically different as indicated by the same letters in Table 2. Insignificant effect of the mineral fertilizers on the accumulation of soil organic C and N is due to quick depletion of mineral fertilizer in the soil either by means of microbial utilization [42] and by plant consumption, or by direct losses via leaching and/or volatilization

3.1.1. Soil total C and N in fallow frequency experiment in Astana, Kazakhstan

Under nearly 50 years of monoculture of wheat, summer bare fallow has been practiced in crop rotation in order to retain moisture, to accumulate nutrients through mineralization and to control weed infestation. Fallowed fields are usually cultivated many times to keep the land bare during the whole cropping season. Of great concern is, however, the adverse effect of fallow, that is, the changes in soil organic matter (SOM) quality and quantity in the context of degradation of the fertility of chernozem soils and subsequent agricultural sustainability. The studies of [43-45] have demonstrated that fallowing significantly exacerbates the depletion of SOM. Organic C and N content of soil after 33 years of cropping decreased with increasing frequency of fallow in a rotation on Canadian soils (53).

Treatments	Fertilization rates kg ha^{-1} N year^{-1}	Fertilizer	N applied per rotation (10 years), kg N ha^{-1} rotation^{-1}	
CON	no	no	no	no
M1	N_{45}	$(NH_4)_2SO_4$	450	450
M3	N_{135}	$(NH_4)_2SO_4$	1350	1350
O	Manure $N_{67.5}$	manure	675	675
MO1	N_{22}+ manure $N_{22.5}$	$(NH_4)_2SO_4$	225	450
		manure	225	
MO3	N_{45}	$(NH_4)_2SO_4$	675	1350
		manure	675	

*Amount of N in manure was calculated as: one ton of cattle manure contains approximately 5 kg of N

Table 1. Fertilization treatments in Uman experimental site from 10-year crop rotation in surface soil of *Argiudolls*, Ukraine

Fertilization rates kg ha^{-1} year^{-1}	Treatments	Organic C	Total N	C/N ratio
		g kg^{-1} soil		
no	CON	20.4a	1.64a	12
N_{45}	M1	19.6a	1.62a	12
N_{135}	M3	20.8a	1.76a	12
Manure $N_{67.5}$	O	21.9b	1.77a	12
N_{22}+ manure $N_{22.5}$	MO1	20.4a	1.71a	12
N_{22} + manure $N_{67.5}$	MO3	20.1a	1.72a	12

Table 2. Effect of fertilization treatments on soil organic C (SOC) and total N (TN) in surface soil of *Argiudolls*, Uman, Ukraine

The objectives of this study were to examine the effects of summer fallow on the characteristics of SOM on a long-term basis (length of crop rotation with a variety of frequencies of fallow) as well as on a short-term basis (pre- and post-fallow phases) with special reference to readily decomposable fractions.

To investigate the impact of bare fallow on soil SOM dynamics the five representatives fallow-spring wheat crop rotation were selected (2-year, 4-year and 6-year with one year of bare fallowing). Soil samples were collected from pre- (2R-pre, 4R-pre and 6R-pre) and post-fallow (2R-post, 4R-post and 6R-post) phases in each rotation. Also, for comparison the continuous cropping of spring wheat (CW) and continuous fallowing (CF) were sampled for comparison.

Soil organic carbon (SOC) content was significantly affected by long-term fallowing. The CF system maintained the least SOC, while 6R and CW stored the most SOC (Table 3). SOC was inversely proportional to fallow frequency, indicating the negative effect of fallow on long-term accumulation of SOM. The effect of the rotations on total nitrogen (TN) paralleled that described for SOC. The highest TN concentrations were observed in the 6R and CW systems and lowest concentrations in the CF system.

To protect the field against weeds and to store more moisture and nutrients in the soil, fallowed field are cultivated 4 to 5 times during the vegetative season. Such intensive mechanical disturbance causes enhanced mineralization of SOM in fallow, firstly, due to better aeration of surface soil, and secondly, particular organic matter occluded within aggregates might become exposed to microbial attack after disruption of aggregates. Additionally, bare fallow does not contribute plant residues for the replenishment of SOM.

In general, distributions of SOC and TN among rotations with different fallow frequencies were comparable to those reported by [50-52] for Chernozem soils. Frequently fallowing systems such as 2R showed less SOM than less frequently fallowing systems, such as 6R. Our results confirmed the findings from North American arable systems that frequently fallowing system accelerates mineralization of SOM [51-53]

Rotation phase	Rotation phase sampled	SOC	TN	C-to-N ratio
		kg Mg^{-1} soil		
CF	Cont. Fallow	21.9a[x]	1.97a	11
2R-pre	(F)[y]-W	25.4b	2.26b	11
2R-post	F-(W)	25.1b	2.16b	12
4R-pre	(F)-W-W-W	26.1b	2.26b	12
4R-post	F-(W)-W-W	24.9b	2.19b	11
6R-pre	(F)-W-W-W-W-W	31.0c	2.57c	12
6R-post	F-(W)-W-W-W-W	30.6c	2.50c	12
CW	Cont. W	27.2c	2.38c	13

[X]*a-c:* values within columns followed by the same letter are not significantly different (*P*=0.05) as determined by *LSD* analysis.

[Y] ( ) denotes rotation phase sampled.

Table 3. Effects of fallow (F) frequency and rotation phase on soil organic C (SOC) and total N (TN) in surface soil of *Haplustolls*, Astana, Kazakhstan

4. Impact of agricultural practices on labile SOM

Labile SOM fractions such as the "light" fraction C [14],microbial biomass carbon [15], mineralizable C [8, 16] are highly sensitive tochanges in C inputs to the soil and will provide a measurable change before any such change in total organic matter [17].In contrast, the more stable (humified) poolsare probably the more appropriate and representative fractions for C sequestration characterization [54].

Carbon mineralization potentials measured via soil respiration in a cascade measurement of the carbon dioxide efflux produced from soil metabolic processes consists of mainly microbial decomposition of soil organic matter and root respiration [55]. A great deal of research money and effort has been invested in studies of soil respiration in recent years because of the potential impacts of this process on the Greenhouse Effect [55]. Measurement of potentially mineralizable C represents a bioassay of labile organic matter using the indigenous microbial community to release labile organic fractions of C. Mineralizable N is also an important indicator of the capacity of the soil to supply N for crops. Individual labile organic matter fractions, such as easily mineralizable carbon and nitrogen, the microbial biomass and activity, are sensitive to changes in soil management and have specific effects on soil function [8]. Together they reflect the diverse but central effects that organic matter has on soil properties and processes.

4.1. Soil labile OM in fertilization and experiment

In the 21st century the mineral fertilizers became determinant for obtaining contented yield of agricultural crops. However, the application of only mineral fertilizer might lead to accelerated mineralization of soil organic matter not mentioning the ecological aspects. Most scientist agree that prolonged application of manure either stabilizes the initial content of humus or increases its content, depending on the rates of applied manure [38, 56-58].

4.1.1. Soil mineral nitrogen in fertilization experiment

The study of the fertilization experiment showed that amount of soil mineral N (min-N) was in direct correlation with added N (Fig 3). The highest content of min-N was recorded in MO3 and M3 treatments, followed by O, MO1and M1 treatments (Figure 3). As shown in Table 1, M3 and MO3 treatments received the highest rate of N that was 1350 kg of N per ha per rotation that was the reason of the increased amount of min-N. Generally, min-N was distributed proportionally to the amount of applied N. But some difference was observed between application of mineral N alone and combination of N applied with mineral fertilizer and manure. For example, M1 and MO1 treatments received the same amount of N in whole rotation, where M1 treatment received only mineral N, and MO1 treatment received 50% N from the mineral fertilizer and 50% N from the manure. Similar pattern was observed in the case of M3 and MO3. Fertilizer N was quickly utilized by plants and microorganisms, while N of manure was decomposed more slowly supplying the soil with N for longer period.

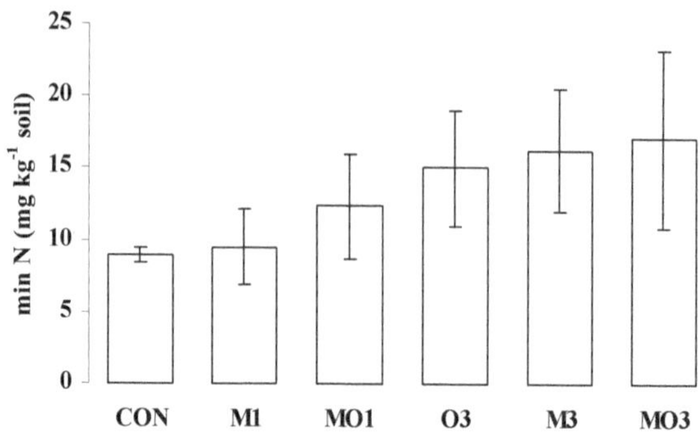

Figure 3. Soil mineral nitrogen at sampling time from fertilization treatments, Uman

The [56] reported that "ready" humic substances applied with cattle manure is thermodynamically non-stable and therefore is subject to faster decomposition and mineralization. Probably, these non-stable organic substances in manure were the main source of mineralized nitrogen resulting in increased accumulation of labile forms of N in manured experiment.

4.1.2. Nitrogen mineralization potentials in fertilization experiment

Mineralization rate constant (*k*) among the treatments varied significantly (P<0.05). The treatments where the high rates of manure wereapplied showed higher mineralization rate (Figure 4). All treatments but MO3 have lowered their mineralization rate by the end of incubation (56-70 days). MO3is the treatment that received mineral fertilizer and high rate of manure.Nitrogen of the mineral fertilizer serves as an easy available substrate for microorganisms at the beginning of the incubation then, after the available mineral nitrogen was depleted by microbial utilization, the nitrogen of the manure was exposed to microbial attack showing high mineralization rate after 70 days of incubation, while in O treatment, manure was attacked from the beginning because no mineral N was added to the soil. Manure consists of labile and non-labile fractions of organic compounds. After the labile fractions of manure were mineralized, the mineralization rate was slowed down thus showing lowered rate after eight weeks of incubation.

PMN content was the highest in the treatments where high rates of manure were applied that are O and MO3 (Table 4). Manure was applied about 19 months before the soil sampling. During about eight months the soil was frozen and no microbial activity was undergoing. It takes about 275 days to start releasing mineral N from manure, and about 391 days for complete mineralization or for reaching the stabilization point [59]. By the time of sampling manure had been releasing N for about 90 days, therefore, during the laboratory incubation manure continued to release mineral N, showing higher PMN in O and MO3 treatments.

Cattle manure contains "ready" humic substances that can be directly and immediately involved into immobilization processes. Probably, these non-stable organic substances in manure were the main source of mineralized N resulting in increased accumulation of labile forms of N under the manured treatments.

Treatment	Fertilization rates kg ha^{-1} year^{-1}	PMN mg kg^{-1}	C/N	Soil total N as PMN, %
CON	No	75.61	12.7	4.6
M1	N_{45}	84.64	8.6	5.2
M3	N_{135}	78.50	15.2	4.5
O	Manure $N_{67.5}$	96.44	12.0	5.4
MO1	N_{22} + manure $N_{22.5}$	88.17	13.7	5.2
MO3	N_{22}+ manure $N_{67.5}$	151.92	6.8	8.8

Table 4. Potentially mineralizable nitrogen as influenced by different fertilization, Uman

Treatment	Fertilization rate kg ha^{-1} year^{-1}	MBC	MBN	C/N	Soil organic C and total N (%)	
		mg kg^{-1} soil			as MBC	as MBN
CON	no	459a	33.3a	14	2.09	2.02
M1	N_{45}	586b	35.5a	17	2.72	2.19
M3	N_{135}	531c	21.3b	25	2.35	1.21
O3	Manure $N_{67.5}$	566bc	85.3c	7	2.40	4.81
MO1	N_{22}+ manure $N_{22.5}$	585b	71.5d	8	2.64	4.19
MO3	N_{22}+ manure $N_{67.5}$	677d	69.6d	10	3.10	4.05

Table 5. Microbial biomass carbon (MBC) and nitrogen (MBN) as influenced by different fertilization, Uman

4.1.3. Microbial biomass carbon and nitrogen in fertilization experiment

Microbial biomass N (MBN) in the treatments where manure was applied amounted from 72 to 85 mg kg^{-1} soil (Table 5). In the treatments where no manure was applied MBN amounted from 21 to 36 mg kg^{-1} soil. Moreover, application of the high rate of mineral fertilizer alone decreased the MBN content. The ratio of MBC to MBN also shows distinctive difference between the treatments where high ratio was observed under the mineral fertilization and lower ratio under the manure application. Noticeably, highest C to N ratio was under the highest rate of mineral fertilization (M3) and the lowest was under the high rate of manure application (O3). This is due to intensive utilization of added nitrogen by microorganisms. Addition of biomass substrate with the content N more than 1.5 to 1,7% do not need additional fertilizer by nitrogen, the soil N satisfies the need of microorganisms during the decomposition. Always, the "requirements" of microorganisms are satisfied first of all, disregarding on the need of plants for nitrogen.

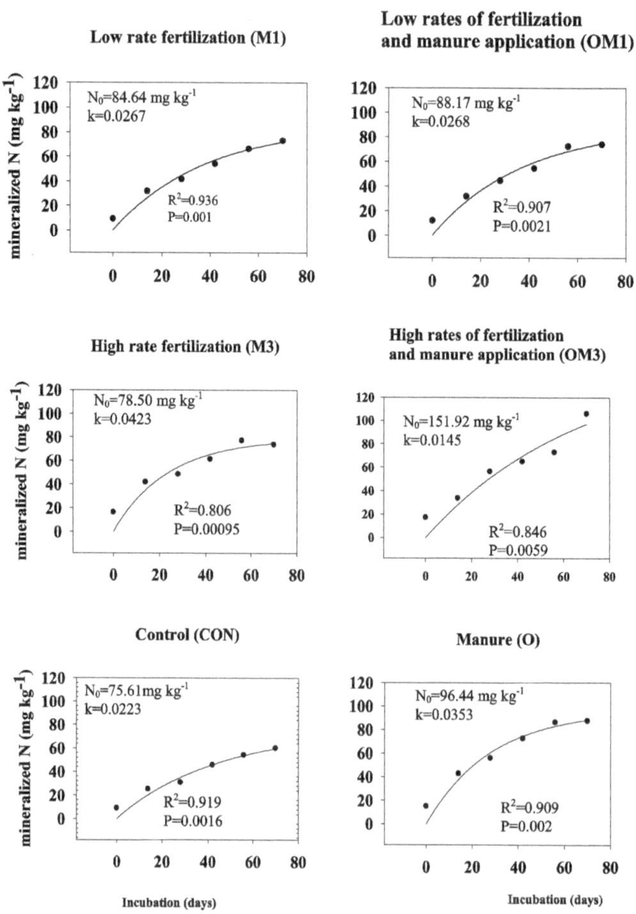

Figure 4. Fitting curves of nitrogen mineralization in fertilization experiment in Uman, Ukraine, as described with the first order kinetic model: $Nmin=N_0(1-e^{-kt})$, where Nmin is the mineralized N at time *t*, N_0 is the potentially mineralizable N (PMN), k is the mineralization rate constant.

4.2. Impact of fertilization and irrigation practices on total soil organic C and N

This study was conducted in long-term experiments with fertilization and irrigation in Kherson, south Ukraine. Sampling scheme of the Kherson experiments were 1) irrigated plus fertilized treatment (I+F), 2) irrigated only (I); 3) fertilized only (F) and 4) control that was neither fertilized nor irrigated (CON).

Analysis of variance showed that soil organic carbon (SOC) and total nitrogen (TN) were not statistically different among treatments. However, there were observed a different trend in accumulating SOM (Table 6). Contents of SOC and TN were for 7.19% and 9.30% respectively, greater in I+F treatments than in the control CON. Higher accumulation of SOC and TN under I+F treatment is due to higher biomass production under irrigation and fertilization. Consequently, higher plant biomass contributes to SOM. Fertilization (F) alone, or irrigation (I) alone maintained similar amount of organic C and total N. Kherson region is characterized with very small amount of precipitation, around 300 mm annually. In such dry conditions, fertilization does not pay off in terms biomass production because the applied fertilizer cannot be dissolved and be available for plant consumption.

Treatments	Applied treatment	Organic C	Total N	C/N
		$g\ kg^{-1}$ soil		
I+F	Irrigated and fertilized	16.7a	1.29a	13
F	Fertilized	16.4a	1.25a	13
I	Irrigated	15.8a	1.24a	13
CON	no	15.5a	1.17a	13

*Fertilizer was applied as $N_{120}P_{120}K_{120}$ for every fertilization treatment at rates indicated as subscripted mark.

*Irrigation water was applied at rate of $3200m^3\ ha^{-1}$ for every irrigated treatment.

Table 6. Organic carbon and total nitrogen concentration in irrigation experiment, Kherson

4.2.1. Soil mineral nitrogen in irrigation experiment

Content of soil mineral N was different among the treatments at P=0.1 (Fig. 5). As expected, the greatest differences were observed between I+F and non-irrigated treatments (F and CON).

Irrigated treatment accumulated higher min-N than the non-irrigated because irrigation of dry soil resulted in flash in microbial growth. During the long dry period the most of microorganisms dies, and then later upon irrigation those microbial necromass serves as easy source of N for survived microorganisms. The microorganism's body has the narrowest C-to-N ratio that is an indicator of the most easily available N source. Thus the flash in microbial growth accelerates mineralization processes in soil and moistening of dry soil causes disruption of organic compounds and soil particles that may contain organic substances. Subsequently, the disrupted organic material is more sensitive for microbial attack thus contributing to N mineralization.

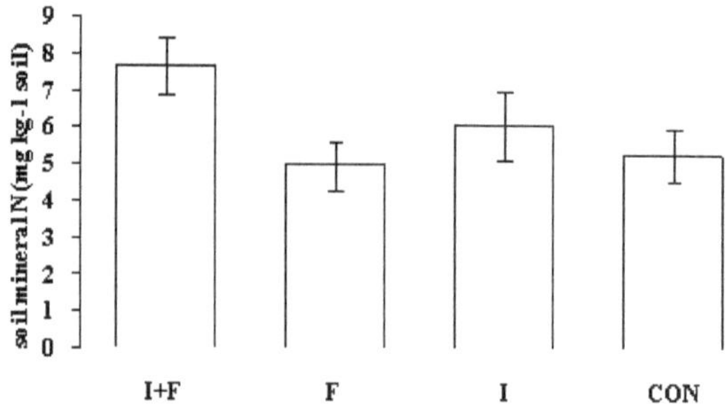

Figure 5. Soil mineral nitrogen in irrigation experiment, Kherson

4.2.2. Nitrogen mineralization potentials in irrigation experiment

Potentially mineralizable nitrogen (PMN) was significantly different (P=0.01) among the treatments with the highest mineralization rate under I+Ftreatment (k=0.0192) (Figure 6). The highest accumulation of mineralizable N (PMN) was also obtained under the I+Ftreatment (171.73 mg kg^{-1}) (Table 7), while all other treatments maintained statistically not different amounts of PMN.

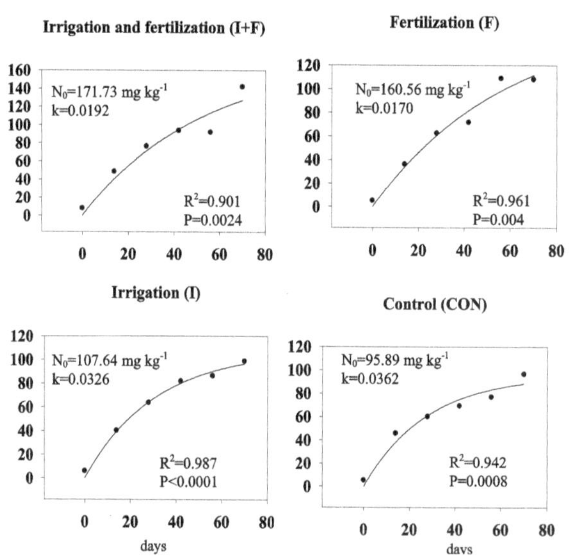

Figure 6. Fitting curves of nitrogen mineralization in fertilization experiment, Kherson, as described by the first order kinetic model: Nmin=$N_0(1-e^{-kt})$, where Nmin is the mineralized N at time t, N_0 is the potentially mineralizable N (PMN), k is the mineralization rate constant

I+F treatment maintained higher plant biomass returned into soil. Later when the soil was placed under the favourable laboratory conditions, those accumulated residues were subjected to mineralization showing higher PMN.

Fertilization alone (F) had suppressed mineralization on the field because of deficiency of water necessary for microbial activity. But when the soil was placed under favourable laboratory conditions the accumulated organic substrate is mineralized, thus giving nearly the same amount of PMN as the irrigated treatment.

4.2.3. Carbon mineralization potentials in irrigation experiment

Irrigated plus fertilized treatment (I+F) showed the highest carbon mineralization rate as well as amount of PMC in 70d (Fig. 7; Table 7). Irrigation of dry soil disrupts soil structure thereby making previously sequestered carbon available for microbial utilization. The [60, 61] found that soil drying destroyed 1/3 to 1/4 of biomass and after remoistening the biomass was progressively restored to approximately the same size as before drying.

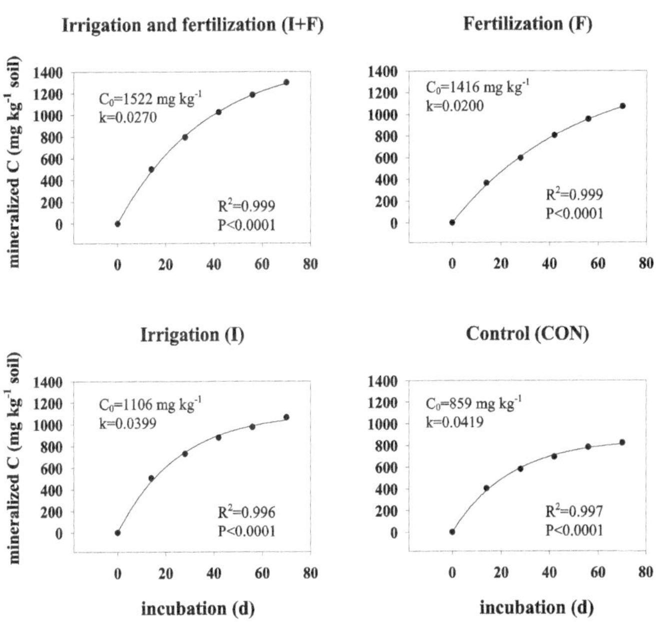

Figure 7. Fitting curves of carbon mineralization in fertilization experiment, Kherson, as described by the first order kinetic model: $Cmin=C_0(1-e^{-kt})$, where Cmin is the mineralized C at time *t*, C_0 is the potentially mineralizable C (PMC), k is the mineralization rate constant

In this study, high temperatures and dry conditions have caused death of microorganisms that were immobilized during desiccation via adsorption on clay surfaces and/or transformation into other forms of organic compounds. Then, the following irrigation revived microbial community and disrupted soil clay particles that released stabilized organic matter. In the study conducted by [62] the similar results were obtained, where extra mineralized ^{14}C, due to soil desiccation came from non-living residues, likely to be those that were stabilized by adsorption to clay surfaces.

The highest percentage of PMC and PMN in soil organic C were under I+F treatment (Table 7). This is in accordance with the earlier discussion and confirms the hypothesis that there are at least two reasons responsible for it: firstly, irrigation of dry soil causes enhanced mineralization of soil organic matter, and secondly, fertilization of irrigated soil provides higher plant biomass that contributes to the accumulation of labile organic matter.

Treatments	Applied treatment	PMC	PMN	C/N	Soil organic C and total N (%)	
		mg kg^{-1} soil			as PMC	as PMN
I+F	Irrigated and fertilized	1522	171.73	8.9	9.11	13.31
F	Fertilized	1416	160.56	8.8	8.63	12.84
I	Irrigated	1105	107.64	10.2	6.99	8.68
CON	no	858	95.89	8.9	5.53	8.20

Table 7. Mineralizable carbon and nitrogen (PMC and PMN) in irrigation experiment, Kherson

4.2.4. "Light fraction" organic matter in irrigation experiment, Kherson

The I+Ftreatment maintained the highest amount of 'light' fraction dry matter (LFDM), carbon (LFC), nitrogen (LFN) and their proportions in soil organic carbon (SOC) and total nitrogen (TN) (Table 8). One of the reasons is, as discussed earlier, higher biomass production in this treatment, hence higher organic substrate was added with residues. Desiccation that caused the death of a large number of microorganisms, followed by immobilization and condensation of their dead tissues thus increasing the amount of recalcitrant, soluble organic C is another reason [60]. Moreover, irrigation of desiccated soil also causes the death of microorganisms due to the osmoregulatory shock [63] that also could contribute to the LFOM.

4.2.5. Microbial biomass in irrigation experiment, Kherson

Microbial biomass carbon (MBC) and nitrogen (MBN) significantly differed among the treatments (Table 9). The highest MBC and MBN were obtained under the irrigation alone (I) treatment followed by the irrigated plus fertilized (I+F) treatment. The least microbial biomass was obtained under the fertilized alone (F) treatment.

Treatment	Applied treatment	LFDM	LFC	LFN	C/N	Soil organic C and total N (%)	
		g mg^{-1} soil	mg kg^{-1} soil			as LFC	as LFN
I+F	Irrigated and fertilized	9.27	1884	122	15	11.3	9.4
F	Fertilized	8.52	1591	104	15	9.7	8.4
I	Irrigated	6.49	1589	104	15	10.0	8.4
CON	no	6.82	1684	108	16	10.8	9.2

Table 8. "Light fraction" dry matter (LFDM), carbon (LFC) and nitrogen (LFN) in irrigation experiment, Kherson

Treatment	Applied treatment	MBC	MBN	C/N	Soil organic C and total N (%)	
		mg kg^{-1} soil			as MBC	as MBN
I+F	Irrigated and fertilized	618	160	4	3.7	12.4
F	Fertilized	450	76	6	2.7	6.1
I	Irrigated	733	175	4	4.6	14.2
CON	no	636	128	5	4.1	10.9

Table 9. Microbial biomass in irrigation experiment, Kherson

Such distribution of microbial biomass was expected because moisture conditions are a major factor controlling survival and activity of microorganisms in the soil [64]. Drying and re-moistening of soils strongly affects microbial growth and activity [61, 65, 66]. After remoistening of dried soil, available C components were assimilated and transformed partly into new biomass C, and partly involved into CO_2 that evolved into the atmosphere [63].

4.2.6. General discussion of fertilization, manure application and irrigation experiments

Many researchers recorded positive effects of manure application on SOM [38, 57, 67-69]. For example, in Nebraska, annual application of 13.5 t ha^{-1} of manure (dry matter) during 31 years on irrigated land has increased content of humus from 0.98 to 1.67%. The [59] found out that increased application of manure resulted in intensification of C mineralization, especially the C that is included in fulvic acids, and in lesser extent in humic acids. Biological analysis showed that application of high rates of manure activates the biochemical processes, which is controlled by particular microbiological community that has ability for active transformations not of only simple organic substances (e.g. fulvic acids), but also of more complex and hardly decomposable substances (e.g. humic acid).

Increased microbial activity in irrigated treatments in Kherson has been ascribed to the rapid metabolization of biomass-derived substrate resulting from the death of part of the microbial community during drying [61-63, 70] and/or rapid rewetting of the desiccated soil material [100]. Alternate drying and re-moistening increases the mobility of organic matter and results in the release of N as ammonium and amides [42].

4.3. Impact of bare fallow on soil labile organic matter in Haplustolls, Astana, Kazakhstan

With increasing cultivation intensity, the SOM of the less stable pools is decomposed, as indicated by decreasing portions of sand-sized SOM (2-0.05 mm) [19, 46, 47], or light fraction C [48, 49). Organic compounds adsorbed to surfaces of clay particles might become exposed to microbial attack after disruption of aggregates due to tillage.

To investigate the impact of bare fallow on soil SOM dynamics the five representatives fallow-spring wheat crop rotation were selected (2-year, 4-year and 6-year with one year of bare fallowing). Soil samples were collected from pre- (2R-pre, 4R-pre and 6R-pre) and post-fallow (2R-post, 4R-post and 6R-post) phases in each rotation. The continuous fallowing (CF) and continuous cropping of wheat (CW) were also sampled to see the effect of fallow impact on SOM.

4.3.1. Soil mineral nitrogen in fallow frequency experiment

On a long-term basis, the CF system accumulated the highest amount of soil mineral nitrogen (min-N). But min-N was strongly affected by summer fallow on a short-term basis as well. Pre- and post-fallow phases showed significant differences with min-N accumulating in post-fallow than in pre-fallow phase (Table 10). Post-fallow phases accumulated 3.0-, 1.9- and 1.9-fold amounts of min-N of pre-fallow phases in 2R, 4R and 6R, respectively

As expected, the CF system maintained the highest amount of soil min-N that was due to enhanced mineralization of SOM compared to the other systems. The short-term effect of fallow on the accumulation of min-N is clearly observed as well. During fallow phase min-N is not subjected to either plant uptake or leaching, thus resulting in a greater accumulation of soil min-N in post-fallow (2R-post, 4R-post and 6R-post) than in pre-fallow phases (2R-pre, 4R-pre and 6R-pre).

4.3.2. Nitrogen mineralization potentials in fallow frequency experiment (PMN)

The pattern of N mineralization showed a different trend between pre- and post-fallow phases in all rotations (Fig. 8). Pre-fallow phases (Fig.8. a, c and e) were characterized by a larger value of PMN (N_0), a smaller mineralization rate constant (k), and a shorter initial delay of mineralization (*c*) than in the post-fallow phases (Fig.8. b, d, and f).

Fallow influenced accumulation of PMN on short-term basis, that is, pre-fallow phases (2R-pre, 4R-pre and 6R-pre) accumulated more PMN than post-fallow (2R-post, 4R-post and 6R-post) phases (Table 10). The lowest PMN was observed under the CF system (69 mg kg^{-1}) and the highest under 6R-pre (124 mg kg^{-1}). Pre-fallow phases accumulated 2.4, 1.3 and 1.5 fold amount of PMN of post-fallow phases in 2R, 4R and 6R, respectively. Larger amounts of mineralized nitrogen (N_0) in the pre-fallow phases indicate larger storage of PMN in these soils than in post-fallow soils. Differences in the rate constant (*k*) between pre- and post-fallow phases indicate that fallowing has caused changes in the quality of the PMN.

Due to multiple cultivations of fallows the soil is subjected to alternating wet-dry cycles. The wet period provided better moisture condition microorganism activity and produced greater biomass than in cropped fields. Then in the subsequent dry period the greater biomass turned into necromass due to drought. This cycle may be repeated several times in a cropping season. Later, during incubation in the laboratory, this microbial necromass as well as living biomass was rapidly mineralized showing a higher mineralization rate constant in the post fallow than in the pre-fallow phases. The soils from the post-fallow phase showed a longer initial delay of mineralization, suggesting that higher concentration of min-N compared to pre-fallow phase probably stimulated microbial activity and resulted in immobilization of mineralized N during the initial stages of incubation.

The long-term effect of fallow was not observed for soil min-N or PMN suggesting that N mineralization is only affected by the substrate added during the previous year or the latest cycle of rotation. Nitrogen in the forms of NO_3^- and NH_4^+ is assimilated by plants and returned into soil whereas C originates from CO_2 in the air and ploughed as organic residue into soil. Nitrogen transformations are closely related to the processes of mineralization of its organic forms in plant-soil system. Therefore, in plant-soil systems N cycling is affected over shorter period than C cycling.

Rotation phase	PMC	min-N	PMN	C-to-N ratio	Organic C as PMC	Total N as PMN
		mg kg^{-1} soil			%	
CF	794b*	46a	69a	11	3.6	3.5
2R-pre	1194ab	14b	166b	7	4.7	7.4
2R-post	1012b	42a	69a	15	4.0	3.2
4R-pre	1224a	13b	86c	14	4.7	3.8
4R-post	1215a	24c	67a	18	4.9	3.1
6R-pre	1524c	16b	124b	12	4.9	4.8
6R-post	1300ac	30c	82c	16	4.3	3.3
CW	1581c	14b	93c	17	5.8	3.9

Table 10. Effect of fallow frequency and rotation phase on labile organic matter content, Astana

4.3.3. Carbon mineralization potentials in fallow frequency experiment, Astana

Differences in PMC among the rotation systems ($P<0.001$) were more clearly shown than for SOC. PMC ranged from 3.6 (CF) to 5.8% (CW) of the SOC. The amount of PMC was more affected by the long-term effect of fallow than by the short-term effect and was inversely proportional to fallow frequency.

Continuous wheat (CW) and 6-y systems (6R) had higher amount of PMC that was inversely proportional to fallow frequency and indicated the long-term effect of fallow. These results corroborate with the study conducted by [73] who found for a silt-loam in southwestern Saskatchewan that mineralized C (measured after 30 days at 21°C) represented 1.06 and 1.45% of SOC in a 2-y fallow-wheat rotation and continuous growing of wheat, respectively. The [74] found that C mineralization was not related to the amount of crop residue from the previous year. In our study PMC was a little higher in the pre-fallow (2R-pre, 4R-pre and 6R-pre) than in the post-fallow (2R-post, 4R-post and 6R-post) phases, probably reflecting the input of crop and weed residues in the preceding year.

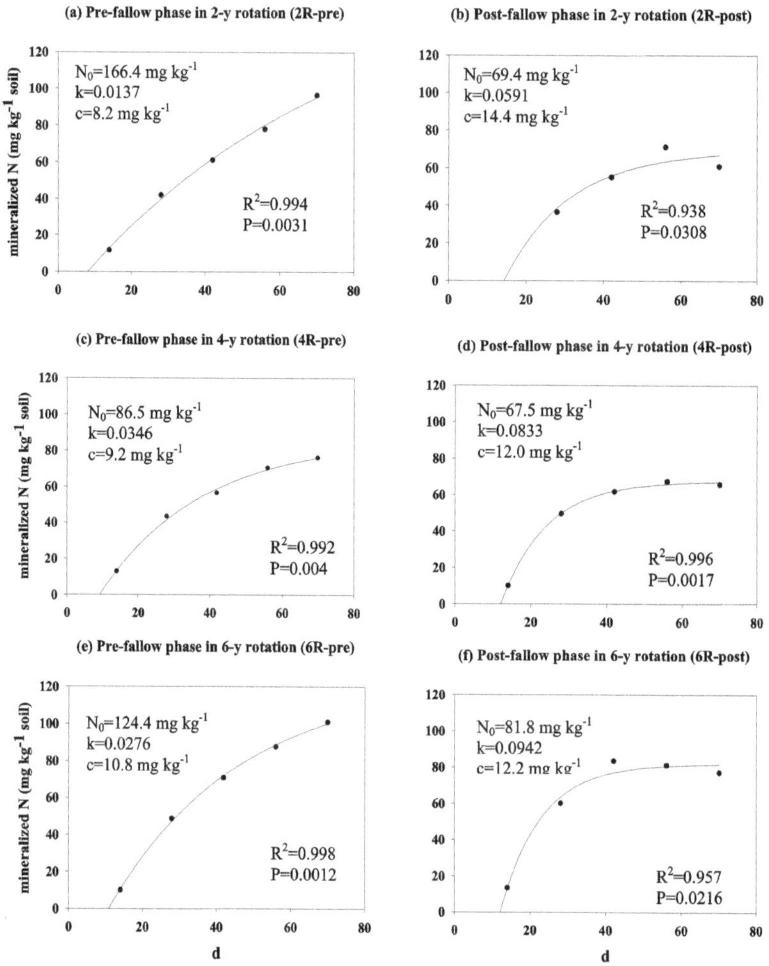

Figure 8. Fitting curves of N mineralization from pre- and post-fallow phases of the 2-, 4-, and 6-year wheat-fallow 40-years rotation experiment, as described by the first order kinetic model with an initial delay of mineralization ($N_{min} = N_0(1-e^{-kt-c})$), where, $N_{min}$is mineralized N at time t,;N_0 is value of potentially mineralizable N (PMN), k is a mineralization rate constant, and c is an initial delay in mineralization.

4.3.4. Light fraction" organic matter in fallow frequency experiment, Astana

The amount LF-OM was highly responsive to fallow frequency, accounting for 1.1(CF)-6.3(CW)% of the SOC and 0.8(CF)-4.3(CW)% of the TN (Table 11). LF-OM, as expressed on the basis of dry matter (LF-DM), C (LF-C) or N (LF-N), was inversely related to fallow frequency. For example, the LF-C content of the CW system was 7.2 times higher than that in the CF system. These results agree with those of other studies [e.g. 36, 75], where LF content was highest under continuous cropping and lowest in those with a high frequency of summer fallow. Additionally, LF-C was affected by the rotation phase, showing larger amounts in pre- than in post-fallow phases in 4R and 6R rotations.

"Light fraction" of SOM (LF-OM) consists mainly of plant residues, small animals and microorganisms adhering to plant-derived particulate matter at various stages of decomposition that serves as a readily decomposable substrate for soil microorganisms and also as a short-term reservoir of plant nutrients [76]. The "light fraction' C (LF-C) was positively correlated with PMC (Fig. 9), and confirm the hypothesis that the reduced fallowing system has more potential to supply soil with easily mineralizable C. However, there was no linear correlation between LF-N and PMN, presumably because the high C-to-N ratio of the LF-OM temporary induced N immobilization [36].

The content of labile OM, which is closely related to LF-OM, may be governed by the degree to which temperature and moisture conditions constrain decomposition of accumulated residues [52]. Under the CW system decomposition of residues during periods of favourable soil temperature was retarded by the depleted soil moisture [77, 78,]. Then, when moisture and temperature constraints were removed during laboratory incubations, soil showed a high respiration rate [36]. On the contrary, residues in the 2R system during the fallow phase were always exposed to an extended period with favourable moisture and temperature. Therefore, labile organic matter was rapidly depleted in the field, and in the laboratory respiration rates were much lower in 2R than in CW [52].

4.4. Grain yield and weed biomass

The main controller of biological activity in soil is the SOM generated from crop residue, while crop residues are in direct correlation with crop yield and sometimes with yield of weeds. In systems with no pesticide application, such as control treatments, or in case of the CW (continuous wheat cropping), where no break in lifecycle of weeds occur, the weed biomass might significantly contribute to the input of crop residue in soil and affect the crop yield output.

The highest grain yield was produced in the first year of the rotations (the year after fallow) (Figure 10). However, the yield decreased sharply with in the second and successive years after fallowing. This trend is, firstly, because plants in a post-fallow phase take advantage of higher soil min-N. Secondly, because when a field is in fallow provides the only break for weed infestation; the amount of weeds was generally least in the first year after fallow and reduced competition for nutrients. In contrast to the grain yield, weed infestation reached its

Rotation phase	LF-DM	LF-C	LF-N	C-to-N ratio	Soil C as LF-C	Soil N as LF-N
	g kg⁻¹ soil	mg kg⁻¹ soil			%	
CF	0.9	240a[x]	15a	16	1.1	0.8
2R-pre	3.6	810b	51b	18	3.2	2.2
2R-post	2.5	660b	38bc	16	2.6	1.7
4R-pre	5.7	1330c	81d	17	5.1	3.6
4R-post	5.3	1250c	73d	17	5.0	3.3
6R-pre	6.4	1560d	74d	20	5.1	3.0
6R-post	6.0	1500d	75d	21	4.9	2.9
CW	7.4	1730e	103e	17	6.3	4.3

[x]a-e: values within columns followed by the same letter are not significantly different ($P<0.005$) as determined by *LSD* analysis.

Table 11. Effects of fallow frequency and rotation phase on the amount of "light fraction" dry matter (LF-DM), "light fraction" C and N (LF-C and LF-N) and their proportions in SOC and TN in surface soil of Haplustolls, Astana

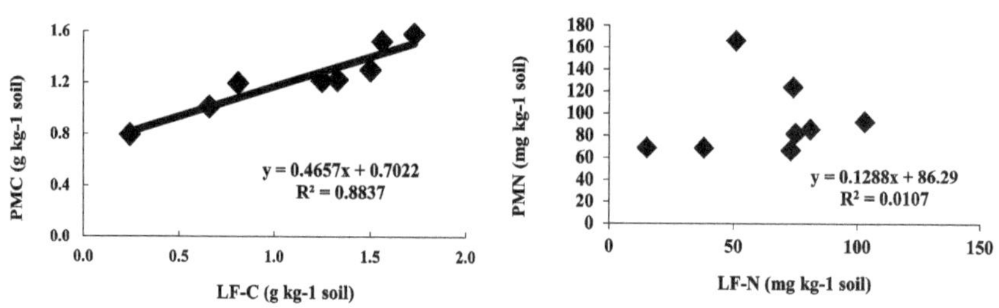

Figure 9. Correlation of potential mineralizable C and N) with "light" fraction C (LF-C) and N (LF-N)

maximum in the second years after fallow in 4R and 6R. Probably, some of the weeds were not destroyed during the fallow and their seeds remained dormant but germinated in the second year after fallow. Correlation between the grain yield and weed infestation (average for ten years) is presented by the following equation of multiple linear regression:

$$Y = 20.82 - 0.189X,$$

where X total amount of weeds, pieces/m^{-2}. The coefficient of determination was also high ($R^2 = 0.78$) or 78% of changes of the yield depend on weed infestation.

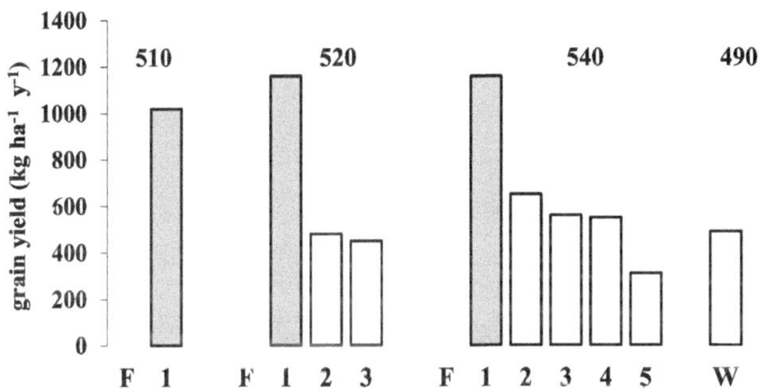

Figure 10. Grain yield (1994-2000) of spring wheat affected by distance from fallowing year. (F is fallow; 1, 2, 3...5 are succession of crops after fallow. (the numbers above bars are average yield per rotation including the fallow year).

The highest grain output per whole rotation, was obtained by 6-year rotation, being 540 kg ha^{-1} that parallels the distribution of soil labile C (PMC and LF-C). These results indicate that longer rotations with fewer fallows contribute more to the accumulation of SOM than shorter rotations with frequent fallowing, and that SOM is replenished continuously under less bare-fallowing system.

The results on *Mollisols* in North Kazakhstan suggested that N dynamics were closely related to the recent input of substrate added as plant residue while C dynamics were more related to long-term substrate addition.

Yearly input of plant residue in a 6-y wheat-fallow rotation system built up more labile OM, especially LF-C or readily decomposable C, whereas 2-y rotation system with a high frequency of fallow depleted SOM via accelerated mineralization. Therefore, with no fertilizer or pesticides application, in the semi-arid regions of northern Kazakhstan, the inclusion of fallow in wheat monoculture every 6 years is the most appropriate farming system in terms of sustainability in both grain production and soil fertility.

The results of this study may provide prediction of SOM response to fallow frequency in wheat-based rotation systems in Chernozem soils of semi-arid regions: the susceptibility of labile fractions of OM and their relationship to fallow frequency suggest the possibility of managing labile OM through controlling the length of wheat-fallow rotation systems.

5. Characterization of soil organic matter dynamics from different climatic zones

Many studies have shown that climatic factors, namely temperature and precipitation are major determinants of microbial diversity and activity in soil at global [79[, regional [5, 80] and local [81] scales [47, 82]. The great meridian and latitudinal extension of Mollisols [83] determines a wide variety of climatic conditions that influence the main genetic characteristics of soils and their natural growth and agronomic properties. The aim of this study was to find out effects of temperature and moisture on soil organic matter accumulation and decomposition.The study conducted by [109] also showed that in dry and cold conditions of dry steppe, soil respiration was mostly controlled by soil temperature while residue input was a function of moisture conditions.

Four types of Mollisols from four different climatic regions were sampled: *Hupludolls* (southern forest-steppe, Kharkov, Ukraine), *Argiudolls* (northern forest-steppe, Uman, Ukraine), *Calciustolls* (southern steppe, Kherson, Ukraine) and *Haplustolls*(northern steppe, Astana, Kazakhstan) (Table 12). The sampling sites represent the most typical soil type and ecosystem for each given region. The selected geographical regions are characterized as wet-frigid (Kharkov; 6.5°C, mean annual temperature, 542 mm, mean annual precipitation), wet-mesic (Uman; 8.5°C, 660 mm), dry-thermic (Kherson; 11°C, 332 mm) and dry-frigid (Astana; 0°C, 324 mm).

Site	Geographical coordinates	Precipitation (mm)	Mean temperature, (°C)		Ecological and climatic region		Soil Taxonomy, USDA
			winter	summer			
Kharkov	50°N, 36°E	515-570	-10	+18	southern forest-steppe	wet-frigid	*Hapludolls*
Uman	48.8°N, 30.2°E	550-770	-5	+17	northern forest-steppe	wet-mesic	*Argiudolls*
Kherson	46.6°N, 32.6°E	315-350	0	+22	southern steppe	dry-thermic	*Calciustolls*
Astana	51.3°N, 71.1°E	300-350	-18	+19	northern steppe	dry-frigid	*Haplustolls*

Table 12. Soil Types and Climatic Characteristics of the Study Sites

5.1. Soil organic C and total N in different climatic zones

The highest amount of soil organic carbon (SOC) and total nitrogen (TN) was observed in wet-frigid (Kharkov) region, 26.8 and 2.50 g kg^{-1} soil, respectively, and the lowest in dry-mesic (Kherson) region with 15.3 and 1.24 g kg^{-1} soil, respectively (Table 13). This is mainly due to the inherently higher humus content in comparison with other studied soils. These results agree with previously reported studies where the stock of SOC was generally greater in colder and wetter compared to hotter and drier climates [28, 84]. In our study the TN content was significantly higher in frigid (Kharkov and Astana) than in mesic and thermic (Uman and Kherson) regions. Higher amounts of precipitation generally lead to a higher plant biomass production and organic C input [e.g. 15]. In addition, lower temperatures, especially in winter when it falls below a threshold for biological activity, limits decomposition of SOM resulting in accumulation over time [84].

Lack of water in the dry-frigid (Astana) region retarded mineralization of plant residues [8] while low winter temperatures conserved plant residues that were partially mineralized when the temperature was favorable, partially immobilized and partially accumulated as a labile organic matter [85].

In wet-mesicUman, the relative temperature sensitivity of decomposition was greater than the net primary productivity [3], where the higher amount of precipitation naturally produced a greater plant biomass that was quickly decomposed due to favorable temperature and moisture conditions. Dry and hot condition in Kherson suppressed the production of plant biomass and limited accumulation of SOM.

Soil type and hydrothermal regime	Total N	Organic C	C/N	pH	$CaCO_3$	Sand (200-20μm)	Silt (20-2μm)	Clay (<2μm)
	g kg^{-1} soil				%		%	
Hapludolls; Wet-frigid (n=24)	2.50	26.8	11	6.3	0	20.4	37.5	42.1
Argiudolls; Wet-mesic (n=24)	1.70	20.5	12	5.6	0	22.9	37.7	39.4
Calciustolls; Dry-thermic (n=12)	1.24	15.3	12	7.5	5.5	43.4	26.9	29.7
Haplustolls; Dry-frigid (n=24)	2.29	20.0	9	8.2	1.6	25.6	30.6	43.8

Table 13. General properties of different types of *Mollisols* from different

5.2. Labile C fractions of soil organic matter from Mollisols in different climatic zones

5.2.1. Potentially mineralizable C and rate constant k from Mollisols in different climatic zones

Drier regions accumulated a higher amount of potentially mineralizable carbon (PMC) that is presented as C_0 in fitting curves (dry-thermic and dry-frigid, 1225 and 1222 mg kg^{-1} soil, respectively) than of wetter regions (wet-frigid and wet-mesic, 754 and 1091 mg kg^{-1} soil, respectively) (Fig. 11). Corresponding rate constant (k) followed a similar trend as PMC (k value of 0.031 and 0.026 in dry-thermic and dry-frigid, respectively; 0.013 and 0.021 in wet-frigid and wet-mesic, respectively). The shape of the fitting curve of Kharkov (wet-frigid) soil greatly differed from others by being the lowest, i.e. reflecting the least amount of mineralized C, and the straightest, i.e. corresponding with the slowest rate of decomposition. On the other extreme, Kherson (dry-thermic) and Astana (dry-frigid) soils showed the highest and the most curved shapes of the fitting curves that reflect the greater amount of mineralized C and faster decomposition rate in 70 days of incubation.

Moisture was the main factor influencing the amounts of soil labile C. In wetter regions (Kharkov and Uman) microbial respiration is always higher [66], and more organic substrate was utilized than in drier regions. In contrast, soil microorganisms from drier regions

(Astana and Kherson) experienced moisture deficiency and were unable to use the existing available organic substrate. Consequently, when microbial activity was not limited by moisture during the laboratory incubation there was enough energy substrate to promote a high respiration rate. Additionally, in dry conditions lethal effects contributed "dead biomass" to the organic substrate pool [66, 86, 87]. This easily available substrate was rapidly taken up and utilized by surviving soil microorganisms, thus contributing to the increased soil respiration observed when soils from dry regions were moistened [66]. Severe moisture conditions in Astana and Kherson, firstly, enhanced turnover of MB and condensation of microbial products, thus increasing the amount of soluble C [60], and secondly, caused disruption of soil aggregates that resulted in the liberation of protected organic C [88].

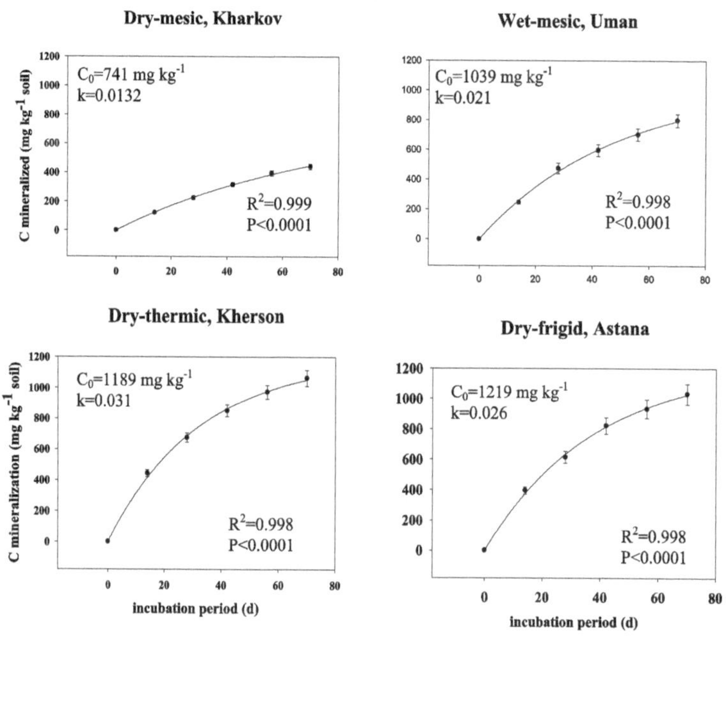

Figure 11. Fitting curves of C mineralization in soils from different climatic regions as described with the first order kinetic model ($Cmin= C_0(1-e^{-kt})$, where, Cmin is a mineralized C at time *t*, C_0 is a potentially mineralizable C (PMC), k is a mineralization rate constant)

Site variation in Kharkov and Uman were not significant (Figure 12). However, dry regions (Kherson and Astana) showed the biggest variation in the amount of potentially mineralizable C (PMC) and *k* values within the site. This indicates that SOM of dry regions are more sensitive to the imposed agronomic treatments than wetter regions. As discussed in previous sections, in drier regions the microbial activity is suppressed by lack of water. Then, when the water limitation is excluded the flash of mineralization take a place, where rapidly growing microorganisms compete for the N thus involving the more stable SOM into the mineralization process. Therefore, the vulnerability of SOM in drier regions should be considered when designing the agronomic treatments.

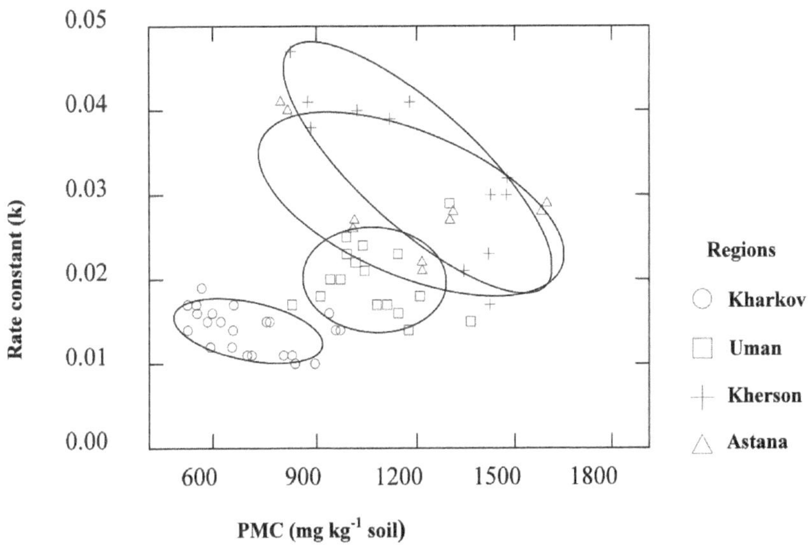

Figure 12. Scatter-plot of mineralized C and decomposition rate constant (k) in *Mollisols* from different climatic regions

5.2.2. "Light" fraction organic matter (LF-C) from Mollisols in different climatic zones

The highest amount of "light" fraction carbon (LFC) was observed in the dry-thermic region (1687 mg kg^{-1} soil, Kherson), followed by the dry-frigid region (1436 mg kg-1 soil, Astana), and the least amount of LFC was observed in wet regions (1180 and 1105 mg kg-1 soil in Kharkov and Uman, respectively) (Table 14). Obviously, drier conditions in Kherson and Astana retarded decomposition of SOM, contributing to accumulation of LFC. Generally, distribution of LFC among sites was well correlated with PMC (r^2= 0.619). The amount of LFC was more affected by precipitation rather than by temperature. These results correspond to studies of, for example, [77, 78] who reported that decomposition of SOM during the period of favourable soil temperature is inhibited by lack of water. In wetter Kharkov and Uman, favorable moisture and temperatures during vegetation season promoted mineralization of "light fraction" OM resulting in less accumulation of LFC.

Site	MBC		PMC		LFC	
	mg/kg	% of SOM	mg/kg	% of SOM	mg/kg	% of SOM
Wet-frigid (n=24)	203±18	0.80	754±39	3.00	1180±31	4.65
Wet-mesic (n=24)	206±16	1.00	1091±34	5.10	1106±49	5.39
Dry-thermic (n=12)	281±31	1.84	1225±56	8.00	1687±67	11.03
Dry-frigid (n=24)	309±36	1.54	1222±50	6.16	1436±58	7.18

Table 14. Labile Carbon Fractions in different types of *Mollisols*

5.2.3. Microbial biomass carbon in Mollisols from different climatic zones

Microbial biomass C (MBC) was significantly higher under drier (281 and 309 mg kg^{-1} soil in Kherson and Astana, respectively) comparing with wetter (203 and 206 mg kg^{-1} soil in Kharkov and Uman, respectively) regions (Table 14). The effect of temperature on microbial biomass was not clearly observed in this study.

As shown by data of LFC, drier regions accumulated a greater amount of organic substrate that favoured accumulation of microbial biomass due to its availability as energy source. However, because microbial activity and survival are in direct physiological dependence on available water [87], the moisture deficiency during the summer period retarded microbial activity and caused death of moisture–sensitive microorganisms. After remoistening of the soils, the inhibition of microbial activity by dry conditions was reactivated and the available necromass was rapidly metabolized by soil microorganisms leading to the higher accumulation of MBC in drier regions.

Changes in the relative contribution of bacteria and fungi to soil respiration occur as soil dries [66] Kharkov and Uman soils normally undergo less severe fluctuations in water potential that Kherson and Astana. [89, 90] have shown that bacterial activity is largely restricted to water films in soil in contrast to fungi activity. Hyphae extension occurs at much lower potentials allowing fungi to bridge air-filled pores and actively explore for nutrients [91].

5.2.4. Relationship between soil organic matter and clay content

The inert carbon is strongly correlated with clay content, while most changes in both carbon and nitrogen occur in the readily decomposable fraction [92]. Firstly, clay minerals can absorb large organic molecules directly, reducing their availability to decomposition. Secondly, organic material may be located in pores too small for microorganisms to enter [93-97).

In this study, clay content was highest in Kharkov and Astana regions (43.1% and 48.8%, respectively) versus Uman and Kherson (39.4% and 29.7%, respectively) (Table 15). It is reasonable to conclude that higher clay content and plant biomass production in Kharkov maintained higher SOM. [92] reported that inert carbon was strongly correlated with clay content, while most changes in both carbon and nitrogen occur in the readily decomposable fraction. [98] determined that "light fraction" (LF) of fine silt and coarse clay was more humified and more aromatic than other LF, concluding that LF represents a continuum of undecomposed to highly humified materials. Sites with higher silt fraction (2-0.2μm) that are Kharkov and Uman (37.5% and 37.7%, respectively) might form organo-mineral complexes with large molecules of LF, where those mineral-associated LF probably were not retrieved from these soils during the separation procedure, whereas, Kherson and Astana contained less silt fraction (26.9% and 26.6%, respectively) that could entrap LF, resulting in higher LFOM in these soils.

Regions	Siol organic carbon	Sand	Silt	Clay
	g kg^{-1} soil	200-20μm	20-2μm	<2μm
Kharkov (n=24)	25.4	20.4	37.5	42.1
Uman (n=18)	20.5	22.9	37.7	39.4
Kherson (n=12)	15.3	43.4	26.9	29.7
Astana (n=24)	20.0	25.6	30.6	43.8

Table 15. Granulometric composition of studied soil from different climatic regions

However, although Astana soil showed the highest content of clay the SOM in this soil was less than in Kharkov. This is explained by the lack of water in dry-frigid Astana that produces less plant biomass, and inhibits mineralization processes contributing to the accumulation of labile OM, which explains higher PMC content in this soil. Also, organic compounds adsorbed to surfaces of clay particles become exposed to microbial attack after disruption of aggregates due to severe dry-wet conditions on soil in Astana [34]. The lowest clay content (29.7%) and lack of water in Kherson can explain the lowest SOM content in this soil, since the dry-thermic conditions don't contribute to high plant biomass.

5.2.5. General discussion of effect of moisture and temperature on SOM in Mollisols

Total SOM among the four regions was distributed as follows: dry-thermic < dry-frigid ≤ wet-mesic< wet-frigid. While the labile OM distributed oppositely, as follows: dry-thermic ≥ dry-frigid > wet-mesic> wet-frigid. In Figure 13 the proportions of labile and stable carbon in the studied regions is presented. The highest amount of stable C and the least amount of labile C was found in wet-frigid (Kharkov) region, while the least amount of stable and the greatest amount of labile C was found in dry-thermic (Kherson) region. Because wet-frigid (Kharkov) region maintained the highest amount of total SOC and the least amount of easily mineralizable organic matter (PMC), the suggestion is: in wet-frigid region transformation of organic substrates into more stable humified forms of OM has taken place more actively

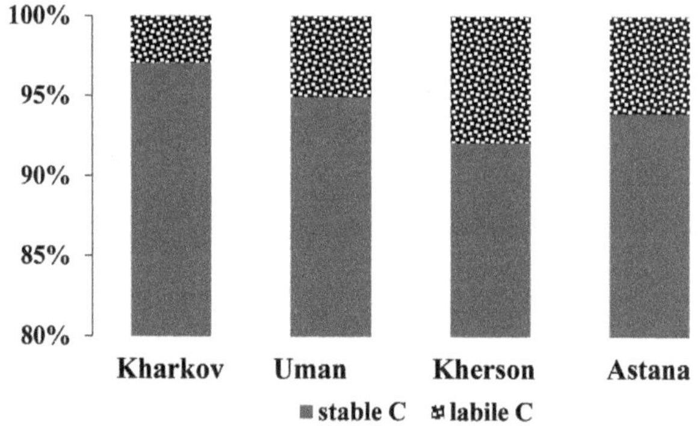

Figure 13. Distribution of labile and stable C in different *Mollisols* from four studied sites

Higher precipitation contributed stronger to the amount of total SOM, while plant biomass production under drier conditions in a lesser degree was subjected to decomposition due to moisture deficiency, thus contributing to the amount of labile SOM. The amount of C reflecting the labile fraction increases with increasing temperature, while the amount of recalcitrant C more controlled by low temperature.

Readily decomposable substrates were also found to originate partially from non-living SOM [62, 63]. This source of non-biomass substrate may become available by aggregate disruption, litter defragmentation and substrate desorption, and redistribution of water, oxygen, substrate and microorganisms resulting from drying and rewetting of soil [62, 63, 65, 99, 100]. Soil drying and rewetting promotes the turnover of carbon derived from added plant material [62, 63]. Drier condition causes more disruptions of entrapped or stabilized organic matter when the soil is rewetted. Also, higher respiration in the soils exposed to wet-dry cycles may have been due to utilization of organic substrate that were gradually built up due to limited microbial activity when the soil was air-dried [66].

6. Carbon to nitrogen ratio (C to N) of plant residues

Quite precise integral index of quality of organic matter, from which the intensity of its decomposition depends, appears the ratio of carbon to nitrogen (C: N). Plant remainders with the wide ratio C to N do not ensure the sufficiency of nitrogen for the metabolism of microorganisms at their high activity. When the rapidly metabolized substrata (carbohydrates) are depleted, the limitation of nourishment is changed from nitrogen to carbon.

"Critical" C to N ratio, which characterizes accessibility to the microorganisms of the nutrients contained in the remainders, and their influence on the soil fertility, varies from 15 to

30 depending on the pool of mineral nitrogen in the soil, the quality of organic matter, duration of their of the decomposition. With an increase in the value indicated the processes of decomposition slow down and the immobilization of nitrogen occurs. With the smaller C:N values the intensive mineralization of plant litter occurs as a result of the invigoration of the activities of the microflora. The studies on the *Mollisols* showed that among the studied crops the maximum speed of decomposition was for the postharvest remainders of alfalfa and pea, which have the narrowest C to N ratio, 19,4 and 26,8 respectively. The percentage of the decomposition of pea in the first two months was 47,2, in seven months - 52,8, in a year - 82,4. The lowest intensity of decomposition process was recorded under the residues of winter wheat. The changes of the ratio in the crop residues occurred in essence due to the content of nitrogen, which increased more rapidly than carbon. In more easily decomposing plant biomass the proteins of microbial synthesis are formed more rapidly thus contributing to the development of humification processes.

During the decomposition process, 55-70% of the carbon in the residue is released to the atmosphere as CO_2, 5-15% of the C is incorporated into the soil microbial biomass, and the remaining 15-40% of C is partially stabilized as new soil humus [101, 102]. Crop residues contain lignin, which is resistant to decomposition and becomes a substrate for soil humus formation. The more lignin in a residue type the slower the decomposition and the larger its contribution to the soil humus formation. Older, mature residues will tend to have more lignin content than young, non-mature residue from the same crop. The simple sugars, amino acids, polysaccharides, proteins, and lipids decompose first in the decomposition process.

Nitrogen (N) is necessary for the decomposition process. The less N the residue contains relative to the C, the slower the decomposition rate. If the residue N content is low, or the C:N ratio high, the decomposition process will require the input of N from either available soil inorganic N or fertilizer. Generally, residues with N concentrations less than 1.5% or C:N ratios greater than 30 will require N from sources outside the residue itself; it will immobilize soil N [107]. Residues with greater N concentration or lower C:N ratios, as is frequently the case with legume residues or non-mature residues, tend to decompose at a more rapid rate and will release or mineralize N. The actual decomposition rate will depend on N content and chemical composition of the residue and the environmental conditions such as soil moisture and temperature. Crop residues such as corn and wheat residue have large C:N ratios and soil N will be immobilized during the decomposition process. However, mineralization of N will generally start to occur after 50-60% of the residue has been decomposed [108] when enough C has been volatilized as CO_2 and N immobilized such that the remaining residue C:N ratio is below 30. Nitrogen immobilization reduces N availability to the growing crop and N mineralization increases N availability. The addition of a large quantity of oxidizable carbon from litter with the content of N less than 1,5% creates microbiological demand on N, thus immobilizing N of plant litter and inorganic N of soil. Effect of substrate composition on decomposition is described by C to N ration, based on the fact that N is the most limiting factor for decomposition. With ongoing decomposition the C to N ratio is narrowing thus the carbon energy supply is decreasing.

7. Conclusions

The studies confirm that the transformations of SOM are generally concentrated within labile pool. The process of mineralization of organic matter in soil is controlled predominantly, by the climatic factors (moisture and temperature) and plant litter quality (mainly by content of N), then by land management.

In the 21st century the mineral fertilizers became determinant for obtaining contented yield of agricultural crops. However, the application of only mineral fertilizer might lead to accelerated mineralization of the more stable soil organic matter. Instead, application of manure in combination with mineral fertilizer [38, 56-58, 103] might contribute in both obtaining the content crop yield and sustaining the soil fertility.

Yearly input of plant residue in less fallowed rotation system built up more labile OM, especially readily decomposable C, whereas frequent fallowing depletes SOM via accelerated mineralization. Therefore, in the semi-arid regions the inclusion of fallow in wheat monoculture every 6 years is the most appropriate farming system in terms of sustainability in both grain production and soil fertility.

Higher precipitation produced higher plant biomass contributing to the amount of SOM with further decomposition upon temperatures and soil texture. While less plant biomass production in drier regions in a lesser degree was subjected to decomposition due to moisture deficiency, thus contributing to the amount of labile SOM. Because wet-frigid region maintained the highest amount of total organic carbon and the least amount of easily mineralizable organic matter (PMC), transformation of organic substrates into more stable humified forms of organic matter might have taken place more actively in this region.

Understanding the processes of SOM changes under the impact of land management and different moisture and temperature regimes would greatly contribute to most ecological problem of C sequestration supplying with valuable information about land use management.

Author details

Elmira Saljnikov[1], Dragan Cakmak[1] and Saule Rahimgalieva[2]

*Address all correspondence to: elmirasal@mail.ru

1 Institute of Soil Science, Belgrade, Serbia

2 West-Kazakhstan Agrarian-Technical University named after Zhangir-Khan, Uralsk, Kazakhstan

References

[1] Sollins, P., Homann, P. and Caldwell, B.A. 1996. Stabilization and destabilization of soil organic matter – mechanisms and controls. Geoderma 74: 65-105

[2] Six J, Bossuyt H, De Gryze S, Denef K. 2004. A history of research on the link between (micro) aggregates, soil biota, and soil organic matter dynamics. Soil Till Res 79:7–31

[3] Baldock, J.A., Nelson, P.N, 2000. Soil organic matter. In Sumner, M.E., (Ed.), Handbook of Soil Science. CRC Press, Boca Raton, pp. B25-B84

[4] Post, W.M., Emmanuel, W.R., Zinke, P.J. and Stangenberger, A.G., 1982. Soil carbon pools and world life zones. Nature 298: 156-159.

[5] Saljnikov E., Cakmak, D., Kostic L., and Maksimovic, S. 2009. Labile fractions of soil organic carbon in Mollisols from different climatic regions. Agrochimica. Vol. LIII, 6

[6] Campbell, C.A. 1978. Soil organic carbon, nitrogen, and fertility. In: Schnitzer M., Khan, S.U., (Eds.) Soil organic matter. Elsevier, New York, pp 173-271

[7] Beiderbeck, V.O., Campbell, C.A. and Zentner, R.P., 1984. Effect of crop rotation and fertilization on some biological properties of a loam in southwestern Saskatchewan. Can. J. Soil Sci. 64: 355-367.

[8] Saljnikov-Karbozova, E., Funakawa, S., Akhmetov, K., and Kosaki, T. 2004. Soil organic matter status of Mollisols soil in North Kazakhstan: effects of summer fallow. Soil Biol.Biochem. 36: 1373-1381

[9] Campbell, C.A. and Souster, W., 1982. Loss of organic matter and potentially mineralizable nitrogen from Saskatchewan soils due to cropping. Can. J. Soil Sci. 62: 651-656.

[10] Rasmussen, P.E., Allmaras, R.R., Rohde, R.R. andRoager, N.C., 1980. Crop residue influences on soil carbon and nitrogen in a wheat-fallow system. Soil Sci. Soc. Am. J. 44: 596-600

[11] Saljnikov E., Hospodarenko H., Funakawa S and Kosaki T. 2005. Effect of fertilization and manure application on nitrogen mineralization potentials in Ukraine. Zemljiste i biljka, vol.54, 3: 221-230.

[12] Couteaux, M.M., Bottner, P. And Berg, B. 1995. Litter decomposition, climate and litter quality. Tree 10: 63-66.

[13] Zou, X.M.; H.H. Ruan, Y. Fu, X.D. Yang and L.Q. Sha.Estimating soil labile organic carbon and potential turnover rates using a sequential fumigation–incubation procedure. Soil Biol. Biochem.37, 10, 2005, pp. 1923-1928

[14] Christensen, B.T. 1986. Barley straw decomposition under field conditions: Effect of placement and initial nitrogen content on weight loss and nitrogen dynamics. Soil Biol. Biochem. 18: 523-529

[15] Sparling, G.P., 1992. Ratio of microbial biomass to soil organic carbon as a sensitive indicator of changes in soil organic matter. Aust. J. Soil Res. 30: 195-207

[16] Franzluebbers, K., Weaver, R.W., Juo, A.S.R. and Franzluebbers, A.J., 1994. Carbon and nitrogen mineralization from cowpea plant parts decomposing in moist and in repeatedly dried and rewetted soil. Soil Biol. Biochem. 26: 1379-1387.

[17] Gregorich, E. G. and Janzen, H.H. 1996. Storage of soil carbon in the light fraction and macroorganic matter. P. 167-190. In M.R. Carter and B.A. Stewart (ed.) Structure and organic matter storage in agricultural soils. CRC press, Boca Raton, FL

[18] Keeney, D.R. 1982. Nitrogen—availability indices. In A.L. Page et al., Eds. Methods of Soil Analysis. Part 2, 2nd ed. Chemical and Microbiological Properties, Agronomy 9. SSSA and ASA, Madison, WI, pp. 711–733

[19] Christensen, B.T., 1996. Carbon in primary and secondary organo-mineral complexes. Advances in Soil Science 24, 97-165.

[20] Elliott, E.T., Paustain, K., Frey, S.D., 1996a. Modeling the measurable or measuring the modelable: A hierarchical approach to isolating meaningful soil organic matter. In D.S.

[21] Elliot, L.F., Lynch, J.M., and Papendick, R.I. 1996b. The microbial component of soil quality. In G. Stotsky and J.M.Bollag, Eds. Soil Biochemistry, Vol. 9. Marcel Dekker, New York, NY, pp. 1–20.

[22] Breland, T.A. and Eltun, R., 1999. Soil microbial biomass and mineralization of carbon and nitrogen in ecological, integrated and conventional forage and arable cropping systems. Biol. Fert. Soils, 30: 193-201.

[23] Nelson, D.W. and Sommers, L.E. 1996. Total carbon, organic carbon, and organic matter. In Page, A.L., Miller, R.H., Keeney, D.R. (eds.) Methods of Soil Analysis, Part 3, SSSA, Madison, WI.

[24] SPSS Inc., SYSTAT 1998. version 8.0, Statistics, Chicago, IL.

[25] SPSS Inc., 2007. Sigma Plot version 6.0, Programming guide, Chicago, IL.

[26] Jalil, A., Campbell, C.A., Schoenau, J., Henry, J.L., Jame, Y.W., and Lafond, G.P. 1996. Assessment of two chemical extraction methods as indices of available nitrogen. Soil Sci. Soc. Am. J. 60: 1954–1960.

[27] Picone, L.I., Cabrera, M.L., and Franzluebbers, A.J. 2002. A rapid method to estimate potentially mineralizable nitrogen in soil. Soil Sci. Soc. Am. J. 66: 1843–1847.

[28] Jenkinson, D.S. 1988. Soil organic matter and its dynamics.. In A. Wild (ed.) Soil Condition and Plant Growth. Longman, New York. pp. 564-607.

[29] Smith, J.L. and Paul, E.A. 1990. The significance of soil microbial biomass estimations. In J.-M. Bollag and G. Stotzky, Eds. Soil Biochemistry, Vol. 6. Marcel Dekker, New York, NY, pp. 357–396

[30] Brookes, P.C., Ocio, J.A., and Wu, J. 1990. The soil microbial biomass: its measurement, properties and role in soil nitrogen and carbon dynamics following substrate incorporation. Soil Microorganisms. 35: 39–51

[31] Powlson, D.S., Brookes, P.C., and Christensen, B.T. 1987. Measurement of soil microbial biomass provides an early indication of changes in total soil organic matter due to straw incorporation. Soil Biol. Biochem. 19: 159–164

[32] Spedding T.A., Hamel C., Mehuys G.R., Madramootoo C.A. (2004): Soil microbial dynamics in maize-growing soil under different tillage and residue management systems. Soil Biology and Biochemistry, 36: 499–512.

[33] Herna´ndez-Herna´ndez, R.M.andLo´pez-Herna´ndez,D. 2002. Microbial biomass, mineral nitrogen and carbon content in savanna soil aggregates under conventional and no-tillage. Soil Biol. Biochem. 34: 1563–1570.

[34] Jenkinson D.S. and Powlson D.S. 1976. The effects of biocidal treatment on metabolism in soil: V. A method for measuring soil biomass. Soil Biol. Biochem. 8: 209-213

[35] Sparling, G.P and West, A. W. 1998. A direct extraction method to estimate soil microbial biomass C: calibration in situ using microbial respiration and ^{14}C labeled cells. Soil Biol. Biochem. 20: 337-343

[36] Janzen, H.H., Campbell, C.A, Brandt, S.A., LaFond, G.P. and L.Townley-Smith. 1992. Light-Fraction Organic Matter in Soils from Long-Term Crop Rotations. Soil Sci.Soc.Am.J. 56: 1799-1806.

[37] Nosko, B.S., 1987. Changes of humus status of Typical Chernozem upon fertilization. Pochvovedenie, 5: 26-31 (in Russian)

[38] Chesnyak, G. Ya., 1981. Changes of humus quantity and quality and the ways of providing of positive humus balance in Typical Chernozems of Ukraine under intensive agricultural use. In publications of VI Meeting of Soil Science Society of USSR, V.2, pp.42-43. (in Russian).

[39] Aleksandrova, L.N., 1980. Soil organic matter and its transformation. Nauka, Moscow, p. 288. (in Russian).

[40] Kononova, M.M., 1956. Humus of the main soils of SSSR, its nature and ways of forming. Pochvovedenie, 3: 18-30. (in Russian).

[41] Mamontov, V.T., 1971. Effect of agricultural use on the agronomical properties of thick Chernozem in west forest-steppes of Ukraine. In doctoral thesis of the author, p. 234. (in Russian).

[42] Ilyaletdinov, A., 1988. Microbiological conversion of nitrogen compounds in the soil. Nauka, Moscow, 119-154pp.

[43] Ferguson, W.S., Gorby, B.J., 1971. Effect of various periods of seed-down to alfalfa and bromegrass on soil nitrogen. Canadian Journal of Soil Science 51. 65-73.

[44] Clarke, A.L., Russell, J.S., 1977. Crop sequential practices. In: J.S. Russell, E.L. Greacen, (Eds), Soil Factors in Crop Production in a Semi-arid Environment. University of Queensland Press, St. Lucia, pp. 279-300.

[45] Dormaar, J.F., 1983. Chemical properties of soil and water-stable aggregates after sixty-seven years of cropping to spring wheat. Plant and Soil 75, 51-61.

[46] Bird, M.I., Chivas, A.R., Head J., 1996. A latitudinal gradient in carbon turnover times in forest soils. Nature 381, 143-146.

[47] Amelung W., Flach, K.W., Zech W., 1998. Climatic effects on soil organic matter composition in the Great Plains. Soil Science Society of America Journal 61, 115-123.

[48] Christensen, B.T., 1992. Physical fractionation of soil and organic matter in primary particles and density separates. Advances in Agriculture 20, 2-90.

[49] Trumbore, S.E., Chadwick, O.A., Amundson R., 1996. Rapid exchange between soil carbon and atmospheric carbon dioxide driven by temperature change. Science 272, 393-396.

[50] Collins, H.P., Rasmussen, P.E., Douglas, C.L., 1992. Crop rotation and residue management effects on soil carbon and microbial dynamics. Soil Science Society of America Journal 56, 783-788.

[51] Campbell, C.A., Zentner, R.P., 1993. Soil organic matter as influenced by crop rotations and fertilization. Soil Science Society of America Journal 57, 1034-1040.

[52] Biederbeck, V.O., Janzen, H.H., Campbell, C.A., Zentner, R.P., 1994. Labile soil organic matter as influence by cropping practices in an arid environment. Soil Biology & Biochemistry 12, 1647-1656.

[53] Janzen, H.H., 1987. Soil organic matter characteristics after long-term cropping to various spring wheat rotations. Canadian Journal of Soil Science 67, 845-856.

[54] Cheng, HH, & Kimble, JM. 2001. Characterization of soil organic carbon pools. In R. Lal, JM Kimble, RF Follett, & BA Stewart (Eds.), Assessment methods for Soil Carbon. CRC/Lewis Press, Boca Raton, FL, 676 pp.

[55] Singh J.S. and Gupta S.R. 1977. Plant decomposition and soil respiration in terrestrial ecosystems. Bot. Rev. 43:449-528.

[56] Chesnyak, G. Ya., 1986. Modification to the determination of coefficient of humification of plant residues in Typical Chernozems of forest-steppe Ukraine in grain-beet crop rotation. Agrochemistry and soil science, Kiev, 49: 77-92. (in Russian).

[57] Voroney, R.P., 1988. Loss of organic matter in Ontario soils. Highlights. V.11, No.3, pp. 25-29.

[58] Pare, T., Dinel, H., Moulin, A.P. and Townley-Smith, L., 1999. Organic matter quality and structural stability of a Black Chernozemic soil under different manure and tillage practices. Geoderma 91: 311-326

[59] Kharin, S.V., 1993. Humification and regulation of humus status of different cropping systems in Typical Chernozems of west forest-steppe of Ukraine. PhD thesis, Institute of Soil Science and Agrochemistry after Sokolovski, Kharkov, Ukraine (in Russian).

[60] Lundquist, E.J., Jackson, L.E. and Scow, K.M. 1999. Wet-dry cycles affect dissolved organic carbon in two California agricultural soils. Soil Biol. Biochem. 31: 1031-1038.

[61] Bottner, P., 1985. Response of microbial biomass to alternate moist and dry conditions in a soil incubated with 14C- and 15N-labelled plant material. Soil Biol. Biochem. 17: 329-337

[62] Van Gestel, M., Merckx, R. and Vlassak, K., 1993b. Microbial biomass responses to soil drying and rewetting: the fate of fast- and slow-growing microorganisms in soils from different climates. Soil Biol. Biochem. 25: 125-134.

[63] Van Gestel, M., Merckx, R. and Vlassak, K., 1993a. Microbial biomass responses to soil drying and rewetting: the fate of fast- and slow-growing microorganisms in soils from different climates. Soil Biol. Biochem. 25: 109-123.

[64] Pulleman, M. and Tietama, A., 1999. Microbial C and N transformations during drying and rewetting of coniferous forest floor material. Soil Biol. Biochem. 31: 275-285.

[65] Lund, V. and Goksoyr, J., 1980. Effects of water fluctuations on microbial mass and activity in soil. Microbial Ecology 6: 115-123

[66] Orchard, V.A. and Cook, F.J. 1983. Relationship between soil respiration and soil moisture. Soil Biol. Biochem. 15: 447-453.

[67] Kononova, M.M., Pankova, N.A. and Belchikova N.P., 1949. Changes in quality and quantity of soil organic matter under cultivation. Pochvovedenie, 1: 28-37. (in Russian).

[68] Beauchamp, E.G., 1980. Nitrogen from liquid dairy cattle manure for corn. Highlights Agr. Res. In Ontario, V.3, No.4, pp.10-12.

[69] Anderson, D.W., Joug, E de, Verity, G.E. and Gregorich, E.G., 1986. The effect of cultivation on the organic matter of soils of the Canada prairies. Transact. 13 Cong. Int. Soc.Soil Sci., Hamburg, 13-20 Aug., V. 4, SI, S.9, pp.1344-1345.

[70] Van Gestel, M., Ladd, J.N., and Amato, M., 1991. Carbon and nitrogen mineralization from two soils of contrasting texture and microaggregate stability: influence of sequential fumigation, drying and storage. Soil Biol. Biochem. Vol.23, No.4, 313-322

[71] Rubinstein, M.I., 1959. Decomposition rate of organic matter of virgin Chernozem in Northern Kazakhstan during their cultivation. Soviet Soil Science 11: 1332-1335

[72] Dzhalankuzov, T D. and Redkov, V.V., 1993. Changes in morphological and agrochemical properties of calcareous Southern Chernozems of North Kazakhstan due to long-term cultivation. In: Proceedings of Academy of Sciences of Republic of Kazakhstan, Biology Series 1: 53-58 (in Russian)

[73] Campbell, C.A., Biederbeck, V.O., McConkey, B.G., Curtin, D., Zentner, R.P., 1999. Soil quality-effect of tillage and fallow frequency. Soil organic matter quality as influenced by tillage and fallow frequency in a silt loam in southwestern Saskatchewan. Soil Biology & Biochemistry 31, 1-7.

[74] Campbell, C.A., Moulin, A.P., Bowren, K.E., Janzen, H.H., Townly-Smith, L., Biederbeck, V.O., 1992. Effect of crop rotation on microbial biomass, specific respiratory activity and mineralizable nitrogen in a Black Chernozemic soil. Canadian Journal of Soil Science 72, 417-427.

[75] Haynes, R.J., 2000. Labile organic matter as an indicator of organic matter quality in arable and pastoral soils in New Zealand. Soil Biology & Biochemistry 32, 211-219

[76] Gregorich, E.G., Carter, M.R., Angers, D.A., Monreal, C.M., Ellert, B.H., 1994. Towards a minimum data set to access soil organic matter quality in agricultural soils. Canadian Journal of Soil Science 74, 367-385

[77] Shields, J.A. and Paul, E.A. 1973. Decomposition of ^{14}C-laballed plant material under field conditions. Can. J. Soil Sci. 53: 297-306, 1973.

[78] Douglas, C.L. Jr., Rickman, R.W., Klepper, B.L., Zuzel, J.F. and Wysocki, D.J. 1992. Agroclimatic zones for dryland winter wheat producing areas of Idaho, Washington, and Oregon. Northwest Science 66: 26-34

[79] Raich, J. W. and C. S. Potter. 1995. Global Patterns of Carbon Dioxide Emissions from Soils. Global Biogeochemical Cycles 9(1)23-36. 10.3334/CDIAC/lue.db1015

[80] Fierer N, B Colman, JP Schimel, RB Jackson. 2006. Predicting the temperature dependence of microbial respiration in soil: A continental-scale analysis Global Biogeochemical Cycles 20, GB3026c

[81] Davidson, E.A., S.E. Trumbore, and R. Amundson. 2000. Biogeochemistry: Soil warming and organic carbon content. Nature. 408:789-790.

[82] Janzen, H.H., 2004. Carbon cycling in earth systems-a soil science perspective. Agriculture, Ecosystems and Environment, 104, 399-417

[83] Soil Taxonomy 2nd ed., 1999. USDA, Washington, DC. pp.555-655.

[84] Franzluebbers, A.J., Haney, R.L., Honeycutt, C.W., Arshad, M.A., Schomberg, H.H. and Hons, F.M. 2001. Climatic influences on active fractions of soil organic matter. Soil Biol.Biochem. 33: 1103-1111

[85] Aleksandrova, L.N. 1972. Study of the humification of plant residues and of the nature of newly formed humic acids. Pochvovedenie 7: 37-45.

[86] Marumoto, T., Kai, H., Yoshida, T. and Harada T. 1977. Drying effect on mineralization of microbial cell walls as a source of decomposable soil organic matter due to drying. Soil Sci. Plant Nutr. 23: 9-19

[87] Mikha, M.M., Rice, C.W. and Milliken, G.A. 2004. Carbon and nitrogen mineralization as affected by drying and wetting cycles. Soil Biol. Biochem. 37: 339-347

[88] Wu, J. and Brookes, P.C. 2005. The proportional mineralization of microbial biomass and organic matter caused by air-drying and rewetting of a grassland soil. Soil Biol. Biochem. 37: 507-515.

[89] Wong, P.T.W. and Griffin, D.M., 1976a. Bacterial movement at high matric potentials-I. In artificial and natural soils. Soil Biol. Biochem. 8: 215-218.

[90] Wong, P.T.W. and Griffin, D.M., 1976b. Bacterial movement at high matric potentials-II. In fungal colonies. Soil Biol. Biochem. 8: 219-223.

[91] Griffin, D.M., 1969. Soil water in the ecology of fungi. Annual Review of Phytopathology, 7: 289-310

[92] Körschens, M., Weigel, A. and Schulz, E., 1998. Turnover of soil organic matter (SOM) and long-term balances-tools for evaluating sustainable productivity of soils. Z. Pflanzenernähr. Bodenk., 161: 409-424

[93] Juma, N.G., 1993. Interrelationship between soil structure /texture, soil biota/soil organic matter and crop production. Geoderma 57: 3-30

[94] Elliott, E.T., 1986. Aggregate structure and carbon, nitrogen and phosphorous in native and cultivated soils. Soil Sci.Soc. Am. J., 50: 627-633

[95] Gupta, V.V.S.R. and Germida, J.J., 1988. Distribution of microbial biomass and its activity in different soil aggregate size classes as affected by cultivation. Soil Biol. Biochem. 20: 777-786

[96] Amelung, W. and Zech, W., 1996. Organic species in ped surface and core fractions along a climosequence in the prairie, North America. Geoderma 74: 193-206

[97] Gregorich, E.G., Kachanoski, R.G. and Voroney, R.P., 1989. Carbon mineralization in soil size fractions after various amounts of aggregate disruption. Journal of Soil Sci. 40: 649-659

[98] Turchenek, L.W. and Oades, J.M., 1979. Fractionation of organo-mineral complexes by sedimentation and density techniques. Geoderma 21: 311-343

[99] Sommers, L.E., Gilmour, C.M., Wildung, R.E. and Beck, S.M., 1981. The effect of water potential on decomposition processes in soil. In: Parr, J.F., Gardner, W.R., Elliot, L.F. (Eds.), Water Potential Relations in Soil Microbiology. Soil Sci. Soc. Am., Madison, pp.97-117.

[100] Kieft, L.T., Soroker, E. and Firestone, M.K., 1987. Microbial biomass response to a rapid increase in water potential when dry soil is wetted. Soil Biol. Biochem. 19: 119-126.

[101] Jenkinson, D.S. 1971. The Accumulation of Organic Matter in Soil Left Uncultivated. Rothamsted Experimental Station Report for 1970, part 2: 113-137

[102] Stott, D.E., and J.P. Marten, 1989. Organic matter decomposition and retention in arid soils. Arid Soil Res. Rehab. 3:115

[103] Kononova, M.M., 1951. Problems of soil humus and contemporary methods of their study. Moscow, pp.390. (in Russian).

[104] Jenkinson, D.S., Adams, D.E. and Wild A. 1991. Model estimation of CO2 emissions from soil in response to global warming. Nature 351: 304-306

[105] Zak, J.C., M.R. Willig, D.L. Moorhead and H.G. Wildman, 1994.Functional diversity of microbial communities: a quantitative approach, Soil Biol. Biochem. 26, 1101–1108.

[106] Gregorich, E.G. and Janzen, H.H., 2000. Decomposition. In Handbook of Soil Science. Sumner M. E. (Ed.), pp. C107-C120

[107] Schomberg, H.H., Steiner J.L., Unger, P.W. 1994. Decomposition and Nitrogen Dynamics of Crop residues: Residue Quality and Water effects. Soil Sci.Soc.Am.J. 58:372-381.

[108] Douglas CL, Allmaras RR, Rassmussen PE, Ramig RE, Roager NC. 1980. Wheat straw decomposition and placement effects on decomposition in dryland agriculture of the Pacific Northwest. Soil Sci.Soc.Am.J. 44:833-837.

[109] Funakawa S., Yanai J., Takata Yu., Karbozova-Saljnikov E., and Kosaki T. 2007. Dynamics of water and soil organic matter under grain farming in Northern Kazakhstan – Toward sustainable land use both from the agronomic and environmental viewpoints. In Lal R., et al., (ed.) Climate Change and Terrestrial Carbon Sequestration in Central Asia. The Netherlands. Taylor&Francis, pp. 279-331.

[110] Six, J. and Jastrow J.D. 2002. Organic Matter Turnover. In Lal R. (ed.) Encyclopedia of Soil Science. Dekker, NY, pp.936-942.

Chapter 10

Effect of Crop Rotation and Nitrogen Fertilization on the Quality and Quantity of Soil Organic Matter

Agnieszka Rutkowska and Dorota Pikuła

Additional information is available at the end of the chapter

1. Introduction

Soil organic matter (SOM) is one of the most important features of the soil. Its characteristic depends on a variety of biotic and abiotic variables of the ecosystem, such as climate, soil texture, mineral composition, quantity of organic residues and other factors. Currently, in an era of rapidly changing civilization, leading to changes in climate and soil conditions, SOM content becomes increasingly important, not only for the proper functioning of ecosystems, but also for socio-economic development of many regions of the world [1, 2]. In the first half of the past century, there were hardly contradictions between the agricultural cultivation and the environment. The substance circulations were closed, animal production comparatively small and mainly regularly allocated. The mineral fertilization was only used to a slight extent. A fundamental change has taken place during the last decades. With the increasing use of mineral fertilizers, yields have increased by more than 100% thus the quantity of roots and harvest residues on the field has increased strongly as a source of organic matter. However, in agricultural practice they are commonly removed from the field after harvest resulting in SOMdecrease.

During the last ten years in EU countries, the progressive degradation of SOM is observed. Thus, this issue was reflected in the EU soil strategy (COM (2002) 179), on which the actual reduction of soil organic matter content was listed as one of the most important problems. In Poland, the reduction of SOM content in soils became a problem particularly significant.

Nearly the whole territory of Poland (99,7%), covering about 313 thousands square kilometres is situated in the Baltic Sea basin. This territory is drainaged by two big rivers Vistula and Odra and seven small rivers flowing directly to the sea. Natural farming condition in Poland are poor, due to prevalence of light, sand-derived soils (60% very light and light soils) and unfavourable climate. Due to soil texture and acid or very acid reaction more than

60% of the soils in Poland can be classified as soils witha relatively low content of organic matter. According to the newest survey 7,6% of arable soils show low content of SOM (below 1%), 47,1% of soils presentmedium content (1,1-2,0%), 29,3%, high content (2,1-3%) and only 3% of soils show very high content of organic matter (above 3%) [3]. The percentage of acid and very acid soils is very high and exceeds 50%, and soil acidity seems to be one of the most important factors leading to degradation of the quantity and quality of organic matter [4]. The summaryof the last four-years period of agrochemical soils monitoring program reveals that 20,2% of soils are vey acid ($pH_{KC}l$ below 4,5) and 29,4% acid (pH_{KCl} 4,5-5,5) [4].

Besides, the intensification of soil use combined with the simplified crop rotation and predominance of cereals together with an expansion of farming systems based on crop production with a reduced number of livestock or without animals intensifies the process of organic matter degradation [5- 7].

The turn over of organic substance are time dependent and become most apparent after decades. Therefore, long-term field experiments are a necessary tool for tracking changes in organic carbon content in the soil. In Europe, the most famous long-term experiences with testing of different mineral and natural fertilisers and/or cultivation of various plant species were held at Rothamsted (England), Halle and Bad Laustädt (Germany), Prague-Ruzyne (Czech Republic) and Skierniewice (Poland) [8].

The Institute of Soil Scienceand Plant Cultivation National Research Institute (IUNG-PIB) in Puławy is also involved in several kinds of long term field studies, including SOM content monitoring, since 1979. A special trait of these experiments is that crop rotations included plants enriching and exhausting soil from humus, have been apermanent factor for the last 33 years.

The objectives of presented paper was to access the impact of mineral nitrogen fertilization, manure application and crop rotation on the quantity and quantity soil organic matter in long term field experiment.

2. Composition of soil organic matter and methods of humus substances fractionation

Humic substances (HS) is the major organic constituents of soil. Humic substances are a mixture of particles that differ in their structure, mass, size, elemental composition and properties. The first record of humic acids extration by Achard dates back to 1786, and the first major study was done by Sprengel in 1820 [in 9]. Despite such a long research history on humic substances, there has been constant dispute over the structure of humus. There is an increasing interest in HS, because of their capacity for complexing metal ions, involvement in the organic geochemical cycle, affecting bioavailability and ecological effects of nutrient in water and stabilizing soil fertility.

The content and quality of humus are directly and indirectly determined by physical, chemical, biological and environmental properties. SOM quality is determined mainly by the percentage share of humic substances in soil. Humic substances can be subdivided into three major fractions: humic acids, fulvic acids and humins. Electron microscope observations reveal the humic acids of different soils to have a polymeric structure, appearing in form of rings, chains, and clusters. The size of their macromolecules ranges from 60 to 500 A and is mainly decided by the occurring humification process, which also exerts an influence on their spatial structure. Some of the main features of humic substances are shown in Figure 1 [10].

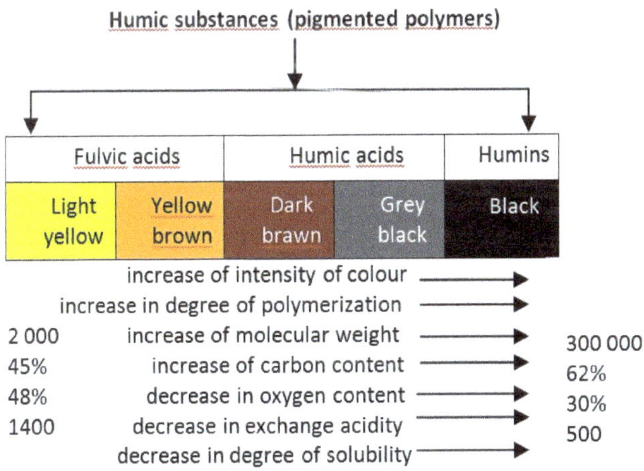

Figure 1. Chemical properties of humic substances.

Humic acids (HA) comprise a mixture of weak aliphatic and aromatic organic acids, containing carboxyl and phenolate groups. HA are not soluble in water under acidic conditions but soluble in water under alkaline conditions and precipitated from aqueous solutions when the pH decreases below 2. HA tend to be more aromatic and more prone to precipitation under the acid conditions common in soils, making them less mobile.

Fulvic acids (FA) are a humic substances soluble in water at any pH. They areless aromatic than humic acids. This part of humus remains in solution after removal of humic acids by acidification. FA have an oxygen content twice that of humic acids thus revealing a more acidic character than that of HA. The exchange capacity of FA is more than double that of HA.

Humins are the fraction of humic substances, not extracted from soil with either a strong base or a strong acid. Humins present within the soil are the most resistant to decomposition (slow to break down) of all the humic substance.This fraction plays a key role in soil fertility by improving structure and soil water capacity or providing a reservoir for the plant nutrients.

There are different ways to separate humic substances into each particular fraction. Every technique has its own advantages and limitations. The reagents used for extraction of humic acids from the soil are listed in table 1.

The most popular method is to use NaOH to extract humic acids from the soil. It is evident that extraction from soil with NaOH solution leads to the recovery of approximately two-thirds of the soil organic matter [6]. The amount of organic matter extracted from soil with the caustic alkali increase with time of extraction. Humic acids extracted by alkali solutions are characterized by high purity what is what is necessary for further physicochemical analyzes (UV-VIS, NMR and IR) [11].

Another reagent used for extraction of soil humusis sodium pyrophosphate ($Na_4P_2O_7$). Its advantage is that the auto oxidation process that occurs during the extraction of humic material is less intensive than with NaOH. However, this reagent allows to isolate only humus compounds loosely linked with the mineral fraction of the soil.

Type of material	Extractant	Organic matter extracted
Humic substances	NaOH Mild extractants	to 80%
	$Na_4P_2O_7$ and other	to 30%
	Organic chelates:acetyloacetone, cupferron, hydroxyquinonline	to 30%
	Formic acid (HCOOH)	to 55%

Table 1. Reagents commonly used for extration of organic constituents from soil [6]

The most common methods of humus fractionation are Schnitzers method and he standard, prior to extraction from the soil is determined by total organic carbon, by one of the recommended methods.

Schnitzer and Turin methods based on NaOH reagent are suitable for soils without carbonate thus the carbonate determination in soil is needed before [12-13]. In both methods, diluted sodium hydroxide is used after decalcification in mineral acid solution, which burnts the conjunctions of humic acids with calcium. Afterwards, humic and fulvic acids pass to the solution during alkaline extraction. On the base of the difference between the quantity of humus fractions before and without decalcification, the calculation of humic and fulvic acids associated with calcium is possible.

The separated fractions of humic substances have been recognized as the main indicators of soil fertility. The quality of SOM can be evaluated by determining the ratio of humic acids to fulvic acids (HA:FA) or carbon of humic acids to carbon of fulvic acids (CHA:CFA). It is widely described that fertile soils are characterized by higher humus content and CHA:CFA ratio>1.On the agricultural lands, soil humus properties are mostly determined by post- harvest residue left after the harvest of crops [8, 14]. Legumes increase organic matter content in soil. This is a consequence of chemical composition of organic material. Cereal straw contains more lignin, and legumes one more cellulose and nitrogen. Thus, the mineralization process of legumes residues occurs faster and the C:N ratio is narrower as compared to other crops [15].

The percentage of humus which occurs in the various humic fractions varies considerably from one soil type to another (Table 2) [16].

Soil	Humic acid/ Fulvic acid ratio	Soil	Humic acid/ Fulvic acid ratio
Chernozemordinary	2.0 – 2.5	Gray forest	1.0
Chernozem deep	1.7	Sod podzolic	0.8
Chestnut dark	1.5 – 1.7	Tundra	0.3

Table 2. Humic acid/fulvic acid ratios of some surface soils

A further attempt is made to investigate the properties of humic acids focusing on their spectroscopic characterization by UV–VIS and fluorescence spectra. This method is an important tool for determining the differences in humic substances structure, maturity and condensation degree [16, 17]. Measuring of optical properties of humic acids (absorbance and absorbance ratio) of alkaline soil extract allows to determine the degree of humification process and humic substances quality. Humic substances from various types of soil differ in the absorbance ratio. The measurement of absorbance is made in the wavelength range 280-665. The absorbance at wavelength 280 nm indicates the high content of lignin, 465 nm - young humic substances components associated with the first phase of humification process. Absorbance at 665 nm is related to well humified components [18].

The intensity of absorbance at wave length 280, 465 and 665 nm is used to calculate the ratiosA 2/4 (the ratio of absorbance in the wave length 280 nm to 465 nm), A 2/6 (280 nm to 665 nm) and A4/6 (465 nm to 665 nm). The absorbance ratio allows to recognize the structure of humus components. Larger values of 465 nm to 665 nm (A4/A6 ratio) are associated with the presence of smaller size organic molecules or more aliphatic structures and usually with higher content of functional groups [19].

2. Materials and methods

The study was conducted on the basis of a three factorial long-term field experiment carried on since 1979 at the Experimental Station Grabów of the Institute of Soil Science and Plant Cultivation in Puławy, on typical soil in Poland, classified as light loamy and sand texture according to USDA soil classification. The experiment was conducted with two crop rotations (I factor): A – recognized assoil exhausting from humus (potatoes, winter wheat, spring barley and corn for silage) and B –considered to enrich soil with humus (potatoes, winter wheat and mustard* as aftercrop for ploughing, spring barley with undersown** and clover with grass mixture). The experiment was performed in the split – block layout in two cycles moved by one year (Table 3).

Rotation	Year	Crop rotation A		Crop rotation B	
		Cycle 1	Cycle 2	Cycle 1	Cycle 2
	1980	potatoes		potatoes	
	1981	winter wheat	potatoes	winter wheat*	potatoes
	1982	spring barley	winter wheat	spring barley**	winter wheat
I	1983	maize for silage	spring barley	clover-grasses mixture	spring barley
	1984		maize for silage		clover-grasses mixture
	1984	potatoes		potatoes	
	1985	winter wheat	potatoes	winter wheat*	potatoes
	1986	spring barley	winter wheat	spring barley**	winter wheat
II	1987	maize for silage	spring barley	clover-grasses mixture	spring barley
	1988		maize for silage		clover-grasses mixture
	1988	potatoes		potatoes	
	1989	winter wheat	potatoes	winter wheat*	potatoes
	1990	spring barley	winter wheat	spring barley**	winter wheat
III	1991	maize for silage	spring barley	clover-grasses mixture	spring barley
	1992		maize for silage		clover-grasses mixture
	1992	potatoes		potatoes	
	1993	winter wheat	potatoes	winter wheat*	potatoes
	1994	spring barley	winter wheat	spring barley**	winter wheat
IV	1995	maize for silage	spring barley	clover-grasses mixture	spring barley
	1996		maize for silage		clover-grasses mixture
	1996	potatoes		potatoes	
	1997	winter wheat	potatoes	winter wheat*	potatoes
	1998	spring barley	winter wheat	spring barley**	winter wheat
V	1999	maize for silage	spring barley	clover-grasses mixture	spring barley
	2000		maize for silage		clover-grasses mixture
	2000	potatoes		potatoes	
	2001	winter wheat	potatoes	winter wheat*	potatoes
VI	2002	spring barley	winter wheat	spring barley**	winter wheat
	2003	maize for silage	spring barley	clover-grasses mixture	spring barley

	2004		maize for silage		clover-grasses mixture
	2004	potatoes		potatoes	
	2005	winter wheat	potatoes	winter wheat*	potatoes
	2006	spring barley	winter wheat	spring barley**	winter wheat
VII	2007	maize for silage	spring barley	clover-grasses mixture	spring barley
	2008		maize for silage		clover-grasses mixture

Table 3. The scheme of the experiment; Soil sampling term for SOM quality evaluation

In each crop rotation, five rates of manure (II factor) were applied under potatoes: 0, 20, 40, 60 and 80 t• ha^{-1} every four years.

These factors were permanent in the history of the experiment.

Since 1984 (the third rotation) in the both crop rotations mineral fertilization has been set (III factor). Each crop was supplied with four rates of nitrogen fertilizers in accordance with table 4.

Crop rotation	**Plant**	**Mineral fertilization [kg ha^{-1}]**					
		N0	N1	N2	N3	P_2O_5	K_2O
A	Potatoes	0	45	90	135	54	160
	Winter wheat	0	40	80	120	54	100
	Spring barley	0	40	80	120	54	85
	Maize for silage	0	45	90	135	54	120
B	Potatoes	0	45	90	135	54	160
	Winter wheat	0	40	80	120	54	100
	Spring barley	0	30	60	90	54	85
	Clover with grass	0	40	80	120	54	115

Table 4. Mineral fertilization for crop rotation A and B

The paper presents the results of permanent experiment in which the quantity of soil organic matter was determined through 33 years. The quality of soil organic matter was evaluated on the basis of the soil samples (0-30 cm) collected after seventh rotation in the both cycles 1 and 2(Table 3). The average values for the cycles of described parameters were considered. SOM was expressed as the percentage content of organic carbon in soil.

In the experiment numerous parameters were evaluated, but for the paper purposes the follows ones were determined: organic carbon content by direct method using Analyzer C-

MAT 5500 and fractional composition of organic matter by Schnitzer method. The content of organic carbon of separated fraction was calculated as follow:

- C_d –carbon in solution after decalcification
- CHA+FA – sum of humic and fulvic acids in extracts obtained with 0,5 m NaOH
- CF – carbon of fulvic acids in solutions, following humic acids precipitation
- CHA – carbon of humic acids calculated from the difference:
- CHA = CHA+FA – CFA

The fractional composition was expressed as the percentage share of respective fraction in the total organic carbon pool (TOC).

Optical parameters of humic acids (HU) were measured in the UV-VIS, and afterwards, A_4/A_6 ratios were calculated.

The absorbance at 280, 465 and 665 nm of a solution (pH 8,3), containing at least 1 mg of in $NaHCO_3$ was measured in a UV –VIS spectrometer Perkin Elmer Lambda 20. The ratio of $A_4/A_6$was used do characterize SOM according to Kononova (1966) [16].

Statistical processing of the results was performed using Statgraphics 5 Plus package.

The data were processed by ANOVA, for each crop rotation, manure and mineral fertilization. There were proofed significant effect of crop rotation and manure application on both organic carbon content in soil and SOM quality. However, these parameters did not been affected by mineral fertilization.

The average values for treatments with different rates of manure and mineral N fertilizers describe the effect of experimental factors on soil organic carbon quantity. SOM quality was evaluated by the average values for crop rotations (A, B), and for the extreme treatments – without mineral nitrogen (N0), and the highest N rate (N3) as well as for manure, rates (1 and 5). Furthermore, the treatment with the highest mineral and manure doses was included (see explanations under table 5).

3. Results

3.1. The quantity of SOM

The analysis of variance demonstrated the importance of main effects and random effect (years of study) as well as the synergies of all the experimental factors on Corg (P-Value 0,0000 for all tests). The content of organic carbon in soil through 33 years of the experiment is presented by Figure 2. Figure 3 illustrates the impact of manure application on Corg. content in soil under cultivation of plants exhausting and enriching soil in organic matter. The results are the average for treatments with different rates of manure and mineral nitrogen fertilizers.

To understand the figure 1, it is needed to consider that for the first two four-years rotations, up to 1984, the experimental scheme included only manure application. At the beginning of the experiment in 1979, the initial organic carbon content amounted to 0, 74% (Fig. 2). After eight years, in crop rotations with clover grass mixture (B), organic carbon content oscillated around the initial value and amounted to 0,78%. Meanwhile, in crop rotation without legumes (A) Corg. value dropped to 0,72%.

In the following years (since the third rotation), in fields under plants exhausting soil with SOM (A), organic carbon quantity decreased regularly through the experiment and after 33 years dropped to 0, 61%. On the contrary, in crop rotation with clover – grass mixture, the tendency to stabilization organic carbon quantity in soil was observed with the highest value 0,79% in 1988 and the lowest one 0,72% in 2004. The disturbances in Corg. content detected through the experiment in both crop rotations could be caused by both climatic conditions and spatial field variability.

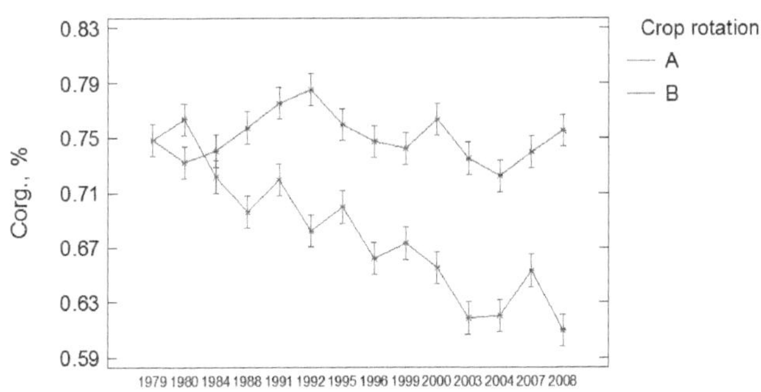

Figure 2. Effect of crop rotation on organic carbon content in soil through 1979-2008

The results illustrated by figure 3 indicated, that the effect of manure on soil organic carbon content was strongly linked to the crop rotation. In the both crop rotations, manure applicationin creased soil organic carbon. However, in crop rotations with plants exhausting soil from SOM even the highest manure rates 60 and 80 t ha^{-1} was not sufficient to secure Corg. content at the initial level (Fig. 3) over the years, what confirms the results presented by Figure 2. Meanwhile, in crop rotation with legumes, only 20 t ha^{-1} manure per hectare was required to obtain such content.

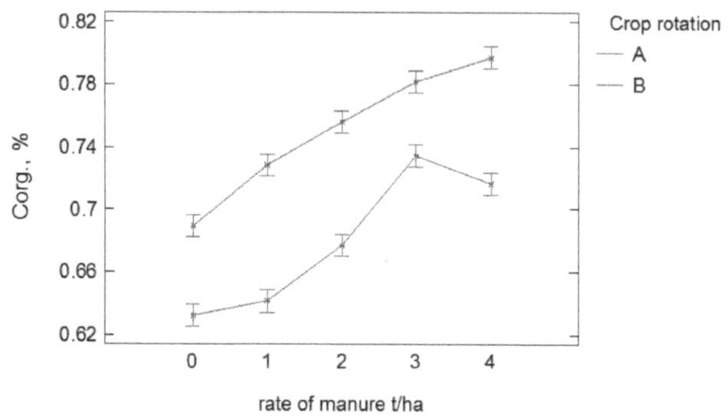

Figure 3. Effect of interaction manure application x crop rotation on soil organic carbon content

3.2. The quality of SOM

The properties of soil organic matter were evaluated on the base of its fractional composition. Table 5 shows the group composition of humus after 29 years of the experiment done by Schnitzer's method.

Combination	C_{org} %	C_d %	CHA+CFA	C_{FA}	C_{HA}	C_H	C_{HA}:C_{FA}
A1N0	0,55	0,033	47,3	20,0	27,3	52,7	1,37
A1N3	0,61	0,034	41,0	16,4	24,6	59,0	1,50
A5N0	0,74	0,039	43,3	14,9	38,4	56,7	1,91
A5N3	0,74	0,039	52,7	18,9	33,8	47,3	1,79
B1B0	0,75	0,035	48,0	20,7	27,3	52,0	1,32
B1N3	0,80	0,035	43,8	18,8	25,0	56,2	1,33
B5N0	0,78	0,042	48,7	21,8	26,9	51,3	1,23
B5N3	0,82	0,033	50,6	22,0	28,6	49,4	1,30

Table 5. Explanations: A – crop rotation with plants exhausting soil from humus B – crop rotation with plants enriching soil with humus1 – treatment without manure5 – treatment with 80 t manure ha^{-1} N0- treatment without mineral fertilizers N3 – 120 kg N ha^{-1} Fractional composition of SOM

As it has been described in Material and Methods, the results for selected treatments were listed. Constantly, the values of organic carbon content in soil samples were presented.

Humus content and its quality in the soil depended on manure application and the choice of plant species for crop rotation. There was found that crop rotation B and manure fertilization generally increased organic C content in the soil. Crop rotation influenced the strong to quality of SOM (Fig.4).

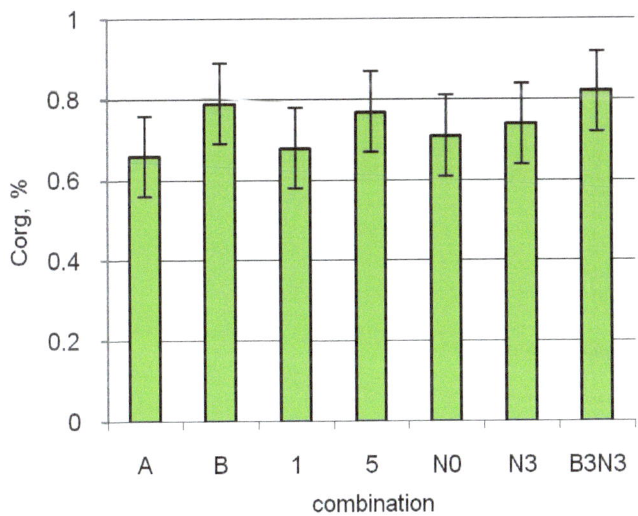

Figure 4. Effect of manure application, mineral fertilization, and crop rotation on Corg. content(explanations as under table 3). Error bars indicate standard deviation

Both manure application and selection of plants with legumes in crop rotation strongly modified the fractional composition of SOM and led to an increase of the sum of humic acids and fulvic acids in soil as well as the decrease in the amount of humins. Manure fertilization increased the ratio of humic acids in soil where as crop rotation with legumes affected a larger proportion of fulvic acids in comparison with humic acids. As a consequence, in treatments with regular, one to four years manure application, the ratio CHA:CFA was higher as compared with only mineral N fertilization. Similarly, soil in fields under clover with grasses cultivation are characterized by the lower ratio of humic acids to fulvic acids.

In the described experiment, the selection of plant species for crop rotation was recognized as crucial for the parameter. Humus under plants which enhance soil with organic matter revealed much lower values of CHA:CFA ratio as compared with the exhausting ones (Fig. 6). These differences will be discussed in the next part of the paper.

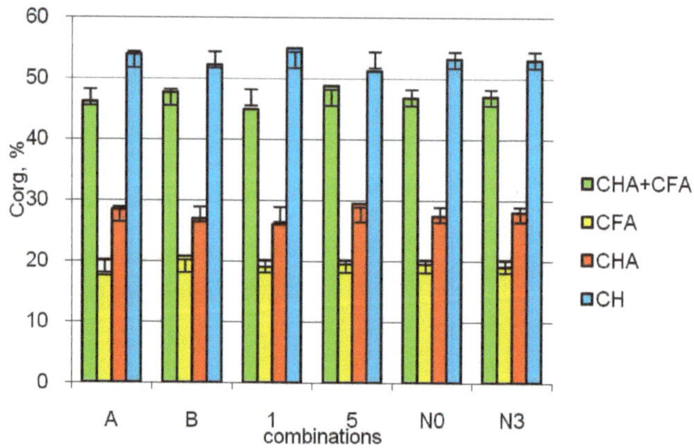

Figure 5. Fractional composition of humic substances (explanations as under table 3).

Error bars indicate standard deviation

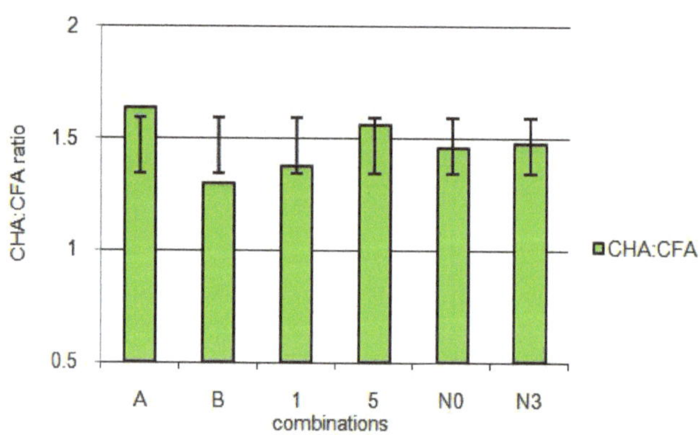

Figure 6. The ratio of humic acids to fulvic acids (CHA:CFA).Error bars indicate standard deviation

The isolated humic acids were treated by 1 mg of in $NaHCO_3$ and the liquid samples were subjected to UV - VIS radiation by the spectrometer for measurement within the range of visible and UV light. Then, the ratio of UV absorbance in different wave-lengths were analyzed. The coefficients of absorbance of humic acids were listed in table 6.

Combination	A280[1)]	A465[1)]	A665[1)]	A280/465[2)]	A280/665[2)]	A465/665[2)]
A1N0	2,16	0,401	0,0871	5,39	24,8	4,60
A1N3	2,09	0,384	0,0864	5,44	24,2	4,44
A5N0	1,85	0,325	0,0695	5,69	26,6	4,68
A5N3	2,03	0,318	0,0657	6,38	39,0	4,84
B1B0	2,08	0,346	0,0736	6,01	28,3	4,70
B1N3	1,96	0,318	0,0650	6,61	32,0	4,89
B5N0	2,04	0,306	0,30585	6,66	34,9	5,23
B5N3	2,09	0,339	0,0657	6,20	31,8	5,16

Table 6. Explanations: A – crop rotation with plants exhausting soil from humus B – crop rotation with plants enriching soil with humus1 – treatment without manure5 – treatment with 80 t manure ha^{-1} N0- treatment without mineral fertilizersN3 – 120 kg N ha^{-1} [1)]– wave-lengths (nm) [2)] – absorption ratioThe coefficients of absorbance of humic acids

The soil with manure application and mineral fertilization revealed the spectrum 280 nm in both crop rotations, which indicates the early stage of humification process characterized by high lignin content. However, there was no clear evidence in relation of fertilization and crop rotation on the intensity of the spectrum which amounted from 2,00 to 2,07 (Fig. 5).

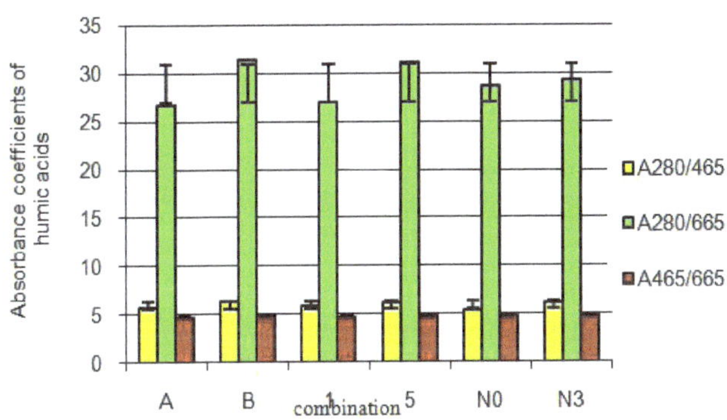

Figure 7. Absorbance coefficients of humic acids. Error bars indicate standard deviation

4. Discussion

Carbon plays a significant role both for agriculture and environment protection. The organic matter is a major component for soil formation. It considerably determines soil fertility and soil properties that are relevant to the yield. In agriculture land, both soil organic matter quantity and quality are determined by natural, agroclimatic condition as well as soil cultivation technology. The relevance of soil organic matter has been long neglected, and for decades the crop production was focused on mineral fertilization. During the last ten years in EU countries, aprogressive lost of soil organic matter has been observed. In Poland, due to simplification in crop production with prevalence of cereal monocultures as well as crop residue removal and reduced manure production intensify the process of natural SOM degradation. An of an adequate farm management relies in maintaining a sustainable balance of soil organic matter. A balance can be achieved by selection of species of cultivated plants, their share in the crop structure, and the quality of organic fertilizers. For different crop species the amount of residues left in filed varies.It can be estimated that the weight of cereals crop residues is about 3-fold greater than the roots, and legumes with grasses by up to 6-fold. Moreover, a different duration and degree of shading the soil surface and the number of tillage performed might affects the mineralization of organic compounds in humus.

The impact of all these factors on soil organic matter transformations could be accounted only in long-term perspective because of gradual character of the processes. The results of described 33 years experiment confirm the positive effect of organic fertilization and crop rotation included legumes – grass mixtures on SOM reproduction. For the research presented, the effect of crop rotation had a larger impact on organic carbon accumulation in soil that the effect of manure applications. Those results could be explained by the fact that clover mixture with grasses results ina greater amount of humus in soil compared with manure application. Moreover, carbon from manure is more prone to undergo mineralization process [20].

The reduction of Corg in soil in the following years relative to the initial value, is not significantly related with manure rates and can be explained by a rather slow depleting of the total pool of humus in agricultural soil as the effect of tillage intensity and regular removal of straw from the field [21, 22]. The results from the long term field experiment indicate that an external source of organic matter such as manure applied regularly might not assure a positive balance of SOM. The results confirmed that including legumes into crop rotation might provide an effective method to preserve the optimal level of humus.

The role of mineral nitrogen fertilization remains unclear. Some authors have described that applicaton of high amounts of mineral fertilizers might accelerate mineralization process and therefore diminish organic carbon. Other authors [23, 24] propose that prolonged application of mineral N fertilisers on lessives and brown soils may cause a decrease of carbon content in soil of about 21% compared with soils not supplied with mineral fertilizers. Numerous papers indicate that the inhance combined application of manure and mineral nitrogen fertilizers enhances the mineralization process of humus [25-27]. However, there exists a strong consensus on the literature that mineral nitrogen has beneficial effect for humus stabilization [8].

In the own research, the significant effect of mineral nitrogen fertilization did not be proven.

A distinctive/significant difference was observed in the evolution of organic carbon content in soil over time for crop rotation that included plants recognized to deplete humic substances or plants that can enrich the humic content in soil.

Soil cultivation and management practice, particular manure application and mineral fertilization, tillage systems and crop rotation induced an important discussion about humus quality. According to stability and decomposability, soil organic carbon is divided into the stable and labile forms. The stable forms are represented by the total carbon content, humic substances sum, humic acids sum and fulvic acids sum [10, 28]. The labile carbon is soluble in water, more active and undergoes short time changes in the soil. Körschnes [29] claims that this part of SOM is a useful tool to characterize the soil management practice. The objectives of own experiment was both quantitative changes in Corg content in soil and quality of humus according to Schnitzer method. That approach enables to point out the most proper soil management regarding to organic and mineral fertilization, and selection of plant species for crop rotations to obtain the positive balance of SOM and soil fertility measured by humus quality [29-31]. SOM quality is determined not by the absolute quantity of humus acids but its percentage share in humus. Humic and fulvic acids content underlies the calculation HA:FA ratio. The ratio describes humus quality and its stability as well. It is commonly accepted that fertile soils are characterized by the ratio HA:FA >1. Soils fertilized with composts, slurry and manure display higher HA:FA ratio than soils supplied with mineral fertilizers [32]. The research presented provides similar results. Furthermore, the strong impact of crop rotations on the ratio was found. Humus of the soil under clover with grasses mixture (crop rotation B) revealed lowest value of the ratio than with maize for silage (crop rotation A) which amounted to 1.30 and 1.64 respectively. The literature reports that the lowest ratio of humic acids to fulvic acids in soil under legumes cultivation, comes off from the great carbon amount left by these plants, which does not impact on the ratio [8]. Besides, legumes affect the rise of fulvic acids fraction and the lower HA:FA ratio as the result.

Fractional composition of humus was modified both by crop rotation and manure application. Legumes left a great amount of residues abundant with carbon and nitrogen - the source energy for soil microorganisms, which promote the process of humus transformation with the prevalence of mineralization one [32-35]. Hence, soils under legumes cultivation are characterized by the lower HA:FA ratio. Nevertheless, the further detailed analyzes of UV-VIS proved the good quality of soil under legumes. It demonstrates that fluorescence, although poorly utilized for humic acids research, is a powerful tool to contribute to the knowledge and the action of the organic matter in soil.

UV-VIS absorption technique is applied for a long time for the characterization of humic substances. It is commonly accepted that absorbance at 280 nm is an indicative of lignin content where as absorbance at 465 nmcan be related to organic matter in early stages of decomposition, and absorbance at 665 nm with advanced humification process of SOM. The data presented in the literature indicate [14, 17, 35] that for a UV- VIS spectrum of "young" humic acids the maximum signal at 280 nm will disappear with advanced humification process. Such spectrum is observed in soils fertilized with slurry and mineral fertilizers. For the research presented, the soil treated with manure revealed a maximum at 280 nm for both crop rotations. However, results on signal intensity at 280 were no conclusive for fertilization and crop rotation.

Higher values of the A465/665 absorbance ratio were recorded in the soil with crop rotation B and in the soil fertilized with manure, especially the highest rates. It is a clear evidence that there were "young" humic acids of a lower condensation level of aromatic structures that predominate in the structure of humic acids as compared with HA in the advanced humification process [14, 18, 36].

Manure applications as compared to mineral nitrogen fertilization increases the content of lignins in the particles of humic acids (higher values of the A280/465 absorbance ratio) (Fig. 5). Manure promoted a formulation of humus with greater share of aliphatic structure (lower absorption values) and higher (higher values of A465/665).

Maintaining or increasing soil organic matter (SOM) is justified both from an agronomic and a climatic perspective because it affects the capacity of the soil to sustain crop growth, is an important factor in decreasing soil compaction and erosion, and is also a source and possible sink of atmospheric CO_2-C [37]. Understanding the processes that control SOM dynamics is the key to SOM management. Land use and agricultural management practices such as crop rotation, soil tillage and organic amendments can affect SOM by influencing both the quantity and quality of crop residues that are returned to the soil; they also influence the rate of decomposition of added residues and native SOM [38-39].

5. Conclusions

1. The results of 32 – years of the field experiment show that the most important factor which stabilizes organic carbon content in agricultural soils is crop rotation with legumes. This effect has not been obtained even by systematic application of the very high rates of manure.
2. Because of intensified soil microbial activity, the ratio of humic acids to fulvic acids was lower in soil under legumes cultivations as compared with manure application. The further UV – VIS spectral analysis of humic acids indicated a high quality of humus of soils under clover with grasses mixture.
3. Mineral fertilization has not modified both soil organic carbon content and humus quality.

Author details

Agnieszka Rutkowska* and Dorota Pikuła

*Address all correspondence to: agrut@iung.pulawy.pl

Department of Plant Nutrition and Fertilization, Institute of Soil Science and Plant Cultivation-State Research Institute, Poland

References

[1] Rusco, E. R. Jones, & Bidoglio, G. (2001). Organic matter in the soils in Europe: Present status and future trends. *European Soil Bureau, JRS Ispra, Italy.*

[2] Komatsuzaki, M., & Ohta, H. (2007). Soil management practices for sustainable agro-ecosystems. Technical Raport. *Sustainability Science*, 2, 103-120.

[3] Terelak, H., Krasowicz, S., & Stuczyński, T. (2000). Srodowisko glebowe Polski i racjonalne użytkowanie rolniczej przestrzeni produkcyjnej. *Pamiętnik Puławski*, 120(2).

[4] Ochal, P. (2012). Wykorzystanie syntetycznego wskaźnika do oceny stanu agrochemicznego gleb w Polsce. *PhD thesis. IUNG-PIB Puławy.*

[5] Mercik, S., Stepień, M., Stępień, W., & Sosulski, T. (2005). Dynamic of organic carbon content insoil depending on long-term fertilization and crop rotation. Roczniki Gleboznawcze; t LVI 3/ , 4-53.

[6] Kusińska, A. (1999). Zasoby i skład humusu glebowego pod niektórymi gatunkami roślin w dwóch systemach uprawy. *Zeszyty Problemowe Postępów Nauk Roliczych*, 465-319.

[7] Pranagal, J. (2004). Wpływ systemu uprawy na zawartość węgla organicznego w glebie. Annales UMCS Sec. E, , 59(1), 1-10.

[8] Gonet, S.S. (1989). Właściwości kwasów humusowych w warunkach zróżnicowanego nawożenia. *ATR Bydgoszcz, Rozprawy*, 33, 4-55.

[9] Susic, M. (2003). Structure and origin of humic acids and their relationsfip to kerogen, bitumen, petroleum ans coal. http://humicacid.wordpress.com/structure-and-origin-of-humic-acids-and-their-relationship-to-kerogen-bitumen-petroleum-and-coal/.

[10] Stevenson, F. J. (1982). Humus chemistry genesis, composition, reactions. *Willey Interscience, New York.*

[11] Dziadowiec, H., & Gonet, S. (1999). Przewodnik metodyczny do badań materii organicznej gleb. *Prace Komisji Naukowej Polskiego Towarzystwa Gleboznawczego*, 42-43.

[12] Griffith, S. M., & Schnitzher, M. (1975). Analytical characteristics of humic and fulvic acids extracted from tropical volcanic soils. *Soil Science Society of America Proceedings*, 89-861.

[13] Chen, Y., Sanesi, N., & Schnitzer, M. (1978). Chemical and physical characteristics of humic and fulvic acids extracted from soils of Mediterranean region. *Geoderma*, 20-87.

[14] Gonet, S. S., & Dębska, B. (1999). Properties of humic acids producted during decomposition of plant residues in soil. *RostlinnaVyroba*, 45(10), 455-460.

[15] Zawiślak, K., Adamiak, J., & Tyburski, J. (1988). Dynamika substancji organicznej i składników mineralnych w warstwie uprawnej gleby pod wieloletnimi monokultur-

ami. Cz. I. *Gatunki o większych wymaganiach glebowych. Zeszyty Problemowe Postępów Nauk Rolniczych*, 331-227.

[16] Kononova, M. (1966). Soilorganic matter. *2ª ed. Pergamon Press Oxford.*

[17] Orlov, D. S. (1985). Humus Acids of Soils. *A.A. Balkema, Rotterdam.*

[18] Kumada, K. (1987). Chemistry of soil organic matter. *Developments in Soil Science 17.Japan Sc. Soc. Press Tokyo, Elsevier Amsterdam.*

[19] Stevenson, F. J. (1994). Humus Chemistry: genesis, composition, reactions. *2nd ed. Wiley, New York.*

[20] Antil, R. (2011). Predicting Nitrogen and carbon mineralization of composted manure and sevage sludge in soil. *Compost Science & Utilization*, 19(1), 33.

[21] Fotyma, M., & Filipiak, K. (2006). The influence of long term application on FYM and nitrogen fertilizers on the yield and uptake of nitrogen by crops grown in two rotations. *Nawozy i Nawożenie- Fertilizers and Fertilization*, 1(26), 71-89.

[22] Pikuła, D. (2012). The yield of winter wheat depending on weather and nitrogen supply. *Nawozy i Nawożenie- Fertilizers and Fertilization*, 41, 187-197.

[23] Janowiak, J. (1995). Wpływ nawożenia obornikiem z dodatkiem słomy i zróżnicowanych dawek azotu na właściwości materii organicznej. *Zeszyty Problemowe Postępów Nauk Rolniczych*, 421, 145-150.

[24] Dziadowiec, H., Jończak, J., Czarnecki, A., & Kejna, M. (2003). Wieloletnia dynamika zawartości węgla organicznego w poziomie ornopróchnicznym gleb intensywnie rolniczo użytkowanych. *[W].Zintegrowany monitoring środowiska przyrodniczego, W. Bochenek, E.Gil (red.). IOŚ, Instytut Geografii i Przestrzennego zagospodarowania PAN, Stacja Naukowo-Badawcza*, 166-172.

[25] Adamus, M., Drozd, J., & Stanisławska, E. (1989). Wpływ zróżnicowanego nawożenia organicznego i mineralnego na niektóre elementy żyzności gleby. *Roczniki Gleboznawcze*, 40(1), 101-110.

[26] Łoginov, W., Andrzejewski, J., & Janowiak, J. (1991). Rola nawożenia organicznego w utrzymaniu zasobów materii organicznej w glebie. *Roczniki Gleboznawcze*, 42(3-4), 19-25.

[27] Panak, H., & Nowak, G. (1989). Wpływ intensywnego nawożenia mineralnego na rozkład materii organicznej w glebie. *Roczniki Gleboznawcze*, 40, 1-39.

[28] Piccolo, A., & Mbagwu, J. S. C. C. (1990). Effects of different organic waste amendments on soil microagregates stability and molecular sitzes of humic substances. *Plant Soil*, 123-27.

[29] Körschens, M. (1996). Long-term data sets from Germany and Eastern Europe. In: POWISON D.S., Smith P., Smith J.N., (eds): Evaluation of soil organic matter models. ,Springer-Verlag, Berlin;, 1

[30] Körschens, M. (1999). Soil organic matter balance, optiumum content in soils. *In: Proc. Oral Presentation Organic Matter Balance and Optimum Organic Matter Stock in Soil. VURV, Praha,* 1-24.

[31] Körschens, M. (2000). Carbon and nitrogen dynamics as well as nitrogen utilization in dependence on soil texture. *K.Skogs-o. Lantbr-akad. Tidskr,* 139, 8-43.

[32] Dębska, B. (2004). Właściwości substancji humusowych gleby nawożonej gnojowicą. *PhD thesis ATR Bydgoszcz.*

[33] Dziamski, A., Żarski, J., & Stypczyńska, Z. (2000). Effect of irrigation and nitrogen fertilization on the mass of roots of spring barley and their distribution in very light soil. *Zeszyty Naukowe Nr 226, Agriculture,* 45-25.

[34] Kondratowicz-Maciejewska, K. (2007). Susceptibility of organic matter to oxidation and soil microbiological activity under conditions of varied crop rotation systems and fertilization. *Polish Journal of soil science,* 15/1, 89-98.

[35] Mc Kenney, D. J., Wang, S. W., Drury, C. F., & Findlay, W. I. (1993). Denitrification and mineralization in soil amended with legume,grass and corn residues. *Soil Science Society of America,* 57-1013.

[36] Kalembasa, D., Kalembasa, S., & Amberger, A. (1999). Spectroscopic characterization of organic compounds extracted from slurries by 0,1 M NaOH. *Humic Substances in Ecosysystems,* 3, 55-57.

[37] Paustian, K., Six, J., Elliot, E. T., & Hunt, H. W. (1999). Management options for reducing CO2 emissions from agricultural soils. *Biogeochemistry,* 48-147.

[38] Gregorich, E. G., Carter, M. R., Angers, D. A., Monreal, C. M., & Ellert, B. H. (1994). Towards a minimum data set to assess organic matter quality in agricultural soils. *Canadian Journal of Soil Science,* 74-367.

[39] Haynes, R. J., & Meare, M. H. (1996). Aggregation of organic matter storage in mesoternmal, humid soils. *In: Structure and Organic Matter Storage in Agricultural Soils (eds M.R. Carter & B.A. Stewart). Advances in Soil Science. CRC Levis, Boca Raton, Florida.*

Chapter 11

Estimation of Soil Carbon Stock in Taiwan Arable Soils by Using Legacy Database and Digital Soil Mapping

Chun-Chih Tsui, Horng-Yuh Guo and Zueng-Sang Chen

Additional information is available at the end of the chapter

1. Introduction

Since the Kyoto Protocol [1] was adopted, recent concerns about global warming have driven many effort been paid to develop methods to quantify both current and future carbon (C) stocks in different ecosystems. In global assessment, the current C stock contained in the plant and microbial biomass is estimated 560 Gt and 110 Gt, respectively, while soil C pool amounts to 2500 Gt, includes 1550 Gt of soil organic carbon (SOC) and 950 Gt of soil inorganic carbon (SIC) [2-6]. The soil C pool is 3.3 times the size of the atmospheric C pool of 760 Gt and 4.5 times of the biotic pool [4]. Therefore, soils are particularly important, as they are the largest reservoir of C in the terrestrial biosphere [2-3]. The SOC pool represent a dynamic equilibrium of C gains from plant production and loss through decomposition [7-8]. Carbon stock in soils is influenced by climate condition, soil properties, vegetation, land use and soil management [9-13].

Vegetation is the only source of carbon to the soils in terrestrial ecosystems. Therefore, land use play a major role in SOC stock built up through organic matter input [14]. At time scale of decades to centuries, changes in land use can exert a major influence on soil C storage [15]. Researches indicated that SOC decreases following the conversion of native ecosystems (forests, shrublands, grasslands) to agriculture [10-11,16-19], for example, losses of SOC from the conversion of prairie to agriculture have resulted in 24 to 89% loss in North America [20-21]. Agricultural activities that result in depletion of the soil C pool include the followings [22]:

1. deforestation, biomass burning and other activities related to conversion of natural to agricultural ecosystems,

2. tillage and other soil disturbances,
3. drainage of wetlands,
4. cultivation of organic soils,
5. removal of biomass for fuel, fodder and other uses, and
6. acceleration of soil erosion.

Besides, the magnitude and rate of SOC loss due to agricultural activities is higher for soils with a high C pool compared to a low C pool, tropics compared to temperature regions, croplands compared to pastures and tree plantations [22].

In agricultural soils, crop sequence, tillage and fertilization change inputs and outputs and, consequently the whole C dynamics. Potential soil C storage of the United States associated with changes in agricultural soil management [23]. In Java, Indonesia, SOC content declined from 2% to 0.75% in 1960 due to rapid conversion of natural vegetation to agricultural farms in the 1930s, but agricultural practices have started to accumulated soil C from 1975 to current, suggesting that the human or management influence on SOC stock can be stronger than the environmental factors [24]. In the United States, conversions of dry land farming to irrigated agriculture may increase SOC content in the soil profile with an average rate of 100 kg ha^{-1} yr^{-1} [23]. Long-term experiments in the Philippines showed that continuous cultivation of irrigated rice with balanced fertilization on submerged soils maintained or slightly increased SOC [25]. The previous study [26] indicated that soils with low to intermediate organic matter levels often exhibit a linear relationship between soil C levels and C inputs from addition of crop residues. The other study [27] found higher SOC content in 0-10 cm topsoil in sugarcane fields than that of Curatella savanna. These inconsistent results can be partly explained by the complexity of SOC, which consists of several pools that have a wide variety range of chemical properties and turnover times and consequently respond differently to land use changes [12].

The island of Taiwan used to be named as Formosa, which means "Beautiful Island". Forest lands cover about 2/3 of the total area of Taiwan. Since the first Han People arriving Taiwan from mainland China in the Qing Dynasty, most lowland forests have been exploited to agricultural lands. At present, a substantial conversion of cropland to urban land and other uses in Taiwan has occurred in recent decades. Consequently, these conversions have great impacts on SOC stock. Previous study [28] indicated that in the forest soils of Taiwan, the average SOC stock, estimated from 63 soil profiles excluding Histosols and Spodosols, was 18.5 kg m^{-2} in the upper 100 cm depth. Meanwhile in the arable land, using 140 soil profiles to estimate the SOC stock indicated that 5.97, 8.06 and 11.0 kg m^{-2} of C storages to the soil depth of 0-30, 0-50 and 0-100 cm, respectively [29]. Therefore, the conversion of forest land to rural soils has resulted in SOC loss in Taiwan.

According to the Kyoto Protocol, national householders need to pay attention on the impacts of land use changes on SOC storage in soils and vegetations. Any national soil-C-monitoring system must incorporate land use change as a key factor controlling changes in soil C [15]. However, monitoring long-term trends in SOC over a large geographical area is rare;

furthermore, the slow rate of changes and large spatial variation in SOC require high sampling densities with sufficient time to observe the changes [24]. This means monitoring SOC for large areas is only available in countries with big monitoring infrastructure. In Taiwan, we still have no formal soil monitoring network yet, but there have been several soil survey projects on agricultural soils for various purposes by Taiwan Agricultural Research Institute (TARI), Council of Agriculture, Taiwan. Therefore, legacy soil survey data is our best resource to monitor the dynamics of soil C [24]. To evaluate the effect of changes in land use on SOC stocks, we use a relatively intensive datasets obtained in 2006 as a reference to former data or to future work. The aims of this study are

1. to estimate the SOC stock at different depths in arable soils of three counties with different soil characteristics and
2. to show the SOC stocks at different depth from surface to 150 cm by digital soil mapping techniques for three important agricultural counties in Taiwan.

2. Materials and Methods

2.1. Pedotransfer function (PTF)

Pedotransfer functions (PTFs) [30] is the predictive functions of some soil properties from other easily, routinely, or cheaply measured properties. PTFs have recently become a popular topic in soil science, and different types of function have been developed to predict physical or chemical soil properties. Soil bulk density (Bd) measurements are often required as an important input parameter for various predictive and descriptive soil models. It is the mass of an oven-dry sample of undisturbed soil per unit bulk volume [31], and it is essential for weight-to volume or area conversions and is indispensable for the assessment of soil carbon stocks and nutrient pools [32]. Field sampling and measurement of bulk density are expensive, labor intensive and time-consuming [33-34]. Bulk density measurements are frequently missing from soil database or have been measured using different procedures [35]. To overcome this problem, PTFs are frequently used to estimate bulk density.

PTFs based on organic matter (OM) and soil texture are often used to estimate bulk density [33-39]. Besides, bulk density has been found to vary with depth, major soil group, water content, land use and vegetation [33,35,37-38,40]. Because published PTFs for bulk density have limited predictive potentials due to their development on specific soils and/or ecosystems [34], and there has been no PTFs for arable soils of Taiwan, we use legacy soil survey data to develop a PTF for estimating bulk density in this study.

2.2. Dataset for developing the pedotransfer function (PTF)

The dataset used for developing the PTF in this study included 230-horizon samples which were collected from arable soils located in Taoyuan and Tainan between 2001 to 2002. Field sampling was conducted following the Soil Survey Manual [41]. Analytical procedures used to measure the soil physical (e.g. bulk density) and chemical (e.g. SOC content) properties were described in the Soil Survey Laboratory Investigations Report No. 42 [42]. According to the location and field description of the profile, we could find its corresponding soil series in Soil Survey Report of Taoyuan and Tainan [43-44], respectively, which were published in 1970s by TARI. Soil texture (sand, silt and clay%) and pH values of horizons were obtained from the Soil Survey Report. Table 1 lists the basic information of 230 soil samples.

Soil property	Data range	Mean ± Standard deviation	Lab method/Description
Profile depth (cm)	20-150	70 ± 33	
Bd (g cm^{-3})	0.92-1.83	1.46 ± 0.20	Core method
SOC (g kg^{-1})	0.30-30.2	8.11 ± 6.35	Walkley-Black method
pH	3.5-8.1	6.1 ± 1.0	Legacy data from Soil Survey Report
Sand (%)	1.67-88.2	38.6 ± 19.4	Legacy data from Soil Survey Report
Silt (%)	5.37-65.7	37.7 ± 12.3	Legacy data from Soil Survey Report
Clay (%)	4.6-56.2	23.7 ± 10.7	Legacy data from Soil Survey Report

Table 1. Description of soil properties of 230 samples used for developing PTFs in this study

The 230 samples were randomly divided into two parts: the training set contained 80% of the data (184 samples) and was used to develop the models, and the validation set contained the remaining 20% of the data (46 samples) and was used to validate the proposed model. In order to determine the relationships between bulk density and soil properties, multiple regression models were developed by the REG procedure with a stepwise variable selection option [45]. Variables in Table 1 were considered for multiple regression modeling because researches have shown significant relationship between these variables and bulk density. R^2 and root mean square error (RMSE) were used to compare the predictive capacities of proposed regression models.

The proposed models for estimating bulk density of arable soils in Taiwan are shown in Table 2. Model A, which included log(d) and $[\ln(SOC)]^2$, can account 40% of the variation of bulk density. Adding pH as a predictor has a minor improvement of predictive quality (44%) in Model B. Besides, we found that pH is not available in part of legacy dataset, thus Model A with less parameters was used to predict bulk density. Although soil texture was used to predict bulk density in many studies [33-34,37-38], it is not significantly correlated with bulk density in this study.

Model	Intercept	Log (d)	$[\ln (SOC)]^2$	pH	R_{adj}^2
A	1.3026***	0.169***	-0.256***	--	0.40
B	1.0386***	0.1447***	0.022***	0.0476***	0.44

Table 2. Coefficients of candidate PTFs for Bd developed in this study. d: depth (cm), SOC: soil organic carbon content (g kg^{-1}). *** $p < 0.001$.

The proposed model was validated by the validation set (46-horizon samples) and the performance was shown in Figure 1. Results shows that only about 15% of the variation (R^2 = 0.15) was explained by the predicted bulk density with a RMSE equivalent to 0.207g cm^{-3}. Due to small available dataset and few variables, the proposed model revealed a very limited predictive potential, however, this model is statistically significant (p = 0.0052). To obtain a high accuracy and great precision in estimating soil bulk density, an equation specific for each range of soils of relevance to a particular research program should be used rather than rely on general PTFs [33]. Because bulk density was rarely measured, and no PTF of Taiwan soils has been developed, we adopted this model to estimate the bulk density of arable soil and, thereafter, to evaluate the SOC stock:

$$\mathrm{Bd} = 1.3026 + 0.169\,\mathrm{Log}\,(\mathrm{d}) - 0.256\left[\mathrm{Ln}\,(\mathrm{SOC})\right]^2 \quad (1)$$

where Bd is the soil bulk density (g cm^{-3}), d is the sample depth (cm), and SOC is the soil organic carbon content (g kg^{-1}).

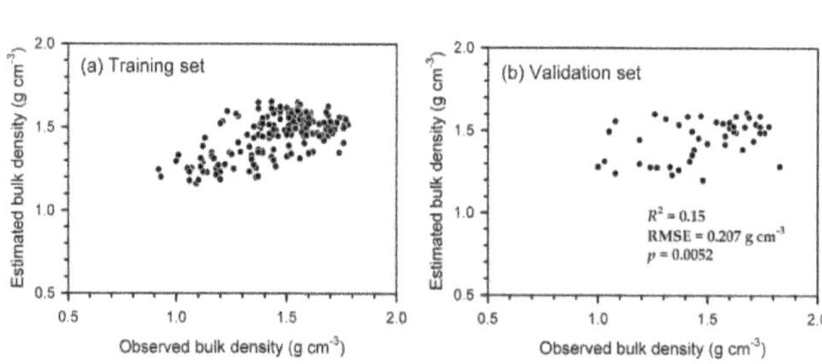

Figure 1. Correlation of observed bulk density and estimated bulk density of (a) training set and (b) validation set.

2.3. Datasets for estimation of SOC stock in arable soils

Dataset for estimating the SOC stock in agricultural soils was obtained from a detailed soil survey which was performed by TARI in 2006. In the field, one pedon was sampled by auger within a 250 m by 250 m grid, that is, every 6.25 ha of the arable land has a representative soil pedon. Sampling depths were 0 to 15, 15 to 30, 30 to 60, 60 to 90, 90 to 120, and 120 to 150 cm, respectively. We use soil data of three important agricultural counties: Taoyuan, Changhua and Tainan County in the northern, central and southern regions of Taiwan, respectively (Figure 2). After removing the outliers and missing data, the extracted database contains the information of 19,024 soil pedons. Only organic matter (OM, %) was available soil properties. Therefore, here we converted the OM content to SOC content by dividing a Van Bemmelen factor of 1.724, on the assumption that SOM contains 58% of organic C averagely. Soil carbon stock for a given depth is calculated as follows:

$$\text{SOC stock}\left(\text{kg m}^{-2}\right) = \left[\text{SOC}\left(\text{g kg}^{-1}\right) \times \text{Bd}\left(\text{g cm}^{-3}\right) \times \text{thickness}\left(\text{cm}\right)\right] / 100 \tag{2}$$

where Bd is bulk density which was calculated by using the Eq (1).

Finally, we estimated the SOC stocks at different depths of three counties and used the kriging method of geostatistics and digital soil mapping techniques.

Figure 2. The location of Taoyuan, Changhua and Tainan Counties in Taiwan.

3. Soil C stock in Taoyuan County (red soil)

3.1. Background of Taoyuan soils

The Taoyuan County is located in northwestern Taiwan, about 50 km southwest of Taipei city. The southeastern part of Taoyuan is mountainous area of high elevation, and the elevation decreases from southeast to northwest into the sea. Except for the mountain area, most of the agricultural lands develop on the terraces which were originally created as an alluvial fan by the paleo-Tahan River. The Tahan River, currently flowing in a northeastern direction, had flowed westwards into the sea before the Taoyuan Terrace formed [46]. Due to tectonic activities, the paleo-Tahan River gradually migrated clockwise and left several terraces behind [47], including Yangmei, Talun, Chungli, and Taoyuan Terraces in the Taoyuan County. Slopes of the terraces are between 1 and 7%, going down gently from the eastern hill land to the western seashore. The soils were developed on Quaternary alluvial deposits and have a minimum thickness of 5 m [48]. Cobbles are overlaid by finer alluvial materials in the terrace, and water usually perches at the contact between these two layers. Most of the terraces mentioned above are covered by red soils and gravels. It is generally believed that the red-colored soils in Taiwan could be developed prior to 30 ka [46].

Figure 3 shows the climatic data from two meteorological stations which are approaching to Taoyuan. Generally, the mean air temperature is 28.3°C in summer and 16.5°C in winter. The mean annual rainfall over the past decades (1981-2010) was 2061 mm. The average monthly evapotranspitation peak is July, but never exceeds the average monthly rainfall. The soil temperature and moisture regimes of Taoyuan soils are hyperthermic and udic.

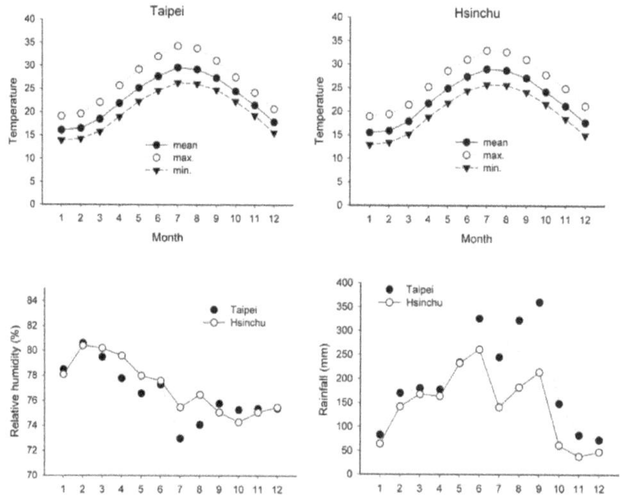

Figure 3. Mean monthly distribution of air temperature, relative humidity and precipitation at Taipei and Hsinchu meteorological stations which are approaching to Taoyuan. Data collected from 1981 to 2010, according to the Central Weather Bureau of Taiwan (http://www.cwb.gov.tw).

3.2. Land use and soil series of Taoyuan

Arable lands of Taoyuan County is 37,544 ha, occupying 30.8% of the total area. About 29% of the arable land were used for rice production and 18% of the arable land were used for upland crops production, including vegetables (10.7%), tea (2.4%), fruit (1.9%), grains (0.3%) and other crops (2.7%). About 53% of the total arable land was fallowed in Taoyuan County. Prior to 1950, the agricultural land in the terrace was used for tea production. After an irrigation system and reservoir were constructed in the 1950s, the terrace soils were used for rice production. Rice was harvested twice per year during the growing season between March and October, and then the soils were fallowed in winter.

A suit of soil profile properties and horizons falling within a particular range are said to belong to the same soil series. According to Soil Taxonomy [49], soil series are specific types of soils named after a geographic feature (town, river, etc.) near where they were first recognized. Based on Soil Survey Report of Taoyuan [43], there are 68 soil series which can be grouped into five Soil Orders: Histosol, Oxisol, Inceptisol, Entisol and Ultisol. About 56%, 32% and 9% of the Taoyuan soils belong to Ultisols, Entisols and Inceptisols, respectively. Description of important soil series in Taoyuan are shown in Table 3. Generally, the most arable soils in Taoyuan County are acidic, well drained, and clayed soil texture.

Soil series code	Occupied Area (%)	Depth class†	Soil texture‡	Drainage¶	pH	SMR§	Subgroup of Soil Taxonomy (USDA)
Pc	15.6	VD	SiC	ED	<5	Udic	Typic Kandiudox
Sk	9.9	VS	SL	W	<5	Udic	Lithic Udipsamment
Lt	6.4	VD	SiCL	W	5.0-7.0	Udic	Oxyaquic Paleudult
Tw	5.8	VD	SiCL	W	<5	Udic	Plinthic Paleudult
Hh	5.1	VD	SiC	P	5.0-7.0	Aquic	Typic Plinthaqult
Tc	4.2	D	CL	W	<5	Udic	Typic Paleudult
Pu	4.2	VD	SiCL	MW	5.0-7.0	Udic	Typic Plinthudult
Hk	4.0	D	SiC	W	5.0-7.0	Udic	Plinthic Paleudult
Tl	3.7	VD	CL	W	5.0-7.0	Udic	Typic Paleudult
Lk	2.2	VD	SiC	W	5.0-7.0	Udic	Plinthic Paleudult
Nc	2.2	D	L	SP	5.0-7.0	Udic	Typic Udipsamment
Lc	2.0	VD	SiCL	SP	<5	Udic	Plinthaquic Paleudult
Total	65.1						

Table 3. Important soil series of Taoyuan County in Taiwan. †Depth class: VS, very shallow (<25 cm); S, shallow (25-50 cm); MD, moderately deep (50-100 cm); D, deep (100-150 cm); VD, very deep (≧150 cm) ‡Soil texture: S, sand; LS, loamy sand; SL, sandy loam; L, loam; SiL, silty loam; Si, silt; SCL, sandy clay loam; CL, clay loam; SiCL, silty clay loam; SC, sandy clay; SiC, silty clay; C, clay. Drainage: ED, excessively drained; W, well drained; MW, moderately well drained; SP, somewhat poorly drained; P, poorly drained; VP, very poorly drained. §SMR: soil moisture regime.

3.3. Estimation of SOC stock and digital soil mapping

Estimation of SOC stocks at different depths of Taoyuan County is shown in Table 4. Available datasets of Taoyuan County are composed of 4,872 pedons. The mean SOC stock at depths of 0-15 cm and 15-30 cm is 3.77 ± 1.29 kg m^{-2} and 3.04 ± 1.14 kg m^{-2}, respectively, and SOC stock decreases with increasing soil depth. The mean SOC stock of arable soils of Taoyuan is 8.17 ± 4.99 kg m^{-2} from surface to 150 cm. It indicates that the red soils of Taoyuan is less soil fertility, and the SOC stock is less than the mean value (11 kg m^{-2}) of all arable soils of Taiwan [29,50]. On average, 83% of the total SOC stock in the upper 150 cm is stocked in the surface 30 cm. It implies the potentially large amounts of CO_2 may be released by changes in land use.

Soil depth	Soil sample	SOC stock (kg m^{-2})				
(cm)	number	Mean ± SD	Median	Skewness	Kurtosis	Data range
0-15	4872	3.77 ± 1.29	3.77	0.07	-0.42	0.45-8.07
15-30	2751	3.04 ± 1.14	2.98	0.25	-0.51	0.55-6.33
30-60	1735	4.73 ± 2.09	4.49	0.36	-0.77	0.77-9.98
60-90	843	3.20 ± 1.31	3.08	0.60	-0.29	0.81-7.08
90-120	552	2.71 ± 0.97	2.65	0.30	-0.86	0.37-4.90
120-150	258	2.52 ± 0.91	2.42	0.54	0	0.43-4.89
0-150	4872	8.17 ± 4.99	6.31	0.78	-0.57	0.65-19.99

Table 4. Estimation of SOC stock at different depths from dataset of Taoyuan County

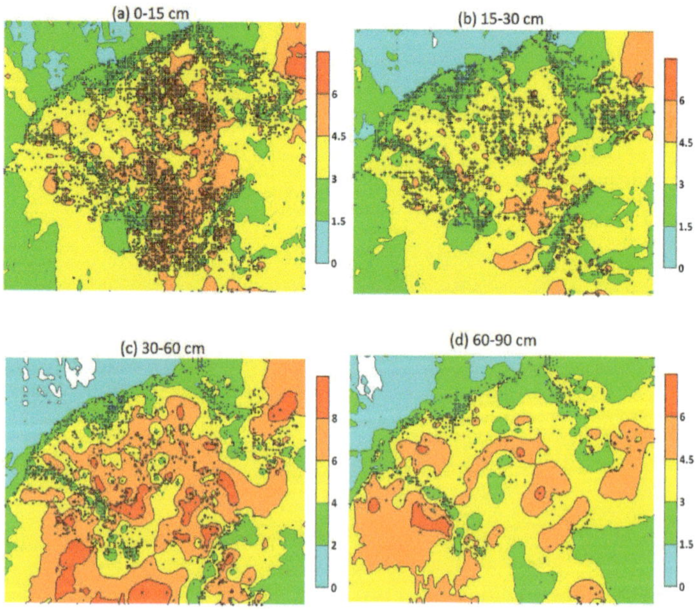

Figure 4. Digital soil mapping of the estimated SOC stock (unit: kg m^{-2}) at different depths in arable soils of Taoyuan County. Cross symbols represent the soil sampling pedons. Each pedon was sampled by auger within a 250 m by 250 m grid (6.25 ha).

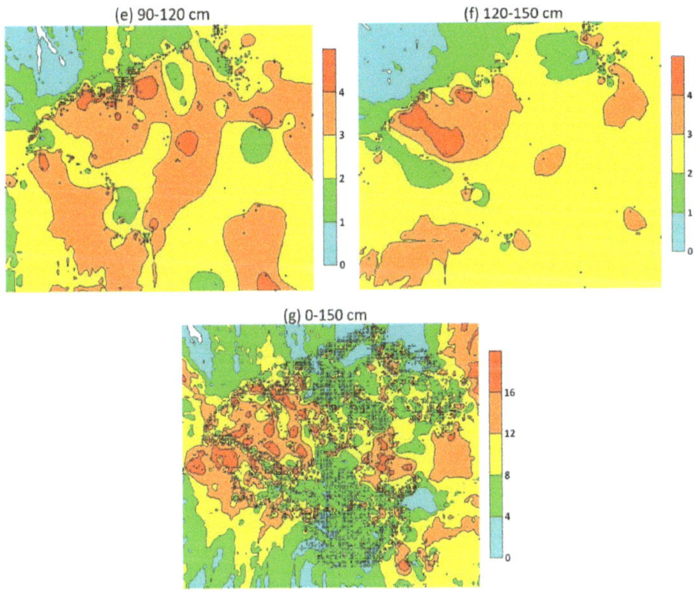

Figure 4. Digital soil mapping of the estimated SOC stock (unit: kg m^{-2}) at different depths in arable soils of Taoyuan County. Cross symbols represent the soil sampling pedons. Each pedon was sampled by auger within a 250 m by 250 m grid (6.25 ha).

Figure 4 shows the digital soil mapping of SOC stock in arable soils of Taoyuan County. Generally, soils in the western parts of Taoyuan have larger SOC in the depth 0-150 cm. In Taoyuan, the rice growing soils are largest in the western parts where Hsinwu and Yangmei are located. Therefore, we suggested that the paddy soils which growing rice can continuously accumulate organic carbon into the soils due to the addition of rice residues and more anaerobic condition during flooding cultivation. On the other hand, the increasing fallow lands with more aerobic condition and without addition of rice residues probably result in the decreases of SOC stock in Taoyuan County.

4. Soil C stock in Changhua County (slate alluvial soils)

4.1. Background of Changhua soils

Changhua County is located in the central Taiwan (Figure 2) and between the Dadu River and the Choushui River, the latter one is the longest river in Taiwan. Hence, there are two sources of soil parent materials in Changhua County, strata under the north part of soils are composed of sandstone and shale deposits from Daua River, while most soils in Changhua County are developed from limestone and slate clay deposited by the Choushui River and its distributaries. Differences in parent materials have great influences on the soil properties of the Changhua County. The elevation decreases gently from the eastern Pakua terrace to the western coastal plain.

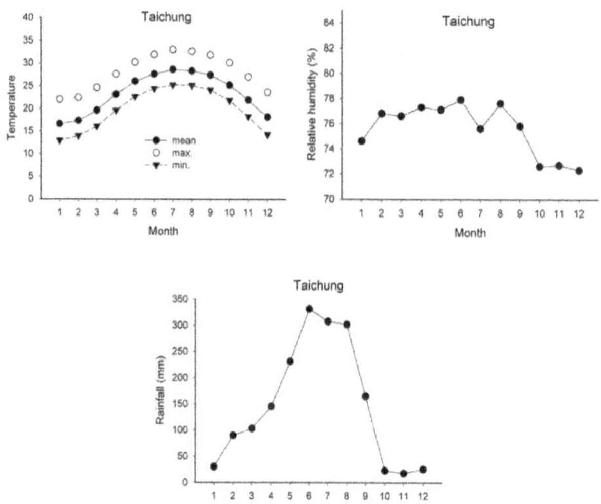

Figure 5. Mean monthly distribution of temperature, relative humidity and precipitation at Taichung meteorological stations which is approaching to Changhua. Data recorded from 1981 to 2010, according to the Central Weather Bureau of Taiwan (http://www.cwb.gov.tw).

Figure 5 shows the climatic data from the Taichung meteorological station which is approaching to Changhua County. The mean air temperature is 28.0 °C in summer and 17.3 °C in winter. The mean annual rainfall over the past decades (1981-2010) is 1773 mm, and the monthly rainfall is relatively low between October and February. The soil temperature and moisture regimes of the study area are hyperthermic and udic, respectively.

4.2. Land use and soil series of Changhua

Arable lands of Changhua is 63,722 ha, occupying 59% of the total area. About 43% of the arable land grows rice, and rice production of Changhua is the highest county in Taiwan. About 50% of the arable land grows crops, including vegetables (21%), fruits (11%), grains (8.5%) and flowers (7.3%). About 7.6% of the total arable land was fallowed in Changhua County. Because of the warm and humid climate, high soil fertility, complete irrigation system and accessibility, most area of the plain are used for croplands. Rice is harvested twice during the growing season, and the agricultural land use of Changhua is most intensive in Taiwan.

Soil series code	Occupied Area (%)	Depth class†	Soil texture‡	Drainage¶	pH	SMR§	Subgroup of Soil Taxonomy (USDA)
Eh	18.7	VD	SiL	SP	>7	Udic	Aquic Dystrudept
Lu	7.4	VD	SiL	SP	>7	Udic	Aquic Dystrudept
Yu	6.5	VD	SiL	SP	>7	Udic	Aquic Dystrudept
Ph	6.0	VD	SiL	MW	>7	Udic	Aquic Dystrudept
Cc	2.8	VD	SiL	SP	5.0-7.0	Udic	Aquic Udorthent
Sp	2.6	S	SiL	W	>7	Udic	Lithic Udorthent
Wh	2.5	VD	SiL	SP	>7	Udic	Aquic Dystrudept
Ct	2.4	VD	SL	SP	>7	Udic	Aquic Udipsamment
Hn	2.4	VD	SiL	SP	>7	Udic	Aquic Dystrudept
Kh	2.1	VD	SiL	SP	>7	Udic	Aquic Dystrudept
Ls	2.1	VD	LS	MW	>7	Udic	Typic Udipsamment
Es	2.1	VD	L	SP	>7	Udic	Aquic Udorthent
Kk	1.9	VD	SiL	MW	>7	Udic	Typic Dystrudept
Cl	1.8	VD	SL	SP	>7	Udic	Aquic Udipsamment
Co	1.8	VD	SL	MW	>7	Udic	Typic Udipsamment
Ha	1.7	VD	S	MW	>7	Udic	Typic Dystrudept
Su	1.7	VD	SL	P	5.0-7.0	Aquic	Typic Endoaquent
Total	66.5						

Table 5. Important soil series in Changhua County †Depth class: VS, very shallow (<25 cm); S, shallow (25-50 cm); MD, moderately deep (50-100 cm); D, deep (100-150 cm); VD, very deep (≧150 cm) ‡Soil texture: S, sand; LS, loamy sand; SL, sandy loam; L, loam; SiL, silty loam; Si, silt; SCL, sandy clay loam; CL, clay loam; SiCL, silty clay loam; SC, sandy clay; SiC, silty clay; C, clay. Drainage: ED, excessively drained; W, well drained; MW, moderately well drained; SP, somewhat poorly drained; P, poorly drained; VP, very poorly drained. §SMR: soil moisture regime.

Based on Soil Survey Report of Changhua [51], there are 65 soil series which can be grouped into two Soil Orders: Inceptisol and Entisol. 65% of the Changhua soils belong to Inceptisols, while the other 35% of soils belong to Entisols. Description of important soil series in Changhua are shown in Table 5. In general, Changhua soils are calcareous, silty-loam textured, and somewhat poorly drained.

4.3. Estimation of SOC stock and digital soil mapping

Available datasets of Changhua are composed of 6,749 soil pedons, and the estimation of SOC stocks at different depths is shown in Table 6. The mean SOC stock at depths of 0-15 cm and 15-30 cm is 2.34 ± 0.90 kg m^{-2} and 1.84 ± 0.68 kg m^{-2}, respectively. The mean SOC stock of arable soils of Changhua is 9.49 ± 5.12 kg m^{-2} from surface to 150 cm. On average, 44% of the total SOC stock of 150 cm is stocked in the surface 30 cm, implying the changes in land use have potentially impacts on CO_2 release.

Soil depth	Soil sample	SOC stock (kg m⁻²)				
(cm)	number	Mean ± Standard deviation	Median	Skewness	Kurtosis	Data range
0-15	6749	2.34 ± 0.90	2.24	0.54	-0.05	0.40-6.25
15-30	5757	1.84 ± 0.68	1.81	0.39	0	0.46-5.84
30-60	5132	2.97 ± 1.11	2.86	0.48	0	0.82-9.19
60-90	4333	2.63 ± 1.01	2.50	0.63	0.60	0.81-8.64
90-120	3516	2.43 ± 0.98	2.26	0.85	0.92	0.80-7.56
120-150	1093	2.31 ± 0.97	2.15	1.26	2.97	0.81-7.82
0-150	6749	9.49 ± 5.12	9.12	0.51	0.62	0.47-41.8

Table 6. Estimation of SOC stock at different depths from dataset of Changhua County

Figure 6 shows the digital soil mapping of SOC stock in arable soils of Changhua. Soils with larger C stocks are located at the high productivity area, including (1) Chutang and Pitou in the southern region, (2) Yungchih and Sheto in the southeastern region, and (3) Hsiushui, Homei and Shenkang in the north region of Changhua. Soils with extremely high SOC content (> 30 kg m^{-2} at 150 cm depth) in the north part of Changhua are belong to Histosols.

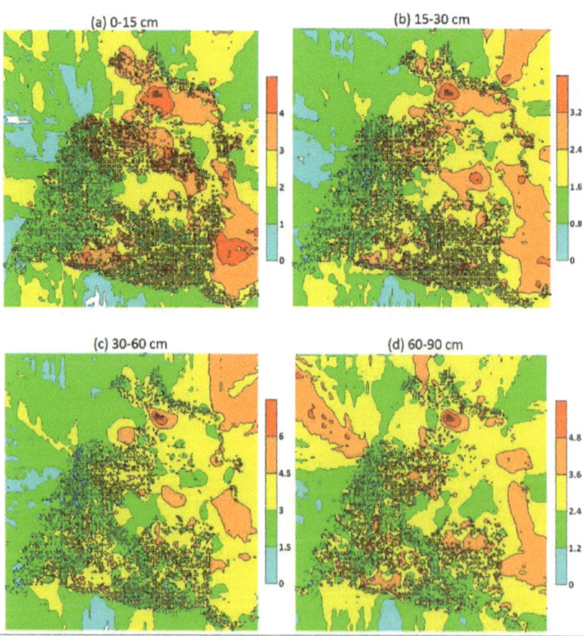

Figure 6. Digital soil mapping of the estimated SOC stock at different depth in arable soils of Changhua (unit: kg m^{-2}). Cross symbols represent the soil sampling pedons. Each pedon was sampled by auger within a 250 m by 250 m grid (6.25 ha).

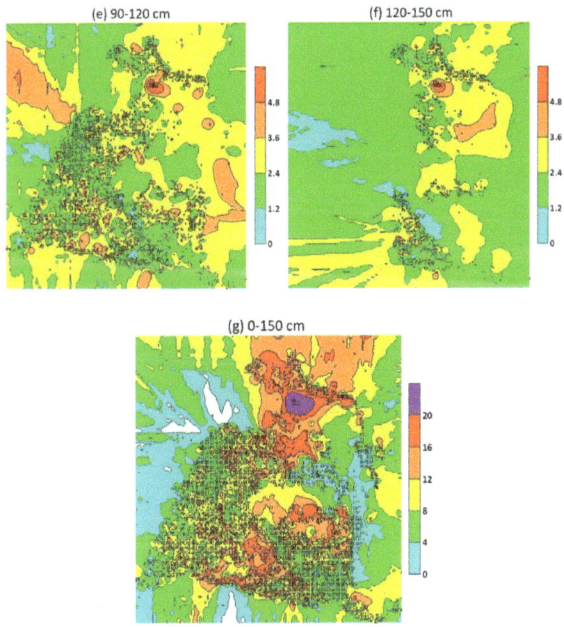

Figure 6. Digital soil mapping of the estimated SOC stock at different depth in arable soils of Changhua (unit: kg m^{-2}). Cross symbols represent the soil sampling pedons. Each pedon was sampled by auger within a 250 m by 250 m grid (6.25 ha).

5. Soil C stock in Tainan County (sandstone and shale alluvial soils)

5.1. Background of Tainan soils

Tainan County is located in the southwestern Taiwan (Figure 2). About one third area is occupied by hill land (30-50 m asl) in the eastern part of Tainan County, and the other two third area is alluvial plain. In general, Tainan County is situated in the central part of Chia-Nan Plain, which is the largest plain with high agricultural production of Taiwan. Most soils of Tainan are developed from sandstone, shale and mudstone deposits of the Zengwun River and the Bajhang River from the eastern hill regions.

Figure 7 shows the climatic data from the Tainan meteorological station. Tainan County is located in south of the Tropic Cancer, thus, the temperature is relatively high. The mean air temperature is 28.7°C in summer and 18.4 °C in winter. The mean annual rainfall over the past decade (1981-2010) is 1698 mm. Except for the raining season beginning from May to September, the monthly rainfall is less than the evapotranspiration. The soil temperature regime of the study area is hyperthermic, and soil moisture regime of most area is ustic.

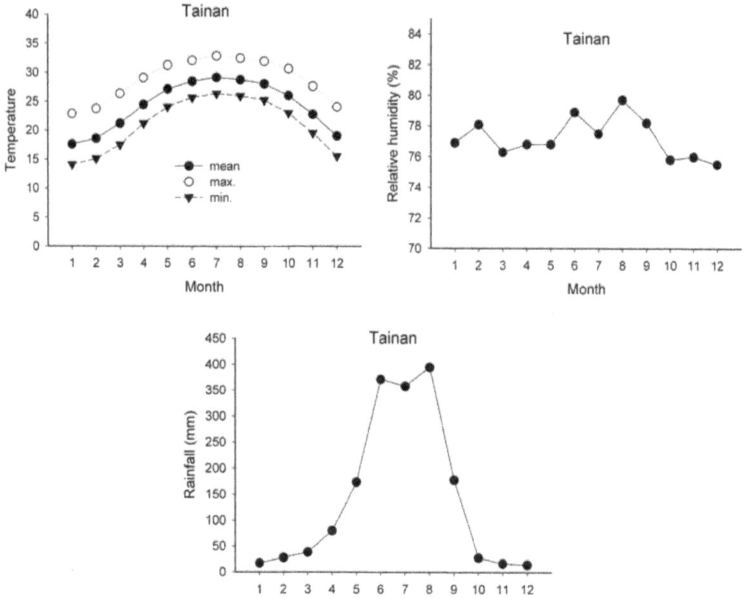

Figure 7. Mean monthly distribution of temperature, relative humidity and precipitation at Tainan meteorological stations. Data recorded from 1981 to 2010, according to the Central Weather Bureau of Taiwan (http://www.cwb.gov.tw).

5.2. Land use and soil series of Tainan

Arable lands of Tainan is 91,974 ha, occupying 45.6% of the total area. About 30% and 25% of the arable land grows fruits and rice, respectively, and 28% of the arable land grows crops vegetables (18%), grains (5.7%), sugarcanes (2.2%) and other crops. About 17% of the total arable land was fallowed in Tainan County. Due to the rainfall is often less than the evapotranspiration, especially when southwestern monsoon occurs during the winter, deficiency of soil water content is the limit factor of crop growing. Strong evaporation and the irrigation from groundwater result in saline soils in the coastal region of Tainan County.

Based on Soil Survey Report of Tainan [44], there are 68 soil series which can be grouped into three Soil Orders: Alfisol, Inceptisol and Entisol. About 57% of the Tainan soils belong to Entisols and 34% of soils belong to Inceptisols. Description of important soil series in Tainan are shown in Table 7. Most Tainan soils are sandy loam to silt loam soil texture, neutral to basic reaction and well drained soils.

Soil series code	Occupied Area (%)	Depth class†	Soil texture‡	Drainage¶	pH	SMR§	Subgroup of Soil Taxonomy (USDA)
Cf	16.1	VD	SL	MW	>7	Ustic	Typic Ustifluvent
An	9.1	VD	SiL	MW	>7	Ustic	Typic Dystrustept
Ts	7.4	VD	SiL	W	>7	Ustic	Typic Dystrustept
Hk	6.4	VD	SL	MW	>7	Ustic	Typic Ustipsamment
Ly	4.7	VD	SiL	W	>7	Ustic	Typic Ustifluvent
Je	4.2	VD	L	W	>7	Ustic	Typic Dystrustept
Lh	3.5	VD	SiCL	W	>7	Ustic	Typic Paleustalf
Tn	3.3	VD	L	W	5.0-7.0	Udic	Typic Dystrudept
Sh	3.2	VD	LS	W	5.0-7.0	Udic	Typic Udipsamment
Kn	2.8	VD	SiL	W	5.0-7.0	Udic	Typic Dystrudept
Sk	2.6	VD	SiC	W	>7	Ustic	Typic Dystrustept
Ku	2.0	VD	SiL	MW	>7	Ustic	Typic Paleustalf
Total	65.2						

Table 7. Important soil series in Tainan County †Depth class: VS, very shallow (<25 cm); S, shallow (25-50 cm); MD, moderately deep (50-100 cm); D, deep (100-150 cm); VD, very deep (≧150 cm) ‡Soil texture: S, sand; LS, loamy sand; SL, sandy loam; L, loam; SiL, silty loam; Si, silt; SCL, sandy clay loam; CL, clay loam; SiCL, silty clay loam; SC, sandy clay; SiC, silty clay; C, clay. Drainage: ED, excessively drained; W, well drained; MW, moderately well drained; SP, somewhat poorly drained; P, poorly drained; VP, very poorly drained. §SMR: soil moisture regime.

5.3. Estimation of SOC stock and digital soil mapping

Estimation of SOC stocks at different depths is shown in Table 8. Available datasets of Tainan are composed of 7,403 pedons. The mean SOC stock at depths of 0-15 cm and 15-30 cm is 2.05 ± 0.61 kg m^{-2} and 1.82 ± 0.50 kg m^{-2}, respectively. The mean SOC stock of arable soils of Tainan is 12.4 ± 5.49 kg m^{-2} in the upper 150 cm. In comparison with Taoyuan and Changhua soils, the total SOC stock in the upper 150 cm is higher in Tainan. On average, 31% of the total SOC stock in the upper 150 cm is stocked in the surface 30 cm.

Soil depth	Soil sample	SOC stock (kg m^{-2})				
(cm)	number	Mean ± SD	Median	Skewness	Kurtosis	Data range
0-15	7403	2.05 ± 0.61	2.03	0.20	-0.32	0.38-3.89
15-30	6259	1.82 ± 0.50	1.80	0.20	-0.39	0.35-3.19
30-60	6152	3.12 ± 0.90	3.05	0.34	-0.13	0.27-5.78
60-90	5880	2.91 ± 0.87	2.86	0.29	-0.28	0.14-5.45
90-120	5635	2.81 ± 0.86	2.75	0.30	-0.36	0.71-5.26
120-150	4711	2.74 ± 0.88	2.68	0.28	-0.53	0.61-4.99
0-150	7403	12.38 ± 5.49	13.82	-0.78	-0.54	0.49-20.00

Table 8. Estimation of SOC stock at different depths from dataset of Tainan County

Fig 8 shows the digital soil mapping of SOC stock in arable soils of Tainan. Generally, the arable lands of Tainan can be divided into eastern and western parts. In the eastern hill regions of Tainan, most soils are used for growing fruits and the SOC stocks are lower than those of the western plains. In the western plains, most soils used for growing rice and crops have larger soil organic carbon pool. Some arable lands are used to grow sugarcane in the western Tainan, and our estimation indicated that the recycling of sugarcane residues may increase the SOC stock in the soil profile.

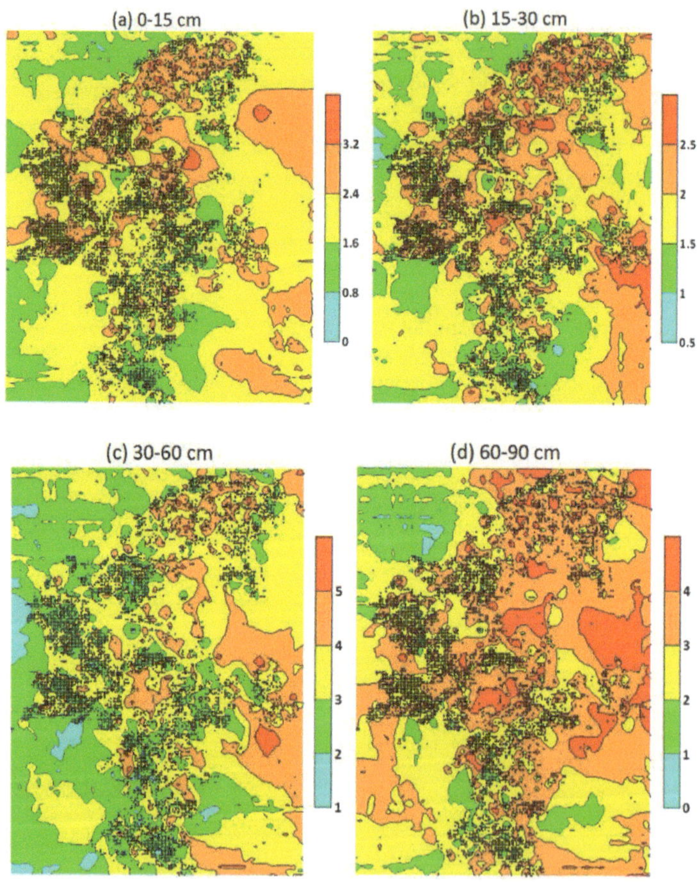

Figure 8. Digital soil mapping of the estimated SOC stock at different depth in arable soils of Tainan (unit: kg m^{-2}). Cross symbols represent the soil sampling pedons. Each pedon was sampled by auger within a 250 m by 250 m grid (6.25 ha).

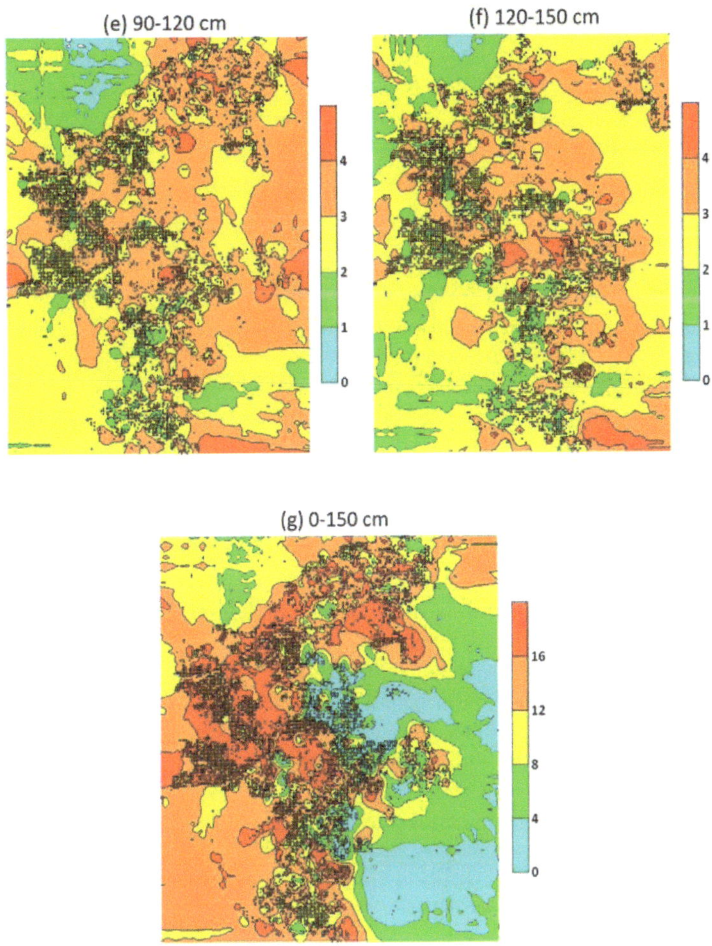

Figure 8. Digital soil mapping of the estimated SOC stock at different depth in arable soils of Tainan (unit: kg m^{-2}). Cross symbols represent the soil sampling pedons. Each pedon was sampled by auger within a 250 m by 250 m grid (6.25 ha).

6. Discussion on SOC stock in arable soils of Taiwan

In the current study, the mean SOC stock in the upper 150 cm of arable soils was highest in Tainan County (12.4 ± 5.49 kg m^{-2}), intermediate in Changhua County (9.49 ± 5.12 kg m^{-2}), and lowest in Taoyuan County (8.17 ± 4.99 kg m^{-2}). Our estimation of SOC stocks was close to those of warm and humid Asian countries, such as Indonesia (1.21 kg m^{-2} in 0-10 cm topsoil of Java) [24] and the Philippines (7.16-10.9 kg m^{-2} in 0-80 cm soils)[30], and was also close to those in semi-arid New Mexico (3.35-3.77 kg m^{-2} in 0-30 cm and 7.68-12.1 kg m^{-2} in 0-100 cm soils) [52]. Mean SOC stocks in the arable soils of Taiwan were lower than those of United Kingdom, Australia and South Africa [53-55], which indicating that the differences in SOC stocks at continental scale may be primarily driven by climate condition [24]. Climate data revealed that the rainfall and air temperature is slightly different among three counties: Taoyuan County has a longer rainy season (Figure 3) while Tainan County has distinctive dry season and rainy season (Figure 7), and Changhua County has a relatively moderate rainy season (Figure 5). Because the depletion of the SOC stocks in cultivated soils is caused by oxidation and mineralization, leaching and erosion [5], it is therefore probable that the consistency and periodicity of rainfall (i.e. effective rainfall) is a more significant factor to affect the quantity of soil carbon accumulation than that of simple value of total annual rainfall [54]. In Taoyuan County, stronger leaching and well-drained condition contributed to the weathering and soil development, and the Ultisols covered over 50% of the arable land of Taoyuan County (Table 3). In Changhua and Tainan Counties, most of the arable soils were classified as Inceptisols and Entisols (Tables 5& 7), which were reported to have larger SOC stocks than those for Ultisols distributed in Taoyuan County [2,16,28-29,50]. Therefore, we suggested that major soil types are also related to the SOC stock of arable soils in this study.

Based on the digital soil mapping (Figure 4,6,8), spatial variation of SOC stock in topsoil was not always parallel to those in 0-150 cm depth of soils, due to major carbon dynamics on topsoil and in subsoil may be controlled by different regulatory mechanisms [9, 56]. For example, land use had strong influence on SOC dynamics in topsoil [54], and the correlation between SOC and soil properties (clay content) was highest in soil deeper interval of the profile [9]. There are many factors and processes that determine the direction and rate of change in SOC content when the vegetation types and soil management practices are changed [17]. The nature and amounts of crop residues can influence the direction and amounts of trends in soil C stocks after the crops are harvested [26]. In this case study, for example, long-term planting and the recovery of sugarcane residues may increase the SOC stock in the western parts of Tainan County (Figure 8). Studies in Java [24] and Philippines [25] indicated that the continuous rice-growing system increased the SOC stocks. Here we have no sufficient SOC data to demonstrate the impact of different crops or land uses changes on arable soils in Taiwan. However, the previous study [29] reported that crop rotation system and fertilizer addition do increase the topsoil SOC content in some counties from 1950 to 1994. Moreover, land use changes from rice-growing soils to fallow or upland cultivation decreased the SOC stock in the upper 30 cm from 1969 to 2002 in Tainan County [29].

Depletion of SOC stock from the root zone has adversely affected the soil productivity and and environmental quality. On the other hand, increasing the SOC stock will increases the crop yield, especially in the soils where it has been depleted for long term. An increase of 1 ton of SOC can increase the wheat grain yield by 27 kg ha^{-1} in North Dakota, USA, and 40 kg ha^{-1} in semi-arid pampas of Argentina, 6 kg ha^{-1} of wheat and 3 kg ha^{-1} of maize in alluvial soils of northern India, 17 kg ha^{-1} of maize in Thailand, and 10 kg ha^{-1} of maize and 1 kg ha^{-1} of cowpea distributed in western Nigeria [4]. A good example of 18-year experiment in Kenya showed that the yield of maize and beans was 1.4 ton ha^{-1} yr^{-1} without external input and 6.0 ton ha^{-1} yr^{-1} when land cover was retained and fertilizer and manure were applied. The corresponding SOC stocks to 15 cm depth were 23.6 tons ha^{-1} and 28.7 tons ha^{-1}, respectively [4]. Therefore, Soil C sequestration is an important strategy to achieve food security through improvement in soil quality.

7. Uncertainties of estimating the SOC stock in Taiwan

In order to obtain a high accuracy and great precision on estimating soil bulk density, an equation specific for each group of soils of relevance to a particular research program should be used, rather than rely on general PTFs [33]. In this study, however, available database used to develop the PTF is very limited (230 horizon-samples), whereas the database used to estimate the SOC stock is larger in sample size (19,024 soil pedons) and composed of soil from different parent materials. The PTF that we proposed here contains only two variables: SOC and the soil depth. Because of the differences of bulk density between topsoil and subsoil, PTFs based on a higher portion of subsoil samples and applied on topsoil and subsoil samples may lead to an overestimation of bulk density [36], subsequently, lead to an overestimation of the SOC stock. Besides, to estimate the SOC stock from the database of 2006, we converted the OM content to OC content by a Van Bemmelen factor of 1.724 on the assumption that SOM contains 58% of organic C averagely. Variation in the ratios of SOC to SOM in different major soil types may result in the errors of SOC stock estimation as well. Finally, quality of the legacy data and the inconsistent measurement of soil properties between independent databases are also possible sources of uncertainties for this estimation.

8. Conclusion

In this study, we used 184 horizon-samples to develop a pedotransfer function (PTF) for bulk density of arable soil in Taiwan. The proposed PTF is $Bd = 1.3026 + 0.169 \text{ Log } (d) - 0.256 [\text{Ln } (SOC)]^2$. Validation by the other 46 horizon-samples in the dataset obtained a $R_{adj}^{\ 2} = 0.15$ and RMSE = 0.207 g cm^{-3}. Database from soil survey by TARI in 2006 was used to estimate the SOC stock in arable soil, and soil bulk densities were estimated by the PTF that we proposed. According to our estimation, the mean SOC stock to the depth of 0-150 cm in arable soils is listed as the decreasing order: Tainan County (12.4 ± 5.49 kg m^{-2}) > Changhua County (9.49 ± 5.12 kg m^{-2}) > Taoyuan County (8.17 ± 4.99 kg m^{-2}). More than 30% to 80% of the total SOC stock in the upper 150 cm is stocked in the surface 30 cm, depending on the soil type and soil management. Based on geostatistics and digital soil mapping techniques, we suggest that land use has great influence on the SOC stock in these arable soils. To obtain a high accuracy and great precision on estimating soil bulk density and SOC stock, databases from various soil types and PTFs for specific soils are needed in the future work.

Author details

Chun-Chih Tsui[1], Horng-Yuh Guo[2] and Zueng-Sang Chen[3*]

*Address all correspondence to: soilchen@ntu.edu.tw

1 Department of Agricultural Chemistry, National Taiwan University, Taiwan

2 Division of Agricultural Chemistry, Taiwan Agricultural Research Institute, Council of Agricultural, Taiwan

3 Department of Agricultural Chemistry, National Taiwan University, Taiwan

References

[1] UNFCCC. (1997). The Kyoto Protocol to the United Nations Framework Convention on Climate Change. *Kyoto: UNFCCC.*

[2] Batjes, N. H. (1996). Total carbon and nitrogen in the soils of the world. *European Journal of Soil Science*, 47(2), 151-163.

[3] Schlesinger, W. H. (1997). Biogeochemistry: An Analysis of Global Change. *San Diego, California: Academic Press.*

[4] Lal, R. (2004). Soil carbon sequestration impacts on global climate change and food security. *Science*, 304(5677), 1623-1627.

[5] Lal, R. (2008). Carbon sequestration. *Philosophical Tran sactions of the Royal Society B*, 363(1492), 815-830.

[6] Jansson, C., Wullschleger, S. D., Kalluri, U. C., & Tuskan, G. A. (2010). Phytosequestration: carbon biosequestration by plants and the prospects of genetic engineering. *Bioscience*, 60(9), 685-696.

[7] Jenny, H. (1941). Factors of soil formation. *New York: McGraw-Hill.*

[8] Schlesinger, W. H. (1977). Carbon balance in terrestrial detritus. *Annual Review of Ecology and Systematics*, 8, 51-81.

[9] Jobbágy, E. G., & Jackson, R. B. (2000). The vertical distribution of soil organic carbon and its relation to climate and vegetation. *Ecological Applications*, 10(2), 423-436.

[10] Guo, L. B., & Gifford, R. M. (2002). Soil carbon stocks and land use change: a meta analysis. *Global Change Biology*, 8(4), 345-360.

[11] Murty, D., Kirschbaum, M. U. F., Mc Murtrie, R. E., & Mc Gilvray, H. (2002). Does conversion of forest to agricultural land change soil carbon and nitrogen? A review of the literature. *Global Change Biology*, 8(2), 105-123.

[12] Paul, S., Veldkamp, E., & Flessa, H. (2008). Soil organic carbon in density fractions of tropical soils under forest-pasture-secondary forest land use changes. *European Journal of Soil Science*, 59(2), 359-371.

[13] Gamboa, A. M., & Galicia, L. (2012). Land-use/cover change effects and carbon controls on volcanic ash soil profiles in highland temperate forest. *Geoderma*, 170, 390-402.

[14] Pandey, C. B., Singh, G. B., Singh, S. K., & Singh, R. K. (2010). Soil nitrogen and microbial biomass carbon dynamics in native forests and derived agricultural land uses in a humid tropical climate of India. *Plant and Soil*, 333(1-2), 453-467.

[15] Scott, N. A., Tate, K. R., Giltrap, D. J., Smith, C. T., Wilde, R. H., Newsome, P. F. J., & Davis, M. R. (2002). Monitoring land-use change effects on soil carbon in New Zealand: quantifying baseline soil carbon stocks. *Environmental Pollution*, 116(1), S167-S186.

[16] Eswaran, H., Van den Berg, E., & Reich, P. (1993). Organic carbon in soils of the world. *Soil Science Society of America Journal*, 57(1), 192-194.

[17] Post, W. M., & Kwon, K. C. (2000). Soil carbon sequestration and land-use change: processes and potential. *Global Change Biology*, 6(3), 317-327.

[18] Lettens, S., van Orshoven, J., van Wesemael, B., Muys, B., & Perrin, D. (2005). Soil organic carbon changes in landscape units of Belgium between 1960 and 2000 with reference to 1990. *Global Change Biology*, 11(12), 2128-2140.

[19] Falloon, P., Smith, P., Bradley, R. I., Milne, R., Tomlinson, R. W., Viner, D., Livermore, M., & Brown, T. A. W. (2006). RothCUK-a dynamic modelling system for extimating changes in soil C from mineral soils at 1-km resolution in the UK. *Soil Use and Management*, 22(3), 274-288.

[20] Knops, J. M. H., & Tilman, D. (2000). Dynamics of soil nitrogen and carbon accumulation for 61 years after agriculture abandonment. *Ecology*, 81(1), 88-98.

[21] Kucharik, C. J., Brye, K. R., Norman, J. M., Foley, J. A., Gower, S. T., & Bundy, L. G. (2001). Measurement and modelling of carbon and nitrogen cycling in agroecosystems of southern Wisconsin: Potential for SOC sequestration during the next 50 years. *Ecosystems*, 4(3), 237-258.

[22] Lal, R. (2004). Agricultural activities and the global carbon cycle. *Nutrient Cycling in Agroecosystems*, 70(2), 103-116.

[23] Lal, R., Follett, R. F., Kimble, J., & Cole, C. V. (1999). Managing US cropland to sequester carbon in soil. *Journal of Soil & Water Conservation*, 54(1), 374-381.

[24] Minasny, B., Sulaeman, Y., & Mc Bratney, A. B. (2011). Is soil carbon disappearing? The dynamics of soil organic carbon in Java. *Global Change Biology*, 17(5), 1917-1924.

[25] Pampolino, M. F., Laureles, E. V., Gines, H. C., & Buresh, R. J. (2008). Soil carbon and nitrogen changes in long-term continuous lowland rice cropping. *Soil Science Society of America Journal*, 72(3), 798-807.

[26] Paustian, K., Cole, C. V., Sauerbeck, D., & Sampson, N. (1998). CO_2 mitigation by agriculture: An overview. *Climatic Change*, 40(1), 135-162.

[27] Jimenez, J. J., Lal, R., Leblance, H. A., Russo, R. O., & Raut, Y. (2008). The soil C pool in different agroecosystems derived from the dry tropical forest of Guanacaste, Costa Rica. *Ecological Engineering*, 34(4), 289-299.

[28] Tsai, C. C., Chen, Z. S., Hseu, Z. Y., Duh, C. T., & Guo, H. Y. (2010). Organic carbon storage and management strategies of the forest soils based on the forest soil survey database in Taiwan. *In: Chen ZS, Agus F. (eds.) Proceedings of International Workshop on Evaluation and sustainable management of soil carbon sequestration in Asian countries, 28,29 September, IPB International Conference Center, Bogor, Indonesia. Food and Fertilizer Technology Center (FFTC) for the Asian and Pacific Region.*

[29] Jien, S. H., Hseu, Z. Y., Guo, H. Y., Tsai, C. C., & Chen, Z. S. (2010). Organic carbon storage and management strategies of the rural soils on the basis of Soil Information System in Taiwan. *In: Chen ZS, Agus F. (eds.) Proceedings of International Workshop on Evaluation and sustainable management of soil carbon sequestration in Asian countries, 28-29September, IPB International Conference Center, Bogor, Indonesia. Food and Fertilizer Technology Center (FFTC) for the Asian and Pacific Region.*

[30] Bouma, J. (1989). Using soil survey data for quantitative land evaluation. In: Stewart BA (ed.) Advances in Soil Science New York: Springer-Verlag;. , 9, 177-213.

[31] ISSS Working Group. (1998). World reference base for soil resources: Introduction. *1st edit. Acco, Leuven, the Netherlands: International Society of Soil Science, International Soil Reference and Information Centre, and Food and Agriculture Organisation of the United Nations.*

[32] Tamminen, P., & Starr, M. (1994). Bulk density of forested mineral soils. *Silva Fennica*, 28(1), 53-60.

[33] Harrison, A. F., & Bocock, K. L. (1981). Estimation of soil bulk-density from loss-on-ignition values. *Journal of Applied Ecology*, 18(3), 919-927.

[34] Kaur, R., Kumar, S., & Gurung, H. P. (2002). A pedo-transfer function (PTF) for estimating soil bulk density from basic soil data and its comparison with existing PTFs. *Australian Journal of Soil Research*, 40(5), 847-857.

[35] Heuscher, S. A., Brandt, C. C., & Jardine, P. M. (2005). Using soil physical and chemical properties to estimate bulk density. *Soil Science Society of America Journal*, 69(1), 51-56.

[36] De Vose, B., van Meirvenne, M., Quataert, P., Deckers, J., & Muys, B. (2005). Predictive quality of pedotransfer functions fro estimating bulk density of forest soils. *Soil Science Society of America Journal*, 69(2), 500-510.

[37] Minasny, B., Mc Bratney, A. B., Mendonça-Santos, M. L., Odeh, I. O. A., & Guyon, B. (2006). Prediction and digital mapping of soil carbon storage in the Lower Namoi Valley. *Australian Journal of Soil Research*, 44(3), 233-244.

[38] Suuster, E., Ritz, C., Roostalu, H., Reintam, E., Kõlli, R., & Astover, A. (2011). Soil bulk density pedotransfer functions of the humus horizon in arable soils. *Geoderma*, 163(1-2), 74-82.

[39] Han, G. Z., Zhang, G. L., Gong, Z. T., & Wang, G. F. (2012). Pedotransfer functions for estimating soil bulk density in China. *Soil Science*, 177(3), 158-164.

[40] Manrique, L. A., & Jones, C. A. (1991). Bulk density of soils in relation to soil physical and chemical properties. *Soil Science Society of America Journal*, 55(2), 476-481.

[41] Soil Survey Staff. (1993). Soil Survey Manual. *Soil Conservation Service. U.S. Department of Agriculture Handbook 18. Washington, DC., US Government Printing Office.*

[42] Soil Survey Staff. (1996). Soil survey laboratory methods manual. *Soil Survey Laboratory Investigations Report No.42 Washington, DC.: US Government Printing Office.*

[43] Chen, C. C. (1976). Soil Survey Report of Taoyuan County. *Taiwan: Taiwan Agricultural Research Institute; (in Chinese).*

[44] Department of Soil, National Chung Hsing University. (1969). Soil Survey Report of Tainan County. *Taiwan: Taiwan Agricultural Research Institute; (in Chinese).*

[45] SAS Institute. (2008). SAS STAT user's guide. 9, *Cary, NC.: SAS Institute.*

[46] Chen, Y. G., & Liu, T. K. (1991). Radiocarbon dates of river terraces along the lower Tahanchi, northern Taiwan: Their tectonic and geomorphic implications. *Proceedings of the Geological Society of China*, 34(4), 337-347.

[47] Wang, Y., Chen, Y., Shyu, J., Chuang, R., Lin, Y., & Chung, L. (2002). The active tectonic features in Taoyuan Area, Northwestern Taiwan. *In: Eos Transactions, American Geophysical Union 83(47) Fall Meeting Supplement, AGU2002 Fall Meeting, 6-10December, Moscone Center, San Francisco, California. San Francisco: AGU.*

[48] Ho, C. S. (1988). An introduction to the geology of Taiwan explanatory text of the geologic map of Taiwan. *Taipei: Central Geological Survey, Ministry of Economic Affairs.*

[49] Soil Survey Staff. (1999). Soil Taxonomy: A basic system of soil classification for making and interpreting soil surveys. *USDA-NRCS, Agricultural Handbook no. 436,Washington, DC.: US Government Printing Office.*

[50] Chen, Z. S., & Hseu, Z. Y. (1997). Total organic carbon pool in soils of Taiwan. *Proceedings of the National Science Council, ROC- Part B: Life Science*, 21(3), 120-127.

[51] Department of Soil, National Chung Hsing University. (1969). Soil Survey Report of Changhua County. *Taiwan: Taiwan Agricultural Research Institute; (in Chinese).*

[52] Jacinthe, P. A., Shukla, M. K., & Ikemura, Y. (2011). Carbon pools and soil biochemical properties in manurebased organic farming systems of semi-arid New Mexico. *Soil Use and Management*, 27(4), 453-463.

[53] Smith, P., Bhogal, A., Edgington, P., Black, H., Lilly, A., Barraclough, D., Worrall, F., Hillier, J., & Merrington, G. (2010). Consequences of feasible future agricultural land-use change on soil organic carbon stocks and greenhouse gas emissions in Great Britain. *Soil Use and Management*, 26(4), 381-398.

[54] Wilson, B. R., Koen, T. B., Barnes, P., Ghosh, S., & King, D. (2011). Soil carbon and related soil properties along a soil type and land-use intensity gradient, New South Wales, Australia. *Soil Use and Management*, 27(4), 437-447.

[55] Mchunu, C. N., Lorentz, S., Jewitt, G., Manson, A., & Chaplot, V. (2011). No-till impact on soil and soil organic carbon erosion under crop residue scarcity in Africa. *Soil Science Society of America Journal*, 75(4), 1503-1512.

[56] Salomé, C., Nunan, N., Pouteau, V., Lerch, T. Z., & Chenu, C. (2010). Carbon dynamics in topsoil and in subsoil may be controlled by different regulatory mechanisms. *Global Change Biology*, 16(1), 416-426.

Chapter 12

Changes in Raised Bog Relief During the Holocene Case Study: Polish Carpathian Mountains

Adam Łajczak

Additional information is available at the end of the chapter

1. Introduction

Less attention has been paid to peat bog growth during the Holocene than to contemporary human impact on peat bogs (e.g. Bower 1961, Mallik et al. 1984, Evans 1989, Shaw et al. 1997, Bragg and Tallis 2001, Bindler 2006, Coggins et al. 2006). The research literature states that in order for peat bogs to grow, certain geomorphological, hydrographic, hydrogeological and climate-related conditions must be satisfied (Tołpa 1949, Maksimov 1965, Grosse-Brauckmann 1974, Lowe and Walker 1997, Tobolski 2000, Chairman 2002, Ilnicki 2002). Research studies have identified several types of peat bogs: limnogenous/river-fed, topogenous, soligenous and ombrogenous, all of which differ in terms of relief (Żurek & Tomaszewicz 1996, Tobolski 2000, Ilnicki 2002). In areas with precipitation barely exceeding evaporation, which includes mountain areas, peat bog development is determined by stable groundwater outflows that foster the continuous expansion of hydrogenic sites (Łajczak 2007, 2011). Groundwater outflows create wetlands that foster the development of low bogs. Once low bogs have formed, minerotrophic contact becomes less significant at the bog surface, which leads to oligotrophication and acidification. Both processes then lead to the development of a raised bog (Gore 1983, Tobolski 2000, Ilnicki 2002). The first researcher to note the difference between a low bog and a raised bog as well as their hydrological determinants was Senft (1862).

The greatest geomorphological differences between peat bogs can be observed in the mountains. Peat bogs can be found on ridges, slopes and valley floors (Bower 1961, Kaule and Göttlich 1976, Rawes 1983, Obidowicz 1985, Carling 1986, Rhodes and Stevenson 1997, Bragg and Tallis 2001, Dykes and Warburton 2007, Łajczak 2007, 2011, Obidowicz and Margielewski 2008). While raised bog relief and extent have not been covered explicitly and extensively in the research literature, certain aspects of bog geomorphology have been covered in paleogeographic research in bog areas and research on peat deposit structure. More papers have focused on historical and modern-day changes in bog relief in areas affected by human activity. What is more rarely encountered is advanced research on modern-day changes in raised bog relief.

Research on raised bog relief in Poland is actually a little more advanced than that in other parts of the world. This is true of northern Poland, which features a large number of bogs, and the Polish Carpathians, which feature just a few bogs. The most thoroughly investigated raised bogs in the Polish Carpathian Mountains are found in the Orawsko-Nowotarska Basin and in valleys in the Bieszczady Range (Fig. 1).

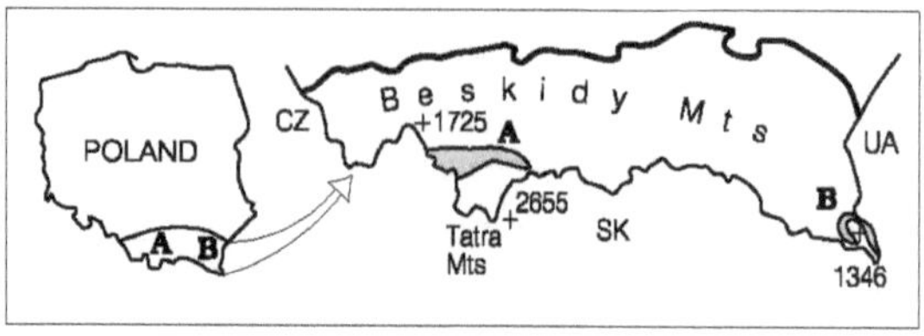

A- Orawsko-Nowotarska Basin, B- Bieszczady Mountains. Fat line shows the northern limit of the Carpathian Mountains in Poland. CZ- Czech Republic, SK- Slovakia, UA- Ukraine.

Figure 1. Location of the study areas in southern Poland.

2. State of the research

While raised bogs in mountain areas have been the subject of research in a number of scientific disciplines, the evolution of raised bog geomorphology has not been adequately covered. Geomorphologists tend to focus on the difference between low bogs and raised bogs – especially those with a dome (e.g. Senft 1862, Gore 1983, Tobolski 2000, Ilnicki 2002). Other key areas of interest include raised bog relief with respect to the different varieties of bogs (ombrotrophic, soligenous, topogenous, river-fed) found across ridges, slopes and valley floors (e.g. Kaule and Göttlich 1976, Obidowicz 1985, Żurek and Tomaszewicz 1996, Tobolski 2000, Ilnicki 2002, Łajczak 2007, 2011, Obidowicz and Margielewski 2008). The following papers cover geomorphological bog classification systems for mountain areas: Früh and Schröter (1904), Sjörs (1948), Gams (1958), Barsiegan (1974), Kaule (1974), Kaule and Göttlich (1976), Ringler (1981), Obidowicz (1990), Łajczak (2007, 2011). Additional analysis of various aspects of geomorphological bog development can be found in: Bower (1961), Rawes (1983), Mallik et al. (1984), Evans (1989), Cooper and McCaan (1995), Rhodes and Stevenson (1997), Shaw et al. (1997), Bragg and Tallis (2001), Dykes and Warburton (2007), Łajczak (2007, 2011).

Four Holocene stages of raised bog development were identified for mountain areas (Łajczak 2005, 2007, 2011): (a) low bog growth, (b) peat dome growth, (c) human impact on raised bogs leading to complete deterioration, (d) revitalisation of remaining bog fragments. Each stage is shorter than the previous stage. A geomorphological analysis of peat bogs during each stage of development may be found in Łajczak (2005, 2007, 2011). Papers on human impact on peat bogs tend to focus on the geomorphological effects of peat extraction, drying and burning as well as the effect of grazing and erosion (e.g. Bower 1961, Rawes 1983, Mallik et al. 1984, Carling 1986, Evans 1989, Cooper and McCann 1995, Shaw et al. 1997, Rhodes and Stevenson 1997, Dykes and Warburton 2007, Łajczak 2007, 2011) but often omit a more detailed analysis of changes in peat bog relief. The issue of raised bog development across valley and basin floors in mountain areas has not been well investigated with respect to local relief and sources of water. In addition, the issue of changes in local relief and surface water drainage patterns resulting from peat dome growth has not been adequately investigated.

The best investigated peat bogs with respect to contemporary changes are those in Great Britain and Ireland, where a lot of attention has been paid to the decline of blanket bogs as a result of sheep grazing, peat burning, new drainage systems and to some extent peat extraction (Bower 1961, Mallik et al. 1984, Evans 1989, Shaw et al. 1997, Bragg and Tallis 2001). Papers on human-induced deterioration of peat bogs often focus on peat erosion and omit the issue of changing bog relief (Cooper and McCann 1995).

The most extensive research on peat bogs in the Polish Carpathians has focused on the Orawsko-Nowotarska Basin and the Bieszczady Range. The first area has been investigated since the early 19th century, while the second area since the 1950s. Until the 1980s, peat bog research focused only on bog paleogeography, peat properties and plant cover. The oldest carbon dated samples obtained from the bottom of peat domes range from about 2,000 to 11,000 BP (Ralska-Jasiewiczowa 1972, 1980, 1989, Obidowicz 1990, Haczewski et al. 1998). This indicates that the Holocene development of raised bogs in the two study areas was non-synchronous. Peat deposits vary in thickness (1.2 m to 3.6 m), which suggests they are of variable age. Peat domes also vary in size from 0.2 km to 6.0 km (Lipka 1999, Łajczak 2007). This is especially true in the Orawsko-Nowotarska Basin. The mean rate of vertical growth in raised bogs in the Polish Carpathians is estimated to be 0.4 to 0.6 mm a^{-1}. This value range is close to that for other European mountain areas (0.3 to 0.7 mm a^{-1}) (Żurek 1987). Low bogs developed first in the study area and filled in local depressions and then developed further into peat domes (Ralska-Jasiewiczowa 1972, 1980, 1989, Horawski et al. 1979, Wójcikiewicz 1979, Haczewski et al. 1998, Kukulak 1998, Lipka 1999, Łajczak 2006, 2007, 2009). Researchers began to address changes in bog relief in the Polish Carpathians in the last 20 years – especially with respect to Holocene evolution (Baumgart-Kotarba 1991-1992, Kukulak 1998, Haczewski et al. 1998, 2007) and human impact (Łajczak 2005, 2006, 2007, 2009, 2011).

3. Study area

The number of raised bogs in the Polish Carpathians is small compared to the northern lowlands of Poland featuring young Glacial relief (Żurek 1983, 1987, Dembek et al. 2000, Dembek and Piórkowski 2007). Most peat bogs in the Polish Carpathians are less than 1 hectare in area and only a few are larger than 100 hectares (Łajczak 2007, 2009, 2011). Polish Carpathian peat bogs are often found on ridges and in spring areas, moraine depressions as well as landslide depressions. However, the largest peat bogs in this region are found in the Orawsko-

Nowotarska Basin and in the largest valleys in the Bieszczady Range (Fig. 1). Peat bogs occur at lower elevations in mountain areas atop local drainage divides (ombrogenous bogs) and across slopes (soligenous or hanging bogs). Topogenous and river-fed bogs are found at the lowest elevations (Kukulak 1998, Haczewski et al. 1998, 2007, Margielewski 2006, Dembek and Piórkowski 2007, Łajczak 2007, 2009, 2011, Obidowicz and Margielewski 2008).

The Orawsko-Nowotarska Basin has an area of 600 km^2 and is the only intra-mountain basin in the Carpathian Mountains where raised bogs developed during the Holocene (Łajczak 2007, 2009). The Basin is located between a high mountain massif (Tatras) and the lower Beskidy Mountains and is tilted to the north. Peat bogs in the Basin developed across glaciofluvial fans and high Holocene terraces at elevations ranging from 590 m to 770 m. Bogs in the region are found between 5 m and 40 m over river channels. The mean peat thickness in domes exceeds 1 m and may reach 11 m. Raised bogs cover 5% of the Basin area. Low bogs cover 7% of the Basin area. The total peat bog area in the Basin may have reached 40% prior to human settlement in the Late Middle Ages. As settlers began to extract peat and dry peat areas, peat bogs began to shrink to a current 70 km^2, which includes dome remnants, post-peat areas and low bogs (Łajczak 2007). The European Drainage Divide runs across the Basin from south to north, separating drainage basins of the Black Sea and the Baltic Sea. The southern and western part of the Basin still experiences upward tectonic shifts, while its remaining area is shifting downward (Vanko 1988, Zuchiewicz 2010).

The bottom of the Upper San Valley and the bottom of the Wołosatka Valley are located at an elevation range of 550 to 700 m and have a total area of 13 km^2. The density of raised bogs in this region is much higher than that in the Orawsko-Nowotarska Basin. However, peat bogs in the Bieszczady Mountains are smaller and less deteriorated due to less peat extraction and less drying (Łajczak 2011). The remaining peat dome fragments and post-peat areas cover 4% of the valley floors in the study area and may be found on postglacial terraces and alluvial fans at heights at 5 to 8 m above river channels. Mean peat thickness in peat domes does not exceed 3 m. Today, the total area of peat domes, post-peat areas and adjacent low bogs does not exceed 1 km^2 (Łajczak 2011).

The parent material of peat bogs in both study areas is a layer of poorly permeable clay about 2 m thick. The clay is located atop water-bearing gravel. The edge zone of virtually every raised bog is recharged by shallow groundwater outflows. Given that precipitation in the study areas barely exceeds evaporation during the vegetation season, minerotrophic recharge must be considered a key determinant of bog development (Łajczak 2009).

4. Purpose of reserach and materials used

The purpose of the paper is to show how raised bogs in mountain valleys and basins develop during each of the four stages of bog development and how this affects local relief. The research was performed in two study areas in the Polish Carpathian Mountains (Fig. 1).

The paper is based on an analysis of maps from the last 230 years (Karte des Königreisches..... 1779-1782, Administrative Karte..... 1855, Die Spezialkarte..... 1894, Tactical Map..... 1937,

Topographic Maps 1965, 1997) and aerial photographs from 1965, 1988 and 2006. The maps and photographs show the shrinking process for each peat bog analyzed in the study area. In addition, extraction scarps and post-peat areas are analyzed. The paper also employs data obtained via fieldwork, which included peat bog and post-peat area mapping using GPS and morphometric measurements. The research was performed over the course of 15 years in the two study areas mentioned earlier (Łajczak 2007, 2009, 2011). Peat deposit thickness was ascertained via drilling. Maximum peat thickness data were obtained from the research literature (Horawski et al. 1979, Wójcikiewicz 1979, Baumgart-Kotarba 1991-1992, Kukulak 1998, Lipka 1999, Haczewski et al. 2007). Fieldwork focused on the location of peat deposit remnants outside of known peat areas, especially in areas where peat extraction was halted before 1850. This type of information makes it possible to make inferences about the previous extent of peat domes, which were often larger than that shown on the oldest maps (Łajczak 2007, 2009, 2011). The analysis of exhumed landforms in post-peat areas helps to identify places with the thickest peat deposits. Such places are understood to be the original peat formation sites. The research results were used to assess the most likely size of peat domes prior to human impact based on local relief and distribution of water phenomena.

5. Results

5.1. Reconstruction of raised bog range for the period prior to human impact

The oldest maps analyzed and the traces of peat found outside of contemporary post-peat areas suggest that 26 raised bogs may have existed in the Orawsko-Nowotarska Basin prior to human settlement (Fig. 2). The total area of raised bogs prior to human settlement has been estimated to be about 4,900 ha (Łajczak 2007). Three of the bogs were completely eliminated in the 19th century. Eighteen became smaller and some became fragmented. Only five of the bogs have remained in their natural state (Łajczak 2011). The raised bogs of the past covered a more topographically diverse landscape than do their fragments today (Horawski et al. 1979, Wójcikiewicz 1979, Baumgart-Kotarba 1991-1992, Lipka 1999, Łajczak 2007, 2009). This makes it possible to assess how raised bogs at the advanced stage of development are able to alter local relief. The two largest peat domes were most likely 1,000 ha in size. Nine peat domes ranged from 100 ha to 1,000 ha in area. The largest peat domes (dimensions: 5 x 2 km and 4 x 2 km) were some of the largest in modern-day Poland (Łajczak 2007). Transit streams flowing around peat bogs, especially in areas beyond the lowest parts of edge zones had a meandering pattern. The streams were recharged primarily by water seeping out of peat bogs. Raised bogs in the Orawsko-Nowotarska Basin sit atop fragments of Quaternary glaciofluvial fans of variable age. Some are found atop Holocene high terraces (Baumgart-Kotarba 1991-1992, Łajczak 2007, 2009). In general, the younger the fragment of Quaternary glaciofluvial fan, the more expansive the raised bogs used to be. This can be explained in terms of neotectonics, local relief and hydrogeological conditions (Fig. 2). Groundwater flows at greater depths in the western and southern parts of the Basin that are being lifted upward and fragmented by erosion. In turn, this does not favor bog growth. Groundwater in the lower part of the Basin can be found at shallow depths and groundwater outflows create wet conditions in the area, which in turn favors bog growth (Łajczak 2009).

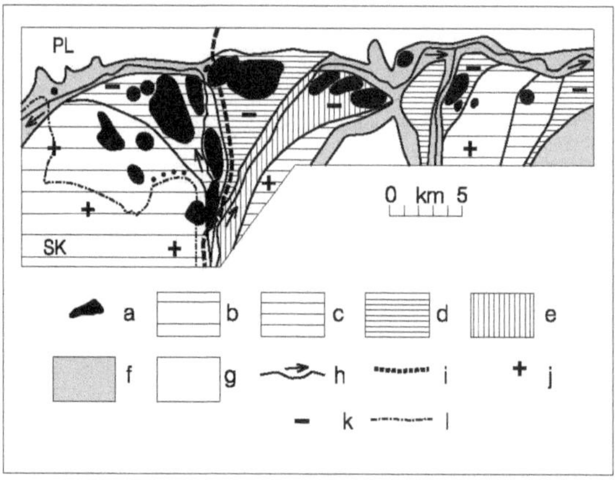

Figure 2. Probable range of raised bogs in the Orawsko-Nowotarska Basin in the period prior to human impact. The raised bogs are presented on the background of Quaternary landforms.a- raised bogs. Quaternary terraces within glaciofluvial fans: b- Mindel terraces, c- Riss terraces, d- Vistulian terraces, e- postglacial terraces, f- Holocene terraces. g- areas located outside the basin, h- main water-courses, i- European Drainage Divide, j- areas shifting upward, k- areas shifting downward, l- state border.

Seventeen raised bogs existed in the Upper San Valley and the Wołosatka Valley in the Bieszczady Mountains prior to human settlement in the 17th century (Kukulak 1998, Haczewski et al. 2007, Łajczak 2011). The 17 bogs had a total area of only about 60 ha and developed across topographically homogenous terrain – often close to streams – on high terraces and alluvial fans (Fig. 3).

5.2. Distribution of peat bogs at different elevations

Eight types of geomorphological situations were identified for raised bogs location at different elevations in the study areas (Fig. 4A). Each type of bog is listed starting at high elevations and ending with low elevations. Their spatial distribution within both studied areas is shown in Fig. 4B. In each geomorphological situation, expanding peat bogs alter relief in a different way (stages "a" and "b" in Łajczak 2005, 2007, 2011). This process is perturbed or halted as a result of human impact – stage "c". While currently almost all of the bogs are classified as ombrogenous or ombrogenous-soligenous using the Kaule and Göttlich (1976) classification system, each group of peat bogs was recharged by water in a variety of ways during its unique development stage.

The first group of peat bogs (I) includes five bogs located atop a drainage divide and are found only in the Orawsko-Nowotarska Basin on ridges 5 to 40 m over adjacent surfaces (Baumgart-Kotarba 1991-1992, Lipka 1999, Łajczak 2005, 2007, 2009). Group I bogs were

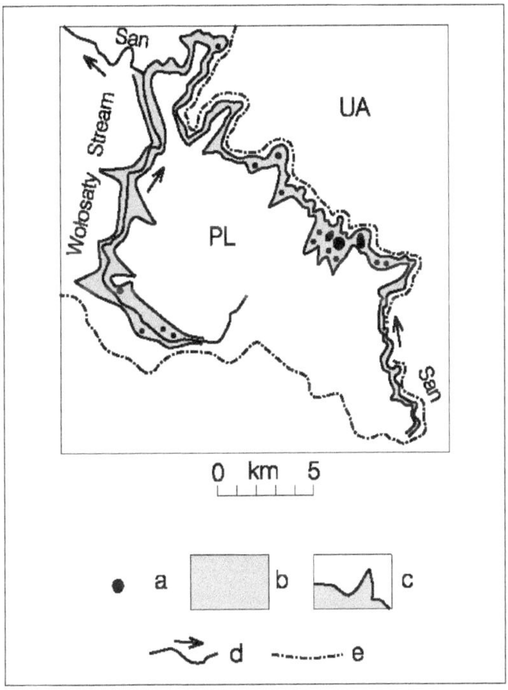

Figure 3. Probable range of raised bogs in bottoms of the Upper San and Wołosatka river valleys in the Bieszczady Mountains. a- raised bogs, b- bottoms of river valleys, c- limit of larger alluvial fans, d- main water-courses, e- state border.

soligenous bogs during the early stage of development. The second group of peat bogs (II) includes eight bogs in spring areas in shallow erosion incisions or at the bottom or on the sides of erosion incisions (Orawsko-Nowotarska Basin) (Łajczak 2007, 2009). Group II bogs were soligenous or river-fed bogs during the early stage of development. The third group of peat bogs (III) includes six bogs in the Orawsko-Nowotarska Basin and one bog in the Bieszczady Mountains. Group III bogs developed in old river channels found on Riss, Vistulian and older Holocene terraces (Baumgart-Kotarba 1991-1992, Kukulak 1998, Łajczak 2005, 2007, 2009, Haczewski et al. 2007). Group III bogs then transformed into river-fed bogs, topogenous bogs, soligenous bogs and finally into ombrogenous bogs. Group IV includes four bogs found on terraces of variable age near the base of the edge of the next higher terrace (Łajczak 2005, 2007). All four are found in the Orawsko-Nowotarska Basin. Group IV bogs were soligenous and later river-fed bogs in the early stage of development. Group V consists of just one bog in the Orawsko-Nowotarska Basin, which had developed on an expansive and uniformly tilted fragment of the Vistulian Terrace. This bog was soligenous at first and then became river-fed. Group VI can be found only in the Bieszczady Mountains and consists of just one bog on an alluvial fan (Łajczak 2011). The bog started

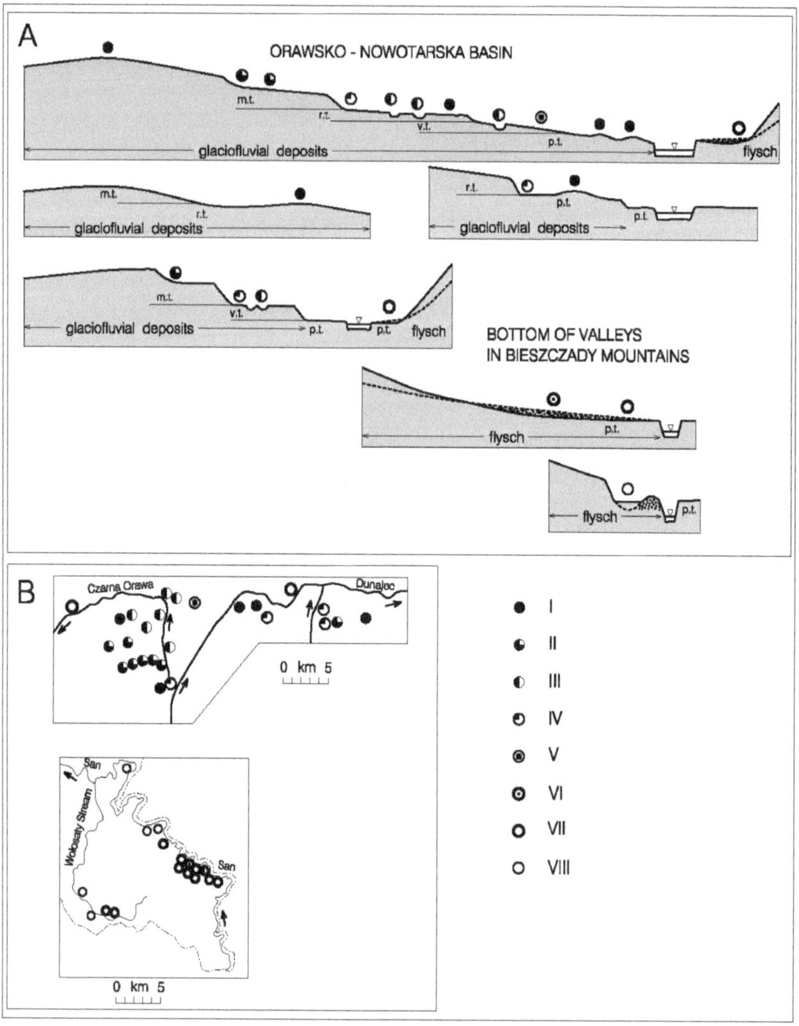

Figure 4. Geomorphological location of identified eight groups of raised bogs in the study areas. A- distribution of peat bogs at different elevations, B- their spatial distribution within both areas. For numbering of peat bog groups (I-VIII) – see the text. Terraces: m.t.- Mindel, r.t.- Riss, v.t.- Vistulian, p.t.- postglacial.

out as a soligenous bog. Group VII is the largest of the groups and includes 12 raised bogs found at the edges of alluvial fans (Łajczak 2009, 2011). Ten of the bogs are found in the Bieszczady Mountains and were initially soligenous. Group VIII is found at the lowest elevations and includes five bogs in the Bieszczady Mountains. The bogs fill in oxbow lakes on the postglacial terrace between an inactive levee and an undercut flysch slope (Kukulak

1998, Haczewski et al. 2007, Łajczak 2011). The bogs were river-fed at first and then remained both ombrogenous and soligenous throughout their period of development.

5.3. Changes in peat bog relief during the first stage of development

The first stage of development of the studied bogs consisted of the formation of a low bog. In the case of bogs located atop drainage divides, the first stage of development included convex landforms, while other types of bogs developed in concave landforms (Ralska-Jasiewiczowa 1972, 1980, 1989, Kukulak 1998, Łajczak 2005, 2007, 2011, Haczewski et al. 2007). At this stage of development, bogs in Group I began to evolve in a way that included increasing differences in local elevation. On the other hand, other groups of peat bogs evolved in a completely different manner by reducing differences in local elevation. This process continued until low bogs filled in concave landforms (Fig. 5). This stage was dominated by soligenous bogs, with some river-fed bogs and topogenous bogs. Even bogs growing on convex landforms were initially recharged by shallow groundwater outflows. As the low bog became thicker and its surface farther removed from minerotrophic waters, oligotrophication and acidification of the site began to occur, leading to the development of a raised bog (Ralska-Jasiewiczowa 1989, Kukulak 1998, Łajczak 2005, 2007).

In raised bogs located atop drainage divides (I), the initial stage of development affected the entire cross section of low ridges. Only the tops of higher ridges were affected. Some Group II bogs were hanging bogs during their initial stage of development. This may be inferred from the presence of modern-day hanging bogs in the area that have not yet proceeded to the raised bog stage. Low bogs developed downstream of springs and expanded around them, although the principal direction of expansion remained downstream (Łajczak 2005, 2007, 2009). Low bogs in Group III began to develop after local streams dried up and became filled with fine-grained sediments featuring shallow groundwater. The initial stage of development of Group IV bogs occurred around spring niches at the base of a scarp of an upper terrace as well as in stream channels fed by these same springs during the Holocene. Such sites became collection points for poorly permeable clayey sediments carried in by sheet wash. Further low bog development encompassed ever larger parts of terraces (Łajczak 2005, 2007, 2009). A Group V raised bog began to develop in an area with numerous springs and over time began to cover the downstream parts of stream channels. A Group VI raised bog began to develop in an area with a gap in the poorly permeable layer of clay sitting atop gravel forming the alluvial fan. This type of situation created the right conditions for shallow groundwater to exit the ground under pressure. Group VII bogs located at lower elevations did not form due to river flooding but due to numerous springs at the base of alluvial fans. The development of these low bogs once again led to the accumulation of peat in various concave landforms situated mainly at lower elevations (Łajczak 2005, 2007, 2009). The first stage of bog development (VIII) at lower elevations was accompanied by the last stage of oxbow lake sediment accumulation (Kukulak 1998, Haczewski et al. 2007, Łajczak 2011). At first, the bogs were periodically flooded. However, the bogs were always recharged to some extent by groundwater from an undercut slope located nearby. This remains true today.

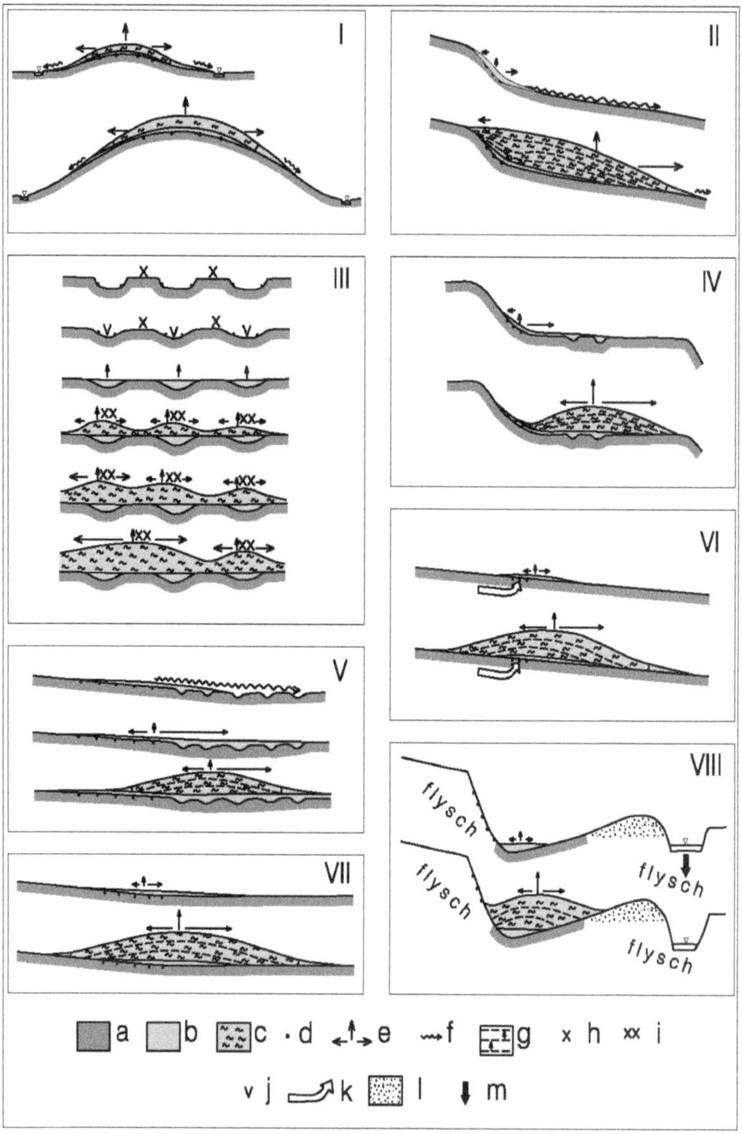

Figure 5. Scheme of growth of distinguished raised bog groups. For numbering of peat bog groups (I-VIII) – see the text. a- sub-peat material, b- low bog material, c- peat typical for raised bog, d- shallow ground water outflow, e- directions of low bog and raised bog expansion, f- surface water outflow, g- vertical peat dome growth, h- local drainage divide lines in Early Holocene, i- shifted local drainage divide lines, j- dried stream channels and filled with fine-grained sediments, k- places of shallow ground water outflows on alluvial fans, l- levee, m- channel deepening during the Holocene.

5.4. Changes in peat bog relief during the second stage of development

The growth of peat domes across low bogs marks the second stage of bog development, which can be interrupted or halted by human impact. The second stage began at different times for different bogs in the study area. Nevertheless, this stage of development of most bogs started during the Atlantic Period or earlier (Ralska-Jasiewiczowa 1972, 1980, 1989, Obidowicz 1990, Kukulak 1998, Haczewski et al. 2007). The second stage produced much larger changes in relief than the first stage (Łajczak 2005, 2007) (Fig. 5). The key change was fossilization of concave landforms, which became filled in by low bogs and then transitioned into raised bogs. Peat dome growth led to the formation of convex landforms atop formerly concave landforms. Other effects included the shifting of local drainage divides and a marked decrease in the density of local streams flowing close to expansive peat domes. Streams flowing in the vicinity of growing peat domes also changed course. Another tendency in raised bog development is the shift towards lower elevations, which now feature thicker peat deposits. This shift started already at the first stage of development. In effect, the thickest peat deposits are found relatively far away from the original peat formation site (Łajczak 2005). Hence, peat dome development creates increasing differences in local elevation. The opposite trend was found to be true for the first stage of bog development. However, each peat bog is different and may exhibit unique changes in relief development.

Growing peat domes covered the tops and sides of ridges found on drainage divides and their edge zones approached nearby stream channels. Higher ridges became covered by peat domes only at the top, while dome edge zones covered the upper parts of gentle slopes. The thickest peat deposits – formerly more than 6 m thick and currently up to 4 m thick – formed atop a drainage divide (Lipka 1999, Łajczak 2005, 2007). As peat domes continued to grow, so did local differences in elevation. In places with low bogs filling spring niches, growing peat domes filled in erosion incisions and created small hills in some places. As peat domes grew, their thickest deposits were to be found downslope. In such cases, the edge zone covered shallow depressions between domes and higher sections of mineral parent material. These areas are recharged by groundwater outflows and possess edge streams and larger transit streams as well. The development of raised bogs in this area also leads to larger local differences in elevation. The path of development for raised bogs in old stream channels was similar. Growing peat domes covered even neighboring erosion incisions and often joined other peat domes to form expansive domes that mask the morphologically diverse parent surface (Baumgart-Kotarba 1991-1992, Łajczak 2007). The thickest peat deposits (up to 11 m) were found at locations where the dome peaks sit atop the deepest old stream channels. In the fourth group of bogs found on high terraces at the base of the edges of even higher terraces, peat domes developed far away from groundwater outflows and cover old stream channels of a rather small size. At these sites, the peat thickness exceeds 6 m. The development of the Group V peat bog followed a similar path. Maps from 1779-1782 and 1855 show that it used to be surrounded by a wide swath of low bogs. Edge streams and larger transit streams beyond the low bogs followed a meandering course. The expansion of raised bog on the alluvial fan was limited by the presence of larger transit streams. On the other hand, the expansion of peat bogs across the lowest parts of the alluvial fans was not limited by any topographic barriers. The growth of the peat dome tends to smooth out the local land surface up to a certain point – peat deposit 5 m thick or more – at which it leads to increasing local differences in elevation. The development of peat domes in the group located at the lowest elevations also leads to increasing local differences in elevation (Kukulak 1998, Haczewski et al. 2007, Łajczak 2011). This group of raised bogs has already reached its maximum extent, as its edge zone runs along the foot of an undercut slope and a levee on the other side.

Growing peat bogs may strongly affect the network of local stream channels. The development of low bogs can affect the course of small streams. Peat also fills in oxbow lakes. At the advanced stage of raised bog development, the stream network becomes substantially reorganized. Peat domes cover some stream channels and some streams are forced to shift away from the dome (Łajczak 2007). Such streams become edge streams flowing around the peat dome. These streams are narrow and cut relatively deep into peat deposits in many cases. As peat domes expand, the thickest peat deposits tend to be found at increasingly lower elevations. This forces edge streams to quickly shift downslope. Larger transit streams are found beyond the edge zone of the peat dome and may limit dome expansion depending on their size. These streams and edge streams were recharged prior to human impact by numerous short tributaries seeping out of peat domes and flowing across the muddy edge zone. In the study area, the edge zones of many bogs approached small streams but remained 300 m or more away from larger rivers. Streams of varying size flowing outside of the peat edge zone, especially at lower elevations, tend to meander. The channels of transit streams flowing near the largest peat bog in the Orawsko-Nowotarska Basin are as much as six meters lower than the old stream channels masked by the expansive peat dome (Baumgart-Kotarba 1991-1992) (Fig. 6). This suggests that these large streams became much deeper during the Holocene in the absence of peat formation.

5.5. Human impact on peat bog relief

Prior to the introduction of agriculture in the Orawsko-Nowotarska Basin towards the end of the Middle Ages, raised bogs most likely occupied about 10% of the Basin, while low bogs may have occupied as much as 30% of the Basin. In the valleys studied in the Bieszczady Mountains, the numbers were closer to 6% and 4% (Łajczak 2007, 2011). Some fragments of the two study areas were already largely covered by peat bogs (Figs 2, 3). In the Orawsko-Nowotarska Basin, incoming settlers began to clear low bogs by burning the peat. In the 18th century, peat extraction began at the edges of peat domes. The peat was used to heat homes. Peat extraction intensified between the mid-19th century and the late 20th century. Peat extraction usually started at the edge of the dome and continued towards the center and normally did not involve the entire dome all at once. Peat dome burning continued until the early 1900s. In the 1950s, industrial-scale peat extraction began at three peat bogs in order to serve the gardening needs of Polish consumers. Drainage work began at the same time around the edges of peat bogs and stream channels became regulated, which led to the drying of large parts of the bogs. This caused a more than three-fold reduction in the low bogs' total area. Raised bogs became reduced 60% (Łajczak 2007, 2011) (Fig. 7). Human impact began to reduce the extent of raised bogs in the Bieszczady Mountains starting in the 19th century. The reductions ended in the

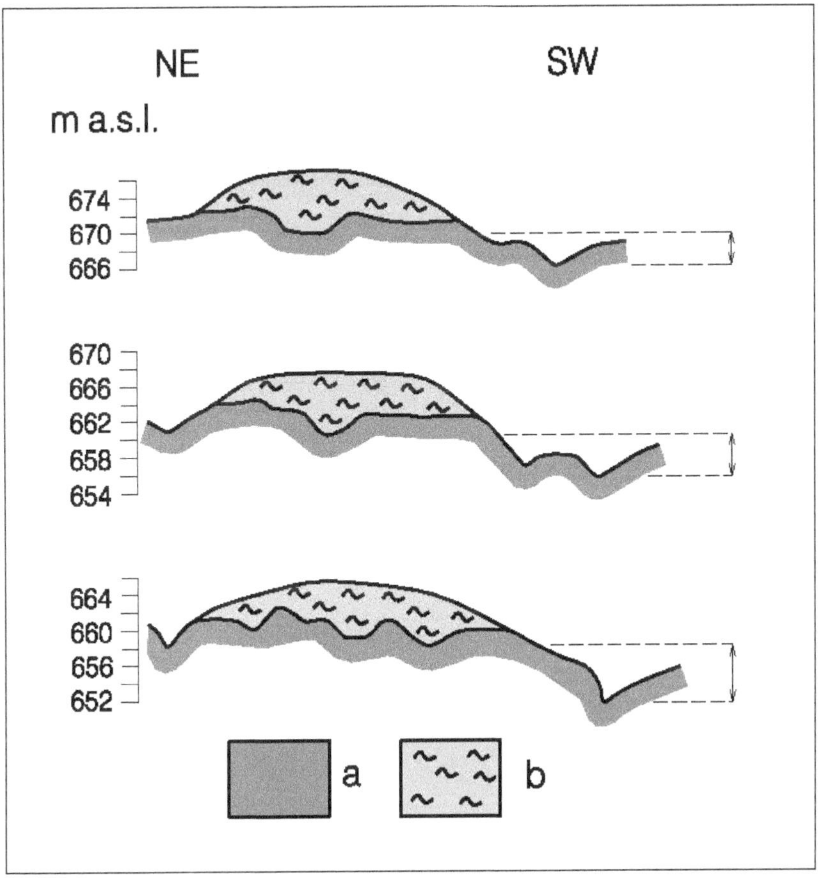

Figure 6. Chosen cross-sections through the largest raised bog in the Orawsko-Nowotarska Basin. a- sub-peat material, b- peat deposit. Differences between elevation of fossilized channels and active stream channels are marked.

1950s. The edges of these bogs were later dried (Łajczak 2011). Today peat bogs in the valleys of the Bieszczady Mountains are protected by law, which makes bog revitalisation possible. Only one large bog in the Orawsko-Nowotarska Basin is protected by law. Almost all others are no longer experiencing human impact and are slowly regenerating.

Human impact on raised bogs helps create the following landforms: 1) shallow hollows after the surface layer of a peat dome has been burned off or after the entire peat deposit in a low bog has been burned off, 2) peat extraction pits of varying size and shape, 3) drainage ditches, 4) regulated and/or straightened stream channels (e.g. Rawes 1983, Mallik et al. 1984, Evans 1989, Cooper and McCaan 1995, Rhodes and Stevenson 1997, Shaw et al. 1997, Bragg and Tallis 2001, Łajczak 2007, 2011, Latocha 2012). Peat extraction alters bog relief in the most visible of

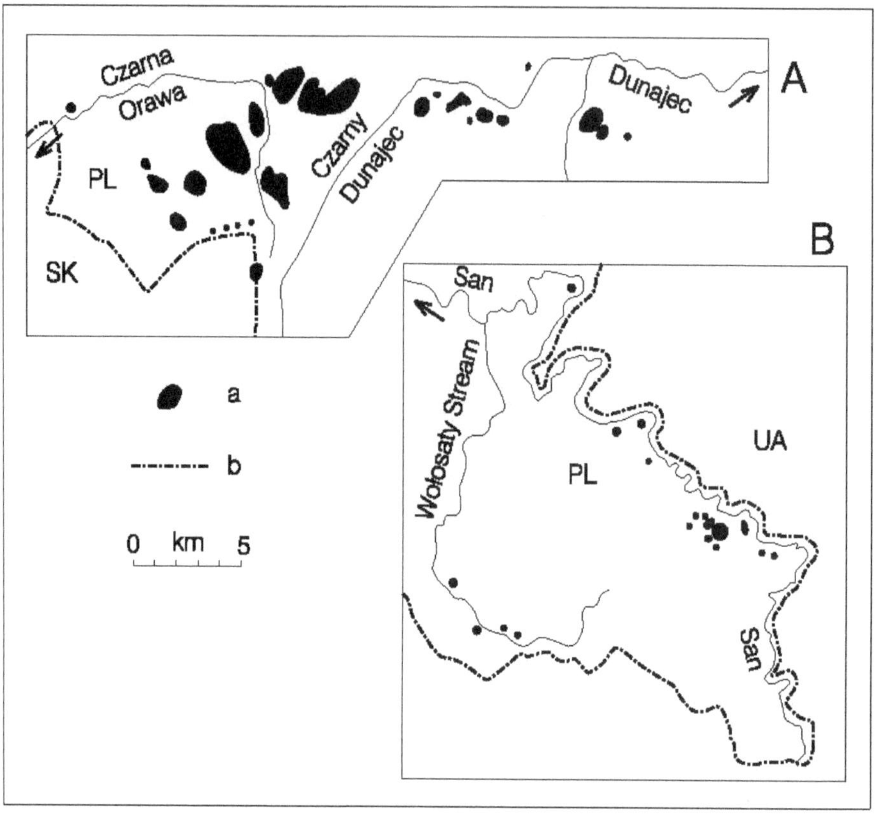

Figure 7. Actual range of remnants of peat domes in the study areas. a- remnants of peat domes, b- state border.

ways (Fig. 8). Extraction from the edges towards the center of the dome produces one type of bog relief, while the opposite direction of extraction produces another type of bog relief. Peat extraction leads to the fragmentation of some peat domes. One dome in the Orawsko-Nowotarska Basin has broken up into three fragments (Łajczaka 2011).

In areas where peat extraction had been taking place for very many years, the following landforms can be observed: 1) older post-peat areas with occasional traces of peat that are used for agricultural purposes, 2) younger post-peat areas with reduced but continuous peat deposits, 3) peat domes reduced to peat remnants, 4) active industrial-scale extraction areas that yield large depressions atop peat domes that usually link with younger post-peat areas, 5) extraction scarps or post-extraction scarps that separate peat dome remnants from younger post-peat areas as well as expansive depressions atop peat domes (Łajczak 2011). In the Orawsko-Nowotarska Basin, older post-peat areas formed not later than the mid-19th century and mark areas previously occupied by low bogs, edge fragments of raised bogs and three

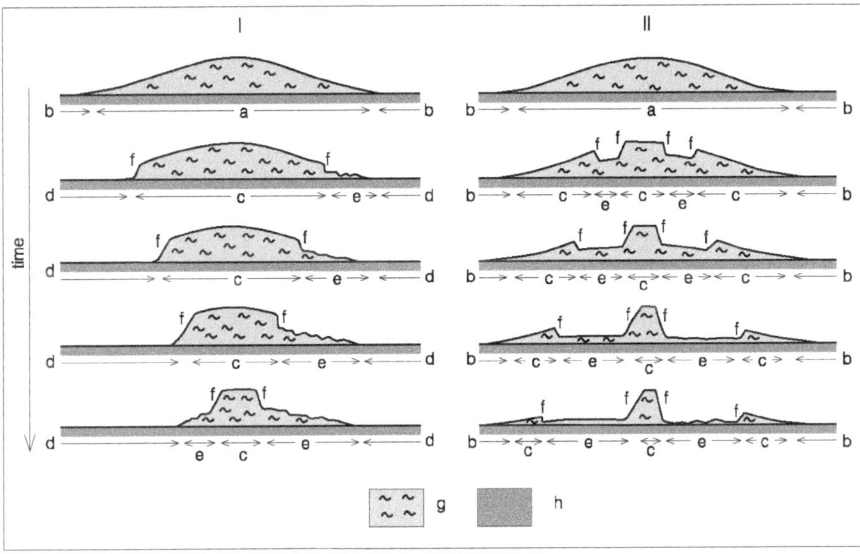

Figure 8. A schematic diagram illustrating the decrease of the range of peat dome as a result of peat extraction. I- peat extraction from the edge towards the center of the dome, II- opposite direction of peat extraction. a- peat dome, b- low bog, c- remnant of peat dome, d- older post-peat area, e- younger post-peat area, f- extraction scarp or post-extraction scarp, g- peat deposit, h- sub-peat material.

entirely destroyed raised bogs. There are no older post-peat areas in the Bieszczady Mountains. Older post-peat areas feature exposed mineral parent material where landforms can be observed that served as potential starting points for peat formation (Łajczak 2006). In younger post-peat areas, the reduced peat layer features a diverse surface with numerous low scarps, pits filled with water and peat deposits overgrown with moss. Younger post-peat areas occupy a much larger area in the Orawsko-Nowotarska Basin than in the Bieszczady Mountains. In the Orawsko-Nowotarska Basin, younger post-peat areas are surrounded by wide older post-peat areas. Existing fragments of peat domes possess virtually fully natural tops and are surrounded by extraction scarps or post-extraction scarps. The scarps can be as high as 6 m and are either fully vertical or stair-shaped. In bogs where most of the peat has been extracted, reduced peat domes take the form of narrow peat remnants. In the Orawsko-Nowotarska Basin, peat dome remnants are much smaller than the original domes. However, in the Bieszczady Mountains, peat dome remnants are only slightly smaller than the original domes (Łajczak 2011). Expansive depressions found atop peat domes have formed only in three peat bogs in the Orawsko-Nowotarska Basin. The depressions occupy no more than 20% of the existing domes' surface and can be as deep as 4 m. Each depression is ringed by vertical scarps and drained by a dense network of drainage ditches (Łajczak 2007, 2011). In the Orawsko-Nowotarska Basin, scarps surrounding peat remnants tend to zigzag, while in the Bieszczady Mountains, the scarp geometry is either bent or circular. In most of the investigated peat bogs where peat extraction had proceeded from the edge towards the center of the peat dome, the aforementioned elements of the morphology of damaged raised bogs tend to form a circular pattern around the peat dome remnants. In peat bogs where peat extraction had proceeded all over the place, the circular pattern does not exist (Fig. 9).

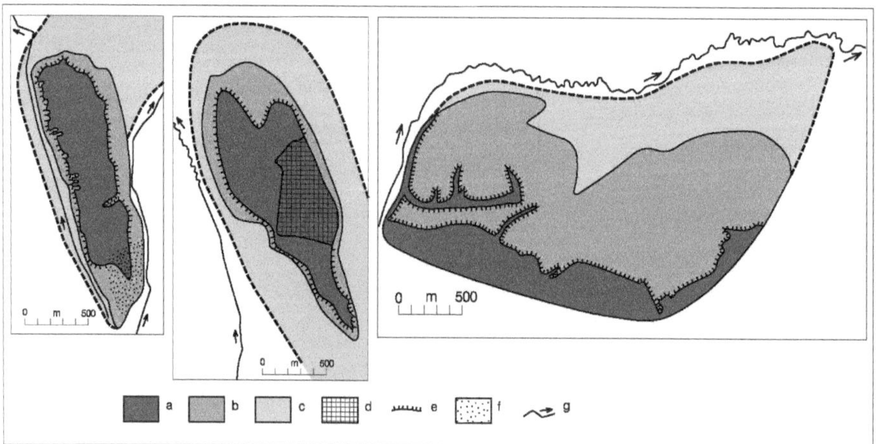

Figure 9. Distribution of main morphological elements within former large raised bogs – examples from the Orawsko-Nowotarska Basin. a- remnant of peat dome, b- younger post-peat area, c- older post-peat area, d- industrial-scale extraction area that yields large depression atop peat dome, e- extraction scarp or post-extraction scarp, f- part of a bog which has preserved natural character, g- main water-courses.

The most visible and most rapidly changing elements of relief in bogs affected by human impact are extraction scarps (Łajczak 2007, 2011, Latocha 2012). The edges of drained areas are also surrounded by scarps but they are lower. The depressed surface with dried peat is often separated from peat saturated with water by a large ditch. When peat extraction comes to an end, the post-extraction scarp changes along its vertical axis, as illustrated over time by Figure 10. The drying of peat on initially vertical walls of the scarp leads to fractures in the peat deposit and to peat sliding downward where it is washed away during snow melting periods, mainly. Peat mud fills numerous pits in younger post-peat areas. Peat hanging over the declining scarp deteriorates over time and the scarp becomes flat. A fully overgrown former scarp assumes a convex-concave shape with a small gradient. This shape becomes even smoother over time as extraction pits become overgrown and new deposits form. Cartographic materials, old photographs, and the opinions of persons involved in peat extraction indicate that post-extraction scarps maintained their vertical walls for ten years after extraction ceased in the Orawsko-Nowotarska Basin. The more time passes since the end of peat extraction, the more a post-extraction scarp resembles a mature scarp. Phase "c" scarp is about twenty years older than phase "b" and phase "d" scarp is between 30 and 60 years old. Scarps in existence more than 60 years since the end of peat extraction are designated "e" or "f". A mature convex-concave peat dome cross section can be found only in the case of one peat dome in the Orawsko-Nowotarska Basin. This peat dome has been protected by law since the 1920s (Łajczak 2006).

Scarp relief transitions from phase "a" to "e" or "f" most rapidly on southern and southwestern "warm" slopes of the peat dome. The slowest rate of change occurs on the opposite slopes. This suggests that peat is washed away during early spring snow melting periods, mainly.

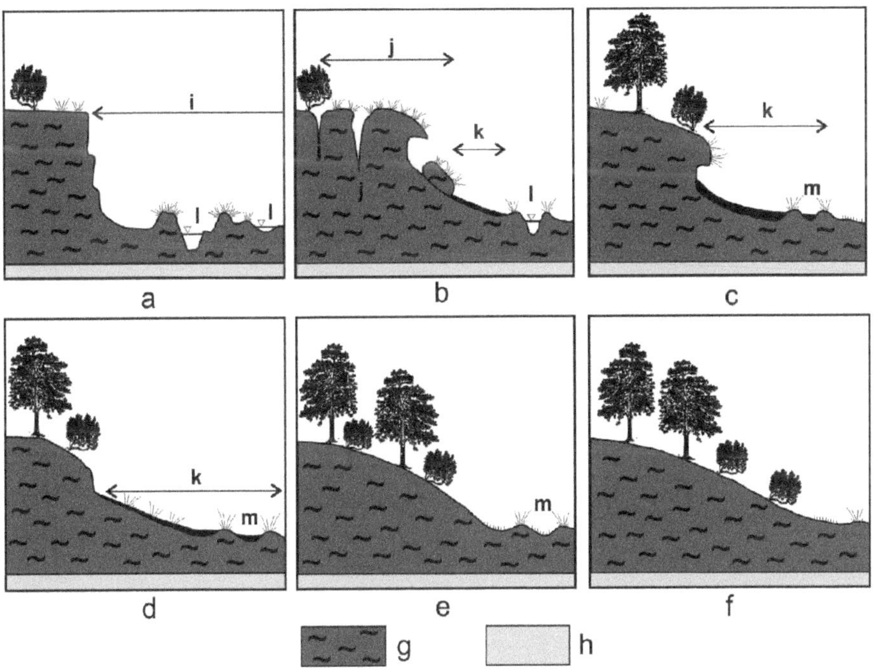

Figure 10. Changes in relief of peat bog scarp since peat extraction is halted. a-f- phases in scarp relief changes, g- peat deposit, h- sub-peat material, i- younger post-peat area, j- bog slides, k- bogflows, l- peat hollows with water, m- shallowed hollows without water.

Edge streams, which used to flow around peat domes, became deeply incised ditches ringing the peat dome and linked with large regulated streams as well as short ditches draining younger post-peat areas and peat dome remnants (Fig. 11). The purpose of the drainage work was to dry the wet edge zone and younger post-peat areas as well as to accelerate water drainage away from the peat bog (Łajczak 2007). The following factors contributed to increasingly abrupt water discharge during flood events: 1) complete extraction of peat deposits across large older post-peat areas, with poorly permeable clayey parent material becoming exposed, 2) some extraction of peat deposits in younger post-peat areas and peat dome remnants, 3) straightening of stream channels, 4) increases in stream gradients. The result is the formation of gravel-bottom braided channels in the case of even small streams with a local tendency to aggradation. This is a sharp contrast to the earlier sinuous stream channels with a stable cross section (Łajczak 2007, 2011).

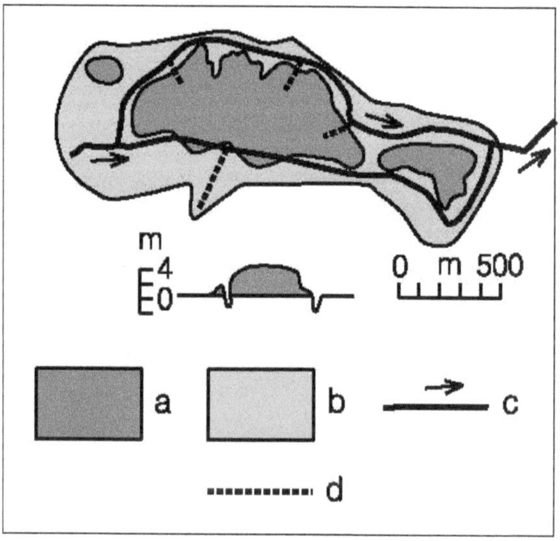

Figure 11. Example of ditches draining anthropogenically disturbed raised bog in the Orawsko-Nowotarska Basin: plan-view and profile. a- remnant of peat dome, b- younger post-peat area, c- girdling ditch, d- other ditch.

5.6. Peat bog revitalisation

Peat extraction has been declining in the Orawsko-Nowotarska Basin for more than two decades. This type of human impact has ceased to exist in the Bieszczady Mountains (Łajczak 2007, 2011). Drainage ditches are overgrown with vegetation due to a lack of maintenance and are effectively retarding the flow of water. This helps create wetlands in younger post-peat areas, which are now becoming a secondary edge zone. Peat moss takes about three years to colonize fresh peat pits filled with water. The increasing sinuosity of stream channels regulated in the past helps to make secondary edge zones more wet. Streams become more sinuous as water undercuts stream channel banks, which leads to more shallow stream channels. Beaver dams built near peat bogs in the Bieszczady Mountains provide another means of retaining water in post-peat areas. Small manmade dams in the region perform the same function (Łajczak 2011). The increasingly wet secondary edge zone and the increasingly flat post-extraction scarp help make peat dome remnants more wet, which prevents the drying of peat and facilitates the growth of peat moss. The cross section of a raised bog at this stage of development is different than that at previous stage of bog development (Fig. 12). Differences in elevation across post-peat areas initially become smaller during the last stage of raised bog development. As the peat dome grows, so do differences in elevation. However, this process may be disrupted once again if more peat is extracted and dried.

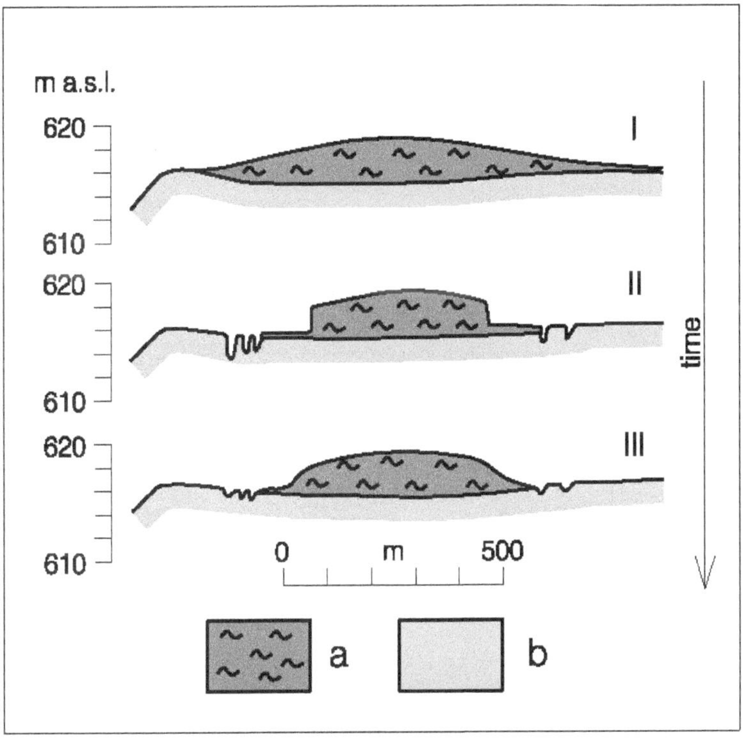

Figure 12. Cross-profiles through Bór na Czerwonem raised bog in the Orawsko-Nowotarska Basin in the periods: I-prior to human impact, II- at the end of peat extraction, and III- at the beginning of revitalisation process. a- peat deposit, b- sub-peat material.

6. Relief development patterns for raised bogs affected by human impact

Figure 13 shows changes in the extent and relief of a raised bog experiencing human impact. Period I shows a pre-human impact state. Period II shows an extraction and drying state. Period III shows the initial bog revetalisation state. Younger and older post-peat areas indicate areas of losses within the peat dome and the edge zone (period II). This was an area of stream channel regulation and drainage ditch construction. Extensive peat extraction primarily along the edges of the peat dome led to major changes in peat bog relief and major losses of water supplies (Łajczak 2007, 2009, 2011). Increases in the density of the drainage network surrounding peat dome remnants led to further drying of peat. An unintended consequence of stream channel regulation was streams becoming more shallow and wider. Another consequence was stream channels evolving into braided stream channels with a local tendency to aggradation. Today peat extraction has ended at most sites and drainage ditches are no longer being maintained and are becoming more shallow. This helps make younger post-peat areas more wet, which helps them evolve into secondary edge zones. Another element of peat dome revitalisation is post-extraction scarps becoming more flat.

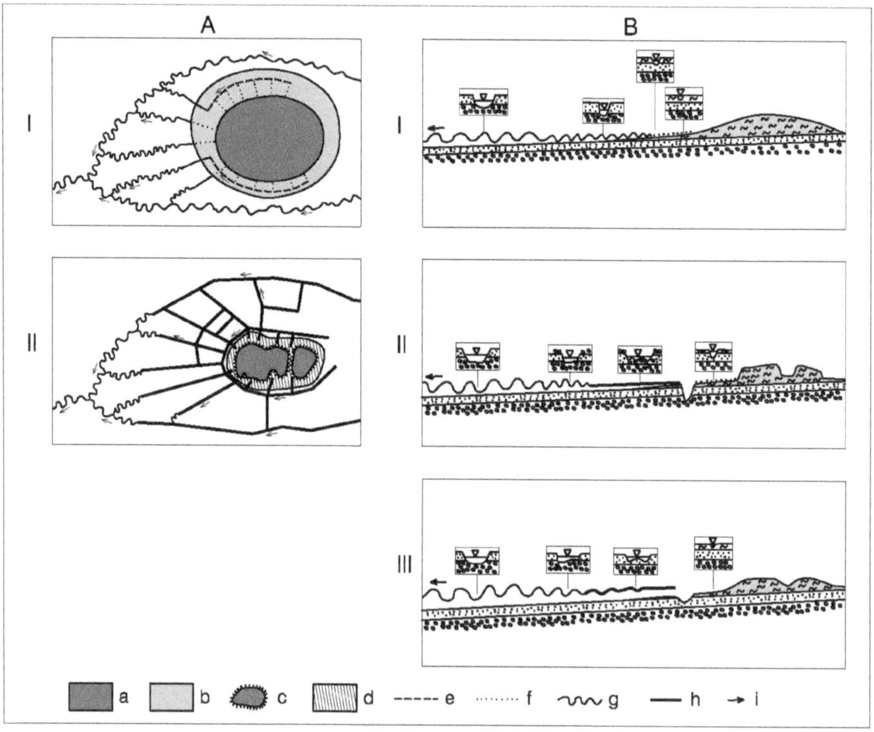

Figure 13. Typical changes in the extent and relief of a raised bog experiencing human impact. A- plan, B- profile. Periods: I- pre-human impact state, II- extraction and drying state, III- initial bog revitalisation state. a- peat dome, b- peat dome edge zone, c- remnant of peat dome surrounded by exploitation scarp, d- younger post-peat area, e- edge stream on outside of dome, f- short stream seeping out of peat dome and flowing across the muddy edge zone, g- meandering stream outside peat bog, h- ditch, i- direction flow. Schematic cross-sections of stream channels and ditches at various stages of their development are presented.

7. Discussion

The paper focuses on changes in raised bog relief in the Polish Carpathian Mountains. It documents bog characteristics that have not been documented before. The investigated peat bogs can be classified as valley-type based on their geomorphology (Ilnicki 2002), although each bog developed in a different mesoform. Raised bogs in the study area are not found exclusively on visible drainage divides, as other researchers seem to indicate (Tobolski 2000,

Ilnicki 2002), but tend to be found at lower elevations. Raised bogs with large peat domes may develop at any elevation in the study area. However, concave landforms are more likely to host peat bogs. This includes spring niches, old stream channels, the base of scarps of higher terraces, and the edges of alluvial fans. Numerous and stable groundwater outflows present within such landforms create the right conditions for low bogs to develop. As raised bogs evolve over time, these outflows maintain a high moisture level in the edge zone (Łajczak 2007, 2009).

Four stages of geomorphological development were identified for raised bogs in the study area. The last two stages are associated with human impact. Stage one is low bog development. Stage two is peat dome development. Peat domes grow depending on the relief of parent material and access to water. Gore (1983) as well as Obidowicz and Margielewski (2008) present a structural scheme of a large raised bog. The paper analyzes the geomorphological development of raised bogs found in a variety of mountain settings (e.g. valleys, basins) as well as analyzes peat bog development prior to human impact. These issues have been discussed only by a small number of researchers thus far (Kaule and Göttlich 1976, Rawes 1983, Obidowicz 1985, Carling 1986, Rhodes and Stevenson 1997, Bragg and Tallis 2001, Dykes and Warburton 2007, Łajczak 2007, 2011, Obidowicz and Margielewski 2008). New knowledge presented in this paper includes trends in bog development during the first and second stage of development relative to stable groundwater outflows facilitating bog formation. Assuming the view of Kaule and Göttlich (1976), raised bogs became ombrogenous-soligenous bogs at this stage, given that edge zones are still largely recharged by groundwater outflows.

The research literature tends to focus on historical and contemporary changes in peat bog relief caused by human impact (Bower 1961, Rawes 1983, Mallik et al. 1984, Evans 1989, Cooper and McCaan 1995, Rhodes and Stevenson 1997, Shaw et al. 1997, Bragg and Tallis 2001, Dykes and Warburton 2007, Łajczak 2007, 2011) in the form of sheep and cattle grazing, peat burning and peat drying. Peat erosion is of particular interest. However, a more in-depth analysis of contemporary changes in peat bog relief is difficult to find. This is especially true of papers published in the British Isles (Bower 1961, Evans 1989, Shaw et al. 1997, Bragg and Tallis 2001). In the study areas covered in this paper, peat extraction and drying are the main determinants of change in raised bog relief caused by human impact. Post-peat areas become larger and peat domes become smaller due to peat extraction by private landowners and industrial companies. Peat extraction, however, is on the decline. The paper also discusses changes related to the third stage of peat bog development by showing how just one form of human impact (e.g. peat extraction) can produce a variety of geomorphological effects based on how and when the impact had occurred.

The Polish research literature rarely covers ongoing changes in raised bog development – classified as stage four in this paper. The most important observations in this respect are the formation of a secondary edge zone in younger post-peat areas featuring shallow overgrown drainage ditches and peat pits as well as post-extraction scarps becoming more flat. Both processes assist in peat dome development (Łajczak 2007, 2011). On the other hand, the British research literature tends to focus on ongoing changes in peat bogs currently used for commercial purposes. In Great Britain and Ireland, both machine-based and manual harvesting of peat produce landforms such as scarps and peat pits that maintain sharp contours for long periods of time (Cooper and McCann 1995, Latocha 2012). In addition, the end of sheep grazing does not lead to a rapid smoothing of landforms produced by trampling (Rawes 1983). While the rate of relief change in post-peat areas in the British Isles is rather slow, the corresponding rate for scarps and peat pits in raised bogs in the Polish Carpathians is rather fast. Latocha (2012) writes about post-extraction depressions in blanket bogs in Ireland, which are still ringed by vertical scarps, even though peat extraction had ended more than 50 years ago at a number of these sites. The scarps in Ireland are stabilized by rapid grass growth. However, older peat pits are much more shallow than younger peat pits, as their bottom is always wet. In the study area in the Polish Carpathians, scarps become overgrown mainly by bushy plants and pine and this takes more time. On the other hand, peat moss first encroaches upon peat pits and drainage ditches (Łajczak 2007, 2011). The burning of peat is a key factor behind the deterioration of upland and mountain blanket bogs in the British Isles (Rhodes and Stevenson 1997). However, this factor ceased to be a key factor in the Polish Carpathians in the early 20th century (Łajczak 2007, 2011).

8. Conclusions

The most important the author`s findings are:

- the younger the fragments of Quaternary accumulation landforms in the studied areas, the more expansive the raised bogs used to be,
- almost of the bogs are classified as ombrogenous or ombrogenous-soligenous,
- the key change during the first two phases of peat bog relief development is fossilization of concave landforms, which become filled in by low bogs and then transitioned into raised bogs,
- another tendency in raised bog development is the shift towards lower elevation which now feature thicker peat deposits,
- among various manners of human impact on the peat bog relief for the last centuries, the peat extraction alters bog relief in the most visible of ways,
- the most visible and most rapidly changing elements of relief in bogs affected by human impact are extraction scarps,
- since the second halt of the 20th century the younger post-peat areas are more wet, which helps them evolve into secondary edge zone of the bogs

One of the most urgent issues affecting Polish environmental conservation policy is the designation as reserves or as sites of ecological interest all the peat bogs studied that form a peatland complex unique at the European scale. A provision of legal protection for peat bogs will require some financial compensation for the local owners. Another way in which the local population should be able to improve their standards of living would be the development of eco-tourism (walking, cycling, horse-riding) promoting the natural qualities of the sprawling mountain bogs.

Well preserved peat domes constitute a valuable component of the studied areas landscape unique at the Carpathian scale. In the past the post-peat areas were converted to pastures, meadows or arable land. Taking into account mountain topography, cool climate and especially low values of local clay soils for agriculture, the post-peat areas should be treated as wastelands or as meadows and pastures.

Author details

Adam Łajczak*

Address all correspondence to: alajczak@o2.pl

Institute of Geography, Jan Kochanowski University, Kielce, Poland

References

[1] Administrative Karte von den Königreichen Galizien und Lodomerien (1855). C. Kummerer Ritter von Kummersberg (ed.), 1:115000, Wien.

[2] Barsiegan, A. M. (1974). Typology of mountainous peat bogs in Armenia. In: Types of peat bogs in the Soviet Union and its classification. Izd. Nauka, Leningrad (in Russian).

[3] The geomorphological evolution of the intermontane Orawa Basin associated with neotectonic movements (Polish Carpathians). Studia Geomorphologica Carpatho-Balcanica, 25-26, 3-28 (in Polish with an English summary). (1991-1992).

[4] Bindler, R. (2006). Mires in the past- looking for the future: Geochemistry of peat and the analysis of past environmental changes. Global and Planetary Changes, 53, 209-221.

[5] Bower, M. M. (1961). The distribution of erosion in blanket peat bogs in the Pennines. Trans. Inst. Br. Geogr., 29, 17-30.

[6] Bragg, O. M., & Tallis, J. H. (2001). The sensitivity of peat-covered upland landscapes, Catena, 42, 345-360.

[7] Carling, P. A. (1986). Peat slides in Teesdale and Weardale, Northern Pennines, July 1983: description and failure mechanisms. Earth Surface Processes and Landforms, 11, 193-206.

[8] Chairman, D. (2002). Peatlands and Environment Change. Wiley, Chichester.

[9] Coggins, A. M., Jennings, S. G., & Ebinghaus, R. (2006). Accumulation rates of heavy metals: lead, mercury and cadmium in ombrotrophic peatlands in the west of Ireland. Atmospheric environment, 40, 260-278.

[10] Cooper, A., & Mccann, T. (1995). Machine Peat Cutting and Land Use Change on Blanket Bog in Northern Ireland. J. of Environmental Management, 43, 153-170.

[11] Dembek, W., Piórkowski, H., & Rycharski, M. (2000). Wetlands on the background of geographical regionalisation of Poland. Publ. IMUZ, Falenty, 97, 135 pp. (in Polish).

[12] Dembek, W., & Piórkowski, H. (2007). Wetlands in landscape of Poland. In: E. Biernacka (ed.) Peatbogs and wetlands. SGGW Publ., Warsaw, in Polish)., 15-48.

[13] Die Spezialkarte der Österreichisch-Ungarischen Monarchie (1894). Wien.

[14] Dykes, A. P., & Warburton, J. (2007). Mass movements in peat: A formal classification scheme. Geomorphology, 86, 73-93.

[15] Evans, R. (1989). Erosion studies in the Dark Peak. North of England Soils Discussion Group, Proceedings, , 24, 39-61.

[16] Früh, J., & Schröter, C. (1904). Die Moore der Schweiz, mit Berucksichtigung der gesamten Moorfrage. Beitr. Geol. Schweiz. Geotechn., Ser. 3, Bern.

[17] Gams, A. (1958). Die Alpenmoore. Jb. Ver. Schutze Alpenpfl. Tiere, 23.

[18] Gore, A. J. P. (1983). Introduction. In: A.J.P. Gore (ed.) Ecosystems of the world 4A. Mires: swamp, bog, fen and moor. Amsterdam-Oxford-New York, 440 pp.

[19] Grosse-Brauckmann, G. (1974). Über pflanzliche Mikrofossilien mitteleuropäischen Torfe. Telma, 4, 51-117.

[20] Haczewski, G., Bąk, K., & Kukulak, J. (1998). Explanations to Geological Map of Poland, 1:50000 (Dźwiniacz Górny). Publ. PIG, Warszawa-Kraków (in Polish).

[21] Haczewski, G., Kukulak, J., & Bąk, K. (2007). Geology and geomorphology of the Bieszczady National Park. Publ. of Pedagogical Academy in Kraków, 468 (in Polish with an English summary).

[22] Horawski, M., Curzydło, J., & Wójcikiewicz, M. (1979). Fluctuations of the level of surface of the Bór na Czerwonem raised bog in 1974 and 1975. Zeszyty Naukowe AR, Kraków, 153, Melioracje, 10: 19-32 (in Polish with an English summary).

[23] Ilnicki, P. (2002). Hoohmors and peat. Poznań, 606 pp. (in Polish).

[24] Karte des Königreisches Galizien und Lodomerien (1779-1782). F. von Mieg (ed.), 1:28000, Wien.

[25] Kaule, G. (1974). Die Űbergangs- und Hohmoore Sűddeutschlands und der Vogesen. Dissertationes Botanicae, 27.

[26] Kaule, G., & Göttlich, K. (1976). Begriffsbestimmungen anhand der Moortypen Mitteleuropas. Moor-und Torfkunde. E. Schweizerbart`sche Verlagsbuchhandlung, Stuttgart, , 1-28.

[27] Kukulak, J. (1998). Origin and development of peat bogs in the upper San River valley in the Bieszczady Mountains. Proc. of the 4th Meeting of Polish Geomorphologists, Lublin (in Polish).

[28] Latocha, A. (2012). Anthropogenically influenced changes in natural environment of mountainous areas in Ireland, an example of Glenveagh National Park. In: A. Łajczak (ed.) Anthropopression in chosen morphoclimatic zones - record in landforms and sediments. Proc. of Faculty of Earth Sciences, Univ. of Silesia, Sosnowiec, 77, 231-241 (in Polish with an English summary).

[29] Lipka, K. (1999). An assessment of the current natural rating of peatlands and their adjacent territories, including the scope for a limited and controlled collection of peat moss in some gminas. Publ. of PEAT-POL, Krakow, 80 pp. (in Polish).

[30] Lowe, J. J., & Walker, M. J. C. (1997). Plant macrofossil analysis. In: J.J. Lowe, M.J.C. Walker (eds) Reconstructing Quaternary Environments. Longman Lim., Edinburgh, 182-190.

[31] Łajczak, A. (2005). The role of biogenic accumulation and peat excavation in geomorphological development of the Orawsko-Nowotarska Basin, West Carpathian Mts. Geomorfologický Sbornik, 4, 97-100.

[32] Łajczak, A. (2006). Raised peatbog Bór na Czerwonem (Podhale, South Poland)- development, degradation, renaturalisation. Nature Conservation, 62, 89-103.

[33] Łajczak, A. (2007). Natura 2000 in Poland, Area PLH120016 The Orawsko-Podhalańskie Peatlands. Publ. of Inst. of Botany, P.A.Sci., Kraków, 139 pp.

[34] Łajczak, A. (2009). Development conditions and distribution of peat bogs in the Orawsko-Nowotarska Basin, Western Carpathian Mountains. Przegląd Geologiczny, 57(8): 694-702 (in Polish).

[35] Łajczak, A. (2011). Contemporary changes in the relief of raised bogs on the example of the Polish Carpathians. Geographia Polonica, 84, Spec. Issue Part 2, 75-92.

[36] Maksimov, A. (1965). Peat deposit and its use in agriculture. Państwowe Wydawnictwa Rolnicze i Leśne, Warszawa, 3-29 (in Polish).

[37] Mallik, A. U., Gimingham, C. H., & Rahman, A. A. (1984). Ecological effects of heather burning: I. Water infiltration, moisture retention and porosity of surface soil, J. Ecol, 72, 767-776.

[38] Margielewski, W. (2006). Records of the Late Glacial-Holocene palaeoenvironmental changes in landslide forms and deposits of the Beskid Makowski and Beskid Wyspowy Mts area (Polish Outer carpathians). Folia Quaternaria, 76, 1-149.

[39] Obidowicz, A. (1985). Mountainous peat-bogs in Europe. Kosmos, 34(2), 299-310 (in Polish).

[40] Obidowicz, A. (1990). Eine Pollenanalytische und Moorkundliche Studie zur Vegetations-geschichte des Podhale-Gebietes (West-Karpaten). Acta Palaeobotanica, 30(1-2), 147-219.

[41] Obidowicz, A., & Margielewski, W. (2008). Classification of mountainous peat bogs. In: S. Żurek (ed.) Peat bogs in mountains, upland and lowlands of Poland. Proc. of Jan Kochanowski University in Kielce, Poland, 103-109 (in Polish).

[42] Ralska-Jasiewiczowa, M. (1972). The forest of the Polish Carpathians in the Late Glacial and Holocene. Studia Geomorphologica Carpatho-Balcanica, 6, 5-19.

[43] Ralska-Jasiewiczowa, M. (1980). Late Glacial and Holocene of the Bieszczady Mts (Polish Eastern Carpathians). PWN, Warszawa-Kraków.

[44] Ralska-Jasiewiczowa, M. (1989). Type region: The Bieszczady Mountains. Acta Palaeobotanica, 29.

[45] Rawes, M. (1983). Changes in two high altitude blanket bogs after cessation of sheep grazing. J. Ecol, 71, 219-235.

[46] Rhodes, N., & Stevenson, A. (1997). Palaeoenvironmental evidence for the importance of fire as a cause of erosion of British and Irish blanket peats. In: J.H. Tallis, R. Meade, P.D. Hulme (eds) Blanket Mire Degradation: Causes, Consequences and Challenges. Macaulay Land Use Research Institute, Aberdeen, 64-78.

[47] Ringler, A. (1981). Die Alpenmoore Bayerns- Landschaftsökologische Grunglagen, Gefährdung, Schutzkonzept. Berichte ANL, 5.

[48] Senft, F. (1862). Die Humus-, Marsch-, Torf- und Limonitbildungen als Erzeunismittel neuer Erdrindelagen (fűr Geonosten, Bergleute, Forst-und Landwirthe), Leipzig.

[49] Shaw, S., Wheeler, B., & Backshall, J. (1997). Review of effects of burning and grazing of blanket bogs- conservation issues and conflicts. In: J.H. Tallis, R. Meade, P.D. Hulme (eds) Blanket Mire Degradation: Causes, Consequences and Challenges. Macaulay Land Use Research Institute, Aberdeen, 174-182.

[50] Sjörs, H. (1948). Myrvegetation i Bergslagen. Acta Phytogeogr. Suec. 21 (in Swedish).

[51] Tactical Map of Poland (1937). Wyd. Wojskowego Inst. Geogr., 1:100000, Warszawa.

[52] Tobolski, K. (2000). Identification guide for peats and lacustrine sediments. Vademecum Geobotanicum, PWN, Warszawa, 508 pp (in Polish).

[53] Tołpa, S. (1949). Peat bogs and peat. Książnica- Atlas, Wrocław-Warszawa, 3-77 (in Polish).

[54] Topographic Maps (1965). Wyd. Państ. Służby Geodez. i Kartogr., 1:25000, Warszawa.

[55] Topographic Maps (1997). Wyd. Państ. Służby Geodez. i Kartogr., 1:10000, Warszawa.

[56] Vanko, J. (1988). Map of recent vertical movements in Western Carpathians in Slovakia until 1952-77. Geodet. a kartogr. Oborz., 76(9), 216-222 (in Slovak).

[57] Wójcikiewicz, M. (1979). The stratigraphy of the Bór na Czerwonem peat bog with the consideration of sub-fossil associations and distribution and differentiation of contemporary plant communities. Part II. The characteristics of the peat bog plant cover. Zeszyty Naukowe AR, Kraków, 153, Melioracje, 10, 133-193 (in Polish with an English summary).

[58] Zuchiewicz, W. (2010). Neotectonics of the Polish Carpathians and Subcarpathian Basins. Publ. Academy of Mining and Metallurgy, Kraków, 234 pp (in Polish).

[59] Żurek, S. (1983). State of stock-taking of peat bogs in Poland. Wiadomości Melioracyjne i Łąkarskie, 7, 210-215 (in Polish).

[60] Żurek, S. (1987). The peat deposits of Poland against the peat zones of Europe. Dokumentacja Geograficzna IGiPZ PAN, 4, 1-84 (in Polish with an English summary).

[61] Żurek, S., & Tomaszewicz, H. (1996). Investigations of marsches. In: M. Gutry-Korycka, H. Werner-Więckowska (eds) Guide-book to field hydrographical investigations. PWN, Warszawa, 190-210 (in Polish).

Section 3

Soil Plant Interactions

Chapter 13

The Role of Soil Properties in Plant Endemism – A Revision of Conservation Strategies

María Luisa Bárcenas-Argüello,
Ma. del Carmen Gutiérrez-Castorena and
Teresa Terrazas

Additional information is available at the end of the chapter

1. Introduction

The soil as the support of all terrestrial ecosystems is distributed as a continuous landscape and varies according to drainage, geomorphology and litho-climatic conditions [1]. There are five major factors that control the formation of soil: parent materials, climate, biota, topography and time [2]. These factors present interdependence; for example contrasting climatic regimes are likely to be associated with contrasting types of vegetation. Nonetheless in certain situation one of the factors had the dominant influence in determining differences among of set of soil [3]. In addition processes of soil genesis are operating under the influence of environmental factors; therefore we can understand the relationship between particular soils and the landscape and ecosystem in which they function [4]. Climate is perhaps the most influential of the four factor acting on large geographical areas (large scale), contributing to the development of specific types of soils and vegetation patterns [5], such as Gelisols (tundra vegetation: lichens, grasses and low shrub), Histosols (water loving plants: pond weeds, cattails, sedges, reed, mosses), Spodosols (mainly coniferous species), Alfisols (deciduous forest), Mollisols (grasses), Aridisols (xerophytic plants), and Oxisols (tropical rain forest). However, at regional level (medium scale) soil variability is often related to small changes in topography and thickness of parent materials or to the effect of organism. In Trans-Mexican Volcanic Belt, is common finding soil sequences or catenas related with specific vegetation species such as Andosols (Pines-volcanic ashes), Cambisols (deciduous trees-colluvial material), fluvisols (crop lands-alluvial material) y solonchack (halophyte grasses-lacutrine material) [6]. Finally at local scales, variation in edaphic characters often provides the best statistical explanation for variation in floristic composition. Frequently, systematic variations in the parent material are closely related to endemism [7]. Endemic species have relatively narrow tolerance to changes in their environment, and can be dependant on certain geologic and edaphic features. There are studies that provide detailed information on species and plant communities, some include relief features, and others report physical and chemical analyses which allow us to infer about soil fertility [8,9] while some other studies have performed statistical analysis to correlate such variables [10-12]. This chapter presents a revision of the current knowledge on the role of soil properties as for pH, H_2O_2 reaction, and reactions to HCl in the distribution of endemic plant species and synthetize in table 1 studies that give details examples of this relationship (plant-soil endemicity).

Species/References	Soi/Substrate type	Technique	Results/Conclusions	Localities
Calochotusobispoensis, *C. tiburonensis*, *C. pulchellus* (Liliaceae) [40]	Serpentine	Concentrations of Ni and Cu in plant tissue	Three species are endemics	Coastal of California USA
Ariocarpus kotschoubeyanus (Cactaceae) [64]	Silty, dry lake beds	Cartographic method by conglomerates	This plant is an edaphic specialist	Chihuahuan Desert, Mexico
Satureja arkansana (Lamiaceae) and *Coreopsis lanceolata* (Asteraceae) [25]	Sandstone glade with alkaline soil	Soil pH	*Satureja arkansana* was absent on soil with a pH less than 6.1 In areas with soil pH greater than 6.1 *C. lanceolata* was absent	Arkansas USA
Hemizonia pungens ssp. *pungens* (Asteraceae) [20]	Alkali pools	Reciprocal transplant greenhouse experiment	This plant grows better on non-alkali soil when grown without competition	Yolo, California, USA
Guaiacum unijugum (Zygophyllaceae) [67]	Coastal dune and arroyo environments	Genetic analysis using microsatellite	The current extraction of gravel its habitat which could pose a direct threat	Baja California Sur, Mexico
Agave bracteosa, *A. victoria-reginae*, *A. albopilosa* *Brahea berlandieri* *Dasylirion berlandieri* *Hesperaloe funifera* var. *funifera* *Yucca filifera*	Litosols stoniness soils, summit, slopes, Stoniness soils depth soils	Sampling of 39 plots (1 Ha each one) all species were recorded and particle size, pH, depth and organic matter	It was observed that soil characteristics drastically alter the conformation of the vegetation and therefore species are not present	Northern Gulf Coastal Plain in northeastern Mexico

Species/References	Soi/Substrate type	Technique	Results/Conclusions	Localities
(All monocotyledons) [71]				
Lupinus subcarnosus (Leguminosae) [72]	Sandy soil	Allozymic variation in enzymes and other proteins	Edaphically restricted species is less genetically variable	East-central Texas USA
Lasthenia californica (Asteraceae) [73,74]	Serpentine soils	Genetic, physiological and phylogenetic studies	Races A and C of *L. californica* coexist on serpentine soil, but inhabit soil of differing physical and chemical properties	Palo Alto, California in the Santa Cruz Mountains USA
Coccoloba cereifera (Polygonaceae), [75]	Sandy soils, gravelly soils and quartzitic outcrops	Each 25 m^2 quadrant was classified according to soil types and were sampled for chemical and granulometric analyses	The spatial distribution of *Coccoloba* was largely related to the arrangement of sandfields	Serra Do Cipó, southeastern Brazil
Erigonum nervulosum (Polygonaceae) *Streptanthus brachatus* and *S. morrisonii* (Brassicaceae) [76]	Serpentine barren soils	Low concentrations of Ca, Mg, P, N	Differences in Ca and Mg between serpentine soils allow distinct species distribution	Lake County, California, USA

Table 1. Studies in which species distribution and soil features are reported.

2. Endemism

Endemic, in botany, means that a plant species is considered native to the country –region- where it can be found [13] and the term is applied to the distribution of organisms [14]. Although climatic factors are the most studied [15,16]; endemism is a non-ecological [17], geological event. Climate limits the flora [18,19], while geological characteristics largely define habitat diversity [4]. Moreover, edaphically severe habitats commonly support edaphic endemics, which are plant species that do not occur elsewhere [20]. Although that might be not enough to recognize the endemic species, edaphological characters (macro and micro) are essential to establish phylogenetic hypothesis of endemic taxa and areas of endemism, looking for consistency with geological models [21-23].

3. Soil

Soil, in soil taxonomy [24] is a natural body comprised of solids (minerals and organic matter), liquid, and gases that occurs on the land surface. Soil occupies space, and is characterized By one of the following: horizons or layers, that are distinguishable from the initial materials a result of additions, losses, transfers, and transformations of energy and matter or the ability to support rooted plants in a natural environment.

We intend to clarify that for some plant species, soil has played an essential role in their evolution and current distribution. Therefore, it is necessary to indicate that according to soil taxonomy is not possible to classify the earthy materials used in pots in greenhouses. In the same sense, plants even grow on trees, but trees are regarded as non soil. Soil covers the earth's surface as a continuum, except on bare rock. Some endemics are restricted to a particular geological formation or to one type of rock or rock outcrops. Endemicity found on rocky outcrops, either calcareous or otherwise, has been reported by several researchers [25-31].

In order to understand the relevance of soil for plant endemism, it is necessary to highlight that the geological processes as genesis of unique soil types may provide the necessary isolation for the genesis of unique biota, and the edaphic factors are used to draw the link between environments and taxa [32]. If the environmental scenario is potentially multi-dimensional [33], some soil properties should be also considered as predictors. Previous research [2] considers that the distinction between soil and environment is arbitrary; and might constitute a theoretical artifact which does not represent natural processes.

3.1. Soil genesis

The transformation of rock into soil is designated as soil formation. Climate, organisms, relief, rocks, and the time are soil forming factors. Therefore, soil can be considered as a particular combination of its forming factors. For a given combination of factors there is only one soil type [2]. Soil properties such as pH, clay content, porosity, etc, are determined by the combination of these factors. The smallest change in any one of the properties, gives rise to a new soil.

Climate is usually considered the dominant soil forming factor, and cannot be described by a single index [4]; for example, the high proportion of smectite in soil, indicate a highly seasonal semiarid subtropical climate [34].

Living organism as bacterial species are able to fix N2, dissolved P, weathered extrusive igneous rock, marble, and limestone, and significantly mobilized useful minerals, such as P, K, Mg, Mn, Fe, Cu, and Zn in rock minerals [35]. Additionally, plant root systems alter the structure of the surrounding soil [36], and the roots of some plants have the ability to exude low-molecular-weight organic acids that produce changes in the availability of nutrients [37].

Relief modifies the water relationships in soils, affected by slope processes such as erosion, landslides and other mass movement [38-39]. However, little is known about the dynamics of soil mosaics formed by slope processes; whereby mountainous regions are characterized by high soil diversity.

The soil formation is regulated by the origin of parent material and by the age of the exposed surface. The nature of the parent material profoundly influences soil characteristics such as the chemical weathering and the quantity and type of clay minerals. While the total composition of the parent rocks is only one of the factors involved in soil formation, it is of considerable interest to analyze some of these rocks, and of soils of similar origin. The time of soils formation refers to the age of the exposed surface. The soils change with time and undergo a process of evolution [2].

3.2. Soil properties

Physical, chemical and mineralogical analyses are used in soil taxonomic criteria. For standard laboratory methods descriptions see Appendix of Keys to Taxonomy of Soils [24]. The diversity of soil properties resulted in the diversity of soil use and soil ecological functions [38,39]. In edaphic islands such as serpentine and limestone outcrops upon which plant grow are necessary specific analyses; per example, in ultramafic soil trace elements as Mn, Cu, Zn, Cr and the heavy metals Ni and Co are extracted, as well as in calcareous soils are measured soluble and exchangeable P [40,41]. Overall, previous authors agree that vegetation differences are strongly associated with differences in the bedrock [29,42].

4. Sampling approach

The existing literature comprises a wide range of sampling techniques to obtain vegetation samples. The plant populations may vary in size and in number of individuals per species. According to our field observations, it is necessary to select populations that adequately represent the endemic area (Figure 1). Soil sampling must be performed in the same locations where vegetation has been previously sampled. The overlap of both sampling activities allows correlating changes in vegetation and soil [43]. Thus, the resulting plots are clearly representative of the surrounding area. Accordingly, geology and topography are essential characteristics that need to be thoroughly examined. For instance, spatial variability studies have indicated that even when the relief of the site is gently rolling, erosion processes can affect soil properties [44, 45], and therefore alter the results of sample analysis. We do not recommend to combined soil samples from which the different edaphic variables are measured [11], this method can give erroneous conclusions because information at the microscale level is lost.

The geology and the topography are factors that influence the formation of particular soil type and the establishment of specific biological forms, [38]. The geological origin of rocks can be identified visually and mineralogical composition by X-ray diffraction or other methods [31]. Undisturbed rock samples must be collected, without showing any chemical or physical weathering. The slope and rock outcrops are some landscape features easy to distinguish. The slope is important by the sediments mobilized by slope processes as for landslides, colluviation, and accumulation of the material that eroded from upper landscape positions [39], this can be recorder in degrees or percent. The geomorphic position can register as summit, shoulder, back slope, toe slope, and floodplain if any (Figure 2).

Figure 1. Delimiting study area in slope or plane environments. (A, B) Nuevo León, México. Habitats for several succulent species endemic to the Chihuahuan Desert.

Figure 2. Geomorphic position of *Cephalocereus* species. (A) *C. nizandensis* on summit; (B) *C. senilis* on summit and shoulder.

The slope aspect would influence insolation, temperature and moisture. Isolation, temperature and moisture also must be into accounted because they have effect on the bedrock or with the vegetation [29, 77]. In the surface it is necessary to test soil and bedrock reactions to HCl as a measure of their calcareousness. Moreover, the soil properties of depth, stoniness, root distribution, H_2O_2 reaction, structure, and Munsell color should be recorded [31,54]. If the characteristics of the site permit, three or more bulk soil or rhizosphere soil and rock samples along the geomorphic position per site should be collected (Figure 3). In the cases where there was no soil, organic debris must be collected [54].

Finally, the abundance of certain minerals in the soil can influence the physiological response and metabolism in plant species, such as heavy metals accumulation and the synthesis and accumulation of biominerals [40, 53]. The most common biominerals in plant tissues are calcium oxalates (Figure 4), and their abundance is associated with calcareous soils [78]. We recommend isolate crystals from the plant tissue for better analyses and carried out X-ray diffraction (XDR), chemical composition and morphology with scanning electron microscopy (SEM). We also suggest use the petrographic microscope to know the optical properties of the crystals for be able to identify in the sand fraction of soil [57]. Additionally to relate the soil elements that plants take up soil and incorporate in their tissues, use energy dispersive X-ray (EDX) on crystals analyses. In order to determine the importance of the biominerals in the soil properties, is necessary that at least three hundred grains from the sandy fraction will count on a grain mount by line counting method using a petrographic microscope [54, 57].

Figure 3. Samples. (A) rock and organic debris for *Cephalocereus apicicephalium*; (B) soil for *C. totolapensis*.

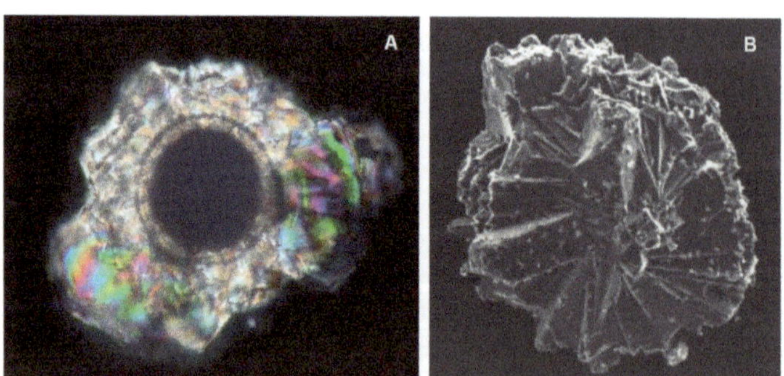

Figure 4. Calcium oxalate crystals isolated from *Cephalocereus* species. (A) petrographic microscope; (B) scanning electron microscopy.

5. Edaphic endemism

The interest for environmental conservation has been growing over the last decades. However, there has been little or no consensus on priority species or conservation strategies Table 2. The conservation of soil diversity greatly overlaps with such for plants, and both become endangered as a result of land use [46]. The fact that soil taxa are geographically restricted is important for planning and conservation efforts. Soil characteristics often play an essential role in determining plant community distributions [47]. The endemism of native plant species in edaphically specialized habitats suggest that these native endemic species are uniquely specialized to survive and grow better under the conditions prevalent in these harsh areas. For instance, several authors [48] reported there are almost five edaphically-restricted or -endemic butterflies, mostly associated with serpentine soils. Some species are absolutely limited by the edaphic restriction of their host plants. These are a better argument for biodiversity preservation [49].

Species	Evidence	Reference
Calochotus (3 spp.)	Soil studies	40
Cephalocereus (5 spp)	Soil-rock studies	54, 57
Calulanthus amplexicaulis var *barbarae*	Soil studies	59
Mammillaria fraileana	Soil-rock studies	31

Table 2. Studies in which species endemism are related to soil features.

Some explanations for the close relationship soil-plant have been looking at morpho-anatomical changes and physiological response to variations in soil parameters. Soil nutrient status determines leaves with glands or without glands: sclerophyllous (leaves without glands) plants exist almost exclusively on oligotrophic soil; whereas orthophyllous (leaves with glands) growth on more or less equally on both oligotrophic and eutrophic soil [50]. In this sense, exceptionally high levels of species turnover were found along all three soil fertility gradients which reflect the high degree of edaphic specialization of the flora [51]. Soil fertility is difficult to quantify, because it dependent not only of the nitrogen (N) and phosphorus (P) status of the soils, but also on their availability. Differences between species in ability to solubilize mineral nutrients could affect the ability or inability of plants to grow in particular soils. In calcareous soils, species suffer lime-chlorosis by Fe deficiency, and their growth is affected by inability to solubilize the native phosphate [52]. In ultramaphic soils, the vegetation accumulates large quantities of heavy metals in their tissues. So, endemic plants have developed strategies to grow successfully in these unusual conditions. For example, to minimize water requirements and excessive water loss, serpentine plants are able to reduce water potentials to levels lower than found on nonserpentine soils, as well as keep stomata closed or nearly closed [53].

Soil constitutes the main source of nutrients for plants. However, for plants growing on bare rock mycorriza have been described as an important factor in promoting edaphic specialization [35, 37]. This high degree of host specificity of symbiotic microbes could enhance nutrient uptake in the infertile soils or rocks. In rhizoplane of cacti, several bacterial species were isolated. This bacterium fixed N_2, dissolved P, weathered extrusive igneous rock, marble and limestone, and significantly mobilized useful minerals, such as P, K, Mg, Mn, Fe, Cu and Zn in rock minerals [35]. Other rock-colonizing cacti usually grow in cracks or fissures that are deeply penetrated by the root system: *Mammillaria fraileana* [31] and *Cephalocereus apicicephalium* and *Cephalocereus nizandensis* (Figure 5) [54]. Little is known about weathering mechanisms, except that the roots of these species can exude low-molecular organic acids (LOAs). The LOAs in root exudates may play an important role in the solubilization and plant availability of mineral nutrients in the rock [55]. Additionally, [20] mentioned that rock outcrop represent refuges from competition with other (often exotic) species.

Figure 5. Rocky habitats. (A) *Cephalocereus apicicephalium* grows in fissures; (B) *C. nizandensis* grows in cracks.

The edaphic endemics are now restricted to unusual and sometimes contaminated soils, but may have been able to withstand large concentrations of metals in their tissues or large quantities of calcium oxalate crystals [40]. The ability to tolerate excessively high levels of nickel and other heavy metals may be a physiological adaptation of the genus *Calochortus* and not necessarily an evolutionary response by several species to life on an ultramafic substrate. The amount of crystalline Ca oxalate in the oldest leaves of *Eucalyptus diversicolor* may be related in part to the high levels of exchange-able soil calcium [56]. *Cephalocereus* species could accumulate great quantities of calcium oxalate crystals even if there is low calcium soluble in the soil [54]. *Cephalocereus nizandensis* and *C. apicicephalium* grow on limestone outcrops, where the Ca is precipitated and *C. totolapensis* preferred acid soils from andesites, siltstones or mica schist (soluble Ca is 19-72 parts per million). The amount of soluble Ca is also very low in where grow *C. columna-trajani* (63-229 parts per million) and *C. senilis* (82-100 parts per million) [57]. The last two species have the larger epidermal crystals of the genus (Figure 6) [58].

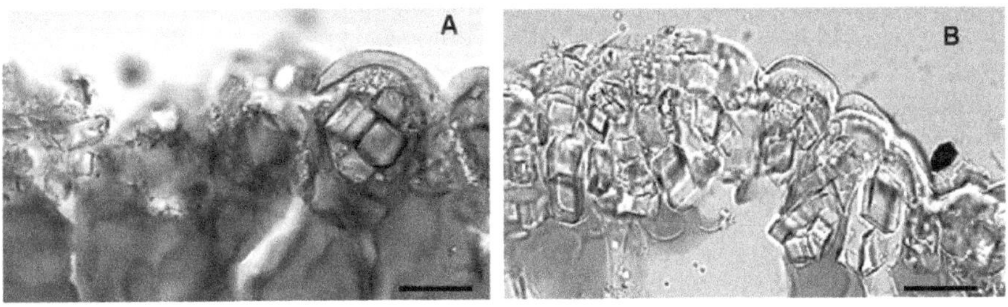

Figure 6. Prismatic calcium oxalate crystals in epidermal cells. (A) *Cephalocereus columna-trajani*; (B) *C. senilis*. Scale A = 40 μm, B = 50 μm.

For studies of evolution of a soil-type adapted endemic species is necessary to perform phylogenetic analysis. The existence of genetically compatible taxa with such distinct edaphic requirements presents a unique opportunity for intensive study of the genetic basis of tolerance to soil-type. A group plant may have evolved in a very dynamic selective context with strong edaphic selective pressures. In [59] they examined phylogenetic relationships of the rare serpentine endemic taxon *Caulanthus amplexicaulus* var. *barbarae*. They found that the serpentine taxa were nonmonophyletic evolving independently at least three times, suggesting that tolerance to serpentine may be gained or lost through relatively few genetic changes. In other case, [50] construct phylogeny for the *Pentaschistis* clade with 82 species in three genera. They investigated the association between leaf anatomy type and soil nutrient type on which species grow. Despite there is little phylogenetic constrain in soil nutrient type. However, only orthophyllous-leaved species diversify on eutrophic soils. Nevertheless, modern phylogenies on a number of phanerogam genera occurring on New Caledonia (*Acianthus, Cupaniopsis, Guioa, Morinda*, and *Oxera*) have shown a shift in soil preference (from non serpentine to serpentine soils and vice versa). Thus [60] concluded that the ability to grown on serpentine soil is either a plesiomorphic or a very homoplasious character and therefore the hypothesis that serpentine soils preserve the indigenous flora in New Caledonia against competition with immigrant species cannot be supported for these groups. Rarely are made specific soil studies, the data are taken often of general charts. We suggest making detailed studies as in *Cephalocereus*, according to plant species and soil type [57]. Therefore, it might be possible to infer the role of soil in the evolution of endemic plant species using the phylogenetic analysis.

6. Conservation strategies

In the landscape, the vegetational differences often serve to delineate the geologic discontinuities of an area even to the casual observer. The remarkable differences often observed in plant cover for different soil types in adjacent areas, have naturally led to attempts to explain these phenomena in terms of the physical or chemical properties of the soil, or of the physiological characteristics of the plants [61]. These areas should be priority sites for conservation to preserve the unique interaction between soil and plant species as well as the microbiota and fauna. For example, the halophytic and gypsophytic vegetation of the Ebro-Basin at Los Monegros [62] or flora of the Coastal Calcareous Hills of the Biosphere Reserve Baconao in Cuba [63] are excellent to demonstrate the varied adaptations of plant types and life-forms as strategies to survive on edapho-climatic harsh conditions of various kinds. In the Chihuahuan Desert region it was found that several Cactaceae species, particularly many members of the Cacteae tribe often inhabit extremely specialized habitats, such as gypsum and other unusual soil formations (Figure 7) [64]. The patches of edaphic endemism also frequently exist as refuges for native species in highly invaded ecosystems [20], ultramafic substrates act as sites in which *Pinus balfouriana* escapes of the competition [65]. Moreover, the work with *Helianthus exilis* showed the need to protect specialized microhabitat found only within the large serpentine outcrops, the species cannot survived outside the narrow conditions proper of its habitat [66]. However, the scarcity of conclusive studies on role of soil to determine the prevalence of endemic plants hampers the efforts of public and private organizations to preserve such areas.

Figure 7. A) *Aztekium ritterii* grows in outcrops of steep slopes of crystalline grypsum; (B) *Turbinicarpus valdezianus* grows in calcarious rocks.

The population size can greatly vary among populations of the same specie generating micro-endemic nature. The small populations of some species consist of adult individuals that may be 3 as in *Cephalocereus totolapensis* [54] or 4 as in *Guaiacum unijugum* [67]. These population sizes are not reported sometimes considerer no significant, but many regions show a unique assemblage of species or a higher level of species richness or other associated species which could serve to protect this ecosystem. In Australian alpine vegetation the analyses of the relationships between physiognomic variation and environment indicate that edaphic factors are more important than climatic factors in differentiating formations [18]. Thus, edaphic discontinuities should be determining the size and population distribution and should be considered when proposing conservation areas.

Frequently, endemic species are less widely distributed and are less well represented in protected areas than other threatened species [68]. Baconao Biosphere Reserve presents high floristic composition and which endemic species represent 21%; however they are restricted at limestone hills, parent material that cover only 6.6% of total area [63]; and the same applies when registering Asteraceae endemic species to the Mexican state of Oaxaca with 53.4% and many of them not found in any Biosphere Reserve [69]; in the same sense, 12 endemic plant species are restricted to serpentine soil in Puerto Rico, all rare, uncommon and very localized within their limited distribution, and only two have been placed on the United States Federal list as threatened or endangered [70], and most species endemic to the state of Nuevo Leon (Northeast of Mexico) present in the submontane scrub have restricted distribution and specially cacti are not located within any protected area [71]. Therefore it is important to reconsider the extension of the sites identified as irreplaceable for various members of the flora endemic worldwide.

The effects of growing human populations on natural communities, on ecosystems, and on some endemic plant populations results in degraded state of sites due to human activity as roadway, tourist development, extraction of mineral as gravel, sand, and others. However, an understanding of the interrelations between soil or bedrock and occurrence of endemics, becomes even more important in the context of restoration ecology and the reversal of land degradation. In addition, the role of soil in the determination of endemic plants has not been sufficiently studied; thus, public and private organizations have not intensified their efforts to preserve such areas. Table 1 show some examples of the different soils supporting endemic species and how through different techniques of study has been able to establish a close relationship between soil and plant. As can be seen, the soil-plant endemism is not exclusive to one type of soil or a plant family.

7. Conclusions

Diverse studies have demonstrated that soil characteristics are correlated with differences in bedrock. Although incipient and timid studies intend to respond to which extent do soil characteristics correlate with vegetation patterns, this is a question that should not be forgotten in plant endemic studies.

We cannot do random sampling hoping to find a relationship between habitat and population size or presence / absence of some species. Efforts should be directed to characterize the habitat or habitats where the species grows in order to determine whether the type of soil, rock, and bedrock are the most important factors for endemic species development. When the limiting factor is the substrate, and abrupt limit in abundance is expected, as well as in population parameters. Narrow endemism in plants is frequently related to soil specificity, and many endemic plants are found in patches of certain soil within a different soil matrix. Additionally as with the soil, vegetation usually changed abruptly at the contact zone.

Edaphic endemic plants are highly vulnerable to extinction due to stochastic events, habitat degradation, climatic change, and invasion by weedy species. Finally, the protected areas may be unable to maintain regional species diversity and representativeness, especially if additional fragments are lost and fragmented landscapes are left unmanaged and they are fragile to soil erosion or degradation by chemical contamination.

Author details

María Luisa Bárcenas-Argüello[1], Ma. del Carmen Gutiérrez-Castorena[2] and Teresa Terrazas[3]

*Address all correspondence to: marialuisabarcenas@gmail.com

1 Botany Program, Postgraduate College, Texcoco, , Mexico

2 Soils Program, Postgraduate College, Texcoco, Mexico

3 Botany Department-Biology Institute, National Autonomous University of Mexico, México City, Mexico

References

[1] Cotler AH. El Uso de la Información Edáfica en los Estudios Ambientales. Gaceta Ecológica 2003;(68) 33-42.

[2] Jenny H. Factors of Soil Formation. A of System Quantitative Pedology, New York: Dover Press;1994.

[3] Boul SW, Hole FD, McCraken RJ: Génesis y Clasificación de Suelos. México: Trillas; 1998.

[4] Brady NC, Weil RR. The Nature and Properties of Soils. New Jersey: Prentice-Hall; 1999.

[5] USDA-NRCS.Global Soil Regions 1:1 million scale. Soil Survey Division, World Soil Resources. 1998.

[6] Cajuste BL, Gutiérrez-Castorena MdelC. El Factor Relieve en la Distribución de Suelos en México. In: Krasilnikov P, Jiménez NFJ, Reyna TT, García CNE. (eds.) Geografía de Suelos de México. México: Las Prensas de Ciencias; 2011. p73-86.

[7] Billings WD. The Environmental Complex in Relation to Plant Growth and Distribution. The Quarterly Review of Biology 1952;27(3) 251-265.

[8] Misra A, Tyler G. Influence of Soil Moisture on Soil Solution Chemistry and Concentrations of Minerals in the Calcicoles *Phleum phleoides* and *Veronica spicata* Grown on a Limestone Soil. Annals of Botany 1999;84(3) 401-410.

[9] Patterson TB, Givnish TJ. Geographic Cohesion, Chromosomal Evolution, Parallel Adaptative Radiations, and Consequent Floral Adaptations in *Calochortus* (Calochortaceae): Evidence from a cDNA Phylogeny. New Phytologist 2004;161(1) 253–264.

[10] Méndez-Larios I, Ortiz E, Villaseñor JL. Las Magnoliophyta Endémicas de la Porción Xerofítica de la Provincia Florística del Valle de Tehuacán-Cuicatlán, México. Anales del Instituto de Biología, Universidad Nacional Autónoma de México. Serie Botánica 2004;75(1) 87-104.

[11] Ruedas M, Valverde T, Zavala-Hurtado JA. Analysis of the Factors that Affect the Distribution and Abundance of Three *Neobuxbaumia* Species (Cactaceae) that Differ in their Degree of Rarity. Acta Oecologica 2006;29(2) 155-164.

[12] Svenning J-C. Microhabitat Specialization in a Species-Rich Palm Community in Amazonian Ecuador. Journal of Ecology 1999;87(1) 55-65.

[13] Font Quer P. Diccionario de Botánica. Barcelona: Labor; 1977.

[14] Esparza-Olguín LG. ¿Qué Sabemos de la Rareza en Especies Vegetales? Un Enfoque Genético-Demográfico. Boletín de la Sociedad Botánica de México 2004;(75) 17-32.

[15] Davies TJ, Barraclough TG, Savolainen V, Chase MW. Environmental Causes for Plant Biodiversity Gradients. Philosophical Transactions of the Royal Society B, 2004;359(1450) 1645-1656.

[16] Giordani P, Incerti G. The Influence of Climate on the Distribution of Lichens: a Case Study in a Borderline Area (Liguria, NW Italy). Plant Ecology 2008;195(2) 257-272.

[17] Luna-Vega I. Aplicaciones de la Biogeografía Histórica a la Distribución de las Plantas Mexicanas. Revista Mexicana de Biodiversidad 2008;79(1) 217-241.

[18] Kirkpatrick JB, Bridle KL. Environmental Relationships of Floristic Variation in the Alpine Vegetation of Southeast Australia. Journal of Vegetation Science 1998;9(2) 251–260.

[19] Yesson C, Culham A. Phyloclimatic Modeling: Combining Phylogenetics and Bioclimatic Modeling. Systematic Biology 2006;55(5) 785-802.

[20] Veblen KE, Young TP. A California Grasslands Alkali Specialist, *Hemizonia pungens* spp. *pungens*, Prefer Non-Alkali Soils. Journal of Vegetation Science 2009;20(1) 170-176.

[21] Andrés HAR, Morrone JJ, Terrazas T, López ML. Análisis de Trazos de las Especies Mexicanas de *Rhus* Subgénero *Lobadium* (Angiospermae: Anacardiaceae). Interciencia 2006;31(12) 900-904.

[22] Morrone JJ, Márquez J. Halffter's Mexican Transition Zone, Beetle Generalized Tracks, and Geographical Homology. Journal of Biogeography 2001;28(5) 635-650.

[23] Webb CO, Ackerly DD, McPeek MA, Donoghue MJ. Phylogenies and Community Ecology. Annual Review of Ecology and Systematics 2002;33 475-505.

[24] Soil Survey Staff. Keys to Soil Taxonomy, 11th ed., Washington: USDA-Natural Resources Conservation Service; 2010.

[25] Jeffries DL. Analysis of the Vegetation and Soils of Glades on Calico Rock Sandstone in Northern Arkansas. Bulletin of the Torrey Botanical Club 1985;112(1) 70-73.

[26] Collins SL, Mitchell GS, Klahr SC. Vegetation-Environment Relationships in a Rock Outcrop Community in Southern Oklahoma. The American Midland Naturalist 1989;122(2) 339-348.

[27] Pérez-García E A, Meave J, Gallardo C. Vegetación y Flora de la Región de Nizanda, Istmo de Tehuantepec, Oaxaca, México. Acta Botanica Mexicana 2001;(56) 19-88.

[28] Pérez-García EA, Sevilla AC, Meave J, Scariot A. Floristic Differentiation in Limestone Outcrops of Southern Mexico and Central Brazil: a Beta Diversity Approach. Boletín de la Sociedad Botánica de México 2009;84(2) 45-58.

[29] Searcy KB, Wilson BF, Fownes JH. Influence of Bedrock and Aspect on Soils and Plant Distribution in the Holyoke Range, Massachusetts. Journal of the Torrey Botanical Society 2003;130(3) 158-169.

[30] Müller JV. Herbaceous Vegetation of Seasonally Wet Habitats on Inselbergs and Lateritic Crusts in West and Central Africa. Folia Geobotanica 2007;42(1) 29-61.

[31] Lopez BR, Bashan Y, Bacilio M, De la Cruz-Agüero G. Rock-Colonizing Plants: Abundance of the Endemic Cactus *Mammillaria fraileana* Related to Rock Type in Southern Sonoran Desert. Plant Ecology 2009;201(2) 575-588.

[32] Kruckeberg AR, Rabinowitz D. Biological Aspects of Endemism in Higher Plants. Annual Review of Ecology and Systematics 1985;(16) 447-479.

[33] Elith J, Leathwick JR. Species Distribution Models: Ecological Explanation and Prediction Across Space and Time. Annual Review of Ecology, Evolution, and Systematics 2009;40 677-697.

[34] Perales R, Serrano H, García BA, Hernández FM. Inferencias Paleoambientales de Mioceno Medio de Somosaguas (Pozuelo de Alarcón, Madrid) Basadas en la Estructura de Tamaños Corporales de su Fauna de Mamíferos. Paleolusitana 2009;(1) 317-325.

[35] Puente ME, Li CY, Bashan Y. Microbial Populations and Activities in the Rhizoplane of Rock-Weathering Desert Plants. II. Growth Promotion of Cactus Seedlings. Plant Biology 2004;6(5) 643-650.

[36] Kearney M, Porter W. Mechanistic Niche Modelling: Combining Physiological and Spatial Data to Predict Species' Ranges. Ecology Letters 2009;12(4) 334-350.

[37] Bar-Yosef B. Root Excretions and their Environmental Effects: Influence on Availability of Phosphorus. In: Waisel Y, Eshel A, Kafkafi U. (eds.), Plant Roots: The Hidden Half. New York: Marcel Dekker; 1996. p.581-605.

[38] Krasilnikov P, García-Calderón NE, Fuentes-Romero E. Pedogenesis and Slope Processes in Subtropical Mountain Areas, Sierra Sur de Oaxaca, México. Revista Mexicana de Ciencias Geológicas 2007;24(3) 469-486.

[39] Krasilnikov P, García CNE, García PMdelS. Soils Developed on Different Parent Materials. Terra Latinoamericana 2007;25(4) 335-344.

[40] Fiedler PL. Heavy Metal Accumulation and the Nature of Edaphic Endemism in the Genus *Calochortus* (Liliaceae). American Journal of Botany 1985;72(11) 1712-1718.

[41] Zohlen A, Tyler G. Soluble Inorganic Tissue Phosphorus and Calciole-Calcifuge Behaviour of Plants. Annals of Botany 2004;94(3) 427-432.

[42] Méndez-Mendoza C, Reyes-Agüero JA, Aguirre-Rivera JR, Peña-Valdivia CB. Distribución Geográfica y Ecológica de *Ephedra* L. en el Altiplano Potosino. Revista Chapingo Serie Horticultura 2000;6(1) 131-138.

[43] Montaño-Arias NM, García-Sánchez R, Ochoa-de la Rosa G, Monroy-Ata A. Relación entre la Vegetación Arbustiva, el Mezquite y el Suelo de un Ecosistema Semiárido en México. Terra Latinoamericana 2005;24(2) 193-205.

[44] Lozano PZ, Bravo C, Ovalles F, Hernández RM, Moreno B, Piñango L, Villanueva JG. Selección de un Diseño de Muestreo en Parcelas Experimentales a Partir del Estudio de la Variabilidad Espacial de los Suelos. Bioagro 2004;16(1) 61-72.

[45] Cristobal AD, Alvarez SME, Hernández AE, Maldonado TR, Pérez GM, Castro BR. Variabilidad Espacial de Propiedades Químicas del Suelo y su Uso en el Diseño de Experimentos. Terra Latinoamericana 2008;26(4) 317-324.

[46] Amundson R, Guo Y, Gong P. Soil Diversity and Land Use in the United States. Ecosystems 2003;6(5) 470-482.

[47] Guo Y, Gong P, Amundson R. Pedodiversity in the United States of America. Geoderma, 2003;117(1-2) 99-115.

[48] Gervais BR, Shapiro AM. Distribution of Edaphic-Endemic Butterflies in the Sierra Nevada of California. Global Ecology and Biogeography 1999;8(2) 151-162.

[49] Raven PH. Catastrophic Selection and Edaphic Endemism. Evolution 1964;18(2) 336-338.

[50] Galley C, Linder HP. The Phylogeny of the *Pentaschistis* Clade (Danthonioideae, Poaceae) Based on Chloroplast DNA, and the Evolution and Loss of Complex Characters. Evolution 2007;61(4) 864-884.

[51] Cowling RM. Diversity Components in a Species-rich Area of the Cape Floristic Region. Journal of Vegetation Science 1990;1(5) 699-710.

[52] Ström L, Olsson T, Tyler G. Differences Between Calcifugue and Acidifugue Plants in Root Exudation of Low-Molecular Organic Acids. Plant and Soil 1994;167(2) 239-245.

[53] Brady KU, Kruckeberg AR, Bradshaw HD.. Evolutionary Ecology of Plant Adaptation to Serpentine Soils. Annual Review of Ecology, Evolution and Systematics 2005;36 243-266.

[54] Bárcenas-Argüello ML, Gutiérrez-Castorena MdelC, Terrazas T, López-Mata L. Rock-Soil Preferences of Three *Cephalocereus* (Cactaceae) of Tropical Dry Forest. Soil Science Society of American Journal 2010;74(4) 1374-1384.

[55] Stöm L. Root Exudation of Organic Acids: Importance to Nutrient Availability and the Calcifuge and Calcicole Behaviour of Plants. Oikos 1997;80(3) 459-466.

[56] O'Connell AM, Malajczuk N, Gailitis V. Occurrence of Calcium Oxalate in Karri (*Eucalyptus diversicolor* F. Muell.) Forest Ecosystems of South Western Australia. Oecologia 1983;56(2/3) 239-244.

[57] Bárcenas-Argüello ML. Distribución Ecológica del Subgénero *Neodawsonia* Backeb. del Género *Cephalocereus* Pfeiff. (Cactaceae), en el Istmo de Tehuantepec, México. PhD thesis. Colegio de Postgraduados México; 2011.

[58] Bárcenas-Argüello ML. Filogenia del género *Cephalocereus* Pfeiff. (Cactaceae) *sensu* Anderson con Base en Caracteres Estructurales. MSc thesis. Colegio de Postgraduados México; 2006.

[59] Pepper AE, Norwood LE. Evolution of *Caulanthus amplexicaulis* var. *barbarae* (Brassicaceae), a Rare Serpentine Endemic Plant: A Molecular Phylogenetic Perspective. American Journal of Botany 2001;88(8) 1479-1489.

[60] De Kok R. Are Plant Adaptations to Growing on Serpentine Soil Rare or Common? A Few Case Studies from New Caledonia. Adansonia 2002;24(2) 229-238.

[61] Whittaker RH. The Ecology of Serpentine Soils. Ecology 1954;35(2) 258-288.

[62] Breckle SW. Halophytic and Gypsophytic Vegetation of the Ebro-Basin at Los Monegros. In Melic A, Blasco-Zumeta J. (eds) Manifiesto Científico por Los Monegros Boletín de la Sociedad Entomológica Aragonesa 1999;(24) 101-104.

[63] Figueredo CLM, Reyes DOJ, Acosta CF, Fagilde EMC. Estudio Florístico de los Cerros Calizos Costeros de la Reserva de la Biósfera Baconao, Cuba. Polibotánica 2009;(28) 69-117.

[64] Hernández HM, Gómez-Hinostrosa C, Hoffman G. Is Geographical Rarity Frequent among the Cacti of the Chihuahuan Desert. Revista Mexicana de Biodiversidad 2010;81(1) 163-175.

[65] Eckert AJ. Influence of Substrate Type and Microsite Availability on the Persistence of Foxtail Pine (*Pinus balfouriana*, Pinaceae) in the Klamath Mountains, California. American Journal of Botany 2006;93(11) 1615-1624.

[66] Wolf A. Conservation of Endemic Plants in Serpentine Landscapes. Biological Conservation. 2001;100(1) 35-44

[67] McCauley RA, Cortés-Palomec AC, Oyama K. Distribution, Genetic Structure, and Conservation Status of the Rare Microendemic Species, *Guaiacum unijugum* (Zygophyllaceae) in the Cape Region of Baja California, Mexico. Revista Mexicana de Biodiversidad 2010;81(3) 745-758.

[68] Aguirre GJ, Duivenvoorden JF. Can we Expect to Protect Threatened Species in Protected Areas? A Case Study of the Genus *Pinus* in Mexico. Revista Mexicana de Biodiversidad. 2010;81(3) 875-882.

[69] Suárez-Mota ME, Villaseñor JL. Las Compuestas Endémicas de Oaxaca, México: Diversidad y Distribución. Boletín de la Sociedad Botánica de México. 2011;88(1) 55-66.

[70] Cedeño-Maldonado JA, Breckon GJ. Serpentine Endemism in the Flora of Puerto Rico. Caribbean Journal of Science 1996;32(4) 348-356.

[71] Estrada-Castillón E, Villareal-Quintanilla JA, Jurado-Ybarra E, Cantú-Ayala C, García-Aranda MA, Sánchez-Salas J, Jiménez-Pérez J, Pando-Moreno M. Clasificación, Estructura y Diversidad del Matorral Sub-Montano Adyacente a la Planicie Costera del Golfo Norte en el Noreste de México. Botanical Sciences 2012;90(1) 37-52.

[72] Babbel GR, Selander RK. Genetic Variability in Edaphically Restricted and Widespread Plant Species. Evolution 1974;28(4) 619-630.

[73] Rajakaruna N, Bohm BA. The Edaphic Factor and Patterns of Variation in *Lasthenia californica* (Asteraceae). American Journal of Botany. 1999;86(11) 1576-1596.

[74] Rajakaruna N, Baldwin BG, Chan R, Desrochers AM, Bohm BA, Whitton J. Edaphic Races and Phylogenetic Taxa in the *Lasthenia californica* Complex (Asteraceae: Heliantheae): an Hypothesis of Parallel Evolution. Molecular Ecology 2003;12(6) 1675-1679

[75] Ribeiro KT, Fernandes GW. Patterns of Abundance of a Narrow Endemic Species in a Tropical and Infertile Montane Habitat. Plant Ecology 2000;147(2) 205-218.

[76] McCarten NF. Rare and Endemic Plants of Lake County Serpentine Soils Habitats. Sacramento:Report for the Endangered Plant Project; 1988.

[77] Nyssen J, Vermeersch D. Slope Aspects Affects Geomorphic Dynamics of Coal Mining Soil Heaps in Belgium. Geomorphology 2010;123(1-2) 109-121.

[78] Garvie LAJ. Decay of Cacti and Carbon Cycling. Naturwissenschaften 2006;93(3) 114-118.

Chapter 14

Effects of Water Stress on Germination and Growth of Wheat, Photosynthetic Efficiency and Accumulation of Metabolites

Rui Guo, Wei Ping Hao, Dao Zhi Gong,
Xiu Li Zhong and Feng Xue Gu

Additional information is available at the end of the chapter

1. Introduction

Especially over the last 100 years, our unbridled exploitation of the world's natural resources has severely damaged its vegetation and has also resulted in worrying accumulations of industrial wastes and greenhouse gases. Together, these have upset natural ecosystem balances and have created many environment and climatic problems, including rising temperatures, increasing desertification, serious soil loss, soil salinization and damaging accumulations of soil nitrogen [39, 31, 37]. In many nations, the recent increased incidences of severe drought and associated desertification are coming into especially sharp focus because of their sudden, long term and devastating consequences for the local human population.

Drought imposes one of the commonest and most significant constraints to agricultural production, seriously affecting crop growth, gene expression, distribution, yield and quality [45, 44, 53]. There are numerous reports on photosynthetic and metabolites characteristics under water stress [22, 52, 25, 5]. Generally, photosynthesis is inhibited by water stress, also affects photosynthetic components and chloroplast stress [54, 52]. Plants have evolved a number of mechanisms to adapt to and survive water stress, Some plant species have evolved mechanisms to cope with the stress, including drought avoidance, dehydration avoidance, or dehydration tolerance. Such adaptive mechanisms are the results of a multitude of morphoanatomical, physiological, biochemical, and molecular changes [1, 2, 6]. But to our knowledge, only a few report about the effects of different level water stress on photosynthetic and metabolites of wheat seedlings.

Wheat is an important crop, with some cultivars tolerant to water stress. The purpose of this study was to investigate the effects of water stress on the growth, chlorophyll fluorescence and accumulations of proline, betaine and carbohydrates of wheat seedlings, by using PEG simulated water stress. It was also desired to elucidate mechanisms of water stress damage and to identify possible adaptive mechanisms to water stress. Understanding how wheat manages water stress is important for the reclamation of drought-prone soils and crop production, and possibly also to discover water-stress resistance genes and hence to develop drought-resistance biotechnology in this crop.

2. Materials and methods

2.1. Design of simulated water stress conditions

Water stress conditions were simulated to polyethylene glycol-6000 (PEG) at one of three concentrations: 0, 5, 15 and 25%. The osmotic potentials of the solutions was measured using a water potential meter (Psypro Wescor Corporation, US) [49]. Table 1 results shows how osmotic potential decreases with increasing PEG-6000 concentration.

PEG-6000 concentration	0%	5%	15%	25%
OP (MPa)	-0.05	-0.09	-0.34	-0.95

Table 1. The osmotic potential (OP) of solutions of polyethylene glycol (PEG).

2.2. Plant materials and growing conditions

Seeds of wheat (*Triticum aestivum*) FengYou-68 were sown 20 seeds in per germination box Seedlings were watered daily with 0.5 Hoagland's nutrient solution [14]. All the boxes were placed in growth chambers [HPG-400, Haerbin, China] with a 16-h photoperiod (Sylvania cool white fluorescent lamps, 200 mmol m^{-2} s^{-1}, 400–700 nm). The temperature was 25 ± 2°C (day) and 21 ± 1.5°C (night).

After three days, 25 boxes containing uniform seedlings were selected and randomly divided into five sets of five replicates. One set was used to determine the seedling growth parameters just prior to treatment, a second set was used as the untreated control (0% PEG-6000, watered with Hoagland's nutrient solution), and the three remaining sets were stressed with one or other of the PEG-6000 solutions. Each PEG subtreatment was applied to a set of five boxes, daily for 7 days.

2.3. Measurement of growth

After the seventh day of treatment, the fresh weights (FW) were recorded after removing surface water by blotting and the dry weights (DW) determined after drying for 15 min in an oven at 80°C and then in a vacuum dryer at 40°C to constant weight. The relative growth rate (RGR) was defined as (ln DW after treatment – ln DW before treatment) / treatment duration. The water content (WC) percentage was calculated as: 100×(FW–DW)/FW [52].

2.4. Measurement of chlorophyll fluorescence and pigments

The maximal photochemical efficiency of PSII (PSII=Fv/Fm), the photosynthetic efficiency of PSII ($Y_{(II)}$=Fm'-F/Fm'), non-photochemical quenching (NPQ=Fm-Fm'/Fm'), non-photochemical quenching coefficient (qN=Fm-Fm'/Fm-Fo'), photochemical quenching (qP=Fm'-F/Fm'-Fo'), the efficiency of excitation energy capture by open PSII reaction centers (Fv'/Fm') and apparent photosynthetic electron transport rate (ETR) were determined between 09:00 and 11:00 h from fully-expanded leaves using an Imaging-PAM (Walz, Effeltrich, Germany], [12, 48]. The leaves were held in the dark for about 20 min before measurement. The intensities of the actinic and saturating light settings were 185μmol/m^2s and 2500μmol/m^2s PAR, respectively. The contents of carotenoids (Car) and chlorophyll (Chl) *a* and *b* were extracted using acetone, and spectrophotometeric determination at 440, 645 and 663 nm of each sample was done three times. The calculations were Chl a = 12.7×OD663-2.69×OD645; Chl b = 22.9×OD645-4.86×OD663 and Car = 4.7×OD440-0.27×(20.2×OD645+8.02×OD663).

2.5. Measurement of metabolites and organic acids

Proline was extracted with 3% sulfosalicylic acid for 30 min at 70°C and measured with ninhydrin [55]. Betaine was extracted with 80% methanol for 20 min at 70°C and measured as described by Grieve and Grattan (1983). Total soluble sugars (SS) were extracted for 30 min at 70°C in 70% alcohol, and measured using anthrone.

2.6. Measurement of germination

One hundred wheat seeds were germinated on filter paper in germination boxes. The dry seeds were submerged in 100 mL of each of the PEG-6000 solutions described above (with distilled water as the control). The boxes were maintained at 20°C in the dark for 10 d, five replicates of each PEG treatment were prepared. Percentages of germinated seeds were scored daily, based on the emergence of the radicles. The germinative Energy(Ge), germinative Percentage(Gp), and germination activity Index(Ai) of wheat seeds were modified using Ge = n/Nx100% (n: the number of germination of seeds in 4 days; N: the total number of seeds); Gp = nl/Nlx100% (n1: the number of germination of seeds at 10 days; N1: the total number of seeds).

3. Statistical analysis

Statistical analysis included one-way analysis of variance (ANOVA) in SPSS (Version 13.0, SPSS, Chicago, IL, USA) and Duncan's method to detect differences in physiological parameters in plants under water stress ($P \leq 0.05$). All measurements represent the means and standard errors (SE) of five replicates.

4. Results

4.1. Growth

The RGR and WC of shoots and roots all decreased with increasing PEG concentration, with the greatest reductions occurring under the highest water stresses (Fig. 1 A - D, $P \leq 0.05$). From the slopes of equations (1) and (2) (Table 2), it was calculated that the RGR for root and shoot increased by 0.229 and 0.231, respectively, per 1% increase in PEG-6000 concentration. Meanwhile, the WC of root and shoot decreased by 24.03 and 21.00, respectively, for each 1% increase in PEG concentration (see equations (3) and ((4)) in Table 2).

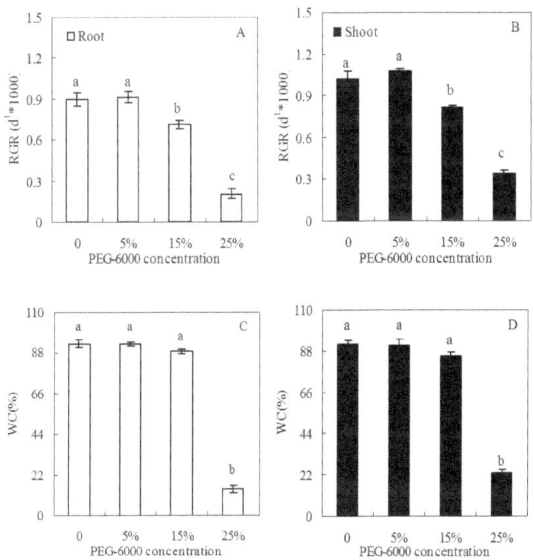

Figure 1. Effects of water stress on shoot (A) and root (B) relative growth rate (RGR) and water content (WC). The values are the means of five replicates. Means followed by different letters in the same stress type are significantly different at $P \leq 0.05$ according to Duncan's method.

	Regression equation	R^2	Decrease in RGR and WC per 1% increment in PEG-6000 concentration
RGR	Y_R=-0.229x+1.254 (1)	0.79	0.229
	Y_S=-0.231x+1.389 (2)	0.80	0.231
WC	Y_R=-24.03x+132.52 (3)	0.64	24.03
	Y_S=-21.00x+125.28(4)	0.68	21.00

Table 2. A regression analysis between RGR, WC and PEG concentration was performed, where Y_R represented the root RGR and WC, Y_S the shoot, and x was PEG concentration.

4.2. Chlorophyll fluorescence and pigments

The Fv/Fm, Y(II), qP and ETR decreased with increasing PEG concentration, while NPQ and qN contents increased significantly, the effects were much more pronounced under high PEG concentration (Table 3; $P \leq 0.05$). The contents of Chl *a* and Chl *b* under PEG induced water stress were less than in the control, each parameter decreased gradually with increasing PEG concentration. The Chl *a*/*b* ratio was higher with PEG than in the control (Table 4, $P \leq 0.05$). The content of Car was scarcely changed by water stress (Table 4, $P \leq 0.05$).

PEG-6000 concentration	Fv/Fm	Y(II)	NPQ	qN	qP	ETR
0%	0.80±0.00^a	0.49±0.00^a	0.29±0.00^b	0.64±0.00^b	0.76±0.00^a	38.30±0.00^a
5%	0.79±0.00^a	0.49±0.00^a	0.27±0.00^b	0.63±0.00^b	0.75±0.00^a	38.00±0.00^a
15%	0.76±0.00^a	0.27±0.00^b	0.48±0.00^a	0.78±0.00^a	0.51±0.00^b	21.20±0.00^b
25%	0.59±0.00^b	0.00±0.00	0.17±0.00^c	0.58±0.00^c	0.00±0.00	0.00±0.00

Table 3. Effects of PEG induced water stress on contents of photosynthetic pigments (g kg^{-1}FM) in seedlings of wheat. The values are the means of five replicates. Means followed by different letters in the same stress type are significantly different at $P \leq 0.05$ according to Duncan's method.

PEG-6000 concentration	Chl*a*	Chl*b*	Chl*a*+ Chl*b*	Chl*a*/Chl*b*	Car
0%	1.14±0.06^a	0.29±0.00^a	1.43±0.08^a	3.93±0.04	0.31±0.00^a
5%	1.13±0.08^a	0.28±0.00^a	1.41±0.06^a	4.04±0.08	0.34±0.00^a
15%	1.02±0.01^a	0.22±0.00^b	1.24±0.02^{ab}	4.64±0.08	0.32±0.00^a
25%	0.86±0.00^b	0.17±0.00^c	1.03±0.01^b	5.06±0.09	0.28±0.00^b

Table 4. Effects of PEG induced water stress on contents of photosynthetic pigments (g kg^{-1}FM) in seedlings of wheat. The values are the means of five replicates. Means followed by different letters.

4.3. Metabolites

The contents of proline increased with increasing PEG concentration, with that in the shoot being significantly higher than that in the root (Fig. 2, A and B; $P \leq 0.05$). In the roots, increasing water stress had a positive effect on betaine content, causing a significant rise at 25% PEG concentration, however, in the shoots a negative effect (Fig. 2, C and D; $P \leq 0.05$). The impacts of water stress on soluble sugar were similar as proline and betaine, it contents significantly increased under high water stress (Fig. 2 E and F; $P \leq 0.05$).

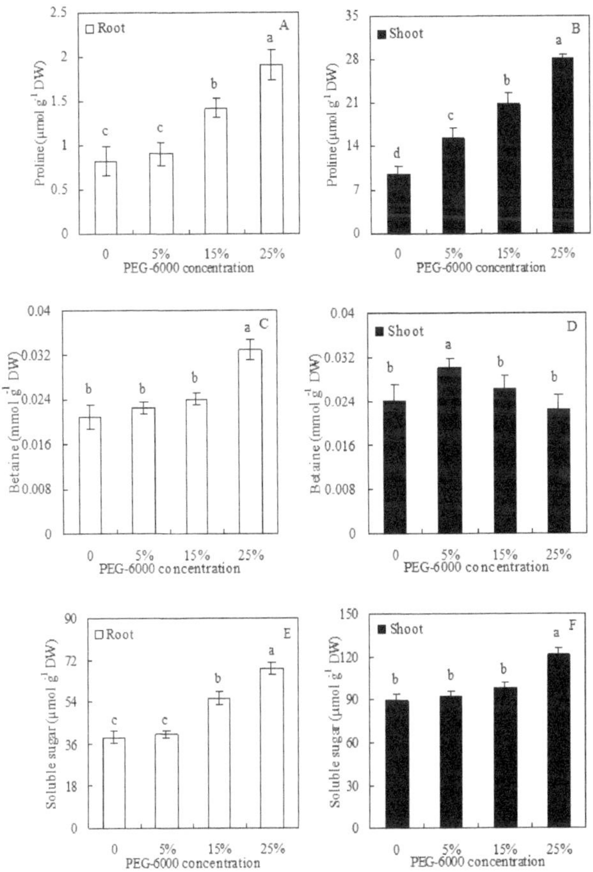

Figure 2. Effects of PEG induced water stress on the contents of proline, betaine and soluble sugar in roots and shoots of wheat seedling. The values are the means of five replicates. Means followed by different letters in the same stress type are significantly different at $P \leq 0.05$ according to Duncan's method.

5. Organic acids

OA, CA, MA, FA, LA and *SA* were all detected in both the shoots and roots of the wheat seedlings under water stress (Fig. 3). In response to water stress, the *OA, MA* and *SA* content of the roots and shoots decreased with PEG-6000 concentration increased, declined significant reduction above 15% PEG-6000 concentration (Fig. 3 A1, A2, C1, C2, F1 and F2, $P \leq 0.05$). The level of *FA* increased under low water stress whereas it decreased under high stress in shoot, but in root it level show completely opposite change (Fig. 3 D1 and D2, $P \leq 0.05$). Water stress had a significant negative impact on CA and LA levels in shoots, but there had no regular impact on the levels of CA and LA in roots (Fig. 3 B and E, $P \leq 0.05$).

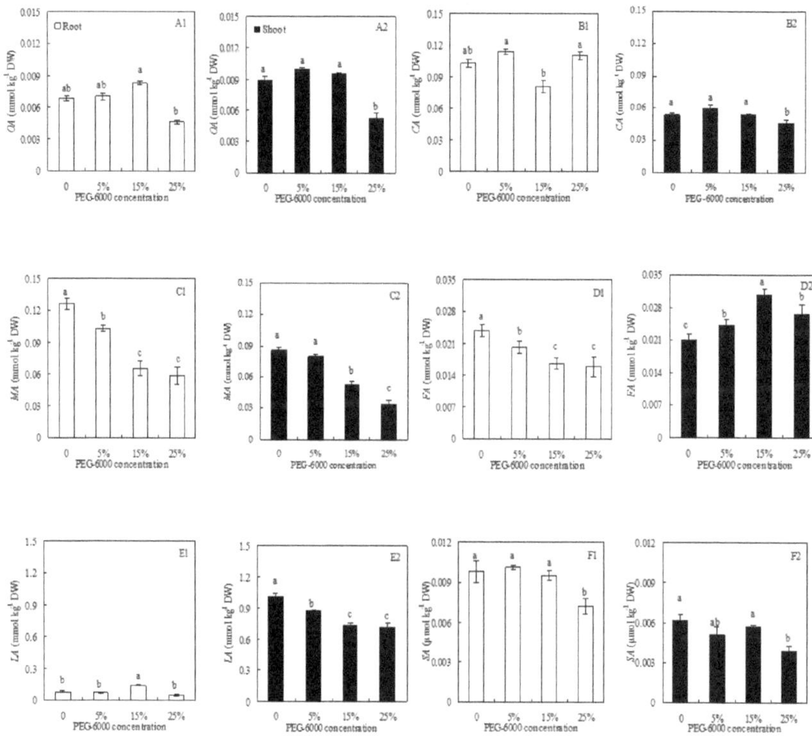

Figure 3. Effects of water stress on the levels of Oxalic Acid (A1), Citric Acid (B1), Malic Acid (C1), Formic Acid (D1), Lactic Acid (E1) and Succinic Acid (F1) in the root, and Oxalic Acid (A2), Citric Acid (B2), Malic Acid (C2), Formic Acid (D2), Lactic Acid (E2) and Succinic Acid (F2) shoot of wheat seedlings. The values are the means of five replicates. Means followed by different letters in the same stress type are significantly different at $P \leq 0.05$ according to Duncan's method.

5.1. Germination

The table 5 shows that the trend in changes in Gp and Ge of wheat seeds under water stress conditions was similar; there was a decreased trend with increased PEG-6000 concentration ($P \leq 0.05$), the reductions were greater when concentration above 15% ($P \leq 0.05$).

Treatment	Gp (%)	Ge (%)
CK	96.66	96.00
5% PEG	95.33	91.67
15% PEG	83.67	80.33
25% PEG	64.00	63.67

Table 5. Effect of water stress on two indices (Gp and Ge) of germination for wheat. Germinative percentage - Gp and germinative Energy - Gp.

6. Discussion

PEG is an osmotic agent, which play an important role in the regulation of mineral elements, hormone, protein metabolism and effects of signal transduction [50, 41]. The main function of PEG is to slow down the moisture rate of import and export seeds, which benefit to reduce membrane system injury in process of seed imbibition and repair impaired membrane system [27, 16]. PEG has been widely used in seed priming and simulated water stress test, the wheat seedlings were treated by three different PEG concentrations.

7. Impact of water stress treatments on growth

In plants in general, an appropriate growth strategy is key to fitness in a competitive situation, so too in wheat seedlings, their growth strategy is critical to survival [10]. The RGR value of a plant reflects its vigour and is considered a good index of its exposure to stresses of all sorts [26, 52]. The RGR response of wheat seedlings exposed to increasing PEG concentrations (Fig. 1 A, $P \leq 0.05$), revealed a decrease for roots and shoots (Table 2, $P \leq 0.05$). This may reflect the impact of water stress on root cell development, which would likely impair nutrient uptake as well as having detrimental effects on photosynthesis, essential for biomass accumulation and therefore on shoot and root elongation. The change trend for WC was similar to that for RGR but the extension of WC in the root was about 1.14-times that of the shoot (Fig 1. B; Table 2;$P \leq 0.05$). Water stress therefore appears to reduce the absorption and utilization of water to such an extent that the tolerance mechanisms employed by these plants in a drought are insufficient to maintain normal growth.

7.1. Impact of water stress treatments on chlorophyll fluorescence and pigments

The chlorophyll fluorescence kinetics react to the "intrinsic" characteristic of photosynthesis and can rapidly and sensitively reflect a plant's physiological status and its relationship with the environment [Huang et al., 2009]. In this study, PSII values decreased with increasing PEG concentration but these began to decline significantly in 15% PEG concentration. The results indicate that photoinhibition occurs under water stress as a result of damage to the reaction center of photosystem II (Table 3, $P \leq 0.05$). The values of Y(II), qP and ETR decreased, while those of NPQ and qN increased with increasing PEG concentrations. These results indicate that electron transport activity and the photosynthetic apparatus of wheat seedlings with certain drought-resistance are damaged.

Chl and Car are the main photosynthetic pigments of plants, so these are good indicators of the photosynthesis capability of a plant. Under water stress, with the exception of Car which barely changed, the contents of Chl *a* and *b* decreased slightly at first but then decreased more sharply at the 15% PEG concentration (Table 4; $P \leq 0.05$). This may be linked with the observation that under -0.34MPa water potential conditions Chl synthesis was severely inhibited with the result that the functioning of the photosynthetic apparatus became seriously impaired [25, 5]. Compared with the control, the water stress effects on Chl *a*/*b* were high and this appears to be closely related to the metabolic regulation of Chl; this possibility is worth further investigation.

7.2. Impact of water stress treatments on metabolites

Proline and betaine are also known to play important roles in osmotic adjustment with their accumulation under water stress being observed in many species [46, 38]. Here, the results show that, along with a decrease in osmotic potential, the accumulation of free proline and betaine increased significantly both in the roots and the shoots. This increase would lower the osmotic potential [i.e. make it more strongly negative] in the cells which would help to maintain turgor and thus sustain the normal physiological and biochemical processes in the face of drought (Fig. 2, $P \leq 0.05$).

Soluble sugars are the main osmotic adjustment substances and so are important indicators of drought tolerance. The results show that the soluble sugars contents of wheat seedlings increases under high PEG concentration. This indicates that they may help to regulate and maintain the activity of physiological processes within the plant in a high water-stress environment by raising the osmotic potential of the cells [14].

7.3. Impact of water stress treatments on organic acids

The accumulation of organic acids is a physiological response of plants to stress, when plants are suffered by water stress, they can through cells apperceive and transmit drought signal [42]. There nearly no impact on the content of organic acids under blew 15% water stress, it decreased significantly under high stress, but in shoot FA completely opposite change (Fig. 3, $P \leq 0.05$). The results confirmed that the organic acids metabolic regulation was closely related to the plant water stress resistance. The change of organic acid may be adaptive mechanism by which wheat seedlings maintain their intracellular osmotic balance under water stress [47, 55].

7.4. Impact of water stress treatments on germination

Germination is one of the most critical periods in the life cycle of plants. Under water stress, low water potential is a determining factor inhibiting seed germination [51, 43]. The inhibiting action of water stress on the wheat germination was increased with PEG-6000 concentration increasing (Table 5).

8. Conclusion

In summary, the growth of wheat seedlings was inhibited by water stress, especially in roots. The function of water regulation occurs outside root, or in apoplast of root, or both outside root and in apoplast of root. Therefore, we propose that the water-potential adjustment of the roots may be a key physiological mechanism for wheat resisting water stress. Proline, betaine and soluble sugar content increase to a greater extent in response to water stress, these data suggest that wheat seedlings may initially sense high drought environments, the harmful effects of water stress on the distribution and accumulation of carbohydrates, it was reflecting the specific detrimental effects of a drought environment. With the extension of PEG-6000 concentrations, wheat seedlings photosynthetic electron transport and photosynthetic primary reaction inhibited, heat disseminate which possess photoprotective effect increased. It implies that there was a closed relationship between the effects of water stress on chlorophyll fluorescence parameters of wheat seedlings. These results provide useful data that will facilitate the development of strategies for the creation of engineered wheat varieties that are more tolerant towards water stress.

Acknowledgements

This work supported by grants from the Project of the National Natural Science Foundation of China (No. 31170303, 30870238, 30871447, 50709040, 31070398). The basic research special fund operations (1610122012001), the international scientific and technological cooperation projects (No. 2010DFB30550).

Author details

Rui Guo*, Wei Ping Hao, Dao Zhi Gong, Xiu Li Zhong and Feng Xue Gu

*Address all correspondence to: guor219@yahoo.com

Institute of Environment and Sustainable Development in Agriculture(IEDA), Chinese Academy of Agricultural Sciences (CAAS)/Key Laboratory of Dry land Agriculture, MOA, Beijing 100081, China

References

[1] Abdalla, M. M., & El -Khoshiban, N. H. (2007). The influence of water stress on growth, relative water content, photosynthetic pigments, some metabolic and hormonal contents of two Triticium aestivum cultivars. *J. Appl. Sci.*, 3, 2062-2074.

[2] Ali, Q., & Ashraf, M. (2011). Induction of Drought Tolerance in Maize (Zea mays L.) due to Exogenous Application of Trehalose: Growth, Photosynthesis, Water Relations and Oxidative Defence Mechanism. *Journal of Agronomy and Crop Science*, 197, 258-271.

[3] Almansouri, M., Kinet, J. M., & Lutts, S. (1999). Compared effects of sudden and progressive impositions of salt stress in three durum wheat (Triticum durum Desf.) cultivars. *J. Plant Physiol.*, 154, 743-752.

[4] Armstrong, D. P., & Westoby, M. (1993). Seedlings from large seeds tolerate defoliation better: a test using phylogenetically independent contrasts. *Ecology*, 74, 1092-1100.

[5] Ashraf, M. (2010). Inducing drought tolerance in plants: Recent advances. *Biotechnology Advances*, 28, 169-183.

[6] Ashraf, M., Ahmad, M. S. A., Öztürk, M., & Aksoy, A. (2012). Crop Improvement Through Different Means: Challenges and Prospects. *Crop Production for Agricultural Improvement*, 1-15.

[7] Bjorkman, O., & Demmi, G. B. (1987). Photon yield of O2 evolution and chlorophyll fluorescence at 77K among vascular plants of diverse origins. *Planta*, 170, 489-504.

[8] Bush, J. K., & Van Auken, O. W. (1991). Importance of time of germination and soil depth on growth of Prosopis glandulosa (Leguminosae) seedlings in the presence of a C4 grass. *Am. J. Bot.*, 78, 1732-1739.

[9] Chen, Y. P., Chen, Y. N., Li, W. H., & Xu, C. X. (2009). Effect of high temperature on photosynthesis in populus euphratica under drought condition. *Journal of desert research*, 29(3), 474-479.

[10] Du, Y., & Huang, Z. L. (2008). Effects of seed mass and emergence time on seedling performance in Castanopsis chinensis. *Forest Ecology and Management*, 255, 2495-2501.

[11] Gale, K. R. (2005). Diagnostic DNA markers for quality traits in wheat. *Journal of Cereal Science*, 41, 181-192.

[12] Genty, B., Briantais, J. M., & Baker, N. R. (1989). The relationship between the quantum yield of Photosynthetic electron transport and quenching of chlorophyll fluorescence. *Biochimiea et Bio Physica Aeta*, 990, 87-92.

[13] Harms, K. E., & Dalling, J. W. (1997). Damage and herbivory tolerance through resprouting as an advantage of large seed size in tropical trees and lianas. *J.Trop. Ecol.*, 13, 617-621.

[14] He, H., Sun, C. H., Du, W., & Li, Y. (2006). Effect and evalution of Entomophthora spp. on controlling Aphis glycines. *Chinese journal of oil crop science*, 28, 76-78.

[15] Huang, H. Y., Dou, X. Y., Deng, B., Wu, G. J., & Peng, C. L. (2009). Responses of different secondary provenances of Jatropha curcas to heat stress. *Scientia silvae sinicae*, 45(7), 150-155.

[16] Jiao, S. Y., Li, Y. Q., Shayila, S. H., & Chen, X. L. (2009). Seeds Germination and Seedling Growth about 3 Pennisetum Ornamental Grasses under Drought Stress. *Acta Bot.Borea l.-Occident. Sin.*, 29(2), 0308-0313.

[17] Jing, J. H., & Ding, Z. R. (1981). Determining organic acid content.]- In: Boqinnoke, X.H. (ed.): Analysis Method of Plant Biochemistry. 264-267.

[18] Johnson, G. N., Youn, G. A. J., Scholes, J. D., et al. (1993). The dissipation of excess excitation energy in British plant species. *Plant, cell and Environment*, 16, 673-679.

[19] Jurado, E., & Westoby, M. (1992). Seedling growth in relation to seed size among species of Arid Australia. *J. Ecol.*, 80, 407-416.

[20] Kerepesi, I., Galiba, G., & Ba′nyai, E. (1998). Osmotic and salt stresses induced differential alteration in water-soluble carbohydrate content in wheat seedlings. *J. Agric. Food Chem*, 46, 5347-5354.

[21] Kerepesi, I., & Galiba, G. (2000). Osmotic and salt stress-induced alteration in soluble carbohydrate content in wheat seedlings. *Crop Sci.*, 40, 482-487.

[22] Krause, G. H., & Weis, E. (1991). Chlorophyll floresence and photosynthesis. *Ann. Rew. Plant Physiol.*, 142, 313-349.

[23] Leishman, M. R., & Westoby, M. (1994). The role of seed size in seedling establishment in dry soil conditions-experimental evidence from semiarid species. *J. Ecol.*, 82, 249-258.

[24] Li, Z. J., Luo, Q. H., Wu, W. M., & Han, L. (2009). The effects of drought stress on photosynthetic and chlorophyll fluorescence characteristics of Populus euphratica and P.pruinosa. *Arid zone research.*, 24(4), 5-9.

[25] Liu, H. Y., Li, J. Y., Zhao, Y., & Huang, K. K. (2007). Influence of drought stress on gas exchange and water use efficiency of Salix psammophila growing in five places. *Arid zone research.*, 24(6), 815-820.

[26] Lutts, S., Almansouri, M., & Kinet, J. M. (2004). Salinity and water stress have contrasting effects on the relationship between growth and cell viability during and after stress exposure in durum wheat callus. *Plant Science*, 167, 9-18.

[27] Ma, H. Y., & Liang, Z. W. (2005). Research progress on improving germination rate of Leymus chinensis. *Grassland of China*, 27(4), 64-68.

[28] Martin, M., Miceli, F., Morgan, J. A., Scalet, M., & Zerbi, G. (1993). Synthesis of Osmotically Active Substrates in Winter Wheat Leaves as Related to Drought Resistance of Different Genotypes. *J. Agric. Crop Sci*, 171, 176-184.

[29] Paz, H., & Martinez-Ramos, M. (2003). Seed mass and seedling performance within eight species of Psychotria (Rubiaceae). *Ecology*, 84, 439-450.

[30] Maxwell, K., & Johnson, G. N. (2000). Chlorophyll fluorescence, a practical guide. *J. Exp.*, 51, 659-668.

[31] Naumann, J. C., Young, D. R., & Anderson, J. E. (2008). Leaf chlorophyll fluorescence, reflectance, and physiological response to freshwater and saltwater flooding in the evergreen shrub, Myrica cerifera. Environ. *Exp. Bot.*, 63, 402-409.

[32] Pawl, B. (1998). Cities and economic development: from the dawn of history to the present. *Chicago: Chicago University Press.*

[33] Pilon-Smits, E. A. H., Ebskamp, M. J. M., Paul, M. J., Jeuken, M. J. W., Weisbeek, P. J., & Smeekens, S. C. M. Improved performance of transgenic fructan-accumulating tobacco under drought stress. *Plant Physiol.*, 107, 125-130.

[34] Qu, Y., Y.,, P., Mu,, X. Q., Li, Y. X., Tian,, F., Wen,, H. L., Zhang, , & Li, Z. C. (2008). QTL mapping and correlations between leaf water potential and drought resistance in rice under upland and lowland environments. *Acta agronomica sinica*, 34(2), 198-206.

[35] Ralph, P. J., & Burchett, M. D. (1998). Photosynthetic response of Halophila ovalis to heavy metal stress. *Environ Pollut.*, 103, 91-101.

[36] Rau, S., Miersch, J., Neumann, D., Weber, E., & Krauss, G. J. (2007). Biochemical responses of the aquatic moss Fontinalis antipyretica to Cd, Cu, Pb and Zn determined by chlorophyll fluorescence and protein levels. *Environmental and Experimental Botany*, 59, 299-306.

[37] Reich, P. B., Tjoelker, M. G., Walters, M. B., Vanderklein, D. W., & Buschena, C. (1998). Close association of RGR, leaf and root morphology, seed mass and shade tolerance in seedlings of nine boreal tree species grown in high and low light. Funct. *Ecol.*, 12, 327-338.

[38] Rengasamy, P. (2002). Transient salinity and subsoil constraints to dryland farming in Australian sodic soils: an overview. *Aust. J. Exp. Agric.*, 42, 351-361.

[39] Rhodes, D., & Hanson, A. D. (1993). Quaternary ammonium and tertiary sulphonium compounds in higher plants. Annu. Rev. Plant Physiol. *Plant Mol. Biol.*, 44, 357-383.

[40] Richards, J. F. (1990). Land transformation. *Turner II, B.L.,et al. (Eds.),The Earth as transformed by human action. New York: Cambridge University Press*, 163-178.

[41] Ross, M. A., & Harper, J. L. (1972). Occupation of biological space during seedling establishment. *J. Ecol.*, 60, 70-88.

[42] Shao, H. B., Liang, Z. S., Shao, M. G., & Wang, B. C. (2005). Impacts of P EG-6000 pretreatment for barley (Hordeum vulgare L.) seeds o n the effect of their mature embryo in vitro culture and primary investigation on its physiological mechanism. *Colloids and Surf aces B: Biointerfaces*, 41, 73-77.

[43] Shen, L. M., David, M., & Joyee, G. F. (1990). Influence of drought on the concent ration and distribution of 2,4- diaminaobutyric acid and other free amino acids in tissues of flat pea (Lathyrus sylvestris L.). *Enviro Expt Bot.*, 30, 497-504.

[44] Shi, G. Y., Liao, W. X., Qin, L. F., & Lu, L. L. (2009). PEG simulated water stress effects on physiological and biochemistry indexes of germination of Toona sinensis seeds. *Journal of Forestry Sci and Tech.*, 4, 142-145.

[45] Shi, Z., Shi, S. Q., Xiao, W. F., & Qi, L. W. (2008). Influence of dehydration on characteristics of chlorophyll fluorescence of detached leaves in Haloxylon ammodendron and Populus euphratica. *Forest Research*, 21(4), 566-570.

[46] Stendle, E., & Peterson, C. A. (1998). How does water get through roots? J Exp Bot. , 49, 775-788.

[47] Stewart, G. R., & Lee, J. A. (1974). The role of proline accumulation in haloplytes. *Planta*, 120, 279-289.

[48] Sun, L., Liu, S. H., Shi, X. D., Xiao, M., Tang, Z. Y., Zhu, H. W., & , J. Z. (2006). Chen-Salt-tolerant physiological and biochemical properties of ten Species of chenopodiaceae halophytes growing in deserts, Xi njiang. *ARI D ZONE RESEARCH*, 23, 209-313.

[49] Van , K. O., & Snel, J. F. H. (1990). The use of chlorophyll fluorescence nomenclature in plant stress physiology. *Photosynthesis Research*, 25, 147-150.

[50] Verslues, P. E., Ober, E. S., & Sharp, R. E. (1998). Root growth an d oxygen relation at low water potentials impact of oxygen availability in polyethylene glycol solution. Plant Physio1. : , 116, 1403-1412.

[51] Wang, J. G., Chen, G. C., & Zhang, C. L. (2002). The effect s of water stress on soluble protein content, the activity of SOD, POD and CAT of two eco types of reeds (Phragmi tescommunis). *ActaBot. Boreal. Occident. Sin.*, 22(3), 561-565.

[52] Xu, S. G., Wang, J. H., & Bao, L. J. (2006). Effect of Water Stress on Seed Germination and Seedling Growth of Wheat. *Journal of Anhui Agri. Sci.*, 34, 5784-5787.

[53] Yang, X. Q., Zhang, S. Q., Liang, Z. S., & Shan, Y. (2004). Effects of water stress on chlorophyll fluorescence parameters of different drought resistance waiter wheat cultivars seedlings. *Acta Bot. Boreal. Occident. Sin.*, 24, 812-816.

[54] Zhao, Y. J., Weng, B. Q., Wang, Y. X., & Xu, G. Z. (2009). Plant physio-ecological responses to drought stress and its research progress. *Fujian science and technology of rice and wheat.*, 27(2), 45-50.

[55] Zhang, Y. Q., Mao, X. S., & Sun, H. Y. (2002). Effects of drought stress on chlorophyll fluorescence of winter wheat. *Chinese Journal of Eco-Agriculture*, 10, 13-15.

[56] Zhou, J. L., Tang, X. Q., & Wang, K. C. (2009). Effect of water stress on content of four organic acids in different cultivated populations of Isatis indigotica. China. *Journal of Chinese Materia Medica*, 34, 127-131.

Chapter 15

Leachability and Vegetable Absorption of Heavy Metals from Sewage Sludge Biochar

Mieko Yachigo and Shinjiro Sato

Additional information is available at the end of the chapter

1. Introduction

Management of industrial wastes has been one of the most challenging problems in most urban municipalities due to increasing waste volume with limited disposal areas, and energy-consuming and high-cost treatment processes for disposal. In Japan, although the total volume of the industrial waste generated has been relatively constant since 1990 being approximately 400 million tons per year, remaining capacity of the final disposal for the industrial waste has decreased by approximately 20% since 1994, reaching to 172 million m^3 in 2007 (Ministry of the Environment Government of Japan, 2011). This disposal capacity is predicted to be filled in the average of 8.5 years for Japan, and 3.6 years for Tokyo Metropolitan areas. Due mainly to high cost of treatment for proper disposal, illegal dumping of the industrial waste has been a new problem despite of severe regulation and monitoring of affected areas. The number and total volume of unsolved cases of illegal dumping and improper disposal of the waste were 2,610 and 1.78 million tons, respectively, in 2010 (Ministry of the Environment Government of Japan, 2011).

Sewage sludge is one of the most produced industrial wastes and comprised approximately 20% (77.2 million tons) of the total volume of industrial wastes generated across Japan in 2010 (Ministry of the Environment Government of Japan, 2011). However, the volume of its final disposal was insignificant (0.37 million tons) since the majority of sewage sludge were reclaimed and/or treated by intermediate processing such as thickening, dewatering, anaerobic digestion, composting, incineration, carbonization, and melting. On the dry solid basis, 78% of the total volume of sewage sludge (1.72 million tons) was recycled in beneficial applications mainly for construction materials (1.39 million tons), energy production (0.02 million tons), and application in agriculture and horticulture as alternative fertilizers and/or soil conditioner (0.31 million tons). Due to new technology and diversification of sewage sludge recycling techniques, further recycling avenues are expected particularly for production of energy, agriculture, forestry, livestock, and fisheries in the future.

Pyrolysis process of sewage sludge is one of the useful approaches for sewage sludge recycling, and its product called sewage sludge biochar (SSB) can be used as supplemental or alternative material to coal for thermal power plants. Some municipalities in Japan have initiated to use SSB as a part of thermal sources for energy production. In Tokyo, since 2007, 8,700 tons of SSB have been produced per year from 99,000 tons of sewage sludge and mixed with coal for energy plant to contribute approximately 1% of the total calorific value by the plant. The SSB produced is pyrolysed at 500°C for one hour and generates 2,000 kcal kg^{-1} of heat calorie, which is equivalent to one third of that of coal. Aichi Prefecture started a similar project as Tokyo in 2012, and uses 2,700 tons of SSB pyrolysed at 500°C from 33,000 tons of sewage sludge every year to generate 4.6 million kWh of energy, which is capable of providing electricity equivalent to annual usages of typical 1,270 households in the area.

Sewage sludge biochar can be also used for supplemental or alternative fertilizer material for crop production. Different feedstock properties and pyrolysis temperatures, however, yield SSB with different physicochemical properties, thus exhibit different degrees of effect when SSB is applied to soil as soil amendments or fertilizers for improving crop production or soil properties (Chan & Xu, 2009). An SSB produced at 800°C in Japan contained moderate amounts of macronutrients for crop production, among which P was relatively high (78.9 g P kg^{-1}), and relatively high amounts of heavy metals (1,100 mg Cu kg^{-1}, 1,630 mg Zn kg^{-1}, 4 mg Cd kg^{-1}, 126 mg Ni kg^{-1}) which were within the limits legislated in Japan (Kawano et al., 2012). Effects of SSB application to soils up to 40% by weight on the growth of Begonia (*Begonia semperflorens*) differed depending on the soil type and were more positively pronounced in soils more infertile such as sand-dune and yellow soils. A wastewater sludge biochar pyrolysed at 550°C in Australia was applied at a rate of 10 t ha^{-1} and improved the production of cherry tomatoes (*Lycopersicon esculentum*) by 64% above the control without SSB application (Hossain et al., 2010). Some heavy metals such as Cd, Cu, and Zn were taken up more by the plant with SSB application (0.04, 6.2, and 22 mg kg^{-1}, respectively) compared with the control, however the plant uptake of all 16 metals and trace elements measured in the SSB itself were below the Australian maximum permitted concentrations for food products. As SSB pyrolysis temperature increased from 300°C to 700°C, while biochar yield, fixed C, volatile matter, total N, and inorganic N contents decreased, pH values, ash, total P, available (Colwell) P, Ca, Mg, Fe, S, Zn, Cd contents increased (Hossain et al., 2011). The total concentration of heavy metals such as Cd, Cr, Ni, and Pb tends to be enriched in the SSB, but the bioavailability (e.g., DTPA-extractable) of many of these trace elements appears to be reduced during pyrolysis process, compared with those in the raw feedstock (Hossain et al., 2011; Méndez et al., 2012). The variability of the micronutrient concentrations in biochar is due to its volatility and pyrolysis temperature effects on both composition and chemical structure of the biochar (Chan and Xu, 2009).

Agronomic effects of SSB application to soils, therefore, widely vary depending on physicochemical properties of the sewage sludge, pyrolysis temperature, and soil properties, which need to be further clarified including the heavy metal dynamics in order for the SSB application to be broadly accepted in agricultural and horticultural practices. Therefore, objectives of this study were to 1) investigate the leachability of heavy metals, particularly Cu, Zn, and Cd from SSB as applied to soil with varying rates, and 2) evaluate the effect of SSB application to soil on plants' growth and absorption of the heavy metals, using SSB produced at two different pyrolysis temperatures.

2. Materials and methods

2.1. Soil and biochar

A soil used in this study was a forest Andisol collected from a mulberry plantation forest in Hachioji, Tokyo, Japan (35.692, 139.316) on Apr. 29, 2011. Surface 15-cm soils were sampled, dried at 45°C overnight, sieved through 2 mm, and used for physicochemical analyses and a pot study.

Sewage sludge biochars were produced in commercial plants at high (approximately 800°C; SSB-H) and low (approximately 300°C; SSB-L) temperatures, crashed by hummer, and sieved through 2-mm sieve for chemical analyses, and homogenized between 2 and 5 mm in size for the pot study.

2.2. Pot study

A pot study was performed on campus of Soka University, Tokyo, Japan. Japanese mustard spinach (*Brassica rapa*) and common bean (*Phaseolus vulgaris*) were planted in 1.3 L planting pots in which the soil and SSB were mixed at different rates. Experimental design was a completely randomized block design with triplicate, two different plants, and four different application rates for each SSB: 0%, 25%, 50%, and 75% (v/v) for SSB-H, and 0%, 5%, 15%, and 25% (v/v) for SSB-L. Each pot received respective amounts of ammonium nitrate, superphosphate, and potassium chloride before planting based on fertilizer application rates recommended for the plants by Tokyo Prefecture: 140-52-100 kg N-P-K ha^{-1} for spinach and 80-39-66 kg N-P-K ha^{-1} for bean. Each plant was sowed on Sep. 23, 2011, and thinned to 4 spinach individuals and 2 bean individuals per pot 2 weeks later. The plants were grown outside and the pots were covered by plastic sheet only when rainfall events occurred. Water was applied periodically to keep 50% of water holding capacity of the soil-SSB mix. Spinach and bean were harvested 46 and 66 days after sowing on Nov. 8 and Nov. 28, 2011, respectively. Throughout the experiment, the daily mean temperature fluctuated between 16°C and 22°C, and the daily rainfall more than 20 mm occurred 5 times on Oct. 5, 15, 22, Nov, 11, and 19, 2011.

After harvest, the plants were divided into shoot, root, and bean sheath (only for common bean), and dried at 70°C overnight. The soil and SSB in the pot were separated using 2-mm sieve, and separately dried at 45°C overnight.

2.3. Soil, biochar, plant analyses

Dry weight (DW) of each part of the plants was determined after drying at 70°C overnight.

The soil, SSBs, and plant tissues after the pot study were ground to pass through 150 μm in size and digested for total elemental analyses for heavy metals. Two grams of the soil sample was heated to 180°C with 30 mL of concentrated nitric acid in a beaker with a glass watch on a hotplate until it turned like a syrup. Then, 30 mL of a perchloric acid-nitric acid mixture (1:4) was added and heated for further 10 min. After cooling, 25 mL of hydrochloric acid (1:5) was added and heated to 130°C for 1 hr. After cooling, the content was quantitatively transferred to a 100 mL flask and filtered through a Whatman No 1 paper (Committee for Analytical Methods for Soil Environment, 1997). Two grams of the SSB sample in crucible was ashed in an electric furnace at 550°C for 8 hr, and the ash was transferred to a beaker with 5 mL of concentrated nitric acid and heated to 120°C on a hotplate for 3 hr. After cooling, 25 mL of hydrochloric acid (1:5) was added and heated at 120°C for further 60 min. After cooling, the content was quantitatively transferred to a 25 mL flask and filtered through a Whatman No 1 paper (Japan Soil Association, 2010). One gram of each of the shoot, root, and sheath plant samples was heated to 140°C with 10 mL of concentrated nitric acid in a beaker with a glass watch on a hotplate until the content is reduced to approximately 1 mL. After cooling, the content was quantitatively transferred to 25 mL flask using 1% nitric acid, and filtered through a Whatman No 1 paper (Committee for Experimental Methods for Plant Nutrition, 1990).

The heavy metals extractable by 0.1 *M* HCl are used as an indicator for environmental pollution in soils. Various chemical extractants are used for single extraction evaluation of heavy metals and may broadly be divided into 3 main classes: (i) weak replacement of ion salts ($MgCl_2$, $CaCl_2$, NH_4CO_3), (ii) dilute solutions of either weak acid (acetic acid) or strong acids (HCl, HNO_3), and (iii) chelating agents (DTPA, EDPA) (Kashem et al. 2007). The unbuffered ion salt solutions are simple procedure to extract bioavailable metals, while a use of 0.1 *M* HCl solution may reflect bioavailability of metals (CSTPA, 1980). Dried 7.0 g of soil samples (< 2 mm) was weighed to a 50 mL centrifuge tube and 35 mL of 0.1 *M* HCl was added. The tube was shaken horizontally for 1 hr at 160 stroke min^{-1}, centrifuged at 5,000 rpm, and the supernatant was filtered through a Whatman No 1 paper (Committee for Analytical Methods for Soil Environment, 1997).

Prior to analytical determination, all filtrates were further filtered through a 0.45 μm membrane. Total and extractable concentrations of Cu, Zn, and Cd in the soil, SSBs, and plant samples were determined using ICP (ICPS-7000 ver. 2.1, Shimazu).

2.4. Statistical analyses

Significant differences of the total and extractable concentrations of Cu, Zn, and Cd in the soil, SSBs, and plant samples among different SSB application rates for each SSB type and each plant were tested by ANOVA using STATISTICA 6.1 (StatSoft Inc., Tulsa, Okalahoma, USA). Unless otherwise stated, the differences were significant at $p \leq 0.05$ level.

3. Results

3.1. Soil and biochar properties

The Andisol used in this study had an almost neutral pH (H_2O) of 7.1 and relatively average TC and TN concentrations (Table 1). The total heavy metal concentrations in the soil were 25 mg Cu kg^{-1}, 43 mg Zn kg^{-1}, and 4.9 mg Cd kg^{-1}. The SSB used in this study had slightly acidic pH levels of 6.6 (SSB-H) and 5.6 (SSB-L). The total Cd concentrations were similar for both SSB types ranging 1.4–1.5 mg kg^{-1}, while the total Cu and Zn concentrations in the

SSB-H were 5 and 1.7 times greater than those in the SSB-L, respectively. The dilute acid (0.1 *M* HCl)-extractable heavy metals varied among elements and SSB types; Cu concentration in SSB-H was 34 times greater than that in SSB-L, while Cd in SSB-H was 5 times less than that in SSB-L.

	pH†	Total C	Total N	Total Cu	Total Zn	Total Cd	HCl Cu‡	HCl Zn‡	HCl Cd‡
		g kg^{-1}		mg kg^{-1}					
Andisol	7.1	45.1	2.8	25	43	4.9	0.23	2.06	0.024
SSB-H§	6.6	532.2	34.1	346	455	1.4	19.2	14.1	0.022
SSB-L§	5.6	528.4	54.8	69	273	1.5	0.56	27.3	0.12

† 1:2.5 soil:solution with H_2O and 1:100 SSB:solution with hot H_2O

‡ 0.1 *M* HCl-extractable

§ SSB pyrolysed at high temperature (800°C; SSB-H) and low temperature (300°C; SSB-L)

Table 1. Basic properties of the soil (Andisol) and sewage sludge biochar (SSB) used in this study.

3.2 Dry weight of the plants

Seeds of both plants with all treatments germinated 100% except for those of the bean with SSB-L 25% treatment being 66% of the germination rate.

The average DW of individual spinach shoot showed a significant decrease with increasing application rates of both SSB-H and SSB-L, respectively (Fig. 1a). The highest DW was 1.1 and 1.4 g with SSB-H 0% and SSB-L 5% treatment, respectively. The average DW of the bean shoot did not show significant differences among the SSB-H treatment, but a significant increase with SSB-L 5% and 15% compared with that of control.

The average DW of the spinach root showed a significant decrease with increasing application rate of SSB-H, but not among the SSB-L treatment except for SSB-L 25% showing the lowest DW among the treatment (Fig. 1b). The average DW of the bean root showed a significant decrease with the increasing SSB-H application rate, and a significant increase only with SSB-L 15% among the SSB-L treatment.

The average DW of the bean sheath was not significantly affected by the SSB application, ranging from 0.04 and 0.18 g among the SSB-H treatment, and 0.04 g for SSB-L 5% rate (the bean did not bear the sheath with SSB-L 15% and 25% rates; Table 2).

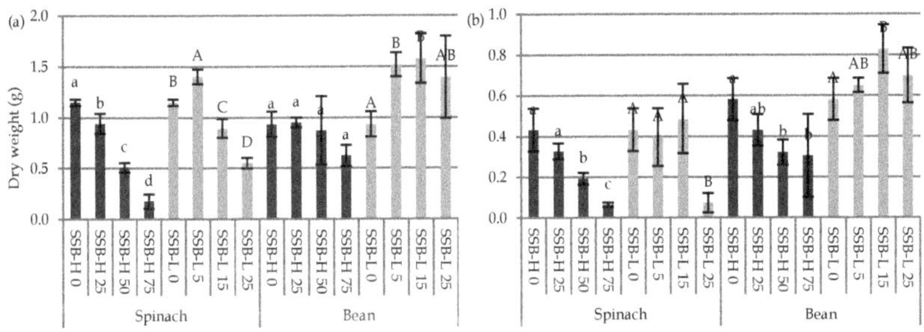

Figure 1. Dry weight of (a) shoot and (b) root parts of Japanese mustard spinach and common bean with different application rates of sewage sludge biochar pyrolysed at high (SSB-H) and low (SSB-L) temperatures. Different letters denote significant differences by Fisher test (p<0.05) among different application rates for each plant and SSB, respectively.

	DW	Cu	Zn	Cd		DW	Cu	Zn	Cd
	g	mg kg^{-1}				g	mg kg^{-1}		
SSB-H 0	0.07ab	5.37a	3.04a	0.58a	SSB-L 0	0.07A	5.37A	3.04A	0.58A
SSB-H 25	0.17ab	1.99a	2.62a	0.28b	SSB-L 5	0.04A	10.75A	5.00A	1.46A
SSB-H 50	0.18a	1.87a	3.36a	0.26b	SSB-L 15	na	na	na	na
SSB-H 75	0.04b	5.77a	3.43a	0.73a	SSB-L 25	na	na	na	na

Table 2. Dry weight (DW), Cu, Zn, and Cd concentrations in the common bean sheath with different application rates of sewage sludge biochar pyrolysed at high (SSB-H) and low (SSB-L) temperatures. Different letters denote significant differences by Fisher test (p<0.05) among different application rates for each SSB and plant, respectively. na denotes non-applicable.

3.3. Total heavy metal concentrations in the plants

Copper concentrations in the spinach shoot significantly increased with both SSB treatments over the control, respectively, while, although not significantly different, those in the bean shoot were lower with both SSB treatments compared with the control, respectively (Fig. 2a). Zinc concentrations in the spinach shoot significantly increased with both SSB treatments over the control, respectively, except for SSB-L 5% rate (Fig. 2b). On the other hand, Zn in the bean shoot was not significantly affected by both SSB treatments, except for SSB-L 5% being significantly lower than the control. The shoot part of each of both plants contained significantly greater Cd concentrations with both SSB treatments over the control, respectively (Fig. 2c).

The concentrations in the plant root were generally greater than those in the shoot for all heavy metals investigated regardless of the plant, SSB type and application rate. The Cu concentration in the spinach root was greatest with SSB-H 25% rate and SSB-L 15% rate among SSB-H and SSB-L treatments, respectively, while Cu in the bean root was greatest with SSB-H 50% rate and SSB-L 25%, among each SSB treatment, respectively (Fig. 2d). The Zn concentrations in the spinach root with SSB-H 25% and 50% rates were significantly greater than that of the control, and although not significant, those with the SSB-L treatment were greater than that of the control (Fig. 2e). The bean root absorbed significantly greater Zn with SSB-H 50% and 75% rates, and not significantly but greater Zn with SSB-L 15% and 25% rates, compared with that of the control, respectively. Cadmium in the spinach root was significantly higher with SSB-H 75% rate than those at other rates, but Cd with the SSB-L treatment was lower than that of the control (Fig. 2f). The Cu in the bean root was significantly lower with SSB-H 75% rate and higher with SSB-L 0% rate compared to other rates, respectively.

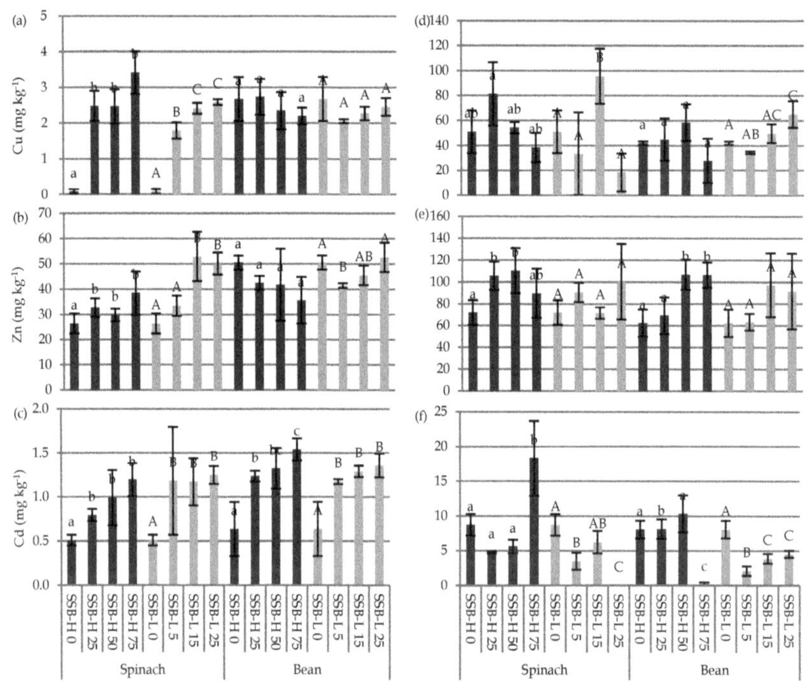

Figure 2. Total concentrations of (a) Cu, (b) Zn, and (c) Cd in the shoot and (d) Cu, (e) Zn, and (f) Cd in the root of Japanese mustard spinach and common bean with different application rates of sewage sludge biochar pyrolysed at high (SSB-H) and low (SSB-L) temperatures. Different letters denote significant differences by Fisher test (p<0.05) among different application rates for each plant and SSB, respectively.

Any of Cu, Zn, and Cd in the sheath was not significantly affected by either application of SSB-H or SSB-L compared with that of the control, respectively, except for Cd with SSB-H 25% and 50% rates being significantly higher than that of the control (Table 2).

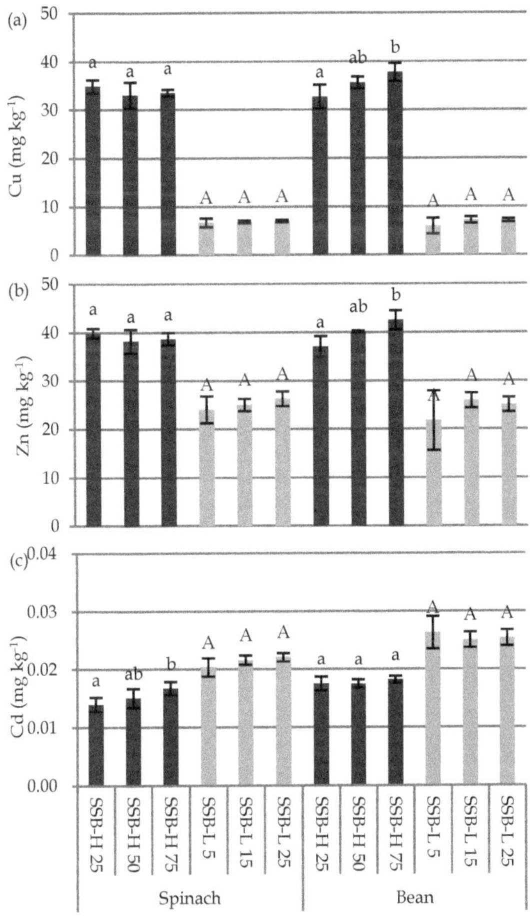

Figure 3. Total concentrations of (a) Cu, (b) Zn, and (c) Cd in the biochar after the pot study with different application rates of sewage sludge biochar pyrolysed at high (SSB-H) and low (SSB-L) temperatures. Different letters denote significant differences by Fisher test ($p<0.05$) among different application rates for each plant and SSB, respectively.

3.4. Total heavy metal concentrations in the biochar after the pot study

The total Cu concentration in the SSB-H after the pot study was not significantly affected for the spinach but significantly increased with SSB-H 75% rate for the bean (Fig. 3a). The total Cu concentrations in the SSB-L were not significantly affected by the application rate for both plants. After harvest of the spinach the total Zn in the biochar was not affected by the application rate for both SSBs, but at the bean's harvest only SSB-H 75% rate caused significantly higher total Zn among the SSB-H treatment (Fig. 3b). The total Cd in the biochar did not significantly change among each of the SSB rates and plants, respectively, except for that with SSB-H 75% rate with the spinach being significantly greater than other SSB-H rates (Fig. 3c).

3.5. Total heavy metal concentrations in the soil after the pot study

The total Cu concentration in the soil after the pot study was significantly increased only with SSB-H 75% rate among the SSB-H treatment for both plants, respectively, but not affected among the SSB-L treatment for both plants (Fig. 4a). The total Zn in the soil tended to increase with the SSB-H application rate and was significantly greater with SSB-H 75% rate compared with those with other rates (Fig. 4b). The total Zn in the soil was not significantly different among the SSB-L treatments for both plants. The total Cd in the soil was significantly increased with SSB-H 25% and 50% rates for the spinach, but not affected for the bean (Fig. 4c). The total Cd in the soil with the SSB-L treatment significantly increased compared with that of the control for both plants, respectively.

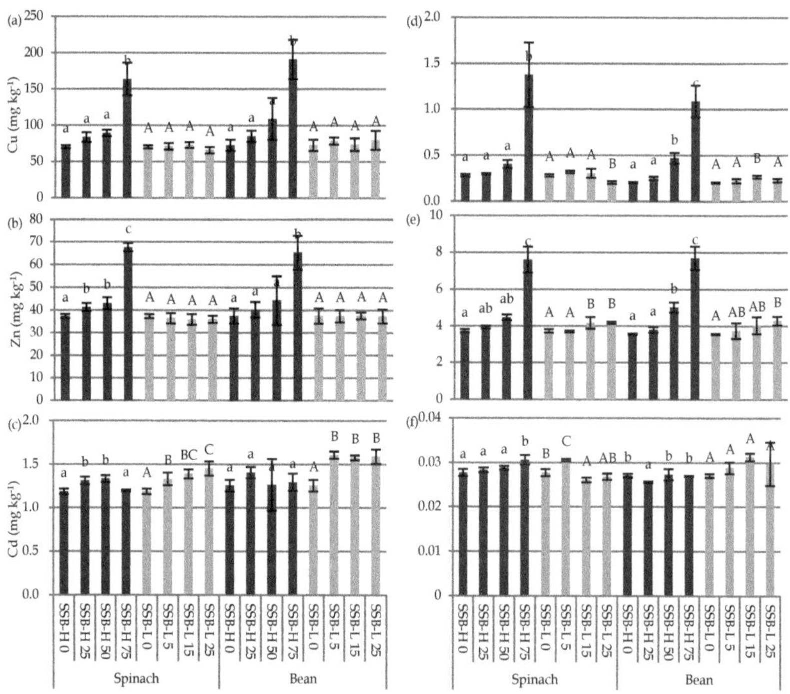

Figure 4. Total concentrations of (a) Cu, (b) Zn, and (c) Cd and dilute acid (0.1*M* HCl)- extractable (d) Cu, (e) Zn, and (f) Cd in the soil after the pot study with different application rates of sewage sludge biochar pyrolysed at high (SSB-H) and low (SSB-L) temperatures. Different letters denote significant differences by Fisher test ($p<0.05$) among different application rates for each plant and SSB, respectively.

3.6. Acid-extractable heavy metal concentrations in the soil after the pot study

The dilute acid (0.1 *M* HCl)-extractable Cu in the soil after the pot study tended to increase with increasing SSB-H application rate for both plants, respectively, and those with 75% for spinach, and 50% and 75% for bean were significantly greater than those with lower rates, respectively (Fig. 4d). The HCl-Cu with SSB-L 25% for spinach and SSB-L 15% for bean was significantly lower and higher, respectively, than those with other rates. Similarly, the HCl-Zn with SSB-H 75% rate for spinach and SSB-H 50% and 75% rates for bean were significantly greater than those with other rates, respectively (Fig. 4e). The HCl-Zn with SSB-L 15% and 25% were significantly greater than those with other rates for spinach, and that with SSB-L 25% rate was significantly greater than those with other rates for bean. The HCl-Cd with SSB-H 75% rate was significantly higher than those with other rates for spinach, and that with SSB-H 25% rate was significantly lower than those with other rates for bean (Fig. 4f). The HCl-Cd with SSB-L 5% was greater compared with those with other rates for spinach, but there were no significant differences among the SSB-L treatment for bean.

4. Results and discussion

4.1. Heavy metal concentrations in the soil and biochar

It appears that the Andisol used in this study contained less but comparable total Cu and Zn, and greater total Cd concentrations compared with the concentration range in similar soils of Japan. The concentrations of Andisols sampled in surrounded subareas of Tokyo ranged from 97 and 140 with the average of 118 mg Cu kg^{-1}, from 121 and 175 with the average of 143 mg Zn kg^{-1}, and from 0.37 and 0.55 with the average of 0.44 mg Cd kg^{-1} (Terashima et al., 2004). Some heavy metals including Cu, Zn, and Cd have tendency of being retained thus accumulated in soils given being relatively immobile after weathering in the soil (Terashima et al., 2004). High concentrations of some heavy metals including Cd, Hg, Pb, and Sb were observed because of possible artificial interventions especially in suburb areas of Tokyo (Terashima et al., 2007). In fact, the area where the Andisol used in this study was sampled was a former forest converted to a plantation area for mulberry (*Morus alba* L.) tree. The conversion and maintenance processes of the plantation could have involved some artificial interferences that might involve Cd deposition to the area.

Many properties including heavy metal concentrations in SSB widely vary primarily depending on those contained in the feedstock (Chan & Xu, 2009). An SSB pyrolysed at 800°C from sewage sludge in Kyoto, Japan contained 1,100 mg Cu kg^{-1}, 1,630 mg Zn kg^{-1}, 4 mg Cd kg^{-1} (Kawano et al., 2012), while one pyrolysed at 500°C using sewage sludge in Madrid, Spain contained 222 mg Cu kg^{-1}, 1,250 mg Zn kg^{-1}, 1.79 mg Cd kg^{-1} (Paz-Ferreiro et al., 2012). It seems that both SSB used in this study fall in the low end of concentration ranges for Cu, Zn, and Cd in the SSB found in the literature. Particularly, Zn concentrations in both SSBs were exceptionally lower than those in the literature. The heavy metal concentration in the SSB appears to accumulate as the pyrolysis temperature increases (Hossain et al., 2011), to which the SSB used in this study followed a similar pattern.

4.2. Effect of biochar on plant growth

The DW of spinach shoot and root and bean root significantly decreased with increasing rates of SSB-H due mainly to a lack of soil volume in the pot. In fact, chlorosis due to N deficiency was observed with the SSB-H treatment and more pronounced on leaves of both plants with higher SSB-H rates. Although the plant was fertilized, it appeared that the lack of soil volume in the pot with high SSB-H rates limited plant root growth thus the plant growth. In addition, although the concentration was not measured in soil or plants, both plants with the SSB-L treatment showed a leave curvature due possibly to excess uptake of boron. When SSB-L was applied, however, the shoot of both plants grew more than that of the control and their DW peaked at 5% rate for spinach and 15% for bean. A similar trend was observed when a begonia (*Begonia semperflorens*) was grown on an Andisol mixed with SSB pyrolysed at 800°C resulting in the DW peak with 25% rate (v/v) (Kawano et al., 2012). In Kawano's study, the effect of SSB application on the plant growth was most pronounced on a most infertile sandy Entisol (Typic Udipsamments) among tested soils with the highest SSB application. It appears that the positive effect of SSB application to soil on the plant may be more prominent in soils which are less favorable for plant growth (Kawano et al., 2012). On the other hand, when an SSB pyrolysed at 700°C was applied to a volcanic ash soil from Kanto area, Japan with 0% (control), 30%, 50%, and 100% (v/v) application rates to grow common bean, the DW at harvest with each treatment was in the order of control<100%<30%<50% (2.2, 3.6, 4.0 times of control, respectively), and rhizobia with all SSB treatments developed more than that with control (Teranuma & Mori, 2002). It appears that the optimum application rates of SSB for maximum plant growth widely depend on varying properties of soil, SSB, and plant, on which requires further investigations.

4.3. Plant absorption of heavy metal

Some heavy metals such as Zn, Cu, and Ni are important for proper functioning of biological systems and deficiency or excess can lead to a number of disorders (Ward, 1995), while others such as Cd are dangerous pollutant due to their high toxicity to plants and inhibit growth at high absorption (Liu et al., 2006). The range of heavy metal concentrations absorbed in both shoot and root of the plants was in the order of Cd<Cu<Zn, respectively, in this study. The same order of the concentration range of heavy metals was observed in various vegetable crops including similar plants used in this study as spinach (*Amaranthus caudatus*), lettuce (*Lactuca sative*), and peas (*Pisum sativum* L.) (Uwah et al, 2011; Singh et al., 2012). The concentrations of Cu, Zn, and Cd in healthy plants can range 4–15, 25–125, and 0.2–0.8 mg kg^{-1}, respectively (National Institute for Agro-Environmental Sciences, 1977). The total Cu and Zn concentrations in the shoot of both plants in this study may have fallen in safe concentration ranges, respectively, whereas the total Cd may have been in excess absorption range which may have caused inhibition of the plant growth. In fact, the DW of the shoot of the plants showed inverse relationships with the total concentrations of the heavy metals absorbed in the plants' shoot, except for the bean DW under the SSB-L treatment with the total Cd (Table 3). The inhibitive effect of heavy metal absorption on the plant growth, judged by the slope of the correlation equation, was in the order of Zn<Cu<Cd for both plants and both SSB types, respectively.

Particularly, the correlation coefficient was highest for Cd among the heavy metals with the SSB-H treatment for the spinach (0.719). A similar trend was observed when heavy metals' effectiveness in producing oxidative damage on spinach, assessed by the manifestation of external visual toxicity effects, was in the order of Zn<Cu<Cd (Pandey et al., 2009). Interestingly, however, the bean shoot showed Cd tolerance to some degree with the SSB-L treatment in this study (its correlation coefficient was positive).

Plant/SSB	Equations	Coefficients
Spinach/SSB-H	(DW) = –0.00097*(Cu) + 0.0048	0.651
	(DW) = –0.00015*(Zn) + 0.0074	0.345
	(DW) = –0.00443*(Cd) + 0.0066	0.719
Spinach/SSB-L	(DW) = –0.00076*(Cu) + 0.0052	0.281
	(DW) = –0.00008*(Zn) + 0.0070	0.422
	(DW) = –0.00129*(Cd) + 0.0051	0.121
Bean/SSB-H	(DW) = –0.00008*(Cu) + 0.0019	0.007
	(DW) = –0.00001*(Zn) + 0.0021	0.038
	(DW) = –0.00022*(Cd) + 0.0020	0.044
Bean/SSB-L	(DW) = –0.00053*(Cu) + 0.0041	0.082
	(DW) = –0.00006*(Zn) + 0.0054	0.219
	(DW) = 0.00151*(Cd) + 0.0010	0.537

Table 3. Correlation equations and coefficients between the dry weight (DW) and total concentrations of Cu, Zn, and Cd absorbed in the shoot of the plants.

Overall concentrations of Cu, Zn, and Cd in the root were greater than those in the shoot of the plants probably because of root's chelating exudates to solubilize heavy metals in the rhizosphere (Inaba & Takenaka, 2000) and rapid absorption by the root and slow translocation to shoot (Nada et al., 2007). Accumulation of heavy metals (Cu, Zn, and Cd) in spinach after exposure to 500 μM supply of the metals was greater in the root than in the leaf and stem at the end of experiment (Pandey et al., 2009). The rate of heavy metal absorption by the plant can be affected by many factors of both soil and plant such as soil pH, plant age, plant species, and nature of soil and climate (Alloway & Ayres, 1997; Uwah, 2009). Although not verified, the application of acidic SSBs used in this study may have decreased the neutral pH of the soil during the pot study, which may have enhanced plant absorption of especially Zn and Cd, as commonly observed irrespective of the vegetable crops and soil types (Kuo et al., 1985; Xue & Harrison, 1991). The effects of SSB application on soil properties, heavy metal bioavailability, and plant growth, therefore, need to be further evaluated for the range of different soils and plants, and over time because heavy metal bioavailability may increase over time in soil (Schauer et al., 1980).

4.4. Heavy metal leachability from biochar

The total concentrations of Cu, Zn, and Cd in SSB after the pot study did not greatly vary among the application rates within the same SSB type and plant, respectively, and 89–91%, 90–92%, and 98–99% of the original concentrations of Cu, Zn, and Cd were lost from the SSB during the pot study, respectively, regardless of the SSB type, application rate, and plant. While there are numerous studies that deal with sorption capacity of biochars for heavy metals when applied to soil in light of phytoremediation perspective (Cao et al., 2009; Namgay et al., 2010; Uchimiya et al., 2011), studies regarding on leachability (through desorption or dissolution) of heavy metals from biochar and their bioavailability are virtually nonexistent. The heavy metals, however, are known to be desorbed by naturally occurring organic acids such as citric, oxalic, acetic, and lactic acids (Nascimento, 2006; Marchi, 2009) and dissolved organic carbon (Antoniadis & Alloway, 2002) from soils amended with sewage sludge. Regardless of equilibrium amounts of Cd adsorbed in an Andisol, more than 80% of Cd was desorbed with one time extraction with citric acid if its concentration was more than 0.1 *M*, and more than 90% was recovered if five soil pore volumes of 0.1 *M* citric acid were continuously run through the Cd-contaminated soil (Abe et al., 2004). An increase of dissolved organic matter in soil by the application of digested dewatered sludge significantly reduced the sorption of Cu on both acidic sandy loam and calcareous clay loam (Zhou & Wong, 2001). The application of SSBs in this study may have increased dissolved organic carbon given the SSBs being acidic, which may have increased Cu desorption (extractability) thus also bioavailability. The heavy metals in the SSB used in this study may have leached due to functions of various interactions between soil and plant roots. Further studies are needed to clarify leachability and bioavailability of heavy metals from biochars whose feedstock contain significant amounts of heavy metals in order to elucidate heavy metal dynamics in soil upon the application of such biochars.

4.5. Heavy metal accumulation and availability in soil

The total concentrations of Cu and Zn in the soil after the pot study, regardless of the treatment, were greater and almost equal compared with those of the original soil (25 and 43 mg kg^{-1}, respectively), while both concentrations significantly increased when SSB-H was applied at 75% rate for both plants. On the other hand, the overall concentrations of the total Cd were smaller after the pot study compared with that at pre-plant condition (4.9 mg kg^{-1}); however those with the SSB-L treatment increased with increasing application rate for both plants. Transfer factors (TF) of heavy metal, as calculated as the ratio of the concentration of heavy metal absorbed in a plant to the concentration of heavy metal in soil, can quantify the relative differences in bioavailability of heavy metals to vegetables or to identify the efficiency of a vegetable species to accumulate a heavy metal (Uwah et al., 2011). The TF values in this study, calculated based on the sum of heavy metal concentrations of the shoot and root as the plant concentration and the sum of heavy metal concentrations in the soil and SSB as the soil concentration, were 0.77, 0.83, and 1.37 for spinach, and 0.69, 0.91, and 1.19 for bean for Cu, Zn, and Cd, respectively, regardless of the SSB application rate. These TF values may explain why Cu appeared to have been accumulated in the soil (TF less than 1 and lowest among the 3 metals for both plants), Zn seemed to have had no changes before and after the pot study

(TF close to 1), and Cd appeared to have been accumulated more in the plant than in the soil (TF more than 1). The TF values widely vary depending on properties of soils and plants, however, those calculated in this study may be comparable with those found by Uwah et al. (2011), which were 0.25–0.95, 0.38–0.55, and 0.42–2.75 for Cu, Zn, and Cd, respectively, for spinach and lettuce grown on tropical soils in Nigeria. Further studies are needed to elucidate heavy metal accumulation in soil and selective absorption by plants with the SSB application.

The heavy metal concentrations extractable by 0.1 *M* HCl acid solution in the soil after the pot study followed similar patterns as the total concentrations in the soil. The Cu and Zn concentrations significantly increased only when SSB-H was applied at 75% rate for both plants, while the Cd concentrations did not show noteworthy differences among the treatment. However, the percentage of the concentration of the acid-extractable Cu, Zn, and Cd to the total concentration (acid solution extractability) was 0.3–0.8%, 9.4–11.8%, and 1.8–2.6%, respectively. When 4 contaminated and 4 non-contaminated soils from a northern part of Japan were extracted for heavy metals using 0.1 *M* HCl, the extractability ranged 12% and 27–33% for Cu, 9% and 12–31% for Zn, and 33% and 73–92% for Cd in the contaminated and non-contaminated soils, respectively (Kashem et al., 2007). Although the extractability of heavy metals from soil by the dilute acid may vary depending on soil type, heavy metal, contamination degrees, and so on, 0.1 *M* HCl extractant may be the best choice for assessment of mobility or bioavailability of heavy metals (Kashem et al., 2007).

The environmental threshold concentrations in soils are 125 mg Cu kg^{-1}, 125 mg Zn kg^{-1}, and 0.01 mg Cd L^{-1} as 0.1 *M* HCl extractable heavy metals, set by the Japanese Environmental Ministry. It seems that neither of SSB types within the application rates and study duration used in this study have caused environmental threads to soils. Based on life cycle assessment analysis on different treatment processes of sewage sludge including anaerobic digestion, pyrolysis, and incineration, it was concluded that the most effective utilization of sewage sludge implied both energy and material reuse, and that the land application of digested sludge was an acceptable and good choice as long as heavy metal contents in the final cake could be minimized (Hospido et al., 2005). Therefore, long-term effects of SSB application on heavy metal dynamics among SSBs, soils, and plants are needed to be evaluated for further acceptance of SSB application in agronomic benefits.

5. Conclusions

Biochars derived from feedstock including heavy metals such as SSB need to be thoroughly evaluated for heavy metal dynamics among SSBs, soils, and plants for environmentally safe and sound application of SSB. Both SSBs used in this study contained a low end of concentration ranges of Cu, Zn, and Cd, respectively, found in the literature, and the heavy metal concentrations accumulated in the SSB as the pyrolysis temperature increased as found in the literature. Both spinach and bean plants suffered from N deficiency due mainly to a lack of soil volume with excessively high application of SSB-H, and possibly from B toxicity especially from SSB-L application. However, when SSB-L was applied, the shoot DW of plants peaked with 5% application rate for spinach and 15% rate for bean. Therefore, it was concluded that the optimum application rates of SSB-H and SSB-L were indeterminate (lower application rates of SSB-H need to be evaluated) and 5–15%, respectively, for the best growth of the plants in this study. The concentration ranges of the heavy metals absorbed in both plants were in the order of Cd<Cu<Zn for both shoot and root. The total Cu and Zn in the shoot may have been in safe concentration ranges, and the total Cd may have been in excess range which may have caused inhibition of the growth. In fact, the plant DW showed inverse relationships with the total concentrations of the heavy metals absorbed in the plants, and the inhibitive effect of heavy metals on the plant growth was in the order of Zn<Cu<Cd. Overall concentrations of the heavy metals in the root were greater than those in the shoot for both plants. The leachability of Cu, Zn, and Cd from SSBs was 89–91%, 90–92%, and 98–99% of the original total concentrations, respectively, during the pot study regardless of the SSB type, application rate, and plant. However, the total concentrations of Cu, Zn, and Cd in the soil after the pot study were accumulated, unaffected, and reduced, respectively, compared with those before the pot study, which could be explained by the TF of each heavy metal which were less than 1, close to 1, and more than 1, respectively. Nevertheless, the percentage of the dilute acid-extractable Cu, Zn, and Cd to the total concentrations in the soil after the pot study was 0.3–0.8%, 9.4–11.8%, and 1.8–2.6%, respectively, which were lower than the environmental threshold concentrations in Japan. However, long-term effects of SSB application on heavy metal dynamics among SSBs, soils, and plants are needed to be evaluated for further acceptance of SSB application in agronomic benefits.

Author details

Mieko Yachigo and Shinjiro Sato

Department of Environmental Engineering for Symbiosis, Soka University, Tokyo, Japan

References

[1] Abe, Y, Yamaguchi, N, Mizoguchi, M, Imoto, H, & Miyazaki, T. (2004). Effect of organic acid on transport of cadmium in soil, *Proceedings for Annual Meeting of the Japanese Society of Irrigation Drainage and Reclamation Engineering,* (in Japanese), 116-117.

[2] Alloway, B. J, & Ayers, D. C. (1997). *Chemical Principles for Environmental Pollution,* Blackie Academic and Professional, London, UK., 190-220.

[3] Antoniadis, V, & Alloway, B. J. (2002). The role of dissolved organic carbon in the mobility of Cd, Ni and Zn in sewage sludge-amended soils, *Environmental Pollution,* 117, 515-521.

[4] Cao, X, Ma, L, Gao, B, & Harris, W. (2009). Dairy-manure derived biochar effectively sorbs lead and atrazine, *Environmental Science & Technology*, 43, 3285-3291.

[5] Chan, K. Y, & Xu, Z. (2009). Biochar: nutrient properties and their enhancement, In: *Biochar for Environmental Management: Science and Technology*, J. Lehmann & S. Joseph, Earthscan, 978-1-84407-658-1, London, UK., 67-84.

[6] Committee for Analytical Methods for Soil Environment(1997). *Analytical Methods for Soil Environment*, Hakuyusha, 978-4-8268-0168-3, Tokyo, Japan. (in Japanese)

[7] Committee for Experimental Methods for Plant Nutrition(1990). *Experimental Methods for Plant Nutrition*, Hakuyusha, 978-4-8268-0118-8, Tokyo, Japan. (in Japanese)

[8] Council on Soil Testing and Plant Analysis (CSTPA)(1980). *Handbook on Reference Methods for Soil Testing*, University of Georgia, Athens, Georgia, US.

[9] Hospido, A, Moreira, M. T, Martín, M, Rigola, M, & Feijoo, G. (2005). Environmental evaluation of different treatment processes for sludge from urban wastewater treatments: anaerobic digestion versus thermal processes, *The International Journal of Life Cycle Assessment*, 10, 36-345.

[10] Hossain, M. K, Strezov, V, Chan, K. Y, & Nelson, P. F. (2010). Agronomic properties of wastewater sludge biochar and bioavailability of metals in production of cherry tomato (*Lycopersicon esculentum*), *Chemosphere*, 78, 1167-1171.

[11] Hossain, M. K, Strezov, V, Chan, K. Y, Ziolkowski, A, & Nelson, P. F. (2011). Influence of pyrolysis temperature on production and nutrient properties of wastewater sludge biochar, *Journal of Environmental Management*, 92, 223-228.

[12] Inaba, S, & Takenaka, C. (2000). Remediation of heavy metal-contaminated soil in the rhizosphere, *Root Research*, (in Japanese), 9, 69-73.

[13] Japan Soil Association(2010). *Analytical Methods for Manure and Organic Matter*, Japan Soil Association, Tokyo, Japan. (in Japanese)

[14] Kashem, M. A, Singh, B. R, Kondo, T, Hug, S. M. I, & Kawai, S. (2007). Comparison of extractability of Cd, Cu, Pb, and Zn with sequential extraction in contaminated and non-contaminated soils, *International Journal of Environmental Science and Technology*, 4, 169-176.

[15] Kawano, Y, Yanai, J, Yamada, H, Nakajima, K, Yasui, D, & Takebe, K. (2012). Physico-chemical properties of carbonized sewage sludge and effects on plant growth, *Japanese Journal of Soil Science and Plant Nutrition*, (in Japanese), 83, 125-132.

[16] Kuo, S, Jellum, E. J, & Baker, A. S. (1985). Effects of soil type, liming, and sludge application on zinc and cadmium availability to Swiss chard. *Soil Science*, 139, 122-130.

[17] Liu, D. H, Wang, M, Zou, J. H, & Jiang, W. S. (2006). Uptake and accumulation of cadmium and some nutrient ions by roots and shorts of maize (*Zea mays* L.), *Pakistan Journal of Botany*, 38, 701-709.

[18] Marchi, G, Guilherme, L. R. G, Chang, A. C, & Nascimento, C. W. A. (2009). Heavy metals extractability in a soil amended with sewage sludge, *Scientia Agricola*, 66, 643-649.

[19] Méndez, A, Gómez, A, Paz-ferreiro, J, & Gascó, G. (2012). Effects of sewage sludge biochar on plant metal availability after application to a Mediterranean soil, *Chemosphere*, 89, 1354-1359.

[20] Ministry of the Environment Government of Japan(2011). *Japan annual report on the environment, the sound material-cycle society and the biodiversity 2011*, Ministry of the Environment Government of Japan, Tokyo, Japan.

[21] Nada, E, Ferjani, B. A, Ali, R, Bechir, B. R, Imed, M, & Makki, B. (2007). Cadmium induced growth inhibition and alteration of biochemical parameters in almond seedlings grown in solution culture, *Acta Physiologiae Plantarum*, 29, 57-62.

[22] Namgay, T, Singh, B, & Singh, B. P. (2010). Influence of biochar application to soil on the availability of As, Cd, Cu, Pb, and Zn to maize (*Zea mays* L.), *Soil Research*, 48, 638-647.

[23] Nascimento, C. W. A. (2006). Organic acids effects on desorption of heavy metals from a contaminated soil, *Scientia Agricola*, 63, 276-280.

[24] National Institute for Agro-Environmental Sciences(1977). Database regarding the concentration of metal elements in plants, National Institute for Agro-Environmental Sciences, Ibaraki, Japan. (in Japanese)

[25] Pandey, N, Pathak, G. C, Pandey, D. K, & Pandey, R. (2009). Heavy metals, Co, Ni, Cu, Zn and Cd, produce oxidative damage and evoke differential antioxidant responses in spinach, *Brazilian Society of Plant Physiology*, 21, 103-111.

[26] Paz-ferreiro, J, Gascó, G, Gutiérrez, B, & Méndez, A. (2012). Soil biochemical activities and the geometric mean of enzyme activities after application of sewage sludge and sewage sludge biochar to soil, *Biology and Fertility of Soils*, 48, 511-517.

[27] Schauer, P. S, Wright, W. R, & Pelchat, J. (1980). Sludge-borne heavy metal availability and uptake by vegetable crops under field conditions, *Journal of Environmental Quality*, 9, 69-73.

[28] Singh, S, Zacharias, M, Kalpana, S, & Mishra, S. (2012). Heavy metals accumulation and distribution pattern in different vegetable crops, *Journal of Environmental Chemistry and Ecotoxicology*, 4, 170-177.

[29] Teranuma, M, & Mori, T. (2002). Applications of carbonized sludge as soil improvement additives, *Journal of Japan Sewage Works Association. Research Journal*, (in Japanese), 39, 167-175.

[30] Terashima, S, Imai, N, Ohta, A, Okai, T, & Mikoshiba, M. (2004). Geochemistry of soils from the southern Kanto district, Japan: Preliminary study for the soil geochem-

ical mapping (Part 5: Generalization), *Bulletin of the Geological Survey of Japan*, (in Japanese), 55, 1-18.

[31] Terashima, S, Imai, N, Tachibana, Y, Okai, T, Mikoshiba, M, Ohta, A, & Kubota, R. (2007). Chemical composition and background evaluation of soils and stream sediments from Kanto district, and marine sediments from Tokyo Bay, *Bulletin of the Geological Survey of Japan*, (in Japanese), 58, 69-91.

[32] Uchimiya, M, Wartelle, L. H, Klasson, K. T, Fortier, C. A, & Lima, I. M. (2011). Influence of pyrolysis temperature on biochar property and function as a heavy metal sorbent in soil, *Journal of Agricultural and Food Chemistry*, 59, 2501-2510.

[33] Uwah, E. I. (2009). Concentration levels of some heavy metal pollutants in soil, and carrot (*Daucus carota*) obtained in Maiduguri, Nigeria, *Continental Journal of Applied Sciences*, 4, 76-88.

[34] Uwah, E. I, Ndahi, N. P, Abdulrahman, F. I, & Ogugbuaja, V. O. (2011). Heavy metal levels in spinach (*Amaranthus candatus*) and lettuce (*Lactuca sative*) grown in Maiduguri, Nigeria, *Journal of Environmental Chemistry and Ecotoxicology*, 3, 264-271.

[35] Ward, N. I. (1995). Environmental analytical chemistry, In: *Trace Elements*, F.W. Fifield & P.J. Haines, Blackie Academic and Professional, London, UK., 320-328.

[36] Xue, Q, & Harrison, H. C. (1991). Effect of soil zinc, pH, and cultivar on cadmium uptake in leaf lettuce (*Lactuca sativa* L. var. *crispa*), *Communications in Soil Science and Plant Analysis*, 22, 975-991.

[37] Zhou, L. X, & Wong, J. W. C. (2001). Effect of dissolved organic matter from sludge and sludge compost on soil copper sorption, *Journal of Environmental Quality*, 30, 878-883.

Section 4

Soil Microbial Processes

Response of Soil Microbial Biomass and Enzyme Activities Under Three Temperate Tree Species to Elevated CO_2 in Changbai Mountain, Northeastern China

Jun Qiang Zheng, Ying Wang and Shi Jie Han

Additional information is available at the end of the chapter

1. Introduction

Owing to fossil fuel combustion, deforestation, and intense agriculture, the concentrations of atmospheric CO_2 [CO_2] has risen by 100ppm since the mid 1800s [1], and it has been predicted to double until the end of this century compared to the pre-industrial value [2]. Numerous studies have shown a greater biomass gain of plants, higher fine root and leaf litter C/N in some species under elevated CO_2 condition [3–7]. Moreover, the rising CO_2 also could alter litter chemistry (e.g., total N, lignin and starch content) and fine root turnover. Because microbial growth is limited by the type and amount of organic substrates entering the soil [8, 9], the changes in above- and below-ground plant input under elevated CO_2 could potentially alter both the substrate availability and microbial activity. Although the effect of elevated CO_2 via plants on soil microorganisms has been few studies investigated [10–12], the detailed plant-mediated effects still are unclear because of the complexity of microbial processes.

Soil microorganisms play an important role in nutrient cycling, CO_2 emission and in formation of soil total organic carbon (TOC) pool. Therefore, any effect of the rising [CO_2] on soil microorganisms might in turn feedback on the response of terrestrial ecosystem to atmospheric CO_2 and the sequestration of extra carbon [9]. Soil enzymes drive soil organic matter decomposition and nutrient transformations. Soil enzyme also was considered as a sensitive indicator, which could be significantly affected by temporal variability [13]. It is evident that the seasonal patterns of temperature and moisture of north temperate ecosystems can affect the activity of soil enzymes [14]. Although several studies have investigated the effects of increased CO_2 on the soil microbial biomass and activity, to our knowledge, only relative few studies have measured the seasonal fluctuations of microbial biomass and soil enzyme activity under higher CO_2 levels [10,11]. At the Oak Ridge FACE site, Sinsabaugh *et al.* (2003) found that soil extracellular enzyme activities, substrate utilization, gross N mineralization and denitrification were all not responsive to CO_2 enrichment [10]. In contrast in an agroforestry ecosystem with typical Mediterranean climate, Moscatelli *et al.* (2005) observed that all soil biochemical properties were significantly affected by the temporal variability, and the interaction between time and CO_2 level significantly influenced β-glucosidase activity and microbial respiration [11]. These studies have produced many conflicting results, with few common themes. Furthermore, the specific seasonal patterns of soil enzyme activities vary with soil properties and types of aboveground vegetations.

Temperate forest ecosystems, which occupy much of the earth's terrestrial surface area, have been considered as the most important C sink for sequestering the increasing atmospheric CO_2 [15]. To understand elevated CO_2 effects on temperate forest ecosystems, a growing number of free air CO_2 enrichment (FACE) and open-top chamber (OTC) research project have been initiated throughout the world. In 1998, we began a long term CO_2-enrichment experiment using open top chamber (OTC) growing three species of trees, Korean pine (*Pinus koraiensis* Sieb. *et* Zucc), Changbai pine (*Pinus sylvestriformis* (Takenouchi) T. Wang ex Cheng) and oak (*Quercus mongolicus* Fisch) in Changbai mountain, China [16]. This mountain ecosystem belongs to the temperate continental climate influenced by monsoon and has the prominent characteristics of mid-latitudinal upland climate. The soil type of this area is upland dark-brown forest volcanic soil [17]. The area has four obvious different seasons as windy spring, hot and rainy summer, cool autumn and cold winter. Our previous studies in this OTC site demonstrated that elevated CO_2 increase aboveground biomass and photosynthesis of the trees (Zhou and Han, unpublished data) and decrease total soil respiration [18]. These results were consistent with other reports from temperate forest ecosystems, e.g., Duke temperate forest FACE site [19]. To date, however, there is still a poor understanding of effects of elevated CO_2 on microbial activity of soil under different tree species.

The objectives of this study were (a) to investigate effects of elevated CO_2 on microbial biomass C/N and the variations in activities of various enzymes throughout the growing season, and (b) to compare the CO_2 response of activities of C, N and P cycling related enzymes in the soils under three different tree species.

2. Materials and methods

Experimental site, design and sampling

The experimental fields were located at Changbai Mountain in Jilin province, northeastern China (42º24′N, 128º06′E, and 738 m elevation). The soil is a dark-brown soil developed from volcanic ash. The topography is basaltic mesa, and the parent rock is loose volcanic ash sand. The mechanical composition of the soil is approximately 29% sand (20 μm – 2 mm), 40% silt (2 – 20 μm) and 31% clay (< 2 μm). Soil organic carbon in 0-10 cm layer is approximately 8.5%, and pH is 5.7 (1:2.5, soil: water). The ecosystem is temperate with a mean annual temperature of 5°C and annual average precipitation of 967.3-1400 mm. A randomized complete block design of ambient and elevated CO_2 was established in an open-top chamber facility at the research station of Changbai mountain forest ecosystems, Chinese Academy of Sciences, in the spring of 1999. Eighteen open-top chambers (each 4.2m in diameter with hexagon and 4 m in height enclosed with a clear glass open-top chamber) were utilized to control CO_2 levels. Korean pine (*P. koraiensis Sieb. et Zucc*) and Changbai pine (*P. sylvestriformis*) seeds were prepared and sowed in May, 1999. CO_2 fumigation treatments began after seeds germination in May 1999. For oak (*Q. mongolica*) experiment, five-year old seedlings were transplanted into open top chambers in autumn 2004, and CO_2 enrichment started at the end of April, 2005. In each year, the exposure started at the end of April and stopped at the end of October (the whole growing season). Half of the chambers were maintained at ambient atmospheric CO_2 concentrations (ca. 350 ppm), others were maintained at elevated levels (ca. 500 ppm) by dispensing 100% CO_2 into the blower fans only in the daytime. Elevated CO_2 concentrations were maintained by continuously monitoring CO_2 concentrations in elevated and ambient-level chambers with an infrared gas analyzer (A-SENSE-D, SenseAir, Sweden) by a computer control system that recorded 10-second averages of CO_2 concentration every 3 minutes and then periodically adjusting the flow of 100% CO_2 into the chambers.

Soil samples were collected seven times: May, June, August and September in 2006, and May, July and September in 2007. At each sampling date, five soil cores (3 cm in diameter and 0–10 cm at deep) were collected within each chamber. The pooled samples were homogenized and roots removed by passing the soil through a 2-mm sieve. The samples used for measurements of soil enzyme activity were kept frozen (-70°C) and microbial biomass were kept cool (4°C) until analysis within 1 week after sampling.

Soil microbial biomass analysis

Subsamples of soil were dried at 105°C for 12 h to determine gravimetric water content. Soil pH was measured in solutions of 50 ml water and 10 g air-dry soil. Microbial biomasses C (C_{min}) and N (N_{mic}) were measured by fumigation-extraction method [20]. Subsamples of sieved soil were fumigated with alcohol-free $CHCl_3$ for 24 h, and then extracted with 0.5 M K_2SO_4 solution. The K_2SO_4 soil extract was analysed for total dissolved organic C (DOC) and total dissolved N (TDN) using a Total Organic Carbon Analyzer (multi N/C 3000, Jena, Germany). Microbial biomass C and N were calculated as differences in extractable DOC and TDN between fumigated and unfumigated soils using a correction factor (Kc) of 0.38 for C_{min} and 0.54 for N_{mic} [20, 21].

Soil enzyme activity analysis

Enzyme activities were determined following Freeman *et al.* (1995) and Kang & Freeman (1999) [22, 23]. We used fluorogenic substrates, 4-methylumbelliferyl-linked (MUB) compounds as substrate analogues (Table 1). Briefly, approximately 0.1 g fresh mass, were homogenized in 1.8 ml of 50 mM tris-maleate buffer (pH 5.0) and 0.2 ml of substrate solution were combined in a 5 ml polypropylene tube. For each sample, there were four analytical replicates plus negative controls (sample plus 4-MUB and buffer, separately) for both sample and substrate color. The tubes were tumbled for 1 h at 30°C, except for phosphatase, which was incubated for 30min. The assay was terminated by adding 2 ml ice cold 96% ethanol. The tubes were centrifuged for 5 min at 4°C (6000 r.p.m.). Fluorescence was determined with a HITACHI 650-60 Fluorescence spectrophotometer at 446 nm emission and 377 nm excitation wavelength. The increase of fluorescence was converted to enzymatic activity with a standard curve was drawn using a series of known concentrations of 4-methylumbellife. Enzyme activities are expressed in units of nmol or μmol of substrate converted per hour per g soil dry mass (nmol h^{-1} g^{-1} or μmol h^{-1} g^{-1}).

Enzyme	Abbreviation	EC	Substrate
1,4-β-glucosidase	bG	3.2.1.21	4-MUB-β-D-glucoside
1,4-α-glucosidase	aG	3.2.1.20	4-MUB-α-D-glucoside
1,4-β-Acetylglucosaminidase	NAG (1,4-β-NAG)	3.1.6.1	4-MUB-*N*-acetyl-β-D-glucosaminide
Cellobiohydrolase	CBH	3.2.1.91	4-MUB-β-D-cellobioside
1,4-β-xylosidase	bX	3.2.1.37	4-MUB-β-D-xyloside
Phosphatase	PA	3.1.3.2	4-MUB-phosphate
Phenol oxidase	PPO	1.10.3.2	L-3,4-Dihydroxyphenylalanine
Peroxidase	PO	1.11.17	L-3,4-Dihydroxyphenylalanine

EC: enzyme commission number; MUB: 4-methylumbelliferyl; Abbreviation: used in this article

Table 1. Enzyme assays conducted on soil samples collected from the open top chamber in this study.

For the phenol oxidase and peroxidase activities assay, l-3,4-dihydroxyphenylalanine (L-DOPA) was used as substrate [24]. We prepared soil slurry solutions of 5.0 g soil in 100 ml of 50 mmol l^{-1} acetate buffer (pH 5.0.) The reaction mixture for the phenol oxidase assay, containing 2 ml, 5 mM l-3,4-dihydroxyphenylalanine (L-DOPA) solution and 2 ml of soil slurry in 5 ml tube, was vortexed for exactly 60 min at 20°C in a shaking incubator and was centrifuged for 5 min at 4°C (6000 r.p.m.). The absorbance of the filtrate was read at 460 nm. For peroxidase activities assay, at the beginning of the incubation, were processed in the same way as phenol oxidase, with L-DOPA substrate and the addition of 200 μl of 0.3 % H_2O_2 [25]. Phenol oxidase activity was subtracted from peroxidase activity to calculate the net peroxidase activities.

Statistical analysis

Data were analyzed using repeated measures analysis of variance (RM-ANOVA) with CO_2 treatment, plant species and their interaction as explaining variables. Prior to analysis, the data was checked for heterogeneity of variance, and when necessary, the variable was transformed to improve normality. RM-ANOVA analyses were performed with SPSS 13.0 statistical software package (SPSS Inc.). Correlation analysis was used to test for correlations between microbial biomass C, N and soil enzyme evaluated as significant at $P<0.05$. Correlations among enzyme activities and microbial biomass as well as dissolve organic C and N were explored using a principal components analysis on the correlation matrix. Principal component analyses (PCA) and canonical discriminate analysis were performed with the STATISTICA 6.0 software package (StatSoft Inc.).

3. Results

Elevated CO_2 significantly decreased the activities of 1,4-β-NAG, 1,4-β-xylosidase, phosphatase, 1,4-β-glucosedase, and phenol oxidase in soil under Changbai pine in 2007 (Fig. 1). On all sampling dates, there was no CO_2 main effect on soil enzyme activity across three tree species (Table 2). Time was a significant factor affecting the activities of hydrolotic enzymes and the two oxidase enzymes in soil (Fig. 1, Table 2). The activity of 1,4-β-NAG and phenol oxidase over the two years for all sample dates showed an interaction between tree species and CO_2 ($P < 0.01$), suggesting the responses of these two enzymes to CO_2 enrichment were dependent on aboveground vegetation. Interactions between CO_2 treatment and sampling time were significant for CBH, peroxidase and phenol oxidase activities.

	βG	αG	βNAG	CBH	βX	PA	PPO	PO
Tests of Between-Subjects Effects								
CO_2	2.69^{NS}	0.70^{NS}	4.06^{NS}	0.63^{NS}	0.06^{NS}	2.59^{NS}	1.75^{NS}	<0.01^{NS}
Species	10.97**	4.03^{NS}	35.05***	2.45^{NS}	120.13***	32.09***	88.93***	410.14***
Species ×CO_2	1.90^{NS}	1.04^{NS}	5.28*	2.10^{NS}	0.39^{NS}	2.94^{NS}	6.69*	3.72^{NS}
Tests of Within-Subjects Effects								
Time	8.89***	2.92*	1.90^{NS}	4.11**	15.19***	5.85***	12.65***	37.86***
Time × Species	7.43***	2.97*	5.84***	2.97**	7.33***	5.06***	10.88***	33.40***
Time × CO_2	1.99^{NS}	0.85^{NS}	2.02^{NS}	2.53*	0.82^{NS}	1.52^{NS}	6.72***	6.02***
Time ×species × CO_2	5.69***	1.16^{NS}	2.66*	3.21**	0.60^{NS}	2.19*	1.91*	5.03***

Note: *F* values are displayed and *, ** and *** represent significane at $P < 0.05$, 0.01 and 0.001 level, respectively. NS represents no significant effect.

Table 2. *F*-values from RM-ANOVA showing effects of elevated CO_2 and time on soil enzyme activity.

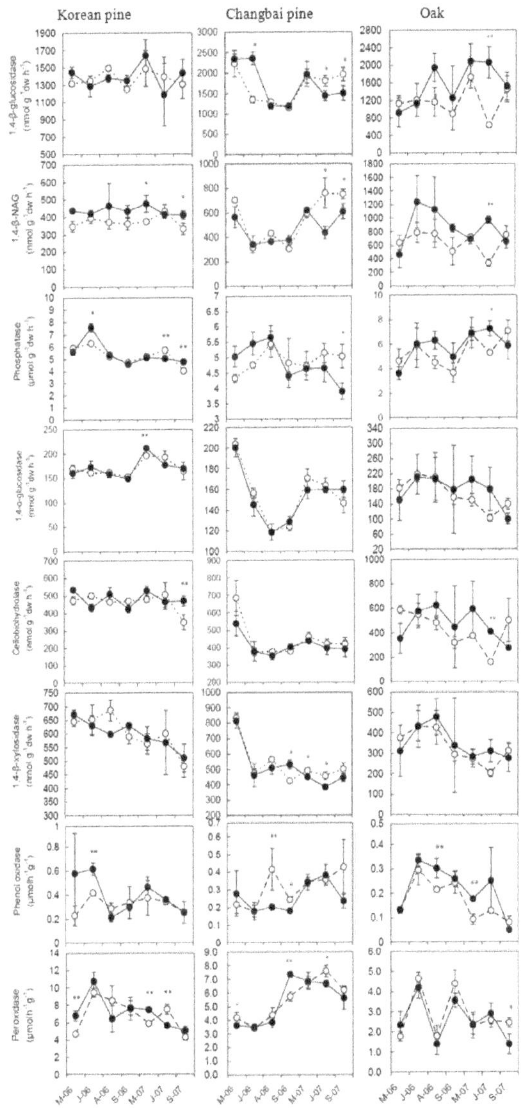

Figure 1. Variations in activities of soil enzymes under Korean pine, Changpai Scotch pine and oak after exposure to ambient (open symbols) and elevated (solid symbols) CO_2 concentration. Error bars represent standard errors of the means; asterisks represent significant differences, e.g. * $p< 0.05$, ** $p< 0.01$.

Correlation analysis indicated that soil moisture was positively correlated with 1,4-α-glucosidase. The TDN content under Korean pine was negatively correlated with DOC (r = -0.6166, P< 0.01) and positively correlated with phosphatase (r = 0.5300, P< 0.01) and peroxidase (r = 0.6559, P< 0.01). TDN from Changbai pine was negatively correlated with DOC (r = -0.7242, P< 0.01), peroxidase (r = -0.7274, P< 0.01) and phenol oxidase (r = -0.5289, P< 0.01), positively correlated with xylosidase (r = 0.6849, P< 0.01).

Mean enzyme activity decreased from highest to lowest in the following order: PPO > AP > βG > βX > CBH > NAG > PO > αG in the Korean pine soil. Mean enzyme activity in the Changbai pine soil decreased from highest to lowest in a very similar order with Korean pine: PPO > AP > βG > βX > NAG > CBH > PO > αG, therefore, tree species is a important factor influencing soil enzyme activities, there were significant interactive effects between species and sampling time.

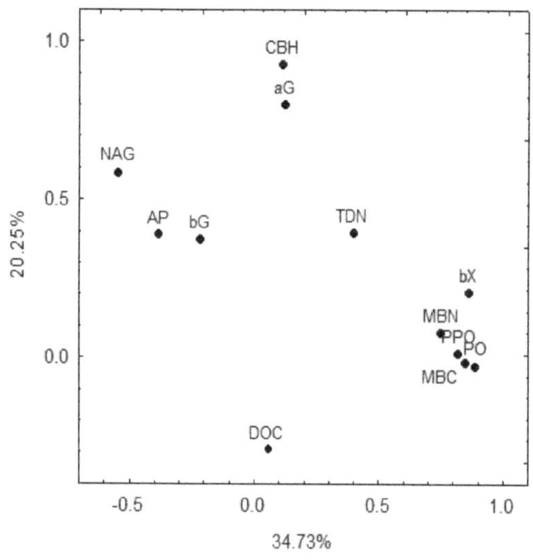

Figure 2. Factor plot from a principal components analysis of soil enzyme activity data for Korean red pine, Changpai Scotch pine and oak exposed to two levels of atmospheric CO_2. PO: peroxidase, PPO: phenol oxidase, AP: acid phosphatase, αG: 1,4-α-glucosidase, CBH: cellobiohydrolase, bX: 1,4-β-xylosidase, NAG: 1,4-β-acetylglucosaminidase, bG: 1,4-β-glucosidase, MBC: microbial biomass C, MBN: microbial biomass N, DOC: dissolve organic C, TDN: total dissolve N.

Soil pH was slightly acidic (5.4–6.1) and did not vary significantly between treatment or sampling dates, and varied among species. Multivariate statistics were used to assess the functional diversity and temporal vary of the soil microbial enzyme. In the principal component analysis of the data of all eight enzyme activities, PC1 and PC2 explained 20.25% and 34.73% of the total variance, respectively (Fig. 2). The first factor (PC1) appears to be associated with labile nutrient acquisition, the second factor (PC2) appears to be associated with lignocelluloses degradation. For the canonical discriminant analysis of the soils under three trees sampled in 2006 and 2007, the following variables were determined: 1,4-β-xylosidase, 1,4-α-glucosidase, 1,4-β-NAG, cellobiohydrolase, 1,4-β-xylosidase, phosphatase, phenol oxidase, peroxidase. Canonical discriminant analysis also showed that the effect of the tree species on soil enzymes during the two experimental years (Fig. 3). Root 1 seems to discriminate mostly between pines and oak (means of the canonical variables: -2.005 and 4.010, respectively). In the vertical direction (Root 2), seems to discriminate the two pines (means of the canonical variables: Korean pine -1.005, Changbai pine 1.264). Correlation analysis indicated that soil moisture was positively correlated strongly with 1,4-α-glucosidase, microbial biomass C and N across three species of tree (Table 3). Peroxidase activity showed strong correlations with N_{mic} (r = 0.54, P< 0.01) and C_{mic} (r = 0.81, P< 0.01), phenol oxidase activity showed a high correlation with N_{mic} (r = 0.48, P< 0.01) and C_{mic} (r = 0.63, P< 0.01).

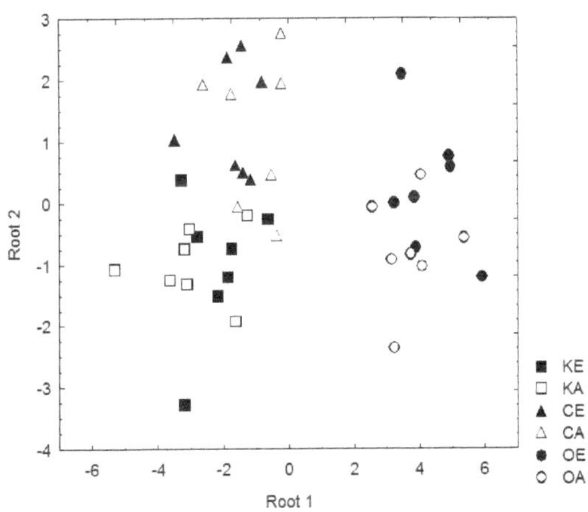

Figure 3. Canonical variates of the microbial properties in the soils under Korean pine, Changpai Scotch pine and oak. KE, Korean pine under elevated CO_2; KA, Korean pine under ambient CO_2; CE, Changpai Scotch pine under elevated CO_2; CA, Changpai Scotch pine under ambient CO_2 ;OE, oak under elevated CO_2; OA, oak under ambient CO_2.

4. Discussioin

Effect of CO_2 concentrations on soil enzyme activities

Elevated atmospheric CO_2 can lead to an increase in the size of the substrate pool utilized by soil microbes and to stimulate the activities of soil enzyme [26]. For example, Larson *et al.* (2002) and Lipson *et al.* (2005) found a significant stimulation of soil extracellular enzyme activities under elevated CO_2 [27, 28]. However, our results, along with other studies [29, 10] have not detected significant increase in C-cycling enzymes across the course of experiment, and only significant at specific time was observed with a higher mean value under elevated

CO_2. Furthermore, the significantly different sampling time point was independent among three types of trees. These results suggest that response of soil enzymes to long-term CO_2 fumigation may be dependent on soil environment and aboveground vegetation type.

1,4-β-NAG is one of the enzymes regulating nitrogen availability in soil, the activity is often used as an indicator of N demand by microbes. Its enhancement under elevated CO_2 (av. +18%) in our study of Korean pine reflected a microbial demand for N nutrient. It is interesting to note that the activities of cellobiohydrolase, 1,4-β-NAG, 1,4-β-xylosidase, phenol oxidase and microbial biomass were decreased significantly in soil of Changbai pine under elevated CO_2 in summer 2007, which can be observed also from mean activity of the enzymes and microbial biomass in 2007. Ebersberger *et al.* (2004) also found a significant reduce in proportion of fungi indicator of PLFA at elevated CO_2 in comparison to ambient CO_2 [30]. Finzi *et al.* (2006) proposed that there is increasing evidence that microbial function is progressively N limited under elevated CO_2 as duration elongation of CO_2 enrichment [31]. The lower microbial activity in the Changbai pine exposed to elevated CO_2 may be attributed to the greater competition for N nutrient by rapidly growth of the tree [32].

Tree species effects on soil enzyme under elevated CO_2

The ecophysiological responses of the three species to elevated CO_2 were significant different (Zhou and Han, unpublished data). The fast-growing Changbai pine tends to show larger growth increases under elevated CO_2 than the slow-growing Korean pine, but the biomass was stored in stem and branch. Many studies showed that above-ground cover plant could determine the composition of the soil microbial community structure and function [33 –35]. In our study, discrimination between the pines and the oak was observed (Fig. 3), but these results must be careful explained, because of variation in duration of elevated CO_2 between the pines and oak. Changbai pine tends to possess characteristics promoting rapid growth generally associated with high competitive ability. This can explain why elevated CO_2 caused a decrease in microbial biomass and extracellular enzyme activity under Changbai pine (Fig. 1), due to the increase in competition for mineral nutrients with microorganisms. We speculate that fast-growing plant may be strongly affect soil N and P cycling compared to the slow-growing plant under elevated CO_2 condition, contributing disadvantage to soil microbial community as well as significant declining the activities of microbial enzyme. These results also imply that tree species might also differ in their influence on soil microbial activity that in turn affects soil properties under elevated atmospheric CO_2 concentration.

Due to Korean pine has a slower growth rate and related physiological characteristics, the tree shows a narrower response to resource levels; whereas Changbai pine has a faster growth rate related to Korean pine shows a broad response to N and P nutrient. Hungate *et al.* (1997) observed a species-specific growth response of species to elevated atmospheric CO_2, that could either increase or decrease soil-N availability [36]. Schimel and Bennett (2004) reported fast-growing species grew significantly better and took up more of the available N and P in the elevated CO_2 condition [32]. Fast-growing Changbai pine required more canopy space and more soil volume than the Korean pine and showed a relative higher demand for soil N, as the high-to-low order of 1,4-β-NAG and CBH shown, compared with CBH-to-1,4-β-NAG order in Korean pine.

Seasonal and temporal dynamic of the activities of soil enzymes under elevated CO_2

Temporal fluctuation of soil moisture, soil temperature, and C input from tree roots, rhizosphere products (e.g., root exudates), and tree residues can have large effects on soil microbial biomass and activity [38]. Most of the enzymes measured in our study (table 2), were significantly affected by the temporal variability, and the interaction between time and CO_2 level significantly influenced CHB, phenol oxidase and peroxidase activities across the three species. Our data also revealed a strong positive correlation between soil moisture and microbial biomass or soil enzyme activity (Table 3), especially the activity of 1,4-α-glucosidase that correlated with soil moisture independent on tree species. These results are consistent with Devi and Yadava (2006), who reported that microbial biomass showed a positive significant correlation with soil moisture [39]. Soil moisture may be a master variable controlling microbial biomass and mineralization of starch (catalyzed by 1,4-α-glucosidase) in this temperate volcanic soil (Table 3).

	Korean pine	Changbai pine	Oak
1,4-β-glucosidase	0.0993	**0.6063***	0.3779
1,4-α-glucosidase	**0.5620***	**0.6385***	**0.6419***
1,4-β-NAGase	0.0551	0.3233	0.5305
Cellobiohydrolase	0.4239	0.3061	**0.5417***
1,4-β-xylosidase	0.2733	0.2320	**0.5856***
Phosphatase	**0.6050***	0.1812	0.3672
Phenol oxidase	0.4411	0.2139	0.2976
Peroxidase	0.2568	-0.2690	-0.1960
Microbial biomass C	0.5843	**0.7042***	0.5018
Microbial biomass N	0.4207	**0.9528****	0.6033

Levels of significance of the coefficients are indicated for $P < 0.05$* or $P < 0.01$**.

Table 3. Correlation analysis of extracellular enzyme and moisture of the soil under Korean pine, Changbai pine and oak after fumigation with elevated CO_2 concentration in 2006 and 2007.

Tree growth often stimulates an increase in the size of microbial biomass during the growing season [40, 41]. The seasonal variations in 1,4-β-glucosidase of Changbai pine we observed were consistent with the seasonal variations in 1,4-α-glucosidase, 1,4-β-NAG, and cellobiohydrodase activity (e.g., greatest activity in spring), which were in agreement with a report of study in a oak forest in USA [42]. Seasonal fluctuations of these enzymes might be strongly influenced by life cycle of Changbai pine. Because spring and early summer is the fast-growing period of the pine, the developing root system could provide enough of easily mineralizable substrate to microorganisms and exhibited a positive effect on their activities. So for Changbai pine, the peak activity of soil microbial community was detected in spring. Seasonal patterns of soil enzyme were controlled by growth also observed from oak, which growth began from mid of May to end of October. This indicates that the activity of soil enzyme may be concomitantly controlled by plant seasonal growth.

In addition, differently seasonal pattern of identical enzyme from elevated and ambient CO_2 was also found, e.g., cellobiohydrolase from Korean pine, 1,4-β-NAG from Changbai pine, 1,4-β-glucosidase and 1,4-β-NAG from oak. This may be due to soil extracellular enzymes can be either induced by the substrate [43], which are often altered by elevated CO_2 [44], and influenced by soil moisture, or controlled by combine effect of there factors and/or other environmental factors [13], e.g., tree species, duration of CO_2 enrichment. Differences between the two coniferous pine and the broadleaf oak results from OTC may be best accounted for by the different seasonal patterns of soil microbial biomass and enzyme in respond to tree species and elevated CO_2. So our results confirm the importance of taking into account the seasonal variation of biochemical parameters when these are used as indicators of soil ecosystem in response to elevated CO_2.

In conclusion, seasonal variations are the factor mostly affecting soil biological properties and nutrients availability in Changbai mountain forest ecosystem. Long-term exposure of elevated CO_2 can alter microbial biomass and the production of enzymes, but the effects are always detectable at specific times and are closely linked to plant processes, soil moisture and aboveground vegetations. Hence, a single sampling of the soil may not fully reveal its response to prolonged elevated atmospheric CO_2. We proposed that it is imperative to assess microbial function for soil ecosystem with one or two year-round sampling regime.

Acknowledgements

We thank XiuXiu Wang, ShiFu Meng, LiHua Niu and GuoZheng Song for their assistance in the open top chamber, and LiHua Xin for help with soil biochemical property analysis. This research was supported by the National Natural Science Foundation of China (NSFC).

Author details

Jun Qiang Zheng[1,2], Ying Wang[1] and Shi Jie Han[1,2*]

*Address all correspondence to: zhjq79@yahoo.com

1 Institute of Applied Ecology, Chinese Academy of Sciences, China

2 State Key Laboratory of Forest and Soil Ecology, Institute of Applied Ecology, Chinese Academy of Sciences, China

References

[1] Neftel, A., Moor, E., & Oeschger, H. (1985). Evidence from polar ice cores for the increase in atmospheric CO_2 in the past two centuries. *Nature*, 315, 45-47.

[2] Meehl, G. A., Stocker, T. F., & Collins, W. D. (2007). Global Climate Projections. *In Solomon S, Qin D, Manning M, Climate Change 2007: The Physical Science Basis. Contribution of Working Group I to the Fourth Assessment Report of the Intergovernmental Panel on Climate Change. Cambridge University Press, Cambridge, UK and New York. USA.*

[3] Catovsky, S., & Bazzaz, F. (1999). Elevated CO_2 influences the responses of two birch species to soil moisture: implications for forest community structure. *Global Change Biol*, 5, 507-518.

[4] Norby, R. J., Long, T. M., & Hartz-Rubin, J. S. (2000). Nitrogen resorption in senescing tree leaves in a warmer, CO2-enriched atmosphere. *Plant Soil*, 224, 15-29.

[5] Norby, R. J., Wullschleger, S. D., & Gunderson, C. A. (1999). Tree responses to rising CO2 in field experiments: implications for the future forest. *Plant Cell Environ*, 22, 683-714.

[6] Oren, R., Ellsworth, D., & Johnsen, K. H. (2001). Soil fertility limits carbon sequestration by forest ecosystems in a CO2-enriched atmosphere. *Nature*, 411, 469-472.

[7] Cotrufo, M. F., & Ineson, P. (1995). Effects of enhanced atmospheric CO_2 and nutrient supply on the quality and subsequent decomposition of fine roots of Betula pendula Roth. and Picea sitchensis (Bong) Carr. *Plant Soil*, 170, 267-277.

[8] Smith, J. L., & Paul, E. A. (1990). The significance of soil microbial biomass estimations. *In Bollag JM, Stotzky G, editors. Soil Biochemistry*, 6, Marcel Dekker, Inc.,, New York., 357-396.

[9] Zak, D. R., Pregitzer, K. S., Curtis, P. S., & Holmes, W. E. (2000). Atmospheric CO2 and the composition and function of soil microbial communities. *Ecol Appl*, 10, 47-59.

[10] Sinsabaugh, R. L., Saiya-Cork, K., Long, T., Osgood, M. P., Neher, D. A., Zak, D. R., & Norby, R. J. (2003). Soil microbial activity in a Liquidambar plantation unresponsive to CO_2 driven increases in primary production. *Appl Soil Ecol*, 24, 263-271.

[11] Moscatelli, M. C., Lagomarsino, A., De Angelis, P., & Grego, S. (2005). Seasonality of soil biological properties in a poplar plantation growing under elevated atmospheric CO_2. *Appl Soil Ecol*, 30, 162-173.

[12] Pinay, G., Barbera, P., Carreras-Palou, A., Fromin, N., Sonie, L., Couteaux, M. M., Roy, J., Philippot, L., & Lensi, R. (2007). Impact of atmospheric CO2 and plant life forms on soil microbial activities. *Soil Biol Biochem*, 39, 33-42.

[13] Nannipieri, P., Kandeler, E., & Ruggiero, P. (2002). Enzyme activities and microbiological and biochemical processes in soil. *In Burns RG, Dick RP, editors. Enzymes in the Environment: Activity, Ecology, and Applications. Marcel Dekker, Inc., New York.*, 1.

[14] Waldrop, M. P., & Firestone, M. K. (2006). Seasonal dynamics of microbial community composition and function in oak canopy and open grassland soils. *Microb Ecol*, 52, 470-479.

[15] Myneni, R. B., Dong, J., & Tucker, C. J. (2001). A large carbon sink in the woody biomass of Northern forests. *Proc Nat Acad Sci U S A*, 98, 14784-14789.

[16] Han, S. J., Zhang, J. H., & Zhou, Y. M. (2002). Dynamics of soil CO2 profiles of Pinus sylvestris var. sylvestriformis seedlings under CO2 concentration doubled. *Acta Botanica Sinica*, 44, 852-857.

[17] Han, S. J., Zhang, J. H., Zhou, Y. M., Wang, C. R., & Zou, C. J. (2001). Influence of CO2 doubling on water transport process at root/soil interface of pinus sylvestris var. sylvestrifornis seedlings. *Acta Botanica Sinica*, 43, 385-388.

[18] Zhou, Y. M., Hart, S. J., & Xin, L. H. (2006). Soil respiration of Pinus koraiensis and Pinus sylvestriformis trees growing at elevated CO2 concentration. *Chinese J Appl Ecol*, 17, 1757-1760.

[19] Hamilton, J. G., De Lucia, E. H., & George, K. (2002). Forest carbon balance under elevated CO_2. *Oecologia*, 131, 250-260.

[20] Vance, E. D., Brookes, P. C., & Jenkinson, D. S. (1987). An extraction method for measuring microbial biomass C. *Soil Biol Biochem*, 19, 703-707.

[21] Sparling, C. P., & West, A. W. (1988). A direct extraction method to estimate soil microbial C: calibration in situ using microbial respiration and 14C labeled cells. *Soil Biol Biochem*, 20, 337-343.

[22] Freeman, C., Liska, G., & Ostle, N. J. (1995). The use of fluorogenic substrates for measuring enzyme activity in peat lands. *Plant Soil*, 175, 147-152.

[23] Kang, H., & Freeman, C. (1999). Phosphatase and arylsulphatase activities in wetland soils: annual variation and controlling factors. *Soil Biol Biochem*, 31, 449-454.

[24] Sinsabaugh, R.L. (1993). Wood decomposition: nitrogen and phosphorus dynamics in relation to extracellular enzyme activity. *Ecology*, 74, 1586-1593.

[25] Sinsabaugh, R. L., Klug, M. J., Collins, H. P., Yeage,r, P. E., & Petersen, S. O. (1999). Characterizing soil microbial communities. *In Robertson GP, Coleman DC, Bledsoe CS, Sollins P, editors. Standard Soil Methods for Long-Term Ecological Research. Oxford University Press, New York.*, 318-348.

[26] Hoosbeek, M. R., & Li, Y. (2006). Free atmospheric CO_2 enrichment (FACE) increased labile and total carbon in the mineral soil of a short rotation Poplar plantation. *Plant Soil*, 28, 247-254.

[27] Larson, J. L., Zak, D. R., & Sinsabaughb, R. L. (2002). Extracellular enzyme activity beneath temperate trees growing under elevated carbon dioxide and ozone. *Soil Sci Soc Am J*, 66, 1848-1856.

[28] Lipson, D. A., Wilson, R. A., & Oechel, W. C. (2005). Effects of elevated atmospheric CO2 on soil microbial biomass, activity and diversity in a chaparral ecosystem. *Appl Environ Microbiol*, 71, 8573-8580.

[29] Dhillion, S. S., Roy, J., & Abrams, M. (1996). Assessing the impact of elevated CO_2 on soil microbial activity in a Mediterranean model ecosystem. *Plant Soil*, 187, 333-342.

[30] Ebersberger, D., Wermbter, N., Niklaus, P. A., & Kandeler, E. (2004). Effects of long term CO_2 enrichment on microbial community structure in calcareous grassland. *Plant Soil*, 264, 313-323.

[31] Finzi, A. C., Sinsabaugh, R. L., Long, T. M., & Osgood, M. P. (2006). Microbial community responses to atmospheric CO2 enrichment in a Pinus taeda forest. *Ecosystems*, 9, 215-226.

[32] Schimel, J. P., & Bennet, J. (2004). Nitrogen mineralization: challenges of a changing paradigm. *Ecology*, 85, 591-602.

[33] Kourtev, P. S., Ehrenfeld, J. G., & Haggblom, M. (2002). Exotic plant species alter the microbial community structure and function in the soil. *Ecology*, 83, 3152-3166.

[34] Kourtev, P. S., Ehrenfeld, J. G., & Haggblom, M. (2003). Experimental analysis of the effect of exotic and native plant species on the structure and function of soil microbial communities. *Soil Biol Biochem*, 35, 895-905.

[35] Montealegre, C. M., van Kessel, C., Russelle, M. P., & Sadowsky, MJ. (2002). Changes in microbial activity and composition in a pasture ecosystem exposed to elevated atmospheric carbon dioxide. *Plant Soil*, 243, 197-207.

[36] Hungate, B. A., Holland, E. A., Jackson, R. B., Chapin, F. S., Field, C. B., & Mooney, H. A. (1997). The fate of carbon in grasslands under carbon dioxide enrichment. *Nature*, 388, 576-579.

[37] Madeira, M., Füleky, G., & Auxtero, E. (2007). Phosphate sorption of European volcanic soils. *In Arnalds O, Bartoli F, Buurman P, Oskarsson H, Stoops G, Garcia-Rodeja E, editors. Soils of Volcanic Regions in Europe. Springer, Berlin.*, 354.

[38] Moscatelli, M. C., Fonck, M., De Angelis, P., Larbi, H., Macuz, A., Rambelli, A., & Grego, S. (2001). Mediterranean natural forest living at elevated carbon dioxide: soil biological properties and plant biomass growth. *Soil Manag*, 17, 195-202.

[39] Devi, N. B., & Yadava, P. S. (2006). Seasonal dynamics in soil microbial biomass C, N and P in a mixed-oak forest ecosystem of Manipur, North-east India. *Appl Soil Ecol*, 31, 220-227.

[40] Kaiser, E. A., & Heinemeyer, O. (1993). Seasonal variations of soil microbial biomass carbon within plough layer. *Soil Biol Biochem*, 25, 1649-1655.

[41] Debosz, K., Rasmussen, P. H., & Pedersen, A. R. (1999). Temporal variations in microbial biomass C and cellulolytic enzyme activity in arable soils: effects of organic matter input. *Appl Soil Ecol*, 13, 209-218.

[42] Boerner, R. E. J., Brinkman, J. A., & Smith, A. (2005). Seasonal variations in enzyme activity and organic carbon in soil of a burned and unburned hardwood forest. *Soil Biol Biochem*, 37, 1419-1426.

[43] Allison, S. D., & Vitousek, P. M. (2005). Responses of extracellular enzymes to simple and complex nutrient inputs. *Soil Biol Biochem*, 37, 937-944.

[44] Billings, S., Schaeffer, S., Zitzer, S., Charlet, T., Smith, S., & Evans, R. (2002). Alterations of nitrogen dynamics under elevated carbon dioxide in an intact Mojave Desert ecosystem: evidence from nitrogen-15 natural abundance. *Oecologia*, 131, 463-467.

Permissions

The contributors of this book come from diverse backgrounds, making this book a truly international effort. This book will bring forth new frontiers with its revolutionizing research information and detailed analysis of the nascent developments around the world.

We would like to thank Dr. M.C. Hernández-Soriano, for lending her expertise to make the book truly unique. She has played a crucial role in the development of this book. Without her invaluable contribution this book wouldn't have been possible. She has made vital efforts to compile up to date information on the varied aspects of this subject to make this book a valuable addition to the collection of many professionals and students.

This book was conceptualized with the vision of imparting up-to-date information and advanced data in this field. To ensure the same, a matchless editorial board was set up. Every individual on the board went through rigorous rounds of assessment to prove their worth. After which they invested a large part of their time researching and compiling the most relevant data for our readers. Conferences and sessions were held from time to time between the editorial board and the contributing authors to present the data in the most comprehensible form. The editorial team has worked tirelessly to provide valuable and valid information to help people across the globe.

Every chapter published in this book has been scrutinized by our experts. Their significance has been extensively debated. The topics covered herein carry significant findings which will fuel the growth of the discipline. They may even be implemented as practical applications or may be referred to as a beginning point for another development. Chapters in this book were first published by InTech; hereby published with permission under the Creative Commons Attribution License or equivalent.

The editorial board has been involved in producing this book since its inception. They have spent rigorous hours researching and exploring the diverse topics which have resulted in the successful publishing of this book. They have passed on their knowledge of decades through this book. To expedite this challenging task, the publisher supported the team at every step. A small team of assistant editors was also appointed to further simplify the editing procedure and attain best results for the readers.

Our editorial team has been hand-picked from every corner of the world. Their multi-ethnicity adds dynamic inputs to the discussions which result in innovative outcomes. These outcomes are then further discussed with the researchers and contributors who give their valuable feedback and opinion regarding the same. The feedback is then collaborated with the researches and they are edited in a comprehensive manner to aid the understanding of the subject.

Apart from the editorial board, the designing team has also invested a significant amount of their time in understanding the subject and creating the most relevant covers. They scrutinized every image to scout for the most suitable representation of the subject and create an appropriate cover for the book.

The publishing team has been involved in this book since its early stages. They were actively engaged in every process, be it collecting the data, connecting with the contributors or procuring relevant information. The team has been an ardent support to the editorial, designing and production team. Their endless efforts to recruit the best for this project, has resulted in the accomplishment of this book. They are a veteran in the field of academics and their pool of knowledge is as vast as their experience in printing. Their expertise and guidance has proved useful at every step. Their uncompromising quality standards have made this book an exceptional effort. Their encouragement from time to time has been an inspiration for everyone.

The publisher and the editorial board hope that this book will prove to be a valuable piece of knowledge for researchers, students, practitioners and scholars across the globe.

List of Contributors

S.E.A.T.M. van der Zee and A. Leijnse
Wageningen University, Environmental Sciences, Wageningen, Netherlands

Diego Antonio França de Freitas, Marx Leandro Naves Silva, Nilton Curi, Mayesse Aparecida da Silva, Anna Hoffmann Oliveira and Sérgio Henrique Godinho Silva
Federal University of Lavras, Brazil

Vito Summa and Maria Luigia Giannossi
Institute of Methodologies for Environmental Analysis of the National Research Council of Italy (IMAA-CNR), Tito Scalo (PZ), Italy

Anna Hoffmann Oliveira, Marx Leandro Naves Silva, Nilton Curi and Diego Antonio França de Freitas
Soil Science Department, Federal University of Lavras, Lavras, MG, Brazil

Mayesse Aparecida da Silva
Soil Science Department, Federal University of Lavras, Lavras, MG, Brazil
CAPES and CNPq (Visiting scholar) scholarships, Brazil

Gustavo Klinke Neto
Vitaramae Environmental Consulting Ltd., Lavras, MG, Brazil

Vito Imbrenda, Mariagrazia D'Emilio, Maria Lanfredi and Tiziana Simoniello
National Research Council of Italy, Institute of Methodologies for the Environmental Analysis, CNR- IMAA, Tito Scalo (Pz), University of Basilicata, Dep. Environmental Engineering and Physics - DIFA, Potenza, DSF-University of Naples Federico II, Naples, Italy

Maria Ragosta
University of Basilicata, Dep. Environmental Engineering and Physics - DIFA, Potenza, Italy

Maria Macchiato
DSF-University of Naples Federico II, Naple, Italy

Mariusz Fotyma and Piotr Ochal
Institute of Soil Science and Plant Cultivation, Puławy, Poland

Jan Łabętowicz
Life Science University, Warsaw, Poland

Raphael Bastão de Souza, Thiago Guilherme Maziviero, Cintya Aparecida Christofoletti, Tamaris Gimenez Pinheiro and Carmem Silvia Fontanetti
Departament of Biology, São Paulo State University, Rio Claro, Brazil

M.C. Hernandez-Soriano and B. Kerré
Division of Soil and Water Management, KU Leuven, Belgium

A. Sevilla-Perea and M.D. Mingorance
Instituto Andaluz de Ciencias de la Tierra, University of Granada - CSIC, Spain, Spain

Elmira Saljnikov and Dragan Cakmak
Institute of Soil Science, Belgrade, Serbia

Saule Rahimgalieva
West-Kazakhstan Agrarian-Technical University named after Zhangir-Khan, Uralsk, Kazakhstan

Agnieszka Rutkowska and Dorota Pikuła
Department of Plant Nutrition and Fertilization, Institute of Soil Science and Plant Cultivation- State Research Institute, Poland

Chun-Chih Tsui
Department of Agricultural Chemistry, National Taiwan University, Taiwan

Horng-Yuh Guo
Division of Agricultural Chemistry, Taiwan Agricultural Research Institute, Council of Agricultural, Taiwan

Zueng-Sang Chen
Department of Agricultural Chemistry, National Taiwan University, Taiwan

Adam Łajczak
Institute of Geography, Jan Kochanowski University, Kielce, Poland

María Luisa Bárcenas-Argüello
Botany Program, Postgraduate College, Texcoco, Mexico

Ma. del Carmen Gutiérrez-Castorena
Soils Program, Postgraduate College, Texcoco, Mexico

Teresa Terrazas
Botany Department-Biology Institute, National Autonomous University of Mexico, México City, Mexico

Rui Guo, Wei Ping Hao, Dao Zhi Gong, Xiu Li Zhong and Feng Xue Gu
Institute of Environment and Sustainable Development in Agriculture (IEDA), Chinese Academy of Agricultural Sciences (CAAS)/Key Laboratory of Dry land Agriculture, MOA, Beijing 100081, China

Mieko Yachigo and Shinjiro Sato
Department of Environmental Engineering for Symbiosis, Soka University, Tokyo, Japan

Jun Qiang Zheng and Shi Jie Han
Institute of Applied Ecology, Chinese Academy of Sciences, China
State Key Laboratory of Forest and Soil Ecology, Institute of Applied Ecology, Chinese Academy of Sciences, China

Ying Wang
Institute of Applied Ecology, Chinese Academy of Sciences, China